Multifunctional Nanomaterials for Energy Applications

Multifunctional Nanomaterials for Energy Applications

Editors

Federico Cesano
Simas Rackauskas
Mohammed Jasim Uddin

MDPI • Basel • Beijing • Wuhan • Barcelona • Belgrade • Manchester • Tokyo • Cluj • Tianjin

Editors

Federico Cesano
Department of Chemistry
University of Torino
Torino
Italy

Simas Rackauskas
Institute of Materials Science
Kaunas University of
Technology
Kaunas
Lithuania

Mohammed Jasim Uddin
Department of Chemistry
University of Texas Rio
Grande Valley
Edinburg
United States

Editorial Office
MDPI
St. Alban-Anlage 66
4052 Basel, Switzerland

This is a reprint of articles from the Special Issue published online in the open access journal *Nanomaterials* (ISSN 2079-4991) (available at: www.mdpi.com/journal/nanomaterials/special_issues/energy_nanomaterials).

For citation purposes, cite each article independently as indicated on the article page online and as indicated below:

LastName, A.A.; LastName, B.B.; LastName, C.C. Article Title. *Journal Name* **Year**, *Volume Number*, Page Range.

ISBN 978-3-0365-4918-7 (Hbk)
ISBN 978-3-0365-4917-0 (PDF)

Cover image courtesy of Federico Cesano

© 2022 by the authors. Articles in this book are Open Access and distributed under the Creative Commons Attribution (CC BY) license, which allows users to download, copy and build upon published articles, as long as the author and publisher are properly credited, which ensures maximum dissemination and a wider impact of our publications.

The book as a whole is distributed by MDPI under the terms and conditions of the Creative Commons license CC BY-NC-ND.

Contents

About the Editors vii

Preface to "Multifunctional Nanomaterials for Energy Applications" ix

Simas Rackauskas, Federico Cesano and Mohammed Jasim Uddin
Multifunctional Nanomaterials for Energy Applications
Reprinted from: *Nanomaterials* 2022, 12, 2170, doi:10.3390/nano12132170 1

Rui D. Oliveira, Ana Mouquinho, Pedro Centeno, Miguel Alexandre, Sirazul Haque and Rodrigo Martins et al.
Colloidal Lithography for Photovoltaics: An Attractive Route for Light Management
Reprinted from: *Nanomaterials* 2021, 11, 1665, doi:10.3390/nano11071665 7

Zhaoxu Song, Kun Fang, Xiaofang Sun, Ying Liang, Wei Lin and Chuanzhong Xu et al.
An Effective Method to Accurately Extract the Parameters of Single Diode Model of Solar Cells
Reprinted from: *Nanomaterials* 2021, 11, 2615, doi:10.3390/nano11102615 45

Mamina Sahoo, Sz-Nian Lai, Jyh-Ming Wu, Ming-Chung Wu and Chao-Sung Lai
Flexible Layered-Graphene Charge Modulation for Highly Stable Triboelectric Nanogenerator
Reprinted from: *Nanomaterials* 2021, 11, 2276, doi:10.3390/nano11092276 61

Arulppan Durairaj, Moorthy Maruthapandi, Arumugam Saravanan, John H. T. Luong and Aharon Gedanken
Cellulose Nanocrystals (CNC)-Based Functional Materials for Supercapacitor Applications
Reprinted from: *Nanomaterials* 2022, 12, 1828, doi:10.3390/nano12111828 79

Snejana Bakardjieva, Jiří Plocek, Bauyrzhan Ismagulov, Jaroslav Kupčík, Jiří Vacík and Giovanni Ceccio et al.
The Key Role of Tin (Sn) in Microstructure and Mechanical Properties of Ti_2SnC (M_2AX) Thin Nanocrystalline Films and Powdered Polycrystalline Samples
Reprinted from: *Nanomaterials* 2022, 12, 307, doi:10.3390/nano12030307 105

Philipp M. Veelken, Maike Wirtz, Roland Schierholz, Hermann Tempel, Hans Kungl and Rüdiger-A. Eichel et al.
Investigating the Interface between Ceramic Particles and Polymer Matrix in Hybrid Electrolytes by Electrochemical Strain Microscopy
Reprinted from: *Nanomaterials* 2022, 12, 654, doi:10.3390/nano12040654 129

Ruowei Cui, Zhenwang Zhang, Huijuan Zhang, Zhihong Tang, Yuhua Xue and Guangzhi Yang
Aqueous Organic Zinc-Ion Hybrid Supercapacitors Prepared by 3D Vertically Aligned Graphene-Polydopamine Composite Electrode
Reprinted from: *Nanomaterials* 2022, 12, 386, doi:10.3390/nano12030386 141

Rosa M. González-Gil, Mateu Borràs, Aiman Chbani, Tiffany Abitbol, Andreas Fall and Christian Aulin et al.
Sustainable and Printable Nanocellulose-Based Ionogels as Gel Polymer Electrolytes for Supercapacitors
Reprinted from: *Nanomaterials* 2022, 12, 273, doi:10.3390/nano12020273 153

Mohammad Bagher Askari, Seyed Mohammad Rozati and Antonio Di Bartolomeo
Fabrication of Mn_3O_4-CeO_2-rGO as Nanocatalyst for Electro-Oxidation of Methanol
Reprinted from: *Nanomaterials* 2022, 12, 1187, doi:10.3390/nano12071187 167

Yinji Wan, Yefan Miao, Tianjie Qiu, Dekai Kong, Yingxiao Wu and Qiuning Zhang et al.
Tailoring Amine-Functionalized Ti-MOFs via a Mixed Ligands Strategy for High-Efficiency CO_2 Capture
Reprinted from: *Nanomaterials* 2021, *11*, 3348, doi:10.3390/nano11123348 179

Yiting Bu, Jiaxi Liu, Hailiang Chu, Sheng Wei, Qingqing Yin and Li Kang et al.
Catalytic Hydrogen Evolution of $NaBH_4$ Hydrolysis by Cobalt Nanoparticles Supported on Bagasse-Derived Porous Carbon
Reprinted from: *Nanomaterials* 2021, *11*, 3259, doi:10.3390/nano11123259 191

Imane Mahroug, Stefania Doppiu, Jean-Luc Dauvergne, Angel Serrano and Elena Palomo del Barrio
$Li_4(OH)_3Br$-Based Shape Stabilized Composites for High-Temperature TES Applications: Selection of the Most Convenient Supporting Material
Reprinted from: *Nanomaterials* 2021, *11*, 1279, doi:10.3390/nano11051279 207

Alvin Orbaek White, Ali Hedayati, Tim Yick, Varun Shenoy Gangoli, Yubiao Niu and Sean Lethbridge et al.
On the Use of Carbon Cables from Plastic Solvent Combinations of Polystyrene and Toluene in Carbon Nanotube Synthesis
Reprinted from: *Nanomaterials* 2021, *12*, 9, doi:10.3390/nano12010009 225

Kirsten I. Louw, Bronwyn H. Bradshaw-Hajek and James M. Hill
Ferric Ion Diffusion for MOF-Polymer Composite with Internal Boundary Sinks
Reprinted from: *Nanomaterials* 2022, *12*, 887, doi:10.3390/nano12050887 243

Natalia Pawlik, Barbara Szpikowska-Sroka, Tomasz Goryczka, Ewa Pietrasik and Wojciech A. Pisarski
Luminescence of SiO_2-BaF_2:Tb^{3+}, Eu^{3+} Nano-Glass-Ceramics Made from Sol–Gel Method at Low Temperature
Reprinted from: *Nanomaterials* 2022, *12*, 259, doi:10.3390/nano12020259 263

Victor Tarasenko, Nikita Vinogradov, Dmitry Beloplotov, Alexander Burachenko, Mikhail Lomaev and Dmitry Sorokin
Influence of Nanoparticles and Metal Vapors on the Color of Laboratory and Atmospheric Discharges
Reprinted from: *Nanomaterials* 2022, *12*, 652, doi:10.3390/nano12040652 279

About the Editors

Federico Cesano

Federico Cesano is an Associate Professor in Industrial Chemistry (Department of Chemistry of the University of Torino, Italy). He obtained his degree in Chemistry in 1999 from the University of Torino. After two years spent at the Italian National Research Council (2000-2002), in 2005 he defended his PhD in Materials Science. Since 2006, he has been working at the Department of Chemistry of the University of Turin. He is the co-author of more than 80 ISI papers, 60 conference presentations, 15 invited talks, and several chapter books published in the journals of Chemistry and Materials Science and he is co-inventor of a few international patents. He has been involved in several EU, national and local research projects. His main research interests are from basic science to applications of 1D, 2D and 3D nanostructured materials (including oxides, carbon nanomaterials, transition metal dichalcogenides, and polymers), either alone or combined to form hybrid structures and composite materials.

Simas Rackauskas

Simas Rackauskas is a Chief Researcher in Kaunas University of Technology (KTU). He defended his PhD in Physics at Aalto University, Finland in 2011. He was a Marie Curie Fellow in University of Turin (Italy). He had fellowships in Swiss Federal Institute of Technology in Lausanne (EPFL, Switzerland), Technical University of Denmark (DTU) and University of Nagoya (Japan). His research interests are mainly focused on non-catalytic growth mechanisms of metal oxide nanowires and application in sensing and multifunctional coatings. He is involved in commercialization of nanoparticle anti-reflection coatings in a spin-off "Zinotech" (CEO).

Mohammed Jasim Uddin

Dr. M. Jasim Uddin (h-index 29) obtained his PhD degree (Materials Science) from University of Turin, Italy. He is currently working as an Associate Professor in the Department of Chemistry, College of Science, University of Texas Rio Grande Valley. He served the Department of Chemistry as Chair for one term. He was awarded $21M external funding within the last ten years. He, along with his team, were recently awarded: CoS Excellence in Research Award 2021, Outstanding and Sustainable Research in Science Award 2016 (UTRGV), High Scholar Research Award (First Prize) 2016, 2018, 2019, 2021 (UTRGV), Texas Governor Award in STEM 2020, ISEF Grand Champion Award 2019, RSEF First Prize Award 2019, United Group Research Award 2016 (International), NASA Texas Space Grant Award (2016), UGC Award in 2010 (International), etc. He has published 70 papers, 100 conference presentations, 21 invited talks, 5 patents, and 6 book/book chapters.

Preface to "Multifunctional Nanomaterials for Energy Applications"

The rapid growth of the world's population has significantly increased energy consumption and environmental impact. The transition from fossil fuels to sustainable energy sources that is needed for a sustainable future demands more efficient materials and improved technologies, but allows us to tackle this great and necessary challenge. This Special Issue highlights some of the latest energy advances in the field of materials, in particular low-dimensional materials, and nanostructured materials. Various topics related to synthesis and characterization methods, properties, and energy application uses are highlighted.

Federico Cesano, Simas Rackauskas, and Mohammed Jasim Uddin
Editors

Editorial

Multifunctional Nanomaterials for Energy Applications

Simas Rackauskas [1], Federico Cesano [2,*] and Mohammed Jasim Uddin [3]

1. Institute of Materials Science, Kaunas University of Technology, 44249 Kaunas, Lithuania; simas.rackauskas@ktu.lt
2. Department of Chemistry, Turin University & INSTM-UdR Torino, 10125 Torino, Italy
3. Photonics and Energy Research Laboratory-PERL, Department of Chemistry, The University of Texas Rio Grande Valley, Edinburg, TX 78539, USA; mohammed.uddin@utrgv.edu
* Correspondence: federico.cesano@unito.it; Tel.: +39-011-6707548

Citation: Rackauskas, S.; Cesano, F.; Uddin, M.J. Multifunctional Nanomaterials for Energy Applications. *Nanomaterials* 2022, 12, 2170. https://doi.org/10.3390/nano12132170

Received: 25 May 2022
Accepted: 16 June 2022
Published: 24 June 2022

Publisher's Note: MDPI stays neutral with regard to jurisdictional claims in published maps and institutional affiliations.

Copyright: © 2022 by the authors. Licensee MDPI, Basel, Switzerland. This article is an open access article distributed under the terms and conditions of the Creative Commons Attribution (CC BY) license (https://creativecommons.org/licenses/by/4.0/).

In the last few decades, global energy requirements have grown exponentially, and increased demand is expected in the upcoming decades. Traditional energy resources have remarkably impacted energy production so far, but the use of renewable energy sources has constantly increased and is gradually substituting fossil fuels. Such non-renewable energy resources are limited in nature, and their use for energy purposes affects climate change. The new paradigm is materials for sustainable energy, and when materials are nanostructured, new key concepts are involved. Nanomaterials exhibit properties very different from their bulk counterparts due to their significant surface boundary and quantum confinement characteristics. Furthermore, the structure (or nanophase assembly) is also relevant for explaining various novel and interesting properties, notably when energy applications are taken into consideration. Remarkably, the aggregation and interface properties of nanostructures, even at lower dimensionality, are expected to boost energy applications.

Nanomaterials and nanotechnologies for energy have been more actively studied and used since the 2000s, as recognized by the number of scientific contributions that are growing exponentially (Figure 1a). As far as the geographical point of view is concerned, the subdivision of the contributions seems unbalanced when considering the continents: North/East/Central Asia (52%), Europe (23%), North/South America (18.3%), Africa (3.8%); Australia (2.5%). In more details, most contributions have been from China (c.a. 26.5%) and the United States (15.0 %), followed by India (9.8%), South Korea (4.3%), UK and Germany (2.9%), Australia, Japan, France (c.a. 2.5%), Italy (2%), and Spain (1.8%) (Figure 1b). This subdivision probably does not reflect the geographical distribution of investments, but it provides an overview of countries providing innovation in the near future in the fields of energy by nanomaterials. However, nanomaterials are not the only materials that have attracted the recent attention of the scientific community. For example, nanostructured materials and compounds are also frequently studied subjects.

As for the global market for energy nanomaterials, recent trends of segment indicate a significantly increasing of demand with a stable growth at a compound annual growth rate (CAGR) of 13.4% over the 2020–2027 period [1]. The global nanotechnology in energy industry was estimated to be at $140 million (2020), and it is estimated near $385 million by 2030 [2]. The geographical distributions of the market prospect a rising product application (especially in North America), the expanding mass production and price reduction of nanomaterials (mainly in Europe) together with the entry for new players due to the government financial support, increasing demand, along with a huge population (Asia-Pacific) [1,2].

From a thematic viewpoint, the field of *"energy materials"* is very wide the research requires a multidisciplinary approach with multifaceted activities from basic and fundamental scientific studies to more applicative works.

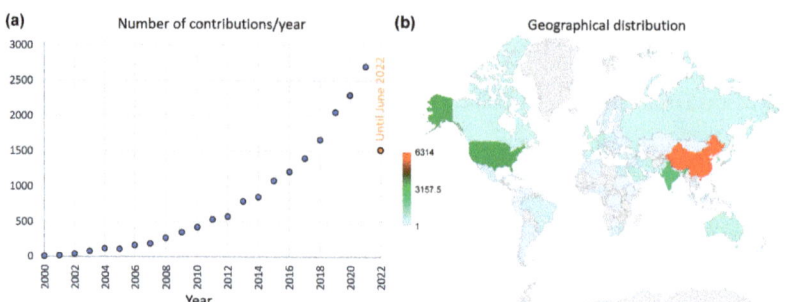

Figure 1. Scientific contributions dedicated to energy nanomaterials: (**a**) documents published since 2000 and (**b**) their geographical distribution. Keywords: "energy" AND "nanomaterials" within: Article title, Abstract, Keywords (Source: Scopus; May 2022).

This Special Issue, comprising two reviews and fourteen research articles, highlights some recent improvements and perspectives in the field of energy nanomaterials, which can be considered to address our current improvements and future challenges. The number and variety of contributions reflect the remarkable interest in topics related to energy nanomaterials. Furthermore, the presence of contributions addressing environmental issues, such as the use of environmentally friendly and recyclable raw materials, indicates an increased interest in sustainability, which is likely to become in the near future an increasingly important issue.

From such a broad and multifaceted item, the reader can expect that the contributions will cover some of the most debated hot topics in the scientific community, including materials and technologies for: (i) light harvesting enabling efficiency improvements [3,4], (ii) triboelectric nanogenerators [5]; (iii) nuclear energy production [6]; (iv) energy (batteries, supercapacitors, and some components, including electrolytes, electrodes, catalysts) [7], and fuel (methanol, H_2) storage [8]; CO_2 capture; (v) thermal energy storage applications [9]; (vi) new wiring based on carbons for electrical applications and electronics [10,11]; and (vii) detectors, optoelectronic devices (i.e., laser technologies, spectral converters, LEDs and three-dimensional displays).

Oliveira et al. [12] reviewed the topic of colloidal lithography (CL) for photovoltaics (PVs). The authors overviewed some of the most promising methods for material micro/nano structuring methods in the field of photovoltaics, where colloidal lithography has been demonstrated to be among the best and preferred patterning techniques for implementation in industrial processes and promise to be a reliable alternative to conventional hard-patterning processes. In this regard, photonic-enhanced PVs have been shown to bypass many of the conventional shortcomings of current solar cell technologies, as follows: (i) reflection losses and (ii) decreased light absorption occurring in thin films. The authors remarked also on the opportunity to enable other attractive functionalities, such as improved transparent contacts and self-cleaning properties due to the high aspect ratio of photonic microstructures.

Song et al. [13] reported a non-iterative method to precisely extract five model parameters for a single diode model of solar cells. The authors' method, overcoming the complexity and accuracy difficulties by employing a simplified calculation process, uses equation parts which are to be dynamically adapted, and thus the five parameters are calculated from the I-V curve. Interestingly, the authors reported that the proposed method is reliable (more than other methods) based on the root mean square error analysis. More interestingly, the authors simulated I-V and P-V characteristics by using the extracted parameters and compared them with the experimental fit of solar cells under different conditions.

Shaoo et al. [14] reported the fabrication of a flexible multilayer graphene triboelectric nanogenerator (TENG) for use as an energy harvester for next-generation flexible electronics. The device was layer-by-layer assembled by introducing a charge trapping layer

(CTL) made of Al_2O_3, which was placed between the conducting electrode and the positive triboelectric layer. The authors reported a 30-fold increase in output power for three layers of graphene TENG (3L-Gr-TENG) with CTL compared to 3L-Gr-TENG without CTL device counterpart. Interestingly, the developed device also showed continuous operations for more than 2000 cycles with remarkable stability. Surprisingly, the same device was capable of powering 20 green LEDs and sufficient to power an electronic timer with rectifier circuits. The authors associated the greatly improved performance with the synergistic effects occurring between the graphene layers and the Al_2O_3 dielectric layer.

Durairaj et al. [15] reviewed the cellulose nanocrystals (CNCs) field for the application as supercapacitors (SCs), from CNC synthesis/surface function to development of conductive CNCs. The authors also remarked and summarized recent perspectives and problems to be addressed including: (i) the importance of fabricating CNCs of millimeter thickness and with a hierarchical porous microstructure; (ii) control/optimization of CNC surface for both hydrophobic and hydrophilic matrices; (iii) bio-precursors allowing the incorporation of biological molecules (i.e., virus, etc.) for making a better impact on their superior properties; (iv) the need to obtain metal-free porous and heteroatom-doped CNC using a low cost, and sustainable approach from abundant biomass; and (v) the need to improve the performance of SCs, energy, and power densities of SCs.

Bakardjieva et al. [16] synthesized layered ternary Ti_2SnC carbides. Among the adopted synthesis methods, an unconventional low-energy ion facility (LEIF) based on Ar+ ion beam sputtering of Ti, Sn, and C targets, was adopted by the authors. The contribution provided insights into the understanding of Sn atom segregation at the surface, highlighting the role played by aberration-corrected STEM techniques, such as high-angle annular dark-field detector (HAADF) analysis combined with simulations of SAED patterns to track atomic paths clarifying the properties of Sn atoms at the proximity of irradiation-induced nanoscale defects and the existence of oxidized species formed during the preparation process.

Veelken et al. [17] investigated the local ionic conductivity in a hybrid electrolyte interface between ceramic particles and polymers in hybrid electrolytes (polyethylene oxide with Li bis(trifluoromethanesulfonyl)imide: PEO_6–LiTFSI; and $Li_{6.5}La_3Zr_{1.5}Ta_{0.5}O_{12}$: LLZO:Ta) by electrochemical strain microscopy (ESM). Interestingly, the results presented by the authors described significant insights to attain an advanced understanding of ionic transport mechanisms in the interior of hybrid electrolytes, and provided two strategies for the hybrid solid-state electrolyte improvements. Firstly, the covering of the particles can help to decrease the interfacial resistance. Secondly, a structure consisting of a continuous percolation path along the ceramic particles may take advantage of the high lithium content at the interfacial regions with the consequent overall ionic conductivity increasing.

Cui et al. [18] reported the fabrication of a graphene-polydopamine electrode (PDA@3DVAG) composite with 3D vertical-oriented macropores by unidirectional freezing and subsequent self-polymerization approach. The authors tested the composite as the positive electrode of zinc-ion hybrid supercapacitors (ZHSCs) and reported excellent electrochemical performances compared with the conventional electrolyte. In this regard, the authors reported that the vertically oriented composite electrode showed better properties compared to performance (48.92% at a current density of 3 A g^{-1}), wider voltage window (±0.8 V voltage drop), better cycle performance with specific capacitance from 96.7 to 59.8 F g^{-1}, and higher energy density (46.14 Wh kg^{-1}).

González-Gil et al. [19] prepared a new nanocellulose-based gel polymer electrolyte (GPE) to be used in supercapacitors. The authors synthesized the GPE from a mixture of an ionic liquid (1-ethyl-3-methylimidazolium dimethyl phosphate) with carboxymethylated cellulose nanofibers at different weight ratios. The addition of nanocellulose-based fibers helped to improve the ionogel properties, including ionic conductivity in the 0.32–0.94 mS cm^{-2} interval and to become easily printable on the electrode surface. Interestingly, the authors reported that the new GPE-based supercapacitor cell showed good electrochemical performances with high specific capacitance (160 F g^{-1}) excellent energy density and good power density (46.14 Wh kg^{-1} at the power density of 393.75 W kg^{-1} and 19.29 Wh kg^{-1} at 2183 W kg^{-1}, respectively). More interestingly, the energy density exceeds that of conventional electrochemical capacitors and of some large-size batteries.

Askari et al. [20] reported the one-step preparation by the hydrothermal method of Mn_3O_4-CeO_2 mixed metal oxides and reduced graphene oxide (rGO) hybrid catalysts to be used in the methanol oxidation reaction (MOR) process. Interestingly, the authors remarked the occurrence of a synergetic effect occurring between rGO and Mn_3O_4-CeO_2. In this domain, Mn_3O_4-CeO_2-rGO material showed an oxidation current density of 17.7 mA/cm^2 in overpotential of 0.51 V and 91% stability after 500 consecutive runs of cyclic voltammetry (CV). Furthermore, the optimal concentrations of methanol for Mn_3O_4-CeO_2 and Mn_3O_4-CeO_2-rGO catalysts were determined in CV experiments to be 0.6 and 0.8 M, respectively.

Wan et al. [21] synthesized a series of amine-functionalized highly stable Ti-based MOFs (MIP-207) by the mixed linkers method from 1,3,5-benzenetricarboxylic acid (H3BTC) and 5-aminoisophthalic acid ($C_8H_7NO_4$) with different weight fractions. Interestingly, the introduction of amino groups demonstrated remarkable CO_2 uptake performance (up to 3.96 and 2.91 mmol g^{-1}), which are 20.7% and 43.3% higher than those of unmodified MIP-207 at 0 and 25 °C, respectively. The CO_2 uptake is attributed by the authors to the introduction of -NH_2 into the framework of MIP-207, leading to the increase of specific surface area and more Lewis basic adsorption sites, thereby enhancing the CO_2 working capacity and CO_2/N_2 selectivity properties.

Bu et al. [22] synthesized Co nanoparticles supported on bagasse-derived porous carbon catalysts for catalyzed hydrolytic dehydrogenation reaction of $NaBH_4$. One of the catalysts prepared by the authors exhibited a remarkable hydrogen generation activity with an optimal H_2 production rate of 11,086.4 mL$_{H2}$·min^{-1}·g$_{Co}$$^{-1}$ and low activation energy (31.25 kJ mol^{-1}). Density functional theory (DFT) results indicated that the metal nanoparticles-supported on porous carbon catalyst structure was advantageous for the dissociation of $[BH_4]^-$, which remarkably enhanced the hydrolysis efficiency of $NaBH_4$. Interestingly, the catalyst presents excellent durability, retaining 72.0% of the initial catalyst activity after several cycling tests.

Mahroug et al. [23] reported a selection of the most promising oxide-based supporting materials for the $Li_4(OH)_3Br$ peritectic compound to be used in thermal energy storage (TES) applications at ca. 300 °C. The authors investigated micro/nanoparticles of MgO, Fe_2O_3, CuO, SiO_2 and Al_2O_3 as candidates for supporting materials. Among all oxides, based on: (i) chemical compatibility of the supporting material with molten $Li_4(OH)_3Br$; (ii) anti-leakage effectiveness and maximum salt loading; and (iii) thermal and microstructural properties and stability of $Li_4(OH)_3B$, MgO nanoparticles were found the most promising oxide also due to the fact that all of the other oxides studied showed more or less pronounced upon heating and upon cycling conditions.

Orbaek White et al. [24] reported the growth of multi-walled car-bon nanotubes (MWCNTs) by the liquid injection chemical vapor deposition (LI-CVD) method at 780 °C from toluene-loaded polystyrene (PS) with different amounts of PS in the presence of ferrocene used as a catalyst. Then, acid-washed Bucky papers were produced, and their DC electrical properties were found to be in the range of 2.2–4.4 Ohm, with no direct correlation with PS loading. In addition, MWCNTs were used by the authors to fabricate MWCNT-based ethernet cables, consisting of tightly packed MWCNT powders. These 2-cm long carbon wires were then tested in server-to-client data transfer operations. Interestingly, the authors measured data transfer rates up to about 99 Mbps for the MWCNT-based wires. More interestingly, the life cycle assessment (LCA) of MWCNT-based wires was compared to that of copper ones for a use-case scenario in a Boeing 747-400 airliner over its lifetime. Due to their lightweight properties, MWCNT wires were found to reduce the CO_2 footprint by 21 kTonnes (kTe) over the overall life of the aircraft.

Louw et al. [25] modelled a new proposed Fe^{+++} ion sensor based on Eu-MOF crystals placed on a polymer surface. In their paper, the diffusion properties of ferric ions through a solution and a polymer layer and the interaction with a MOF crystal located at the interface between the solution and the polymer were investigated. The authors adopted a 2D diffusion model to predict the progress of Fe^{+++} through the solution and the polymer, and the association of Fe^{+++} with a MOF crystal at the interface. In the paper, a facile 1D model was reported to find the most appropriate values for the dimensionless parameters required to optimize the time for a MOF crystal to reach a steady state. Interestingly, a large non-dimensional diffusion coefficient was obtained and a small effective flux association reducing the time to reach a steady state was predicted by the authors' model.

Pawlin et al. [26] reported the synthesis of multicolor light-emitting nanomaterials based on Tb^{3+} and Eu^{3+} rare earth co-doped oxyfluoride glass-ceramics containing BaF_2 nanocrystals. Excitation and emission spectroscopy along with decay analysis from the $^5D_4$ level of Tb^{3+} was performed and the authors observed that co-doping with Eu^{3+} caused the decrease in decay times of the $^5D_4$ state from 1.11 ms to 0.88 ms and from 6.56 ms to 4.06 ms for xerogels and glass-ceramics, respectively. Hence, based on lifetime values, the Tb^{3+}/Eu^{3+} energy transfer (ET) efficiencies were estimated to be ca. 21% for xerogels and 38% for nano-glass-ceramics. The authors explained the increase in energy transfer efficiency by shortening the separation between interacting Tb^{3+} and Eu^{3+} cations embedded into the BaF_2 nanocrystal lattice.

We hope that contributions collected in this Special Issue may benefit researchers and experts in various fields for growing their knowledge in the fields of energy nanomaterials, thus stimulating new relevant studies. We also convey our truthful appreciation to the editorial staff, authors, and referees for their constant and rapid support, beneficial contributions, and appropriate comments.

Author Contributions: The Editorial was written through the contributions of all authors (S.R., F.C. and M.J.U.). All authors have read and agreed to the published version of the manuscript.

Funding: This research received no external funding.

Acknowledgments: This work was supported by MIUR (Ministero dell'Istruzione, dell'Università e della Ricerca), INSTM Consorzio and NIS (Nanostructured Interfaces and Surfaces) Inter-Departmental Centre of University of Torino. The Guest Editors express their gratitude to the valuable contributions by all authors. MJU gratefully acknowledge the support received by NSF PREM award under grant No. DMR-2122178. SR gratefully acknowledges European Regional Development Fund, project No. 01.2.2-LMT-K-718-02-0011.

Conflicts of Interest: The authors declare no conflict of interest.

Abbreviations

CL: colloidal lithography; CNCs: cellulose nanocrystals; CTL: charge trapping layer; CV: cyclic voltammetry; DC: direct current; DFT: density functional theory; ESM: electrochemical strain microscopy; ESR: equivalent series resistance; GPE: gel polymer electrolyte; HAADF: high-angle annular dark-field detector; LCA: lifecycle assessment; LEIF: low-energy ion facility; LI-CVD: liquid injection chemical vapour deposition; LLZO:Ta: Li6.5La3Zr1.5Ta0.5O12; MIP-207: Ti-based metal-organic framework, MOFs: metal-organic frameworks; MOR: methanol oxidation reaction; MWCNTs: multi-walled carbon nanotubes; PCM: phase change material; PEO_6–LiTFSI: polyethylene oxide with Li bis(trifluoromethanesulfonyl)imide; PV: photovoltaic; rGO: reduced graphene oxide; SAED: selected area electron diffraction; SCs: supercapacitors; TENGs: triboelectric nanogenerators; TES: thermal energy storage; ZHSCs: zinc-ion hybrid supercapacitors.

References

1. Inshakova, E.; Inshakova, A.; Goncharov, A. Engineered nanomaterials for energy sector: Market trends, modern applications and future prospects. *IOP Conf. Ser. Mater. Sci. Eng.* **2020**, *971*, 032031. [CrossRef]
2. Jain, P.; Prasad, E. *Nanotechnology in Energy Market by Material Type (Nanostructured Material, Carbon Nanotubes, Fullerene, Others), Application (Photovoltaic Film Coating, Fuel cells and Batteries, Thermoelectric Materials, Aerogels) and End Use (Electrical, Manufacturing, Renewable & Non-Renewable Energy and other Applications): Global Opportunity Analysis and Industry Forecast, 2021–2030*; Allied Market Research: Portland, OR, USA, 2021.

3. Scholes, G.D.; Fleming, G.R.; Olaya-Castro, A.; van Grondelle, R. Lessons from nature about solar light harvesting. *Nat. Chem.* **2011**, *3*, 763–774. [CrossRef]
4. Xu, Q.; Zhang, L.; Yu, J.; Wageh, S.; Al-Ghamdi, A.A.; Jaroniec, M. Direct Z-scheme photocatalysts: Principles, synthesis, and applications. *Mater. Today* **2018**, *21*, 1042–1063. [CrossRef]
5. Wu, C.; Wang, A.C.; Ding, W.; Guo, H.; Wang, Z.L. Triboelectric Nanogenerator: A Foundation of the Energy for the New Era. *Adv. Energy Mater.* **2019**, *9*, 1802906. [CrossRef]
6. Hosemann, P.; Frazer, D.; Fratoni, M.; Bolind, A.; Ashby, M.F. Materials selection for nuclear applications: Challenges and opportunities. *Scr. Mater.* **2018**, *143*, 181–187. [CrossRef]
7. Wu, F.; Maier, J.; Yu, Y. Guidelines and trends for next-generation rechargeable lithium and lithium-ion batteries. *Chem. Soc. Rev.* **2020**, *49*, 1569–1614. [CrossRef]
8. He, Y.; Zhou, W.; Qian, G.; Chen, B. Methane storage in metal–organic frameworks. *Chem. Soc. Rev.* **2014**, *43*, 5657–5678. [CrossRef]
9. Alva, G.; Lin, Y.; Fang, G. An overview of thermal energy storage systems. *Energy* **2018**, *144*, 341–378. [CrossRef]
10. Cesano, F.; Uddin, M.J.; Damin, A.; Scarano, D. Multifunctional Conductive Paths Obtained by Laser Processing of Non-Conductive Carbon Nanotube/Polypropylene Composites. *Nanomaterials* **2021**, *11*, 604. [CrossRef]
11. Cesano, F.; Uddin, M.J.; Lozano, K.; Zanetti, M.; Scarano, D. All-Carbon Conductors for Electronic and Electrical Wiring Applications. *Front. Mater.* **2020**, *7*, 219. [CrossRef]
12. Oliveira, R.D.; Mouquinho, A.; Centeno, P.; Alexandre, M.; Haque, S.; Martins, R.; Fortunato, E.; Águas, H.; Mendes, M.J. Colloidal Lithography for Photovoltaics: An Attractive Route for Light Management. *Nanomaterials* **2021**, *11*, 1665. [CrossRef]
13. Song, Z.; Fang, K.; Sun, X.; Liang, Y.; Lin, W.; Xu, C.; Huang, G.; Yu, F. An Effective Method to Accurately Extract the Parameters of Single Diode Model of Solar Cells. *Nanomaterials* **2021**, *11*, 2615. [CrossRef]
14. Sahoo, M.; Lai, S.-N.; Wu, J.-M.; Wu, M.-C.; Lai, C.-S. Flexible Layered-Graphene Charge Modulation for Highly Stable Triboelectric Nanogenerator. *Nanomaterials* **2021**, *11*, 2276. [CrossRef]
15. Durairaj, A.; Maruthapandi, M.; Saravanan, A.; Luong, J.H.T.; Gedanken, A. Cellulose Nanocrystals (CNC) based functional materials for supercapacitor applications. *Nanomaterials* **2022**, *12*, 1828. [CrossRef]
16. Bakardjieva, S.; Plocek, J.; Ismagulov, B.; Kupčík, J.; Vacík, J.; Ceccio, G.; Lavrentiev, V.; Němeček, J.; Michna, Š.; Klie, R. The Key Role of Tin (Sn) in Microstructure and Mechanical Properties of Ti_2SnC (M_2AX) Thin Nanocrystalline Films and Powdered Polycrystalline Samples. *Nanomaterials* **2022**, *12*, 307. [CrossRef]
17. Veelken, P.M.; Wirtz, M.; Schierholz, R.; Tempel, H.; Kungl, H.; Eichel, R.-A.; Hausen, F. Investigating the Interface between Ceramic Particles and Polymer Matrix in Hybrid Electrolytes by Electrochemical Strain Microscopy. *Nanomaterials* **2022**, *12*, 654. [CrossRef]
18. Cui, R.; Zhang, Z.; Zhang, H.; Tang, Z.; Xue, Y.; Yang, G. Aqueous Organic Zinc-Ion Hybrid Supercapacitors Prepared by 3D Vertically Aligned Graphene-Polydopamine Composite Electrode. *Nanomaterials* **2022**, *12*, 386. [CrossRef]
19. González-Gil, R.M.; Borrás, M.; Chbani, A.; Abitbol, T.; Fall, A.; Aulin, C.; Aucher, C.; Martínez-Crespiera, S. Sustainable and Printable Nanocellulose-Based Ionogels as Gel Polymer Electrolytes for Supercapacitors. *Nanomaterials* **2022**, *12*, 273. [CrossRef]
20. Askari, M.B.; Rozati, S.M.; Di Bartolomeo, A. Fabrication of Mn_3O_4-CeO_2-rGO as Nanocatalyst for Electro-Oxidation of Methanol. *Nanomaterials* **2022**, *12*, 1187. [CrossRef]
21. Wan, Y.; Miao, Y.; Qiu, T.; Kong, D.; Wu, Y.; Zhang, Q.; Shi, J.; Zhong, R.; Zou, R. Tailoring Amine-Functionalized Ti-MOFs via a Mixed Ligands Strategy for High-Efficiency CO_2 Capture. *Nanomaterials* **2021**, *11*, 3348. [CrossRef]
22. Bu, Y.; Liu, J.; Chu, H.; Wei, S.; Yin, Q.; Kang, L.; Luo, X.; Sun, L.; Xu, F.; Huang, P.; et al. Catalytic Hydrogen Evolution of $NaBH_4$ Hydrolysis by Cobalt Nanoparticles Supported on Bagasse-Derived Porous Carbon. *Nanomaterials* **2021**, *11*, 3259. [CrossRef]
23. Mahroug, I.; Doppiu, S.; Dauvergne, J.-L.; Serrano, A.; Palomo del Barrio, E. $Li_4(OH)_3Br$-Based Shape Stabilized Composites for High-Temperature TES Applications: Selection of the Most Convenient Supporting Material. *Nanomaterials* **2021**, *11*, 1279. [CrossRef]
24. Orbaek White, A.; Hedayati, A.; Yick, T.; Gangoli, V.S.; Niu, Y.; Lethbridge, S.; Tsampanakis, I.; Swan, G.; Pointeaux, L.; Crane, A.; et al. On the Use of Carbon Cables from Plastic Solvent Combinations of Polystyrene and Toluene in Carbon Nanotube Synthesis. *Nanomaterials* **2022**, *12*, 9. [CrossRef]
25. Louw, K.I.; Bradshaw-Hajek, B.H.; Hill, J.M. Ferric Ion Diffusion for MOF-Polymer Composite with Internal Boundary Sinks. *Nanomaterials* **2022**, *12*, 887. [CrossRef]
26. Pawlik, N.; Szpikowska-Sroka, B.; Goryczka, T.; Pietrasik, E.; Pisarski, W.A. Luminescence of SiO_2-BaF_2:Tb^{3+}, Eu^{3+} Nano-Glass-Ceramics Made from Sol-Gel Method at Low Temperature. *Nanomaterials* **2022**, *12*, 259. [CrossRef]

Review

Colloidal Lithography for Photovoltaics: An Attractive Route for Light Management

Rui D. Oliveira †, Ana Mouquinho *,†, Pedro Centeno, Miguel Alexandre, Sirazul Haque, Rodrigo Martins, Elvira Fortunato, Hugo Águas and Manuel J. Mendes *

CENIMAT/I3N, Departamento de Ciência dos Materiais, Faculdade de Ciências e Tecnologia, FCT, Universidade Nova de Lisboa, and CEMOP/UNINOVA, 2829-516 Caparica, Portugal; rdd.oliveira@campus.fct.unl.pt (R.D.O.); p.centeno@campus.fct.unl.pt (P.C.); m.alexandre@campus.fct.unl.pt (M.A.); s.haque@campus.fct.unl.pt (S.H.); rfpm@fct.unl.pt (R.M.); emf@fct.unl.pt (E.F.); hma@fct.unl.pt (H.Á.)
* Correspondence: a.mouquinho@campus.fct.unl.pt (A.M.); mj.mendes@fct.unl.pt (M.J.M.)
† Those authors contributed equally to this work.

Abstract: The pursuit of ever-more efficient, reliable, and affordable solar cells has pushed the development of nano/micro-technological solutions capable of boosting photovoltaic (PV) performance without significantly increasing costs. One of the most relevant solutions is based on light management via photonic wavelength-sized structures, as these enable pronounced efficiency improvements by reducing reflection and by trapping the light inside the devices. Furthermore, optimized microstructured coatings allow self-cleaning functionality via effective water repulsion, which reduces the accumulation of dust and particles that cause shading. Nevertheless, when it comes to market deployment, nano/micro-patterning strategies can only find application in the PV industry if their integration does not require high additional costs or delays in high-throughput solar cell manufacturing. As such, colloidal lithography (CL) is considered the preferential structuring method for PV, as it is an inexpensive and highly scalable soft-patterning technique allowing nanoscopic precision over indefinitely large areas. Tuning specific parameters, such as the size of colloids, shape, monodispersity, and final arrangement, CL enables the production of various templates/masks for different purposes and applications. This review intends to compile several recent high-profile works on this subject and how they can influence the future of solar electricity.

Keywords: colloidal lithography; thin-film photovoltaics; photonics; light-trapping; self-cleaning

1. Introduction

Highly efficient renewable energy sources and storage devices are needed to deal with the increasingly expensive energetic demands of our society. Considering the depleting fossil fuel stock and the devastating effects of global warming, technologies like photovoltaics (PV) have become one of the leading contenders in this field, as PV offers a broad diversity of devices—each with their potential use and functionality [1–4]. Recent reports [5,6] show that, despite its current small output, about half of the growth in the electric production capacity worldwide is now held by solar energy systems (Figure 1), with the technology costs decreasing largely. These are indicators of a clear energy transition with large investment, highlighting a tremendous growth potential.

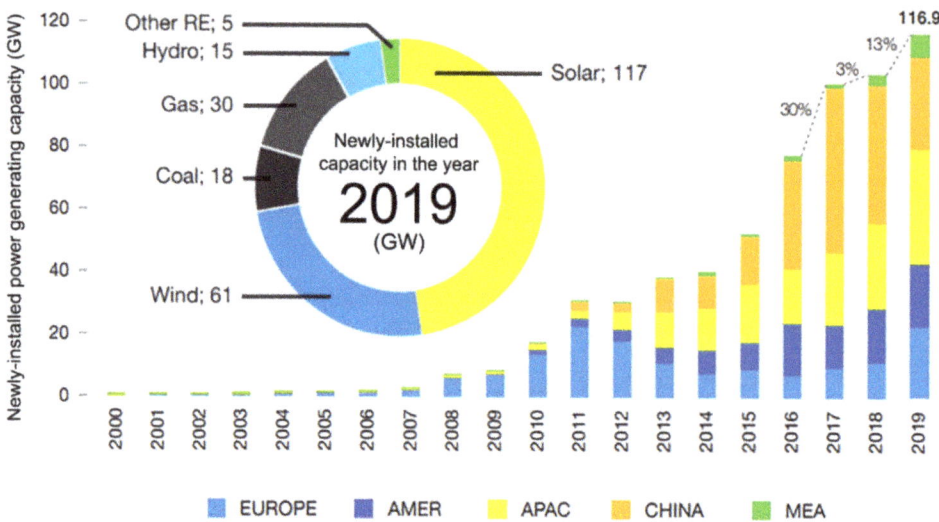

Figure 1. Annual growth of the solar energy market in each development area/country. Although Europe is slowly increasing its production, China is decreasing its yearly rate—overall, the global market is growing largely throughout the years. The inset pie plot presents the net power generating capacity (in GW) added in 2019 for several energy-generating sources, showing that the biggest market growth belongs to solar energy. Adapted with permission from [5]. Copyright 2020 SolarPower Europe.

First-generation solar cells based in mono-crystalline silicon wafers convert a large fraction of the incident sunlight energy with an efficiency of up to ~26 %, being still the most commercially used PV technology with a widespread application on rooftops and solar farms. Second-generation solar cells, based on thin-film technologies, have shown signs of becoming a competitive PV class due to potential advantages in low cost, large area, lightweight, solution-process fabrication, and mechanical flexibility [7]. In this category, we can additionally include semitransparent organic and hybrid (organic-inorganic) PV devices, which also tend to be suitable for indoor applications as they work relatively well with diffuse visible light [8–10].

Thin-film solar cells enable fast and cheap production methods, such as flexible roll-to-roll processes [9,11]. Moreover, thin devices benefit from lower material usage and thence further cost reduction, also with the potential of increase in the open-circuit voltages, V_{OC} (and consequently efficiencies), due to lower bulk recombination. These are crucial factors at the industrial level for cost-effective production, making the technology attractive for application in affordable solar-powered consumer products such as mobile electronics (e.g., wearable PV), intelligent packaging (e.g., smart labels), electronic devices for Internet of Things (IoT) applications (e.g., smart buildings), and portable medical diagnostic services [12–18].

The efficiency of the solar cells is inherently limited by the absorber material's bandgap, as it sets the lower energy limit for absorption [19]. Moreover, thin-film PV suffers from additional absorption losses from the smaller travel path of light within the thin absorber layer [20]. One method to circumvent the first problem would be shifting the incident lower energy photons into higher energy photons [21]. To overcome the second drawback, advanced light management techniques must be used to improve the optical response of the devices. From these, light-trapping (LT) schemes that create challenging conditions for light to escape the device have been the topic of many studies [22–31]. The development of photonic structures, implemented via nano-patterning methods [32,33], has been fundamental for PV performance enhancements, for instance by extending the absorption onset to the near-infrared (NIR) region of the solar spectrum [34].

As a production technique, microfabrication has been essential to modern science and technology through its role in microelectronics and optoelectronics. Photolithography is the current state-of-the-art patterning technique, and as such, it is the most well-established microfabrication method [35,36]. However, it is primarily limited by its low diffraction-limited resolution, high-cost, and low-throughput. The sizes of the features that it can produce (i.e., patterning resolution) are mostly determined by the wavelength of the radiation used. As such, small features require high-energy radiation and thence complex and expensive facilities and technologies [37,38]. Moreover, photolithography cannot be easily applied to nonpolar surfaces, as it tolerates little variation in the materials that can be used, and it provides almost no control over the chemistry of the patterned surfaces.

With these disadvantages in mind, many alternative techniques have been developed to fabricate nanostructures [39]. For instance, electron beam lithography is capable of fabricating designs with <10 nm resolution, with a high level of acceptance for application in more sophisticated devices where costs are not critical, but is even more limited in terms of patterning speed [40]. Focused ion beam (FIB) lithography has similar advantages and drawbacks, but permits patterning without the use of a resistor or mask [41]. Nevertheless, these techniques do not entirely mitigate the previously mentioned issues of photolithography, such as the cost, low throughput, and the requirement for highly sophisticated equipment. Hence, such disadvantages prompted research for unconventional soft-lithography fabrication techniques [39] as: nanoimprint lithography (NIL) [42,43], hot embossing [44], thermal injection molding [45], light-initiated polymerization (with ultraviolet, UV-NIL, and step-and-flash NIL [43,46]), solvent-based processing [47], and colloidal lithography (CL) [48]. These cost-effective soft techniques have brought focus to the patterning world, as they can be used in nonpolar surfaces—increasing the range of allowed materials, enable large scale patterning, and most importantly, employ industrially attractive fabrication methods due to their ease of use and low manufacturing cost [49].

Among several soft-lithography techniques developed in the last decade, NIL and CL sparked the highest research interest for micro and nanostructuring in photovoltaics [42]. In NIL [50], a pattern is created by pressing a mold into the resist, thus printing the inverse design of the mold. This technique is severely limited when applied to large areas, due to sticking issues from the large contact area between the mold and the imprinted structure, as well as the low pattern fidelity over large areas since the polymer chains in the stamping materials tend to relax elastically. NIL depositions are also strictly reserved for materials that can be molded and cured at moderated temperatures, resulting in a limited range of materials that can be effectively NIL patterned. As such, NIL may not provide the best solution for photonic applications, which usually require dense dielectric or metal oxide materials with a high refractive index for stronger interaction with light [26,51].

The main focus of this study, CL, is presently considered the most promising nano/micro-structuring method for photonic and PV applications [52]. It is an especially interesting technique since it can pattern almost any material, as it is not affected by the aforementioned limitations of NIL. It uses low-temperature steps (<100 °C), therefore not limiting the usage of temperature-sensitive materials (e.g., polymeric-based flexible substrates) or devices (e.g., perovskite solar cells, PSCs) [52], tolerating a wide range of materials and surface chemistries. Properties such as processing simplicity, low cost, and substrate agnostic patterning make CL a highly desirable method [53]. It can also produce well-ordered two-dimensional (2D) and three-dimensional (3D) periodic arrays of nanoparticles from various materials on many substrates. Three-dimensional layers are of tremendous interest for photonic crystal-based applications [49], whereas two-dimensional layers can be used as etching or lithographic masks that can be used for nanofabrication of several structures, specially photonic-enhanced PV devices [29,31,52].

In terms of the CL resolution, it is solely dependent on the colloidal particle sizes that can be deposited, thus allowing nanoscale patterning. However, the smaller the particles the more they become affected by destabilizing Brownian forces which prevent their ordered arrangement in the self-assembly process. Consequently, thus far the minimum feasible resolution of CL ranges between 50 to 200 nm [22,54], which is comparable to that of state-of-the-art hard-lithography (more costly) systems, such as photolithography (set by the diffraction limit of UV light), but not as low as the resolution of E-beam or FIB (order of nanometers). Nevertheless, research means are underway to further improve the CL resolution, for instance by operating at low temperature to hinder the Brownian diffusion [55].

This article provides an overview of the present panorama of CL, exploring its working concept and the patterning materials, with a focus on its last-generation PV-related applications.

2. Colloidal Lithography (CL) Methodologies

The use of colloids in lithography has been studied for about 35 years, with the continuous development of nano and microfabrication methods reaching increasing potentialities [39]. The methodology (Figure 2) generally comprises two main stages: the patterning mask preparation (Figure 2a,b), followed by the nano/micro-structure production (Figure 2c,d). The process starts with a colloidal deposition technique which is further described in Section 2.2, where we present a large set of procedures that use self-assembled colloidal arrays for surface patterning. The variety of methods that can be used for the colloidal array formation, as well as for the subsequent structure production, shows the high versatility of this method for implementation in various applications.

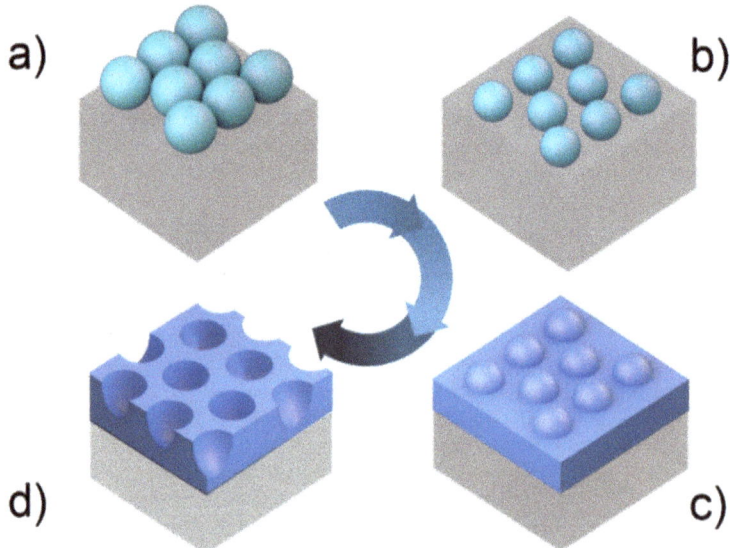

Figure 2. Illustration of the colloidal lithography (CL) main steps, depicting the sequence of (**a**) the deposition of colloidal particles on a surface, (**b**) reactive ion etching (RIE) for particle shaping, (**c**) material deposition, and (**d**) lift-off of the colloids leaving only the patterned material on the surface.

2.1. Colloidal Self-Assembly

Colloid particles are an important class of materials, sharing properties with bulk and molecularly dispersed systems. Their behavior is mostly governed by the particles' size, shape, surface area, and surface charge density [56]. Several techniques and protocols have been developed to synthesize highly monodispersed colloidal spheres with diameters ranging from a few to thousands of nanometers.

For CL applications, any particle material can potentially be used to create the self-assembled colloidal mask in the first 2 steps of Figure 2. However, the preference lies in colloidal materials that: (1) can be synthesized with precise monodispersed particle sizes; (2) allow highly-selective etching (RIE) in step b) of Figure 2; and (3) can be easily removed by chemical lift-off in step d) [54,57–64]. In view of that, the most synthesized particle materials for use in CL have been polystyrene (PS), polymethyl methacrylate (PMMA), and silica [65]. Good phase stability, together with a narrow colloidal size distribution (less than 5% for the typically employed microspheres), has been achieved by suspension [40], emulsion [66], and dispersion polymerization [64,67] synthesis techniques.

The self-assembly of colloidal particles, crucial for next-generation surface and volume nanostructuring applications, consists of their spontaneous arrangement into ordered superstructures (Figure 3) [68].

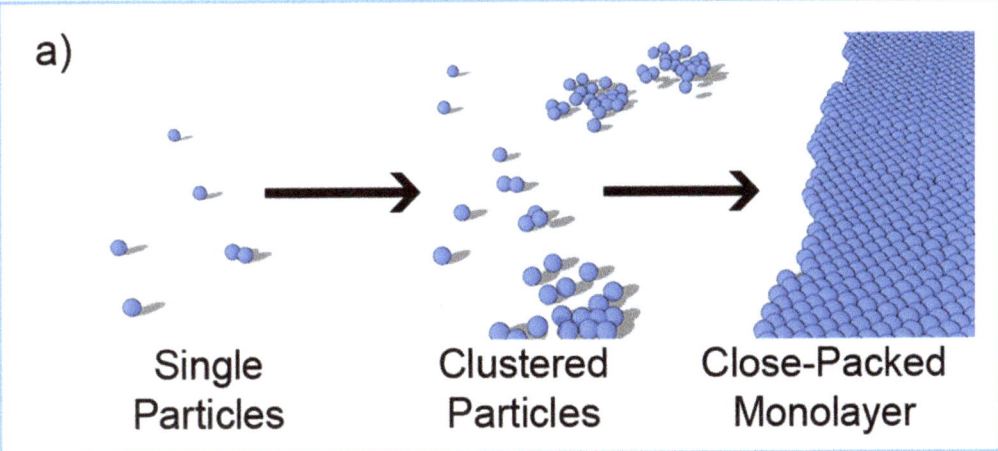

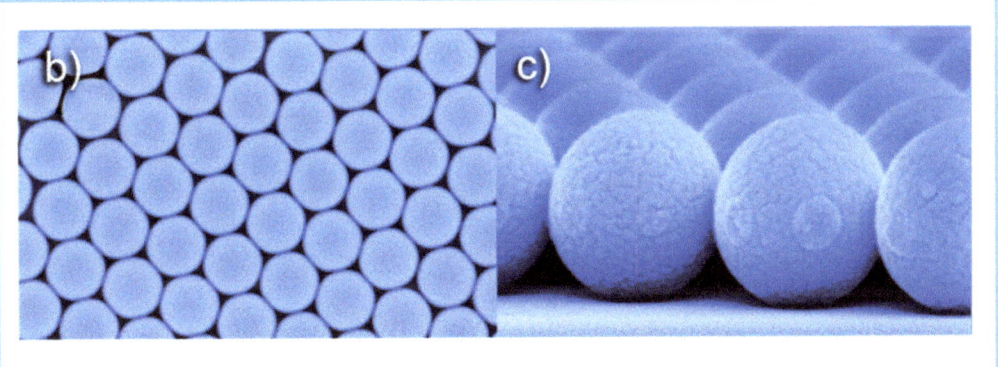

Figure 3. (**a**) Illustration of the self-assembly of colloidal particle structures forming a hexagonal close-packed array (also known as honeycomb), which results in the highest in-plane packing density; (**b**) Top-view and (**c**) cross-sectional scanning electron microscopy (SEM) images of honeycomb arrays of 800 nm PS spheres. Adapted with permission from [69]. Copyright 2021 Elsevier.

Based on the type of dominant force driving the self-assembly, these methods can be organized into four classes: physical (process dominated by shear forces, adhesion, and surface structuring), fluidic (by capillary forces, evaporation, surface tension), external fields (by electric and magnetic fields) and chemical (by chemical interaction, changing the surface charge or creating binding sites) [70].

In 1981, a lithographic method using self-assembled PS monolayers as a mask was first proposed by Fisher and Zingsheim [71]. Afterwards, Deckman and co-workers successfully increased the mask area for patterning [72]. Owing to the size, shape, and monodispersity, colloidal particles can self-assemble into 2D or 3D extended periodic arrays, but the 2D colloidal crystals are those that captured the most attention for PV application [53,73,74].

The production of self-assembled arrays of colloidal particles is the starting point of the CL process, which utilizes the close packing of such colloidal crystals to fabricate long-range ordered nano/micro-structures in/with any material [68,70,73].

Interesting examples are the fabrication of nanoporous templates [75], 3D photonic bandgap structures [76], and thin-film nanocrystal solids for electronic devices [77]. For instance, it has been shown that spherical colloidal particles coated with liquid crystals, or other materials having nematic degrees of freedom, can form composite materials that exhibit point defects with sp and sp^3 valences. For future applications, the most intriguing aspects of colloidal particles are their potential utility as building blocks, capable of mimicking molecular self-assembly through covalent and non-covalent interactions, to created artificially-designed materials [78].

2.2. Deposition of Colloidal Arrays

Three techniques should be emphasized when considering the initial step of colloidal monolayer deposition on the substrate of the CL method (see Figure 2a). These are spin-coating, doctor blade, and Langmuir–Blodgett sketched in Figure 4 and described in this sub-section.

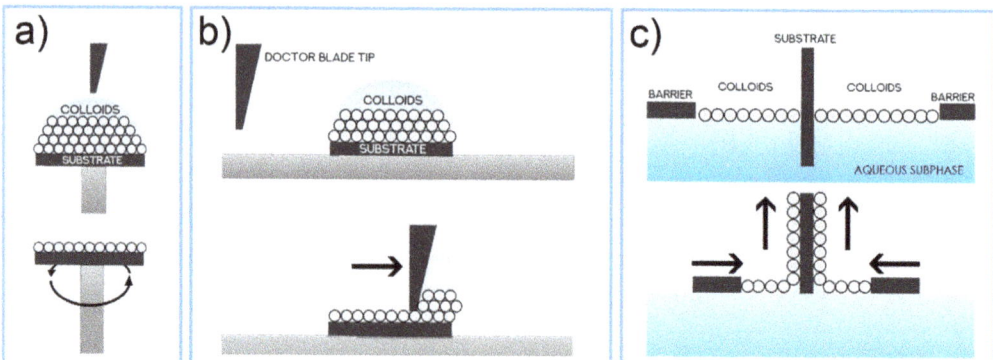

Figure 4. Production of a colloidal monolayer using (**a**) a spin-coating technique, (**b**) doctor blade trough, (**c**), and a Langmuir–Blodgett trough.

The spin-coating technique (Figure 4a) can be considered a simple process for rapidly depositing thin coatings onto relatively flat substrates [54,79–81]. A spinning fixture holds the substrate (often using vacuum to clamp and position the substrate in place), and the coating solution/dispersion is then dispensed onto the surface. The revolving action causes the solution to spread out and leave behind a uniform coating of the chosen material on the surface. Due to its short time of production, combined with its simplicity and low cost, this method is useful in industrial conditions as far as small-area batch coating processes are concerned. However, it is not compatible with large-area deposition, and the resulting colloidal films tend to be less uniform than those produced by dip-coating methods such as Langmuir-Blodgett [82].

Spin-coating experiments have mainly been designed to deposit small-area nanosphere monolayers, severely limiting the application of these films as physical masks. Therefore, important research parameters have been optimized to prepare high-ordered colloidal films with different diameter nanospheres on larger scales. For instance, by adjusting the spin speed and acceleration, Chen et al., (2013) [54] spin-coated long-range ordered colloidal crystal films of PS spheres with diameters of 223 nm, 347 nm, 509 nm, and 1300 nm. Furthermore, for the 509 nm of spheres' diameter, the team also used these conditions to inspect the relation between the monolayer coverage area and spin parameters. It was found that with the increase of the spin speed and acceleration, the monolayer coverage areas oscillated, with the largest ordered areas (near 100% of 25 mm $\times$ 25 mm $\times$ 0.5 mm quadrate and 3-inch circular silicon substrates) being achieved at a speed of 1700 rpm and acceleration of 600 rpm/s. Chen's results thus revealed the successful preparation of monolayer and bilayer films of PS nanospheres with four different diameters. In the only structure that was considered to be with reasonable hexagonal close-packed ordering, both monolayers and bilayers could be found, which is not suitable for colloidal lithography applications.

Another impactful development was reported by Park et al. [81], who introduced polyoxyethylene (12) tridecyl ether (PEO-TDE) as a surfactant for the spin-coating of PS nanosphere monolayers onto Si wafers and glass substrates, under ambient laboratory conditions, with optimal surfactant properties, as opposed to the conventional highly toxic Triton X-100 surfactant. Low viscosity and surface tension cause this mixture to show excellent wettability, which results in superior coverage and uniformity [81].

Another simple, but highly scalable process for nano/microspheres deposition is the doctor blade coating or blade coating technique (Figure 4b). This method is widely used in the textile, paper, photographic film, printing, and ceramic industries to create highly uniform flat films over large areas [83]. An immobilized blade (or rod) applies a unidirectional shear force to a slurry that passes through a small gap between the blade and the substrate. This is a roll-to-roll compatible method that has played a crucial role in ceramic processing to produce thin, flat ceramic tapes for dielectrics, fuel cells, batteries, and functionally graded materials. A simplified doctor blade coating process was developed by Velev et al. [84], based on an evaporative colloidal assembly technology that relies on capillary forces to drive and merge colloidal particles into crystalline structures with thicknesses ranging from a single monolayer to a few layers. Inspired by this technology, Yang et al., (2010) [85] reported a roll-to-roll compatible doctor blade technology for producing highly ordered colloidal crystals (mainly polymer nanocomposites) and macroporous polymer membranes. The resulting 3D-ordered structures exhibited uniform diffractive colors, and Yang has shown that the templated macroporous membranes with interconnected voids and uniform interconnecting nanopores can be directly used as filtration membranes to achieve the size-exclusive separation of particles.

Lastly, the Langmuir–Blodgett (LB) method [86] (Figure 4c) consists of the compression of nanoparticles, floating in an air-liquid interface, into monolayers—Langmuir films—and its transferal onto immersed solid substrates via vertical dipping [87]. This technique offers the possibility to obtain highly ordered, well-defined, controlled mono/multilayers, ultimately serving the patterning purposes for CL applications [88]. Common LB-deposited materials have amphiphilic molecules with two distinct regions: a hydrophilic head group (water affiliation) and a hydrophobic tail group (water repulsion). They must be soluble in organic nonpolar and water-immiscible solvents (ethanol, diacetone, chloroform, benzene, among others [89,90]), with water-insoluble amphiphilic molecules forming a floating monolayer at the air-water interface. Long-chain fatty acid and lipid molecules are examples of typical LB-deposited materials, but the method has also been found successful (highly precise) for the patterning of close-packed monolayers of colloidal spheres, as for CL.

The two main steps of the LB technique are the preparation of a floating self-assembled colloidal monolayer at the air-water interface (Langmuir-film) and its deposition on a solid substrate [91]. At first, the colloidal particles are dispersed in a volatile and preferably water-insoluble solvent to prepare the colloidal dispersion. Then, small amounts of this solution are carefully deposited and spread onto the air-water interface at the LB trough. Afterward, the volatile solvent evaporates, and the LB barriers are compressed accordingly to force the formation of a self-assembled close-packed colloid monolayer at the interface [88]. Finally, the immersed substrate is withdrawn vertically from the aqueous subphase, while the lateral barriers continue to close in towards the substrate, at controlled rates, therefore transferring the colloids stabilized at the air/water interface to the upwards moving substrate [86], resulting in a successfully deposited monolayer colloidal film. Multilayer films can also be engineered by successively subjecting the previously deposited substrate to further cycles of LB deposition. These multilayers have been considered model membranes due to their remarkable 3D uniformity; and offer potential application as photonic waveguides and in breakthrough optoelectronic, nonlinear optical, and sensory devices [92,93].

Surface and interface chemistry is of paramount importance for defining how the colloidal particles float and are packed [32,33,81,94]. For instance, relatively small PS colloids with sizes close to visible wavelengths (under 1 μm) tend to sink into the aqueous subphase, contrasting with larger ones that typically float [94]. The fabrication of monolayers by interface coating methods as LB has been subject to numerous studies, varying the size of colloids, the amount of solution, temperature, deposition angle, and others [22,95–97].

It is also believed that the use of surfactants in the aqueous subphase may enhance the floating and Langmuir-film production of colloidal particles at the air/water interface, with larger areas and mechanical strength, similarly to the solutions presented by Vogel et al., (2011) [94] for the spin-coating method. Surfactant molecules tend to occupy the media interfaces and join the incoming colloidal particles together, thus opposing their dispersion caused by the Brownian motion. The surface assembly forces tend to enlarge the array area, increasing the monolayer order and coverage [95]. There is also a reduction of surface tension at the interface due to the presence of the surfactant, which favors colloidal particle movement along the interface to find their lowest energy configuration, resulting in an optimally-ordered hexagonal close-packed monolayer [98]. Adding the surfactant, however, may introduce undesired contamination to the interface. Therefore, care must be taken to avoid the transfer of substantial amounts of contaminants to the substrate during the LB lift-up process, to avoid imperfections in the deposited colloidal array [79].

From the three methods mentioned above, the Langmuir-Blodgett method combines quality, versatility, and scalability, being the headmost characteristics for the fabrication of high-quality 2D or 3D crystalline films. Furthermore, the capacity to precisely control the deposition of each layer in a layer-by-layer process, the ability to choose different particle sizes for each deposited layer, and the possibility for this method to be adapted to fast industrial production techniques such as roll-to-roll processing [22,99–101] make it an outstanding candidate for the first step of the CL process. These advantages are demonstrated in the research of O. Sanchez-Sobrado et al. [29,33,48] that has revealed outstanding results of thin-film solar cells enhanced with photonic front structures that were patterned via CL using highly-uniform LB-deposited colloidal masks.

2.3. Colloidal Masking for Surface Patterning

After achieving a good monolayer of close-packed colloidal particles, as previously described, it is then necessary to define how one can effectively use such array as a mask to achieve the desired microstructures in the targeted material (recall Figure 2). The final structures and properties achieved by the CL process are highly dependent on the prepared mask of material and packing [102,103]. Although attaching functional molecules or coating materials to colloids offers various possibilities for additional tuning of their properties [104], most polymeric or silica colloids end up being straightforwardly assembled into monolayers without functionalization, for further processing in CL.

The use of the originally deposited close-packed colloids (Figure 2a) as a mask allows only a limited exposure area in the interstitial spaces of the hexagonal array. Therefore, increasing the inter-particle distance in non-close-packed arrays (Figure 2b) is an important tool to optimize the masked area on the surface, at the expense of requiring an extra intermediate step of physical etching of the colloids.

Still, using the simpler CL version with close-packed 2D colloidal crystals as etching masks, triangular nanoparticles [105], nanodots [106], and thin-films with nanohole arrays [107], nanotips [108], or nanopillars [109] have been fabricated on several substrates (such as polymer-based, silica, and silicon) [65,69,97].

Regarding the formation of non-close-packed monolayers (Figure 2b), dry etching methods have been used (such as reactive ion etching, RIE) to reduce the size of the spheres after their deposition, and therefore increase their inter-space distancing in the array (Figure 5).

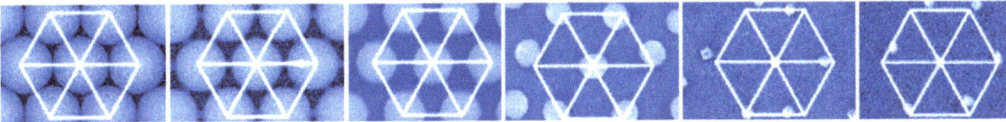

Figure 5. Schematic illustration of the reduction of the colloidal spheres dimension (with 0.5 µm initial diameter) and increase of their spacing in the array, via low-frequency plasma etching, with increasing etching time (left to right). Adapted with permission from [60]. Copyright 2021 MDPI.

A recent study from Yun Chen et al., (2019) [60] has shown that low-frequency plasma etching (40 kHz) can be used to produce PS nanospheres-based arrays with smooth surfaces, doubling the etching rate when compared to high-frequency systems. This study revealed that low-frequency RIE processes are dominated by a thermal evaporation etching mechanism, different from the atom-scale dissociation mechanism that underlies the high-frequency etching. It was found that the PS features size can be precisely controlled by adjusting the etching time and/or power. By introducing oxygen as the assisting gas in the low-frequency RIE system, one can achieve a coalesced PS particle array and use it, for instance, as a template for metal-assisted chemical etching which can significantly improve the aspect ratio of silicon nanowires to over 200 due to their improved flexure rigidity.

RIE has also been used in the CL fabrication of optimized photonic front structures for light-trapping in thin-film solar cells, which is crucial for increasing light absorbance in the absorber layer and subsequently the performance of the devices. Recently, two efficient four-step approaches were described by Mendes et al., (2020) [52] that can produce two types of geometries based on arrays of semi-spheroidal voids or domes, as shown in Figure 6. Briefly, for both cases, this method starts with the deposition of a monolayer of close-packed colloidal PS microspheres (LB method), followed by RIE producing a non-close-packed array.

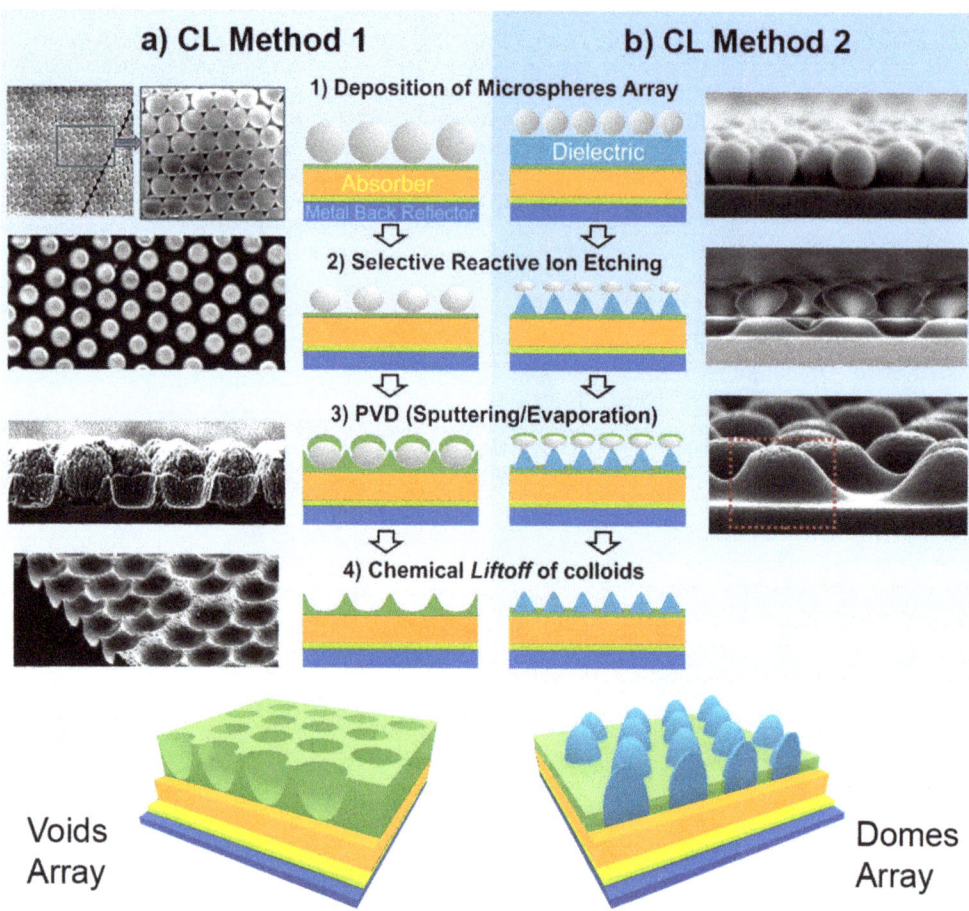

Figure 6. Depiction of two different CL methods used to create distinct geometries of photonic microstructures for light-trapping, integrated in the front contact of thin-film solar cells, arranged in non-closed-packed hexagonal (honeycomb) arrays of semi-spheroidal voids (**a**) or domes (**b**) Adapted with permission from [52]. Copyright 2021 Elsevier.

The main difference between both approaches occurs in the first two steps. In the first method (Figure 6a), the selective RIE only acts on the particles, so the final void-like structures are defined by the subsequent deposition of material in the inter-spaces between particles (as in Figure 2c,d). Although in the second method (Figure 6b), the less-selective RIE process also etches the underlying layer, ultimately defining the final dome/cone-like structures obtained.

Using etched nanospheres as molds/masks in processes such as metal deposition, infiltration, or imprint, it has been possible to produce ordered arrays of spherical voids [110] and nanoshells [111,112]. Through dewetting around nanospheres, nanorings of polymers [113], carbon nanotubes, or nanoparticles [63] can be obtained. Resorting to site-selective deposition or etching, nanospheres with asymmetric shapes or functional features have been produced [114], difficult or impossible to obtain by other synthetic routes.

After the deposition of the intended material onto the RIE-shaped colloidal mask in step c) of Figure 2, a lift-off treatment removes the colloids (step d) leaving only the microstructured material on the front surface. After this process, undesirable colloidal residues may be found in the areas previously occupied by the particles, due to incomplete removal. In such cases, besides the chemical removal (e.g., with toluene), both oxygen plasma [115] and thermal annealing [116] treatments can also be used to remove the polymeric particle residues which are usually quite volatile at temperatures around 100 °C.

3. Photonic Crystals

The previous section showed that there is a wide range of nano/micro-structure designs that can be engineered with CL techniques for various technologies, whose dimensions are chiefly set by the size of the masking particles. The colloidal particles' size propinquity to visible light's wavelengths will therefore grant the fabricated structures excellent interaction properties with this type of radiation (Figure 7) [57,70,117–119].

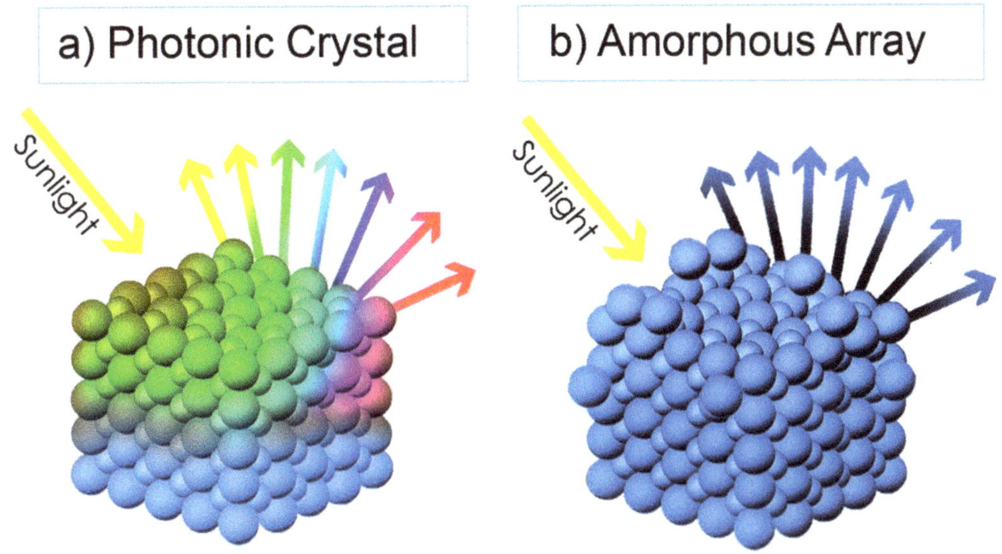

Figure 7. Schematic optical interaction of (**a**) an ordered colloidal crystal and (**b**) an amorphous colloidal array under white light. The structural color from the colloidal crystal changes depending on the viewing angle, while that of the amorphous array remains nearly unchanged.

It is important to note that photonic crystals have structural similarities with common crystals but have no direct relationship with crystalline materials. The dimensionality of the photonic crystal is defined by the length(s) in which the dielectric constant varies periodically, and they can be represented by basic 1D, 2D, and 3D crystals. However, only 3D photonic crystals allow for omnidirectional photonic bandgaps [120] that are an optical analog of the energy bandgap of the crystalline network.

In nature, one can find many examples of natural photonic crystals (Figure 8), such as in wings of butterflies and natural opals (Figure 8a,b). These natural crystals are composed of periodic microstructures whose scattering and transmission properties strongly depend on the incident light frequency, thus displaying brilliant colors, which have inspired artificial designs (Figure 8c,f).

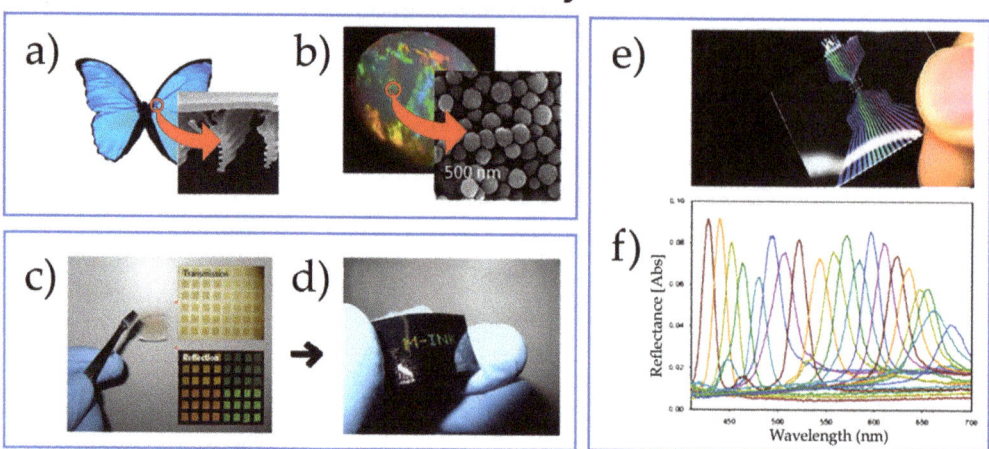

Figure 8. Natural photonic crystals (left): (**a**) photo showing the blue iridescence and SEM image of the 1D structure of the Morpho butterfly, (**b**) photo of an opal gemstone, and SEM image of the silica sphere structure within it. Adapted with permission from [121]. Copyright 2021 The Royal Society of Chemistry. Artificial fabricated photonic crystals deposited on flexible substrates made of semitransparent: (**c**) photonic crystal films and (**d**) photonic crystal films with anti-transmission black tape as the transferred substrate, which blocks backlight transmission [122], as well as (**e**) photo and (**f**) reflectance spectra of stripe-patterned composite photonic crystals with 20 different optical bandgaps. Adapted with permission from [123]. Copyright 2021 John Wiley and Sons.

Due to their unique characteristics, photonic crystals fabricated via colloidal assembly have attracted much interest for various prospective applications, ranging from gas sensing to optical filters [124,125], photonic papers [126], inkless printing [127], flat reflective displays [128], optical devices, photochemistry, and biological sensors [70]. Recent developments have further enhanced their complexity using non-spherical particles [129,130], binary colloidal dispersions [131], as well as controlled production of 3D defects (acting as optical cavities) within the crystals [132,133]. Several approaches have been developed enabling defect engineering controlled to a great extent, such as surface micromachining which allows for symmetries other than face-centered cubic [117,134].

Self-assembly has a crucial role in the fabrication of photonic crystals with a photonic bandgap in the visible and near-infrared region [135]. For instance, freestanding films have been fabricated by the layer-by-layer assembly [136], solution casting [137], surfactant-assisted deposition [138], and filtration of dispersions of materials using membrane filters [61]. A facile approach to fabricate such large asymmetric free-standing 2D array films is by forming 2D colloidal particle arrays at the air−water interface, as in the LB method [139]. Nevertheless, besides LB, colloidal photonic crystal growth via self-assembly of monodispersed colloids can involve, as previously mentioned, various fabrication methods such as controlled evaporation, spin coating, shear growth, among others.

Although the fabrication of 1D or 2D photonic crystals is relatively straightforward, adapting the conventional patterning techniques to fabricate 3D crystals remains a challenge. This originates from the stringent constitutional quality, and functional requirements. Several methods have been proposed in this respect. The most economical and direct approach to fabricating 3D photonic crystals is also by the self-organization of colloidal particles. The most inexpensive and direct approach to fabricating 3D photonic crystals is also by the self-organization of colloids that can be used as a template. Inverse opals materials with a high degree of periodicity in three dimensions are important templates for the design of photonic crystals. One method used to prepare these photonic crystals consists in infiltrating the void spaces between spheres in a colloidal crystal template with the desired material in solution phase (sol-gel), which is subsequently solidified. The subsequent removal of the templating spheres leaves a structured photonic crystal [140]. The sequential passivation reactive ion etching (SPRIE) method, [117] as the name suggests, relies on sequential passivation and reactive ion etching reactions using C_4F_8 and SF_6 plasma chemistries. It allows the addition of the third dimension using a simple and robust protocol for direct structuring of silicon-based 3D photonic crystals. Through a single processing step, SPRIE transcribes 2D colloidal crystal arrangements into well-ordered 3D architectures. The lateral etch extent controls various 3D topologies, useful in the delamination of 3D photonic crystals slabs or for the insertion of structural defects [117].

Alternatively, direct writing or single-step processing techniques have emerged as powerful tools for rapid and scalable 3D photonic crystals fabrication. Multiphoton polymerization lithography seems an attractive scenario, as it allows for unprecedented control of the crystal geometry and the defect incorporation, although it suffers from low throughput due to the serial writing procedure [117]. By infiltrating the interstices of polymer latex colloidal crystals with inorganic materials, and subsequently burning out the polymer latex, ordered macroporous films known as 'inverse opals' can be prepared. Inverse opals can have higher reflectivity over wider optical stop gaps (which prohibits light from propagating in only some directions) [121] due to the higher refractive index mismatch between the spheres and the medium. Such structures can also have 3D photonic bandgaps in the very high-frequency regions. Despite their bandgaps being very narrow, these can be reduced even further with the introduction of small defects, since they exhibit wider stop gaps and broader mechanical stability. For the production of inverse opal structures, colloidal templates of inorganic colloids (silica) can be used. Within this process, composites with polymeric materials are formed and then converted to polymeric inverse opal structures via the removal of the inorganic particles using selective etchants such as hydrofluoric acid [141].

Non-crystalline colloidal arrays—photonic glasses (see Figure 8b)—have also much interest for certain applications. These structures consist of aggregates of monodisperse colloids with short-range order, over a range of a few particles, that can be detected from the structure's diffraction pattern. Although photonic crystals can be used to manipulate ballistic photons, photonic glasses are useful in controlling light diffusion. The random structures of designed uniform colloids can interact strongly with light and produce unusual diffusion phenomena, including random lasing, angle-independent color, and light localization. These disordered monodisperse and short-range ordered particle structures have been produced by destabilizing the colloids with the control of the salt concentration, or by adding particles of different sizes to the colloidal solution [141].

4. Photovoltaics Enhanced with Micro-Structuring

The amazing light-interaction properties of wavelength-sized structures overviewed in the previous section have motivated their development for optical manipulation in PV devices, aiming for maximum sunlight conversion to electrical power. In particular, thin-film solar cells suffer from significant absorption losses, relative to thicker wafer-based cells, due to their diminished absorber thickness. As such, advanced light management techniques are necessary to compensate for such losses and ensure that high efficiencies are achieved [20,33]. As was already introduced, nanophotonic elements in the wave-optics regime are seen as a promising method to efficiently trap light inside the thin absorber material, thus boosting its broadband absorption, as further discussed in Section 4.1 [142].

The self-assembled templates from colloidal spheres can provide monolayers with long-range order throughout large device areas, thus providing an inexpensive and easily scalable mask to engineer materials with the physical parameters appropriate for efficient light-trapping (LT) in solar cells. This has led to a significant interest in CL methods for photovoltaic devices, which is the focus of this review. Nevertheless, the CL applications in the PV field are not limited to the integration of LT structures. In the next sub-sections, we will highlight two other promising nano/micro-structuring solutions that have also been demonstrated with CL: namely for transparent electrodes (Section 4.2) and self-cleaning functionality (Section 4.3).

A review article by Wang (2018) [73] described that CL can produce several different patterns and geometries that could form an anti-reflective LT mechanism. It may include Janus particles, hexagonal and non-close-packed single layers, double layers, free-standing films, and template-induced arrangements. The nanostructures obtained by this process can already be promptly applied to many different areas. Furthermore, this technique can also be adapted by tweaking the experimental parameters, such as the dimension of the spheres, the morphology of the surface, and chemical composition, thereby increasing the spectrum of possible applications.

Micro-meshed electrodes (MMEs) obtained by CL have been one of the most promising approaches to produce industrial-compatible transparent conducting materials (TCMs), with excellent optical transmittance and electrical conductance, composed of TCO/metal/TCO multilayers (TCO = transparent conductive oxide). In particular, replacing the intra-layer metallic film (usually silver) with a micro-grid of the same material has allowed additional degrees of freedom to optimize the TCM performance, namely enabling much higher transparency in the red-NIR (near-infrared) spectral range while maintaining highly conductive TCMs [62].

Besides efficiency, the outdoor reliability of PV systems is another crucial factor necessary for their widespread deployment. Solar panels tend to lose efficiency with time mainly due to unavoidable environmental degradation. Phenomena such as the formation of hot-spots (areas of large heat dissipation) caused by partial shading of solar cells (e.g., due to debris/dirt deposits) can be responsible for pronounced efficiency losses that, for instance, have reached 11% in three days and 65% after six months in certain power plants [74,143]. Therefore, decreasing unwanted processes that block the amount of sunlight reaching the cell, such as dust and other accumulation of particles, merits particular attention due to their inevitability and ubiquity [74].

Most current solutions to this include the mechanical cleaning of the devices [31] to expel dust specks via four types of techniques: the robotic method, air-blowing method, water-blowing method, and ultrasonic vibration method. Nevertheless, such type of active mechanical methods requires a power source for enabling the self-cleaning mechanism. Moreover, manual cleaning can also create cracks on the PV panel surface due to harsh brushing which will further deteriorate PV performance. Moreover, very small particles cannot be removed effectively by a manual cleaning process. On the other hand, the use of a self-cleaning coating, as no PV panel movement is required for its working function, is way simpler and more fitting to PV applications [74]. Another active method—the electrostatic cleaning method [144]—expels the surface dust through electrostatic standing and traveling waves, due to an existing electric curtain. The electric curtain consists of a series of parallel electrodes embedded in a dielectric surface, across which are transmitted oscillations in the electrode potentials. During this process, the standing wave oscillates the dust particles up and downward while a traveling wave does the same process in a horizontal direction.

Concerning passive coating methods, they employ either a superhydrophilic or superhydrophobic film on the outer PV surface. Superhydrophilic coatings reduce the amount of dirt through photocatalytic reactions, while superhydrophobic coatings potentiate the formation of water droplets and their roll-off, carrying away the dirt from the surfaces with minimum water usage [31,74]. To allow water (or other liquids) droplets to effectively roll down a surface, different superhydrophobic-oriented strategies are being investigated mainly via surface micro-structuring, mimicking natural processes such as the skin of certain plant leaves with self-cleaning capability [145], as further detailed in Section 4.3.

The aforementioned applications reveal that CL offers a wide range of promising possibilities for the advancement of PV-related technologies, as illustrated in Figure 9 and elaborated in the following sections.

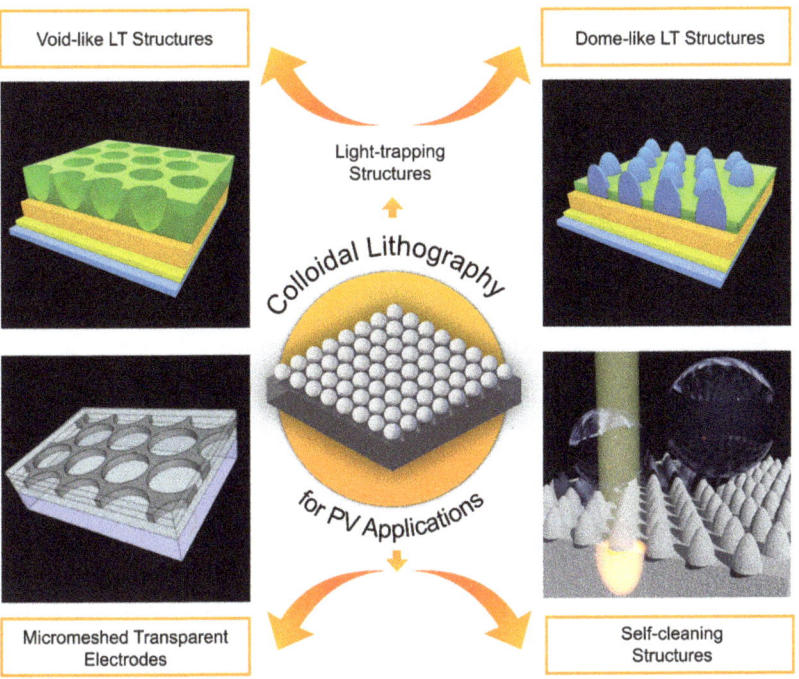

Figure 9. Illustration of the distinct applications of CL for micro-structuring in photovoltaics. Adapted with permission from [62]. Copyright 2021 Elsevier. Adapted with permission from [52]. Copyright 2021 Elsevier. Adapted with permission from [31]. Copyright 2021 John Wiley and Sons.

4.1. Light-Trapping in Photovoltaics

Light-trapping (LT) structures are a critical enabling factor in PV technology, as they improve the absorption of incident photons, therefore impacting its conversion efficiency [23,29,31,51,52,62,146,147].

On one hand, reflection losses are an unavoidable shortcoming in all types of PV technologies. Proper index matching in the surface—using materials with an index between that of the absorber material and the light incidence medium—can help mitigate this problem. Photonic structures can further diminish these losses in a broad wavelength range—by providing geometric index matching.

On the other hand, these structures also bring about LT mechanisms to help with light management within the device. As previously mentioned, this is particularly important for thin-film solar cells, where the short optical path is not enough to absorb all the incoming solar radiation, and it can be useful also for other emerging solar technologies [148]. This shortcoming has also been severely hindering flexible thin-film PV technology from achieving its market potential. As a matter of fact, many thin solar cells have so far only reached modest efficiencies (~14%) compared to those of conventional cells based on rigid silicon wafers (22–25%). Therefore, there is much improvement potential in thin-film PV with the implementation of effective LT techniques, mainly as a means to make the cells optically thicker but without increasing their physical thickness (to allow efficient charges collection) [52].

Here, a detailed analysis is presented of different types of photonic structures integrated via CL methods for LT in thin-film solar cells, particularly based in silicon and perovskite PV materials, enabling the development of high-efficiency flexible devices.

4.1.1. Computational Design and Optimization of Photonic Solutions

The resonant nature of wavelength-sized photonic structures substantially limits the parameter space in which their optical effects can provide exceptional absorption improvements across the relevant sunlight spectrum. Specifically, the wave-optical front features (as those integrated via the CL methods of Figure 6) need to provide a gradually varying effective refractive index, from the air towards the absorber layer to minimize reflection. Simultaneously, their geometry must interact with the incoming light to produce strong scattered fields preferentially directed into the higher index absorber layer. Therefore, before experimental implementation, it is crucial to perform a rigorous screening based on modeling, to understand the influence that the parameters of the LT structures have on such effects, and then appropriately search for the best parameter set that allows the highest photocurrent enhancement in the devices [142,149,150].

Theoretical studies reported by Mendes et al., (2016 and 2018) [142,149] have sought for designs of wavelength-sized structures optimized towards maximum broadband light absorption, with the double aim of enabling the reduction of the absorbing layer thickness (potentiating flexibility) while improving the efficiency of thin solar cells (Figure 10). Two geometries of front-located photonic structures were pointed out, compatible with the CL fabrication methods of Figure 6, and computationally optimized to maximize absorption without degrading the electrical performance of the devices (by avoiding increased recombination since the absorber layer remains flat, i.e., it is not corrugated as occurs with texturing).

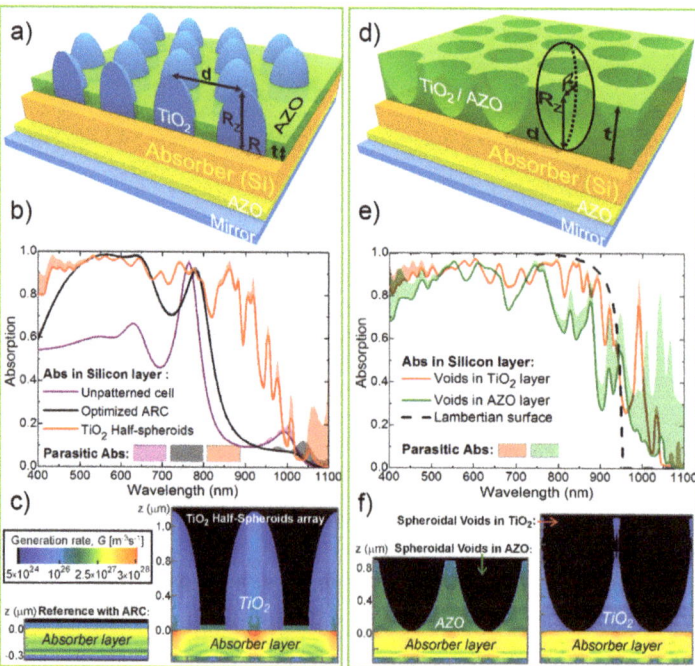

Figure 10. Electromagnetic modeling results of two types of LT geometries (sketched in **a**,**d**) composed of hexagonal arrays of TiO$_2$ half-spheroids (**a**–**c**) or semi-spheroidal voids in a TiO$_2$ or AZO layer (**d**–**f**), both integrated into thin-film (300 nm) Si solar cells. The results show the light absorption, Abs spectra (**b**,**e**) and photogeneration rate, G, profiles (**c**,**f**) of the optimized photonic front structures, compared with flat reference cell structures without LT or with a standard AZO antireflection coating (ARC). The G profiles reveal much higher carrier generation in the cells with the photonic elements due to the enhanced broadband absorption. Adapted with permission from [142]. Copyright 2021 Elsevier.

These LT structures are composed of honeycomb arrays of semi-spheroidal features either made of domes (TiO_2 half-spheroids separated by a flat aluminum-doped zinc oxide—AZO—layer—Figure 10a) or voids (semi-spheroidal holes in a layer of either TiO_2 or AZO—Figure 10d). The electromagnetic field distribution simulations in the thin-film solar cell structures were carried out using a 3D Finite Difference Time Domain (FDTD) method.

The modeling results show that the structures patterned on the front of the cells drastically reduce reflection losses at short wavelengths—Figure 10b,e (at energies above the absorber bandgap)—via geometrical refractive index matching with the cell media. They also boost the absorption of longer wavelengths by increasing their path length via light bending and coupling with wave-guided modes confined in the absorber layer.

These combined effects (antireflection and light scattering) lead to a substantial broadband absorption enhancement in the absorber material, which allows reducing its thickness without lowering the output current.

When evaluating the different LT geometries, it was found that the optimized absorption spectra attained with the two types of structures do not differ significantly. Nevertheless, the highest photocurrent gains were generally provided by the void geometry due to higher NIR absorption enhancement, as it allows a higher degree of angular spreading of the scattered light. Simultaneously, the domes tend to act as micro-lenses that instead focus the scattered light in localized hot-spots located beneath them, as observed in the optical generation profiles of Figure 10c,f.

Pronounced photocurrent enhancements, up to 37%, 27%, and 48%, are demonstrated with honeycomb arrays of semi-spheroidal dome or void-like elements front-patterned on the cells with ultrathin (100 and 300 nm thick) amorphous, and thin (1.5 µm) crystalline silicon absorbers, respectively [142].

The geometrical optics limits for photocurrent enhancement via LT are known as the Lambertian or Green broadband absorption limits. Isabella et al. [147] purposed an advanced LT scheme applied to thin-film silicon-based solar cells, capable of actually overcoming such limits. They showed that optimized 3D optical modeling of thin-film hydrogenated nanocrystalline silicon (nc-Si:H) solar cells endowed with decoupled front and back textures, result in pronounced photocurrent densities (>36 mA/cm^2), thus developing a suitable base for the fabrication of high-efficiency single and multi-junction thin-film solar cells. The simulated enhancements result from a gain in light absorption, especially in the NIR part of the spectrum close to the bandgap of nc-Si:H. Within this wavelength region, the material is weakly absorbing, whereas with the LT design of Isabella et al., significant absorption peaks are observed that can only be explained by the simultaneous excitation of guided resonances by front and rear textures. Using the same advanced LT employed for nc-Si:H, one would obtain a very high implied photocurrent density of 41.1 mA/cm^2 for a device with a 2-µm thick absorber.

A recent breakthrough contribution by Li et al. [151] showed that the Green light absorption limit can also be approached without the need for complex LT geometries. With simple (yet smartly designed) grating structures, composed of checkerboard and/or penta arrangements, the photocurrent of thin-film silicon cells can be realistically doubled, thus revealing performance improvements at the level of the most sophisticated LT grating structures but here attained with much simpler (hence industrial-friendly) geometries.

4.1.2. Thin-Film Solar Cells Improved with Front-Located Photonic Structures

Guided by the modeling studies, in the past few years, LT mechanisms have been successfully implemented in both rigid and flexible solar cell devices with promising experimental results. In this section, we summarize some of these results and the progress that has been achieved in practice. The LT mechanisms, in addition to allowing remarkable enhancement of the optical density of ultrathin PV films, can be fabricated with low-cost materials and integrated by industrial-scale procedures via inexpensive and large-area soft-lithography processes. Accordingly, colloidal lithography has been applied to pattern thin-film solar cells on a photonic length scale with low manufacturing costs.

The simplest CL-related LT implementation consists of using the close-packed array of colloidal spheres (Figure 2a), self-assembled on the front contact of the solar cells, not as a mask but as the photonic front structure itself. This allows forming the photonic structures with a single step performed as a post-process on the cells, as it avoids the additional colloidal masking steps. Such an approach was tested by Grandidier et al. [26], employing a monolayer of silica nanospheres deposited by Langmuir-Blodgett on the front TCO of hydrogenated amorphous silicon (a-Si:H) solar cells. In this design the dielectric colloids act as resonant Mie scatterers, coupling light into the absorber via their near-field proximity. At the same time, it acts in the far-field as a graded-index antireflection coating to further improve the photocurrent. Overall, even though the simple approach of Grandidier et al. allowed significant enhancements in the photocurrent (average of 6.3%) and efficiency (1.8%, reaching the absolute value of 11.1%), the attained gains are relatively low in comparison with those shown in Figure 10, predicted with optimized LT structures. One of the main reasons for this has to do with the low refractive index (n~1.5) of the silica spheres, while higher n values are desired for stronger antireflection and light scattering action, as indicated by the theoretical models [149].

As such, photonic structures composed of high-index dielectric materials, (e.g., TiO_2) as suggested in the simulation works reported in Section 4.1.1, should be capable of yielding much more pronounced enhancements. Given that, Sanchez-Sobrado et al. [33] (Figure 11) developed a CL technique aimed at engineering the TiO_2 semi-spheroidal void-like array geometry of Figure 6b. The TiO_2-based wave-optical structures were first integrated by CL on the front surface of a-Si:H thin-film absorbers, to optimize the parameters of the fabrication method while looking at the absorption enhancement caused in the films.

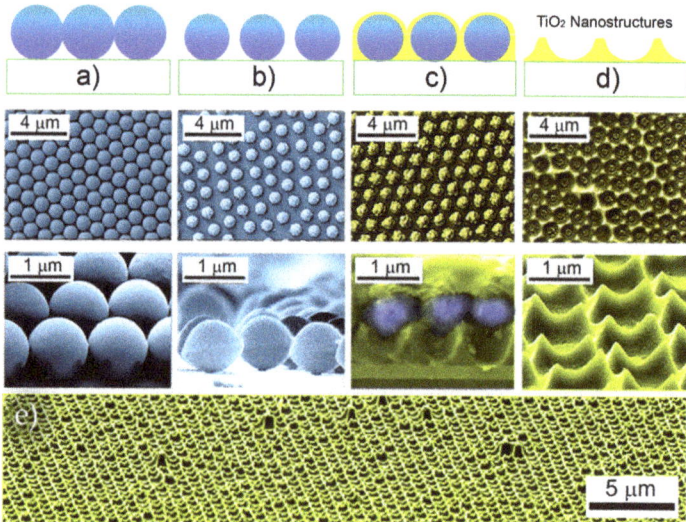

Figure 11. Schematic drawings and SEM pictures of the samples obtained after the different steps of the TiO_2 nanostructure construction via CL on glass substrates: (**a**) a hexagonal array of colloidal PS spheres is patterned on the cell front, (**b**) O_2 dry etching, (**c**) TiO_2 is deposited, filling the inter-particle spaces, (**d**) the spheres are removed leaving an array of semi-spheroidal void-like features. The final TiO_2 structure (**e**) uniformly covers the entire sample area. Adapted with permission from [33]. Copyright 2021 The Royal Society of Chemistry.

The method developed by the authors employs the four main steps illustrated in Figure 11: (a) deposition of periodic close-packed arrays of PS colloids (original diameter of 1.0 µm, 1.5 µm or 2.0 µm) which act as the mask, (b) shaping the particles and increasing their spacing via dry etching, (c) infiltration of TiO$_2$ in the inter-particles spacing and (d) removal of the PS particles to leave only the nanostructured TiO$_2$ layer [33].

It was demonstrated that when directly deposited on a-Si:H absorber films, such LT microstructures provided pronounced broadband absorption enhancement (27.3% on spectral average) in the a-Si:H medium relative to the unpatterned sample.

In subsequent work [48], the authors advanced to integrating this type of photonic structure in actual solar cell devices by CL. These were implemented as a top coating (see Figure 12), with two different materials tested for the photonic coatings in this work: TiO$_2$, due to its high refraction index, and indium zinc oxide (IZO), for better optical and electrical coupling with the front contact of the cells composed of a flat IZO layer.

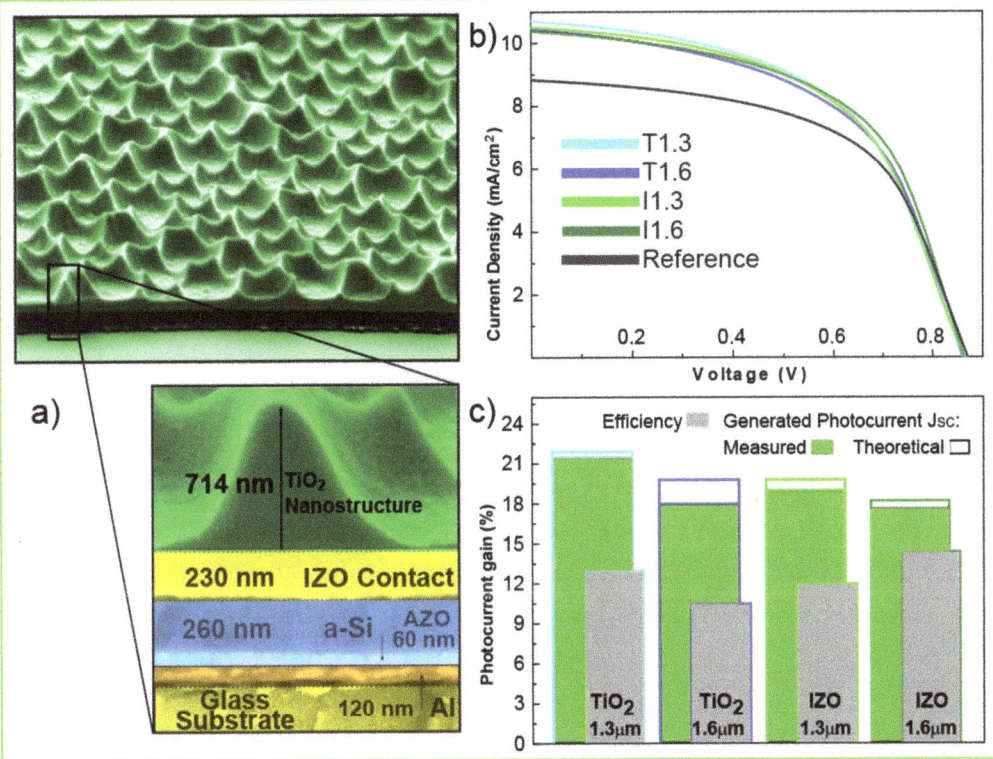

Figure 12. (a) SEM images of the cross-section of an a-Si:H solar cell coated with a TiO$_2$-based LT structure (patterned by CL) over its front flat IZO contact; (b) Measured J-V curves of solar cells incorporating four different photonic coatings made of either TiO$_2$ or IZO and formed with either 1.3 or 1.6 µm diameter PS spheres, and compared with the curve of the planar reference cell with no LT structure over the IZO contact; (c) measured and theoretically modeled enhancement, relative to the reference, of the generated current density (dark grey and empty bars, respectively), and the measured conversion efficiency (light grey bars) of the devices in (b). Adapted with permission from [48]. Copyright 2021 The Royal Society of Chemistry.

It was observed that all the different photonic structures applied on the a:Si:H cells produced significant broadband absorption enhancement, leading to systematic increases in the current (17.6–21.5%, Figure 12c), relative to that of the planar reference cells without the LT features. The TiO$_2$ structures achieved higher optical performance than the IZO ones, as expected, mostly due to the higher refractive index and lower optical absorption of TiO$_2$. Nevertheless, the extra electrical benefits (reduced sheet resistance) on the front electrode caused by the IZO structures allowed for the highest efficiency enhancement (Figure 12c) [48].

Given the promising results attained with the photonic-structured IZO front contacts, in a subsequent work Sanchez-Sobrado et al., (2020) [29] further improved the device architecture and CL process parameters to further optimize the geometry of the IZO structures and, importantly, their location concerning the Si absorber. This was achieved by optimization of the thickness (investigating from 30 to 250 nm) of the flat IZO layer beneath the LT structures, to allow an electrically effective front contact ideally coupled with the geometry of the CL-patterned IZO structures (Figure 13).

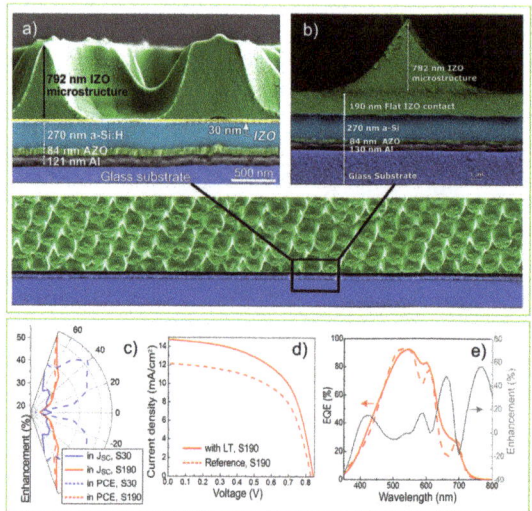

Figure 13. SEM cross-sections of a-Si:H solar cells with two different thicknesses of the flat IZO layer (30 (**a**) and 190 nm (**b**), respectively) located between the a-Si: H absorber and the front LT structures; (**c**) Polar plot representing the angular response of the solar cells with 30 nm, S30 (blue lines), and with 190 nm of IZO thickness, S190 (red lines), in terms of the gains attained in the short-circuit current density, (J$_{SC}$, solid lines) and power-conversion efficiency (PCE, dashed lines) of the cells; (**d**) J-V curves and (**e**) external quantum efficiency obtained for S190 (190 nm of IZO thickness). Adapted with permission from [29]. Copyright 2021 Elsevier.

Here, the CL fabrication method consisted of first dispersing a colloidal suspension of 1.6 μm PS spheres in a water-ethanol mixture (1:3) at a solution concentration of 2.5% wt. Using the Langmuir–Blodgett technique [152], a close-packed monolayer was then deposited onto the flat IZO-coated surface of the a-Si:H cells.

The final IZO structure is revealed after removing the PS particle mask by a combination of Ar/CF$_4$ reactive ion etching and a toluene bath.

As previously mentioned, the main parameter optimized in this work was the thickness of the flat IZO layer deposited on the cell before the CL process of integrating the top microstructured IZO. Such thickness defines the separation between the a-Si:H absorber and the photonic features, and the best results were attained with 30 nm and 190 nm thicknesses, shown in Figure 13a. These results reveal that the LT structures lead to a remarkable broadband enhancement of the total light absorbed by the devices, therefore leading to photocurrent gains (relative to reference planar cells) up to 26.7% with the 30 nm flat IZO space—Figure 13c. However, the best efficiency enhancement (23.1%) was attained with the optimized thickness of 190 nm for this layer (Figure 13d,e), as it provides the most favorable combination of optical and electrical gains.

Another important finding of this work was the remarkable LT gains attained at oblique light incidence. The angular response of the devices was evaluated by measuring the JV curve for a range of light incidence angles from 0° to 90°. The highest gain reaches 53.2% enhancement in photocurrent at 70° incidence angle for the cell with the optimized 190 nm thick flat IZO, and 52.2 in efficiency (at 40° incidence) with the ultrathin (30 nm) flat IZO (Figure 13e) [29].

4.1.3. Photonic-Enhanced Perovskite Solar Cells and CL Compatibility

Perovskite-based PV materials, as thin-film absorbers for perovskite solar cells (PSCs) [153], have received unprecedented attention in both academia and industry due to the exceptionally rapid efficiency advancement from ~3.8% to >25.2% over the last decade [30,154]. As usual, for most PV technologies in the early stages of development, the progress was mainly accomplished by exploring the composition of perovskite-based materials and optimizing the fabrication process of the perovskite layer, as well as the quality of the cells' interfaces to ensure efficient charge collection and to suppress unwanted recombination routes [155]. As such, PSCs have reached a stable point in the development phase, and optical strategies are now paramount to advance beyond its current record efficiency, particularly for the thinner perovskite layers which are attractive to enable device flexibility [156].

The rapid progress in this PV technology has enabled improved PSC deposition methods allowing conformal coating of the cell layers onto microstructured substrates, resulting in improved efficiencies. This is a promising path that has only recently started being unraveled for nano/micro-structuring in the PSCs field. Photonic microstructures can improve the cells' absorption beyond the standard values achieved with planar devices, facilitating thickness reduction without compromising the output current and without degrading the electrical performance. On the other hand, the operational stability of the devices can also be improved by blocking the harmful higher-energy photons of UV radiation. This is particularly advantageous to assist in the stability of this less-matured PV technology which quickly degrades with UV exposure [52].

Recently, several LT schemes have been shown to improve the performance of PSCs, such as disordered micro-pyramids [155,156], nanojets, corrugated substrates, self-cleaning nanostructures, and micro-cones [157], as well as other approaches such as plasmonic nanoparticles, surface plasmon resonances, down and/or upconversion, etc. [158–160]. It is also observed that simple grating structures in the front [161] and back electrodes [162] enable enhancement in the light absorption and PSCs stability.

CL can play an incredible role in thin-film PV process technology due to its compatibility with large-scale (even roll-to-roll) manufacturing. The developed low-cost CL processes commented in the previous sections, for the integration of LT structures in silicon-based solar cells [29,31], can be adapted for photonic-structuring in PSCs. Previous theoretical contributions presented novel LT designs [27,28], operating in the wave-optics regime that demonstrated record photocurrent gains in PSCs via the incorporation of wavelength-sized features in the front electron transport layer (ETL) of the cells with a substrate configuration, similarly to the integration discussed in detail in Section 4.1.2. Interestingly, it was shown that the proposed front-patterned TiO_2 LT coatings also lead to UV stability improvements [27], with even better results being achieved by incorporating another front-located luminescent down-shifting layer that can convert the harmful UV photons to non-harmful visible photons [28]. These structures can be straightforwardly fabricated using the same procedures of CL (Figure 6) as those developed for silicon-based solar cells [29,31], as further detailed next.

The initial simulation work by Haque et al., (2019) [27] considered an inverted (substrate-type) PSC architecture, allowing the integration of the LT structures as a post-process on the front contact of the cells, in a similar way as in Figure 12 for thin-film silicon cells. The wave-optical structures optimized by the authors (Figure 14) for the inverted PSCs are also based on high-index dielectric (TiO$_2$) micro-scale features, with semi-spheroidal geometries.

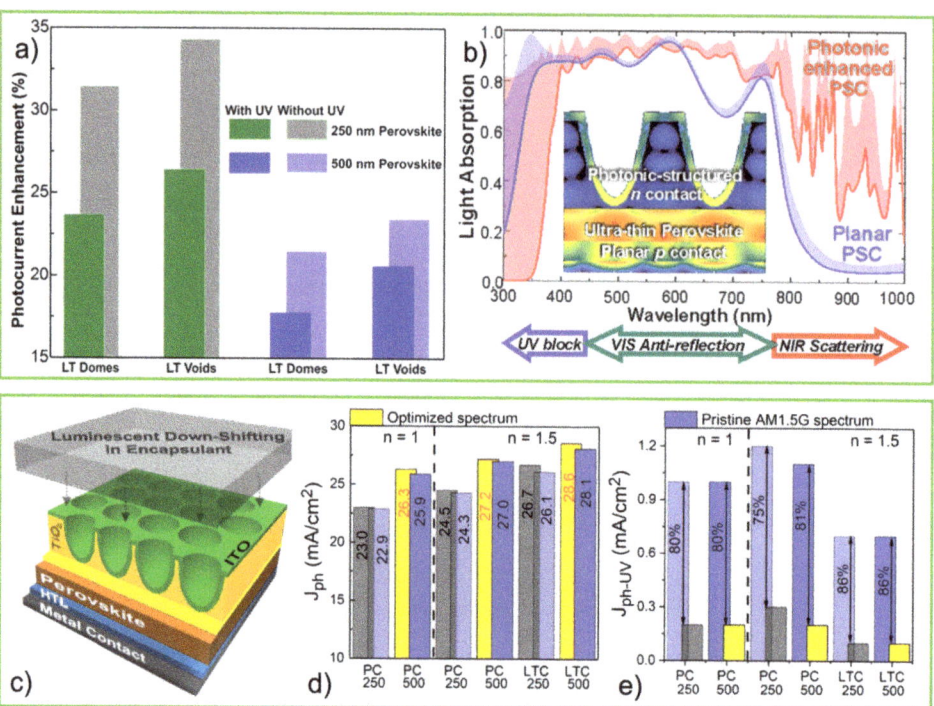

Figure 14. (a) Photocurrent enhancements attained with the optimized photonic structures presented in Haque's work. (b) Light absorption spectra of PSCs with (photonic-enhanced) and without (planar) the optimized LT structures integrated into the ETL of the cell—a low absorption in the UV region indicates the desired blocking effect caused by the photonic structure. Adapted with permission from [27]. Copyright 2021 Elsevier. (c) Sketch of the luminescent down-shifting material encapsulating the photonic-structured PSC. (d) J_{ph} values obtained considering the full UV-Visible-NIR wavelength range (300–1000 nm), for two different refractive indexes (n) values of the encapsulating media. (e) UV photocurrent ($J_{ph\text{-}UV}$) values for wavelengths ranging from 300 to 400 nm (the more transparent bars refer to the devices with 250 nm perovskite thickness, while the others refer to those with 500 nm). Adapted with permission from [28]. Copyright 2021 American Chemical Society.

The optically lossless TiO$_2$ material allows the structures to be patterned in the final processing steps, integrated into the top n contact of the cell. This, in turn, avoids structuring the cell layers, thence avoiding increased roughness and consequent electrical losses due to higher recombination. The electromagnetic field distribution simulations in the PSC structures were also carried out using the same FDTD method of Figure 10, and the main results are summarized in Figure 14a. In particular, the optimized array of TiO$_2$ voids, which was shown to be optically favorable when compared with the domes, enables a photocurrent enhancement of 21% and 27% in PSCs with conventional (500 nm thick) and ultrathin (250 nm) perovskite layers, respectively.

The photocurrent enhancements attained with the optimized LT designs are mainly due to absorption improvements for wavelengths above 600 nm, which the authors attribute to strong antireflection (of visible light) and light scattering effects (of near-infrared), as shown in Figure 14b.

Furthermore, the TiO$_2$ material of the structures advantageously acts as a UV blocking layer, as also shown in Figure 14b, protecting the perovskite from known degradation mechanisms caused by UV penetration [163–165]. Here the UV light is absorbed in the front TiO_2 features and does not reach the perovskite, while the light at the longer visible plus NIR wavelengths is coupled and trapped within the cells, generating an overall enhanced photocurrent. UV blocking mechanisms as this one can enhance the operational stability of PSCs upon solar exposure, but inevitably they prevent the conversion of the UV energy by the cell, so they are not the ultimate solution when aiming for maximizing efficiency.

Alexandre et al. [28] presented an interesting approach to exploit such, otherwise lost, UV energy via the combined effects of LT and luminescent down-shifting (LDS). Simply put, this latter effect shifts higher energy light into lower-energy light. Considering the extensively studied degradation problems of PSCs with UV radiation [166,167], this method can bypass these unwanted mechanisms while also recurring to the energy coming from the UV photons.

The use of optimized LDS materials in the front encapsulation of the previous photonic-enhanced PSCs designed by Haque et al. (Figure 14c) led to an increase in photocurrent of at best 2% (~0.6 mA/cm^2), which is almost half of the theoretical maximum current (1.4 mA/cm^2) that could be gained from all the UV range. The optimum spectral down-shift was found to be from a central 350 nm UV wavelength to around 500 nm visible wavelength, matching well with the electrical performance peak of PSCs, which could imply that an increased device's efficiency surpassed the projected gains. The LT cells also revealed a decrease in the harmful TiO_2 photogeneration near the perovskite/TiO_2 interface due to the LT structures' UV shading effect. By assessing the UV penetration in the perovskite material (given by the UV generated photocurrent by the PSC) for the different simulated cells, reductions up to 86% (Figure 14d) were obtained when comparing photocurrent values for the original (pristine) and optimized down-shifted spectrum. Therefore, these analyses show that the use of LDS provides a more effective way to eliminate the unwanted effects of UV radiation in the perovskite, demonstrated by the hefty decrease in UV absorption coupled with the diminished TiO_2 photoactivity from lower photogeneration, while also enabling additional power generation from the UV portion of the sunlight [28].

Although promising and compatible with CL structuring, the aforementioned LT implementations in Figure 14 require the photonic elements to be patterned on top of the planar cell layers during the final processing stages, which brings the risk of degrading the highly sensitive PSC materials located underneath [52]. Specifically, the fast degradation of PSCs with humidity exposure may render this class of devices incompatible with immersion or coating with aqueous solutions, as required in the first steps of the CL process (see Section 2.2).

Given this, Haque et al., (2020) [30] investigated a more industrially viable LT strategy consisting of the development of photonic substrates used for subsequent PSC deposition. The authors have shown that it is possible to achieve pronounced broadband absorption enhancement provided by the LT-patterned substrates, relative to planar cells, which was also observed for a broad range of incidence angles (0–70 degrees), as seen in Figure 15a. Apart from substantial light absorption enhancement, this approach has a significant advantage for its practical compatibility with PSC technology over the previous ones, as the PSC layers can be wet coated by traditional methods onto a substrate already patterned with the designed LT structure.

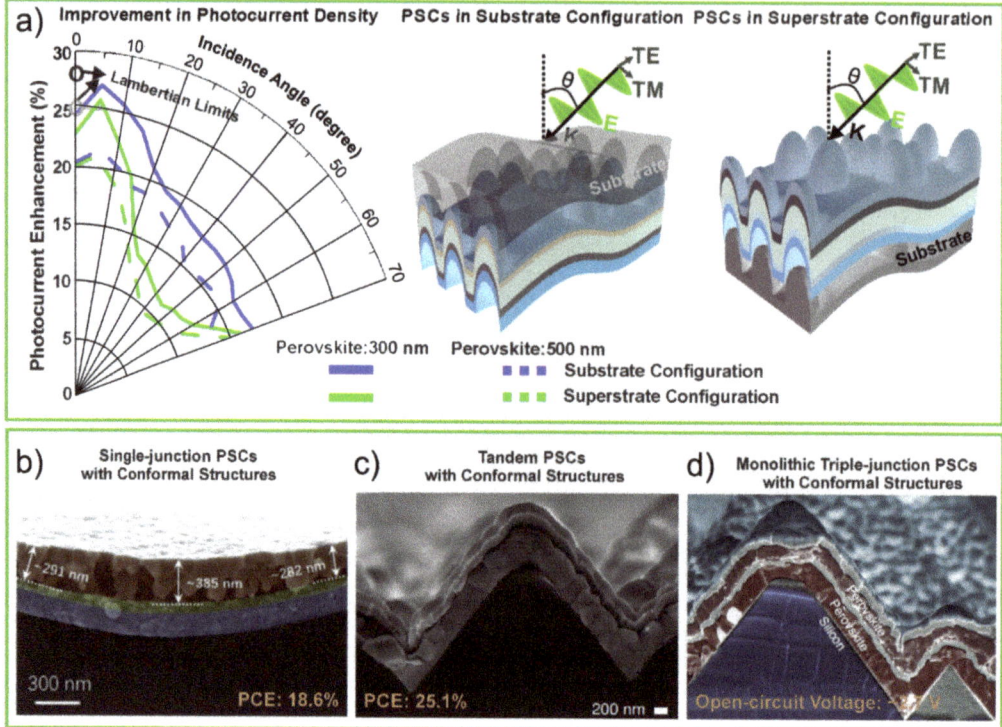

Figure 15. (**a**) The polar plot shows the photocurrent density gain as a function of the incidence angle (θ), attained with the optimized photonic-structured PSCs in a conformal architecture, for both the substrate and superstrate cell configurations illustrated in the middle and right sketches. The Lambertian limits of LT in PSCs, in the geometrical optics regime, are also indicated for normal incidence angle (θ = 0°) in the left plot. Adapted with permission from [30]. Copyright 2021 Elsevier. (**b**) SEM image of a single-junction PSC conformally coated onto a textured glass substrate. Adapted with permission from [168]. Copyright 2021 Elsevier. (**c**) SEM image of a tandem perovskite/silicon cell with ~25.1% efficiency, in which the PSC is conformally coated onto the textured silicon wafer-based bottom cell [169]. (**d**) SEM image of a triple-junction perovskite/perovskite/silicon cell in which the top and middle PSCs are also conformally deposited onto textured silicon. Adapted with permission from [170]. Copyright 2021 American Chemical Society.

This makes the full photonic integration independent of the fabrication of the PSCs. Moreover, this is a potentially more cost-effective approach since it requires no extra materials (coatings) for the LT structuring.

However, the LT designs of Figure 15a are only achievable with highly conformal PSC deposition methods, capable of coating the cell layers onto the micro-patterned substrate surfaces without defect formation. In that respect, the work of Wang et al. [168] is an important contribution, as the authors developed a recrystallization treatment that enabled the conformal coating of high-quality perovskite layers onto textured glass with features in the order of the micrometer, resulting in 18.6% PSC efficiency with a ~300 nm thin perovskite absorber, as seen in Figure 15b.

Conformal deposition methods have also shown remarkable potential in monolithic tandem PV cells, in which perovskite-based top cells are coated onto fully textured crystalline silicon bottom cells. That is the case of the perovskite/silicon double-junctions developed by Aydin et al. [169] with 25.1% efficiency (Figure 15c) and the perovskite/perovskite/silicon triple-junctions (Figure 15d) of Werner et al. [170] reaching ~2.7 V.

Within this section, we went through some of the latest trends on LT strategies for PSCs, revealing promising modeling results with optimized designs for efficiency and stability improvement. We also discuss the first experimental steps that have been recently undertaken to circumvent the challenges associated with the practical realization of photonic-structured PSCs, which is a new avenue for CL implementation with enormous potential in the PV field.

4.2. Micro-Meshed Transparent Electrodes

Another highly promising solution offered by nano/micro-structuring is the development of high-performing transparent electrodes, with strong interest not only for the illuminated contacts of solar cells but also for many other optoelectronic applications. Nano/microstructured metallic films (termed micro-mesh electrodes, MMEs) offer an exciting alternative to flat TCO (transparent conductive oxide) layers, and CL was shown to be one of the most promising approaches to produce industrial-compatible MMEs with excellent properties [62].

State-of-the-art indium tin oxide (ITO) based TCOs are ubiquitously spread in optoelectronic technologies as they are in the illuminated contacts of solar cells, in displays, touch screens, and light-emitting diodes, with low sheet resistances between 8 and 12 $\Omega/\square$ (for commercial ITO) [171] and optical transmittance above 80% in the visible range [62,172]. However, as with most TCO materials, their free electrons cause strong parasitic absorption losses in the NIR range, limiting their performance for solar cell applications. This is further compounded by the rising cost of In, a rare material, while alternative TCOs made with Earth-abundant materials (e.g., based in ZnO [173]) offer reduced figures-of-merit in terms of transmittance-over-resistance ratio. Furthermore, ITO is usually deposited by costly DC-magnetron sputtering involving high temperatures (up to 300 °C) [174], making it unsuitable for deposition on thermal-sensitive materials. Moreover, ITO is brittle, making it also less attractive for flexible electronics as well as resistive touch screens [175].

A solution to mitigate the issues of state-of-the-art TCOs is to use a metallic micro-grid (MMEs) sandwiched between ultrathin layers of TCO, which allows a pronounced increase in transparency (especially in the red-NIR region) while maintaining high sheet conductance; thus, offering an exciting alternative to flat TCOs. CL provides a promising approach to produce industrial-compatible MMEs, as demonstrated with the copper (Cu) nanomeshes developed by Gao et al., (2014) [175] exhibiting an excellent diffuse transmission of 80% and sheet resistance of 17 $\Omega/\square$ (Figure 16a,b).

More recently, Torrisi et al., (2019) [62] have shown the use of CL to create silver (Ag) micro-grids that were sandwiched between TCO layers. First, a PS colloidal microsphere monolayer (deposited using the Langmuir–Blodgett technique) is submitted to plasma etching to serve as a deposition mask. The subsequent evaporation of Ag throughout this mask, followed by lift-off of the colloids, creates a metallic grid as shown in Figure 16c,d. Compared to conventional transparent electrodes (i.e., ITO), excellent electrical and optical characteristics have been accomplished with such TCO/Ag micro-grid/TCO multilayers, such as sheet resistances below 10 $\Omega/\square$ and pronouncedly higher near-infrared transmittance, using different layer thicknesses and mesh dimensions (Figure 16e).

The structural parameters of the produced mesh (openings, line width, and thickness) play a vital role in the electrical and optical performance of the transparent electrodes. Larger hole structures, attained with large sizes of the PS colloids, result in excellent transmittance values at the cost of increased sheet resistance. Alternatively, meshes with smaller holes are less transparent but have a lower resistance, permitting higher currents (Figure 16f). To reduce the sheet resistance values, one can also increase the metal grid thickness, by evaporating more Ag material, without having a large effect on transparency [62]. Finally, the best performing correlation between optical and electrical properties (Figure 16g) is attained with 5 µm spheres and 17 nm Ag thickness, either (120 or 240s RIE).

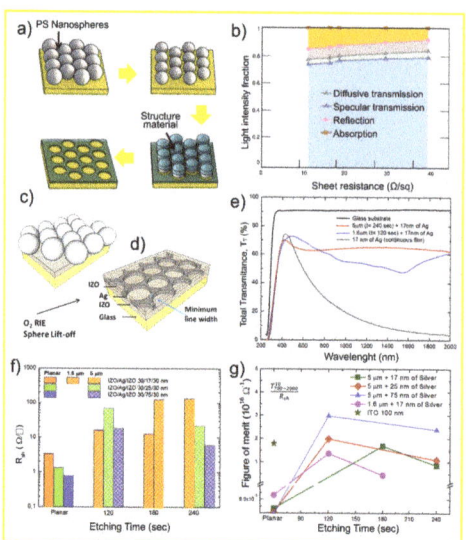

Figure 16. (a) Schematic of the Cu MME fabrication process by CL. (b) Plot of the measured diffuse transmission, specular transmission, reflection, and absorption at λ = 550 nm versus sheet resistance for a variety of Cu MMEs on quartz. Adapted with permission from [172]. Copyright 2021 American Chemical Society. (c,d) Schematic of the CL fabrication of Ag MMEs, showing the initially deposited PS sphere mask (c) and the final structure of the IZO/Ag grid/IZO electrode (d), in which the top and bottom IZO layers are ultrathin (30 nm). (e) Transmittance of samples with 17 nm thick Ag grids (blue and red lines), produced with distinct sphere size and etching time, compared to a continuous Ag film (grey line)—high transparency can be noted in the NIR region. (f) Sheet resistance of the electrodes fabricated with different diameters of PS spheres (1.6 or 5 µm), etching time, and Ag thickness. (g) Haacke's figure-of-merit (expression in inset) for the main MMEs attained in this work, in comparison with a state-of-the-art TCO film made of ITO and with continuous (planar) Ag layers. Adapted with permission from [62]. Copyright 2021 Elsevier.

It is also believed that the use of metallic grids for electrodes when compared to continuous layers, can bring several advantages for flexible transparent materials. In particular, honeycomb grids as produced via CL allow the highest flexural robustness with thinner MME structures [175], due to the fact that the honeycomb lattice enables the highest packing density in 2D arrays. Overall, the use of metallic micro-grids opens new and promising paths for transparent electronics, offering additional degrees of freedom for further electro-optical improvements. Specifically, their much higher transmission if the red-NIR region can be crucial for applications such as smart windows, low-energy photovoltaic devices, and as intermediate contacts in multi-terminal multi-junction (tandem) solar cell architectures [176,177].

4.3. Self-Cleaning with Photonic Structuring

Unavoidable environmental degradation is a major cause of efficiency losses in solar panels over time. Phenomena such as the formation of hot-spots (areas of large heat dissipation) caused by partial shading of solar cells can be responsible for large efficiency losses [74,143]. Among different solutions mentioned at the beginning of Section 4, the use of a self-cleaning coating appears to be the simplest and most fitting approach for large-scale PV installations [74], as no mechanical structure is required for its cleaning function.

Through superhydrophobic coatings, the concept of self-cleaning glass materials showed a maturing research trend between 2009 and 2017, with 1125 research articles being published (ScienceDirect keywords: "self-cleaning glass, superhydrophobic glass"). These coatings can be fabricated using top-down approaches such as template-based, photolithographic, and surface plasma treatment [178], or top-down methods such as chemical modification [179], colloidal assembly [180], layer-by-layer deposition [181], and sol-gel methods [182] can also be applied [74].

An important application for PV technology was presented by Centeno et al., (2020) [31], showing a simple, low-temperature, low-cost, and scalable CL method (Figure 17a) to engineer parylene-C (poly(chloro-p-xylylene)) coatings with encapsulating properties that endow effective light-trapping (LT) and water-repelling functionality when applied in thin-film solar cells (Figure 17b–e). Parylene-C was used as the preferred coating material, as it is a polymer with excellent barrier properties for encapsulation [183,184] and low water-adhesion surface energy.

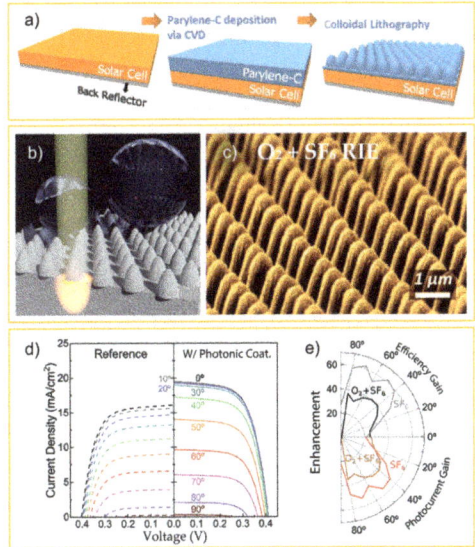

Figure 17. (a) Depiction of the patterning of the photonic-structured parylene-C coating on the transparent front contact of thin-film silicon solar cells, via chemical vapor deposition (CVD), followed by CL with a patterning approach similar to that of Figure 6b. (b) Artistic image of the photonic coating on the solar cell, illustrating its superhydrophobic surface where water droplets easily roll-off. The microstructured parylene-C provides LT while also acting as a water-repellent protective layer, which allows an effective self-cleaning functionality. (c) SEM image of the parylene-C surface microstructured via the CL process ($O_2 + SF_6$ RIE). (d) 1-Sun JV curves of the solar cells before (uncoated reference, left curves) and after (right curves) coating with the photonic-structured parylene-C shown in images (c,d), for illumination angles varying from 0° (normal to cell's surface) to 90° (parallel to the surface). (e) Polar plot presenting the angular dependence of the gain in efficiency (top) and photocurrent (bottom) of the LT-enhanced solar cells relative to the planar references. Adapted with permission from [31]. Copyright 2021 John Wiley and Sons.

It also benefits from optical transparency, adequate refractive index, outstanding flexibility, and mechanical strength. Moreover, parylene is deposited via chemical vapor deposition, which allows for the deposition of more uniform coatings at room temperature, thus making it usable with non-flat temperature-sensitive materials [183,185,186].

In this work, the hydrophobicity of parylene was controlled by simultaneously adjusting the surface corrugations (i.e., roughness, patterned features) and surface chemical composition with the CL patterning process following Figure 17a. In this way, a parylene-C film was micro-patterned with a hexagonal array of cones (see SEM in Figure 17c) after being coated on the front TCO of nc-Si:H solar cells [31].

These microstructured parylene coatings also enable outstanding antireflection and light scattering properties that were shown to enhance the solar cells' photocurrent by up to 23.6% with normal incident light. Furthermore, the cells' angular response was also improved, similarly to the previous work of Sanchez-Sobrado et al., (2020) [29] shown in Figure 13. Figure 17d displays the JV curves with an illumination angle ranging from 0° to 90°. The polar plot of Figure 17e shows the influence of the illumination angle on the efficiency and J_{SC} enhancement relative to the flat reference cell without the front microstructures.

These enhancement values increase with the angle until ~50°, with peak enhancements up to 52% and 61% in J_{SC} and efficiency, respectively, while both V_{OC} and FF (fill-factor) were only marginally reduced with increasing angle. Such gains for oblique incident light translate to an estimated average daily enhancement of ~35% in the generated energy, as compared with the daily energy delivered by the uncoated reference cells.

4.4. Overview of Achievements

The following tables list the best results concerning the CL-patterned PV devices (relative to the unpatterned references) commented along this section (see Figure 9 schematic). Table 1 summarizes the optimized computational designs and Table 2 the experimental outcomes.

Table 1. Overview of simulation results on photonic microstructures for thin-film PV.

Authors	Year [Ref.]	Simulated Structure	Best Results
Mendes et al.	2016 [149]	Front-located TiO$_2$ domes on planar Si solar cells	Photocurrent gain of 50% for thin c-Si cells with 1.5 µm thick absorber
	2018 [142]	Front-located TiO$_2$/TCO wavelength-sized dome/void arrays	Photocurrent gains of 37%, 27%, and 48% in 100 nm a-Si, 300 nm a-Si, and 1.5 µm c-Si, respectively
Isabella et al.	2018 [148]	Thin-film nc-Si:H solar cells with decoupled front and back textures	Photocurrent densities reaching 41.1 mA/cm2 with 2-µm thick nc-Si:H absorber.
Haque et al.	2019 [27]	TiO$_2$ LT structures integrated on the front contact of flat Perovskite solar cells (PSC)	Photocurrent gains up to 27% with planar ultrathin (250 nm) perovskite layers, together with UV blocking effect granting improved stability.
	2020 [30]	Microstructured PSCs on transparent and opaque photonic substrates	Photocurrent gains of 22.8% and 24.4%, respectively with transparent and opaque substrates, with 300 nm perovskite.
Alexandre et al.	2019 [28]	Combined luminescent down-shifting and front-located photonic structures on PSCs	Photocurrent gains similar to Haque et al., together with up to 86% reduction of UV penetration in the perovskite layer.

Table 2. Overview of experimental results on CL-fabricated microstructures for thin-film PV.

Authors	Year [Ref.]	Experimental Structure	Best Results
Grandidier et al.	2013 [26]	Close-packed monolayer of silica microspheres on the front of a-Si:H solar cells	Spheres acting as resonant Mie scatterers provide photocurrent and efficiency gains up to 8.9% and 11.1%, respectively.
Sanchez-Sobrado et al.	2017 [33]	Front-located TiO$_2$ microstructures on a-Si:H thin-film absorbers	Absorption enhancement of 27.3% on spectral average in 300 nm a-Si:H absorbers
	2019 [48]	TiO$_2$ or IZO structures integrated into front contact of a-Si:H solar cells	Photocurrent gains up to 21.5% with TiO$_2$ (better optically) and efficiency gain up to 14.4% with IZO (better electrically) LT structures.
	2020 [29]	Front-located IZO structures on a-Si:H solar cells with optimized parameters	Photocurrent and efficiency gains up to 26.7% and 23.1%, respectively, while gains >50% are attained at an oblique incidence.

Table 2. Cont.

Authors	Year [Ref.]	Experimental Structure	Best Results
Centeno et al.	2020 [31]	Micro-patterned parylene-C film with micro-cone array coated on the front TCO of nc-Si:H solar cells.	Water repellant (self-cleaning) surface, granting photocurrent and efficiency gains up to 23.6%, and up to ~60% at oblique incidence.
Gao et al.	2014 [172]	Cu micro-mesh electrode as a transparent conductive material	Diffuse transmission of 80% in the visible range; sheet resistance of 17 $\Omega/\square$ on quartz; high flexural robustness.
Torrisi et al.	2019 [62]	Ag micro-grids sandwiched between ultrathin TCO layers	Sheet resistances below 10 $\Omega/\square$ and near-infrared transmittance over 50%.

5. Other Applications of Colloidal Lithography: Biological Cell Studies

Apart from the engineering of photonic solutions for optoelectronic-related technologies, such as photovoltaics, several nano/micro-structuring colloidal lithography (CL) approaches have been used to develop breakthrough advances in most various scientific areas. Among those, the biology community has been gaining increasing interest in CL, as overviewed next to show the bigger picture of promising applications.

Micro and nanofabrication techniques have revolutionized the pharmaceutical and medical fields. Complex geometries can be reproduced with technologies such as photolithography and particularly colloidal lithography [187]. The interaction of cells with material surfaces has been widely studied, having relevance for the function of medical devices, as well as to gain a better understanding of cellular action in vivo and in vitro [188]. In this field, nano-topography has shown that nanoscale features can strongly influence cell morphology, adhesion, proliferation, and gene regulation, but the mechanisms mediating this cellular response remain unclear [189]. Nano-topographies produced by CL may be of great importance when considering how cells respond to their environment, allowing for the elucidation of nano-feature effects on cell behavior. This appears to significantly alter the morphology of endothelial cells, epithelial cells, epitenon cells, macrophages, osteoblasts, and fibroblasts [190]. The cell reactions can be controlled by the surface characteristics morphology, chemistry, and viscoelastic properties. Investigations on the influence of nano-topographies in the cell response require surfaces patterned over large areas with high reproducibility, high throughput, and fabricated in biocompatible materials [189,191].

For instance, the ability of fibrinogen molecules to specifically bind to receptors in platelet membranes has been correlated with both nanoscale chemistry (hydrophobic and hydrophilic) and surface topography (nanopits and planar surface) [192]. Thus, CL was used to create a continuous thin-film of TiO_2 with a distribution of nanopits with 40 nm diameter and 10 nm depth.

Similarly, collagen adsorption in terms of both adsorbed amount and the supramolecular organization was investigated concerning the influence of substrate surface topography (smooth surface versus nano-protrusions) and surface chemistry (CH_3 versus OH groups). CL and functionalization with alkanethiol were used to create model substrates with these controlled topographical and chemical surface properties [193] (Figure 18a). The surface chemistry was found to control collagen protein adsorption, while chemistry and topography control collagen protein conformation.

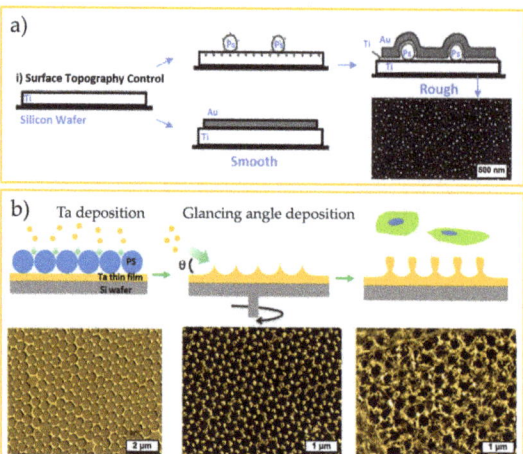

Figure 18. (a) Schematic illustration of prepared substrates with controlled nanotopography using CL and surface chemistry by alkanethiol self-assembly: i) smooth substrates were obtained by depositing titanium and gold layers onto silicon wafers; rough substrates were obtained by adhesion of negatively charged polystyrene colloidal particles to a smooth positively charged surface, followed by coating with a thin layer of gold. Adapted with permission from [193]. Copyright 2021 American Chemical Society. (b) Schematic structure of fabrication of ordered tantalum (Ta) nanotopographies using a combination of CL and glancing angle deposition techniques; and corresponding SEM images. Adapted with permission from [194]. Copyright 2021 American Chemical Society.

The application of nanotechnology in the field of regenerative medicine has opened a new realm of advancement in this field. The human fibroblasts have a pivotal role during the initial phases of implant integration and resultant healing processes. Concerning the tissue–implant interface, in vitro investigations indicate that micro-topography can be used to control cell behavior, including that of fibroblasts, via initial adhesive interactions [191,195]. In this respect, CL was used to fabricate irregular nanotopographies with features of either 20 or 50 nm diameter pillars. These nano-pillar topographies modified fibroblast adhesion, morphology, and behavior relative to those deposited on planar substrates. The fibroblast adhesion is observed to be greater on substrates patterned with 20 and 50 nm diameter colloidal topographies, which result in increased fibroblast adhesion at 20 min and 60 min. Although the colloidal topographies alter the fibroblast adhesion, morphology, and behavior when compared to control conditions (planar substrates), for 180 min the adhesion is similar to planar surfaces [191].

Andersson et al. [188] investigated how pancreatic epithelial cell lines (AR4-2J) respond to surface structures (titanium pillars) of systematically increasing size. For this study, a surface with the same surface chemistry and similar surface roughness (but with increasing size of column diameters from 60 to 170 nm) was produced via CL. The changes in the epithelial cells were studied by evaluating cell area and cell shape. It was observed that the larger the feature, the more spread the cells became [188].

The behavior of primary human adipose-derived stem cells was studied on highly ordered bowl arrays (Figure 18b). PS colloids (722 nm diameter) were self-assembled into a hexagonally close-packed crystal array at the water-air interface, transferred onto a biocompatible Ta surface, and used as a mask to generate an ordered Ta pattern. The results showed that ordered Ta nanotopographies inhibited cell spreading, focal adhesion formation, and filopodia extension when the surface roughness and feature height increased [194]. It was also demonstrated that by changing the feature size, the ordered topographies could alter stem cell adhesion and differentiation.

These remarkable investigations have been expanding our knowledge in cell-surface interactions, which is of great interest in biomaterials, tissue engineering, and cell therapy applications; being CL a preferential industry-compatible technique for the needed surface structuring.

6. Conclusions

As nano/microstructures fabrication grows in importance in a wide range of areas from electronics through optics, microanalysis, combinatorial synthesis, displays, cell biology, etc., the research on ever-more cost-effective patterning methods will undoubtedly continue to increase.

In the past decade, colloidal lithography has grown to become one of the most attractive soft-patterning techniques, and a solid alternative to conventional hard-patterning processes. Its main characteristic is that it embodies a simple and inexpensive way of producing the lithographic mask, using monodisperse colloidal particles self-assembled on the target surface. This method offers advantages in applications in which photolithography falters, for instance, manufacturing below the (sub-wavelength) scale of 100 nm, patterning on non-planar surfaces, fabrication of three-dimensional structures, patterning functional materials other than photoresists, among other types of surface modifications. Another key advantage is that colloidal lithography allows the formation of nano/microscale structures of virtually any material, via scalable process steps that are compatible with industrial mass-production requirements. As such, it is highly attractive for crafting photonic schemes, such as those discussed here for photovoltaic applications, since it enables the precise engineering of long-range ordered wavelength-sized features, with the materials and geometries appropriate for efficient light-trapping.

This article overviewed some of the most promising nano/micro-structuring solutions in the growing field of photovoltaics, where colloidal lithography was shown to be a preferential patterning technique for the practical realization of the concepts in industrially attractive ways. In particular, photonic-enhanced photovoltaics has been shown to circumvent many of the conventional shortcomings of current solar cell technologies, such as the reflection losses and impaired light absorption especially in thin-film devices; as well as enabling other interesting functionalities such as improved transparent contacts and self-cleaning due to the high aspect ratio of the photonic microstructures.

Author Contributions: Original draft preparation—R.D.O. and A.M.; review and editing—P.C., M.A., S.H., H.Á. and M.J.M.; project and funding management—R.M., E.F., H.Á. and M.J.M. All authors have read and agreed to the published version of the manuscript.

Funding: This work received funding from the European Union's Horizon 2020 research and innovation program under the projects APOLO (H2020-LCE-2017-RES-RIA), grant agreement n° 763989 and Synergy (H2020-Widespread-2020-5, CSA), proposal n° 952169. This publication reflects only the author's views, and the European Union is not liable for any use that may be made of the information contained therein. The work was also funded by FCT (Fundação para a Ciência e Tecnologia, I.P.) under the projects UIDB/50025/2020, SuperSolar (PTDC/NAN-OPT/28430/2017), TACIT (PTDC/NAN-OPT/28837/2017) and LocalEnergy (PTDC/EAM-PEC/29905/2017). M. Alexandre, P. Centeno, and S. Haque acknowledge funding by FCT-MCTES through the grant SFRH/BD/148078/2019, DFA/BD/7882/2020, and the AdvaMTech Ph.D. program scholarship PD/BD/143031/2018, respectively.

Data Availability Statement: Not applicable.

Conflicts of Interest: The authors declare no conflict of interest.

References

1. Green, M.A. Commercial progress and challenges for photovoltaics. *Nat. Energy* **2016**, *1*, 15015. [CrossRef]
2. Tawalbeh, M.; Al-Othman, A.; Kafiah, F.; Abdelsalam, E.; Almomani, F.; Alkasrawi, M. Environmental impacts of solar photovoltaic systems: A critical review of recent progress and future outlook. *Sci. Total Environ.* **2021**, *759*, 143528. [CrossRef]
3. Kannan, N.; Vakeesan, D. Solar energy for future world: A review. *Renew. Sustain. Energy Rev.* **2016**, *62*, 1092–1105. [CrossRef]

4. Breyer, C.; Bogdanov, D.; Aghahosseini, A.; Gulagi, A.; Child, M.; Oyewo, A.S.; Farfan, J.; Sadovskaia, K.; Vainikka, P. Solar photovoltaics demand for the global energy transition in the power sector. *Prog. Photovolt. Res. Appl.* **2018**, *26*, 505–523. [CrossRef]
5. Schmela, M. *Global Market Outlook for Solar Power 2020–2024*; Solar Power: Brussels, Belgium, 2020.
6. International Renewable Energy Agency. *Renewable Power Generation Costs in 2019*; IRENA (2020), Renewable Power Generation Costs in 2019; International Renewable Energy Agency: Abu Dhabi, United Arab Emirates, 2020.
7. Liu, J.; Yao, M.; Shen, L. Third generation photovoltaic cells based on photonic crystals. *J. Mater. Chem. C* **2019**, *7*, 3121–3145. [CrossRef]
8. Hagfeldt, A.; Boschloo, G.; Sun, L.; Kloo, L.; Pettersson, H. Dye-Sensitized Solar Cells. *Chem. Rev.* **2010**, *110*, 6595–6663. [CrossRef]
9. Yan, N.; Zhao, C.; You, S.; Zhang, Y.; Li, W. Recent progress of thin-film photovoltaics for indoor application. *Chin. Chem. Lett.* **2020**, *31*, 643–653. [CrossRef]
10. Li, M.; Igbari, F.; Wang, Z.; Liao, L. Indoor Thin-Film Photovoltaics: Progress and Challenges. *Adv. Energy Mater.* **2020**, *10*. [CrossRef]
11. Lee, T.D.; Ebong, A.U. A review of thin film solar cell technologies and challenges. *Renew. Sustain. Energy Rev.* **2017**, *70*, 1286–1297. [CrossRef]
12. Edoff, M. Thin Film Solar Cells: Research in an Industrial Perspective. *Ambio* **2012**, *41*, 112–118. [CrossRef]
13. Lizin, S.; van Passel, S.; de Schepper, E.; Vranken, L. The future of organic photovoltaic solar cells as a direct power source for consumer electronics. *Sol. Energy Mater. Sol. Cells* **2012**, *103*, 1–10. [CrossRef]
14. Hoffmann, W. PV solar electricity industry: Market growth and perspective. *Sol. Energy Mater. Sol. Cells* **2006**, *90*, 3285–3311. [CrossRef]
15. Schuss, C.; Rahkonen, T. Photovoltaic (PV) energy as recharge source for portable devices such as mobile phones. In Proceedings of the 2012 12th Conference of Open Innovations Association (FRUCT), Oulu, Finland, 5–9 November 2012; pp. 1–9. [CrossRef]
16. Vicente, A.T.; Araújo, A.; Mendes, M.J.; Nunes, D.; Oliveira, M.J.; Sanchez-Sobrado, O.; Ferreira, M.P.; Águas, H.; Fortunato, E.; Martins, R. Multifunctional cellulose-paper for light harvesting and smart sensing applications. *J. Mater. Chem. C* **2018**, *6*, 3143–3181. [CrossRef]
17. Vicente, A.; Águas, H.; Mateus, T.; Araújo, A.; Lyubchyk, A.; Siitonen, S.; Fortunato, E.; Martins, R. Solar cells for self-sustainable intelligent packaging. *J. Mater. Chem. A* **2015**, *3*, 13226–13236. [CrossRef]
18. Águas, H.; Mateus, T.; Vicente, A.; Gaspar, D.; Mendes, M.J.; Schmidt, W.A.; Pereira, L.; Fortunato, E.; Martins, R. Thin Film Silicon Photovoltaic Cells on Paper for Flexible Indoor Applications. *Adv. Funct. Mater.* **2015**, *25*, 3592–3598. [CrossRef]
19. Shockley, W.; Queisser, H.J. Detailed Balance Limit of Efficiency of p-n Junction Solar Cells. *J. Appl. Phys.* **1961**, *32*, 510–519. [CrossRef]
20. Karg, M.; König, T.A.; Retsch, M.; Stelling, C.; Reichstein, P.M.; Honold, T.; Thelakkat, M.; Fery, A. Colloidal self-assembly concepts for light management in photovoltaics. *Mater. Today* **2015**, *18*, 185–205. [CrossRef]
21. Balling, P.; Christiansen, J.; Christiansen, R.E.; Eriksen, E.; Lakhotiya, H.; Mirsafaei, M.; Møller, S.; Nazir, A.; Vester-Petersen, J.; Jeppesen, B.; et al. Improving the efficiency of solar cells by upconverting sunlight using field enhancement from optimized nano structures. *Opt. Mater.* **2018**, *83*, 279–289. [CrossRef]
22. Parchine, M.; Kohoutek, T.; Bardosova, M.; Pemble, M.E. Large area colloidal photonic crystals for light trapping in flexible organic photovoltaic modules applied using a roll-to-roll Langmuir-Blodgett method. *Sol. Energy Mater. Sol. Cells* **2018**, *185*, 158–165. [CrossRef]
23. Brongersma, M.L.; Cui, Y.; Fan, S. Light management for photovoltaics using high-index nanostructures. *Nat. Mater.* **2014**, *13*, 451–460. [CrossRef] [PubMed]
24. Battaglia, C.; Hsu, C.-M.; Söderström, K.; Escarré, J.; Haug, F.-J.; Charrière, M.; Boccard, M.; Despeisse, M.; Alexander, D.T.L.; Cantoni, M.; et al. Light Trapping in Solar Cells: Can Periodic Beat Random? *ACS Nano* **2012**, *6*, 2790–2797. [CrossRef] [PubMed]
25. Rim, S.-B.; Zhao, S.; Scully, S.R.; McGehee, M.D.; Peumans, P. An effective light trapping configuration for thin-film solar cells. *Appl. Phys. Lett.* **2007**, *91*, 243501. [CrossRef]
26. Grandidier, J.; Weitekamp, R.A.; Deceglie, M.G.; Callahan, D.M.; Battaglia, C.; Bukowsky, C.R.; Ballif, C.; Grubbs, R.H.; Atwater, H.A. Solar cell efficiency enhancement via light trapping in printable resonant dielectric nanosphere arrays. *Phys. Status Solidi Appl. Mater. Sci.* **2012**, *210*, 255–260. [CrossRef]
27. Haque, S.; Mendes, M.J.; Sanchez-Sobrado, O.; Águas, H.; Fortunato, E.; Martins, R. Photonic-structured TiO_2 for high-efficiency, flexible and stable Perovskite solar cells. *Nano Energy* **2019**, *59*, 91–101. [CrossRef]
28. Alexandre, M.; Chapa, M.; Haque, S.; Mendes, M.J.; Águas, H.; Fortunato, E.; Martins, R. Optimum Luminescent Down-Shifting Properties for High Efficiency and Stable Perovskite Solar Cells. *ACS Appl. Energy Mater.* **2019**, *2*, 2930–2938. [CrossRef]
29. Sanchez-Sobrado, O.; Mendes, M.J.; Mateus, T.; Costa, J.; Nunes, D.; Águas, H.; Fortunato, E.; Martins, R. Photonic-structured TCO front contacts yielding optical and electrically enhanced thin-film solar cells. *Sol. Energy* **2020**, *196*, 92–98. [CrossRef]
30. Haque, S.; Alexandre, M.; Mendes, M.J.; Águas, H.; Fortunato, E.; Martins, R. Design of wave-optical structured substrates for ultra-thin perovskite solar cells. *Appl. Mater. Today* **2020**, *20*, 100720. [CrossRef]
31. Centeno, P.; Alexandre, M.; Chapa, M.; Pinto, J.V.; Deuermeier, J.; Mateus, T.; Fortunato, E.; Martins, R.; Águas, H.; Mendes, M.J. Self-Cleaned Photonic-Enhanced Solar Cells with Nanostructured Parylene-C. *Adv. Mater. Interfaces* **2020**, *7*. [CrossRef]
32. Ye, X.; Qi, L. Two-dimensionally patterned nanostructures based on monolayer colloidal crystals: Controllable fabrication, assembly, and applications. *Nano Today* **2011**, *6*, 608–631. [CrossRef]

33. Sanchez-Sobrado, O.; Mendes, M.J.; Haque, S.; Mateus, T.; Araujo, A.; Aguas, H.; Fortunato, E.; Martins, R. Colloidal-lithographed TiO$_2$ photonic nanostructures for solar cell light trapping. *J. Mater. Chem. C* **2017**, *5*, 6852–6861. [CrossRef]
34. Dimitrov, S.D.; Durrant, J.R. Materials design considerations for charge generation in organic solar cells. *Chem. Mater.* **2014**, *26*, 616–630. [CrossRef]
35. Levenson, M.; Viswanathan, N.; Simpson, R. Improving resolution in photolithography with a phase-shifting mask. *IEEE Trans. Electron Devices* **1982**, *29*, 1828–1836. [CrossRef]
36. Xia, Y.; Whitesides, G.M. Soft Lithography. *Annu. Rev. Mater. Res.* **1998**, *28*, 153–184. [CrossRef]
37. Stepanova, M.; Dew, S. *Nanofabrication: Techniques and Principles*; Springer: Vienna, Austria, 2011; ISBN 9783709104248.
38. Cui, Z. *Nanofabrication: Principles, Capabilities and Limits*, 2nd ed.; Springer: Gewerbestrasse, Switerzland, 2017; ISBN 9783319393612.
39. Ai, B.; Möhwald, H.; Wang, D.; Zhang, G. Advanced Colloidal Lithography Beyond Surface Patterning. *Adv. Mater. Interfaces* **2016**, *4*. [CrossRef]
40. Vieu, C.; Carcenac, F.; Pépin, A.; Chen, Y.; Mejias, M.; Lebib, A.; Manin-Ferlazzo, L.; Couraud, L.; Launois, H. Electron beam lithography: Resolution limits and applications. *Appl. Surf. Sci.* **2000**, *164*, 111–117. [CrossRef]
41. Melngailis, J. Focused ion beam technology and applications. *J. Vac. Sci. Technol. B Microelectron. Nanom. Struct.* **1987**, *5*, 469–495. [CrossRef]
42. Liu, H.; Ding, Y.; Jiang, W.; Lian, Q.; Yin, L.; Shi, Y.; Lu, B. Novel imprint lithography process used in fabrication of micro/nanostructures in organic photovoltaic devices. *J. Micro/Nanolithogr. MEMS MOEMS* **2009**, *8*, 021170. [CrossRef]
43. Glinsner, T.; Lindner, P.; Mühlberger, M.; Bergmair, I.; Schöftner, R.; Hingerl, K.; Schmidt, H.; Kley, E.-B. Fabrication process of 3D-photonic crystals via UV-nanoimprint lithography. In Proceedings of the 2007 NSTI Nanotechnology Conference and Trade Show, Santa Clara, CA, USA, 20–24 May 2007. [CrossRef]
44. Pourdavoud, N.; Hu, T.; Chen, Y.; Marianovich, A.; Kowalsky, W.; Heiderhoff, R.; Scheer, H.; Riedl, T.; Wang, S.; Mayer, A. Photonic Nanostructures: Photonic Nanostructures Patterned by Thermal Nanoimprint Directly into Organo-Metal Halide Perovskites. *Adv. Mater.* **2017**, *29*. [CrossRef]
45. Nazirizadeh, Y.; von Oertzen, F.; Plewa, K.; Barié, N.; Jakobs, P.-J.; Guttmann, M.; Leiste, H.; Gerken, M. Sensitivity optimization of injection-molded photonic crystal slabs for biosensing applications. *Opt. Mater. Express* **2013**, *3*, 556–565. [CrossRef]
46. Miller, M.; Brooks, C.B.; Lentz, D.; Doyle, G.; Resnick, D.; LaBrake, D.L. Step and flash imprint process integration techniques for photonic crystal patterning: Template replication through wafer patterning irrespective of tone. In Proceedings of the Advanced Fabrication Technologies for Micro/Nano Optics and Photonics, San Jose, CA, USA, 21–23 January 2008. [CrossRef]
47. Hong, L.; Yao, H.; Wu, Z.; Cui, Y.; Zhang, T.; Xu, Y.; Yu, R.; Liao, Q.; Gao, B.; Xian, K.; et al. Eco-Compatible Solvent-Processed Organic Photovoltaic Cells with Over 16% Efficiency. *Adv. Mater.* **2019**, *31*, e1903441. [CrossRef]
48. Sanchez-Sobrado, O.; Mendes, M.J.; Haque, S.; Mateus, T.; Águas, H.; Fortunato, E.; Martins, R. Lightwave trapping in thin film solar cells with improved photonic-structured front contacts. *J. Mater. Chem. C* **2019**, *7*, 6456–6464. [CrossRef]
49. Acikgoz, C.; Hempenius, M.A.; Huskens, J.; Vancso, G.J. Polymers in conventional and alternative lithography for the fabrication of nanostructures. *Eur. Polym. J.* **2011**, *47*, 2033–2052. [CrossRef]
50. Chou, S.Y.; Krauss, P.R.; Renstrom, P.J. Imprint Lithography with 25-Nanometer Resolution. *Science* **1996**, *272*, 85–87. [CrossRef]
51. Grandidier, J.; Atwater, H.A.; Deceglie, M.G.; Callahan, D.M. Simulations of solar cell absorption enhancement using resonant modes of a nanosphere array. In Proceedings of the Physics, Simulation, and Photonic Engineering of Photovoltaic Devices, San Francisco, CA, USA, 23–26 January 2012.
52. Mendes, M.J.; Sanchez-Sobrado, O.; Haque, S.; Mateus, T.; Águas, H.; Fortunato, E.; Martins, R. Wave-optical front structures on silicon and perovskite thin-film solar cells. In *Solar Cells and Light Management*; Enrichi, F., Righini, G.C., Eds.; Elsevier: Amsterdam, The Netherlands, 2020; pp. 315–354. [CrossRef]
53. Yu, Y.; Zhang, G. Colloidal Lithography. In *Updates in Advanced Lithography*; Hosaka, S., Ed.; InTechOpen: London, UK, 2013; ISBN 978-953-51-1175-7.
54. Chen, J.; Dong, P.; Di, D.; Wang, C.; Wang, H.; Wang, J.; Wu, X. Controllable fabrication of 2D colloidal-crystal films with polystyrene nanospheres of various diameters by spin-coating. *Appl. Surf. Sci.* **2013**, *270*, 6–15. [CrossRef]
55. You, J.; Wang, J.; Wang, L.; Wang, Z.; Wang, Z.; Li, J.; Lin, X. Dynamic particle packing in freezing colloidal suspensions. *Colloids Surf. A Physicochem. Eng. Asp.* **2017**, *531*, 93–98. [CrossRef]
56. Birdi, K.S. *Surface and Colloid Chemistry: Principles and Applications*; CRC Press Taylor & Francis Group: Boca Raton, FL, USA, 2010; ISBN 9781420095036.
57. Vlad, A.; Huynen, I.; Melinte, S. Wavelength-scale lens microscopy via thermal reshaping of colloidal particles. *Nanotechnology* **2012**, *23*, 285708. [CrossRef] [PubMed]
58. Chandramohan, A.; Sibirev, N.V.; Dubrovskii, V.G.; Petty, M.C.; Gallant, A.J.; Zeze, D.A. Model for large-area monolayer coverage of polystyrene nanospheres by spin coating. *Sci. Rep.* **2017**, *7*, 40888. [CrossRef] [PubMed]
59. Technique, M.S.L.; Wei-dong, R.; Zhi-cheng, L.; Nan, J.I.; Chun-xu, W.; Bing, Z.; Jun-hu, Z. Facile Fabrication of Large Area Polystyrene Colloidal Crystal Monolayer via Surfactant-free Langmuir-Blodgett Technique. *Chem. Res. Chin. Univ.* **2007**, *3*, 712–714.
60. Chen, Y.; Shi, D.; Chen, Y.; Chen, X.; Gao, J.; Zhao, N.; Wong, C.-P. A Facile, Low-Cost Plasma Etching Method for Achieving Size Controlled Non-Close-Packed Monolayer Arrays of Polystyrene Nano-Spheres. *Nanomaterials* **2019**, *9*, 605. [CrossRef] [PubMed]

61. Peng, X.; Jin, J.; Ericsson, E.M.; Ichinose, I. General Method for Ultrathin Free-Standing Films of Nanofibrous Composite Materials. *J. Am. Chem. Soc.* **2007**, *129*, 8625–8633. [CrossRef]
62. Torrisi, G.; Luis, J.S.; Sanchez-Sobrado, O.; Raciti, R.; Mendes, M.J.; Águas, H.; Fortunato, E.; Martins, R.; Terrasi, A. Colloidal-structured metallic micro-grids: High performance transparent electrodes in the red and infrared range. *Sol. Energy Mater. Sol. Cells* **2019**, *197*, 7–12. [CrossRef]
63. Chen, J.; Liao, W.-S.; Chen, X.; Yang, T.; Wark, S.E.; Son, D.H.; Batteas, J.D.; Cremer, P.S. Evaporation-Induced Assembly of Quantum Dots into Nanorings. *ACS Nano* **2008**, *3*, 173–180. [CrossRef] [PubMed]
64. Ober, C.; Lok, K.P.; Hair, M.L. Monodispersed, micron-sized polystyrene particles by dispersion polymerization. *J. Polym. Sci. Part C Polym. Lett.* **1985**, *23*, 103–108. [CrossRef]
65. Kallepalli, L.D.; Constantinescu, C.; Delaporte, P.; Utéza, O.; Grojo, D. Ultra-high ordered, centimeter scale preparation of microsphere Langmuir films. *J. Colloid Interface Sci.* **2015**, *446*, 237–243. [CrossRef]
66. Zou, D.; Aklonis, J.J.; Salovey, R. Model filled polymers. XI. Synthesis of monodisperse crosslinked polymethacrylonitrile beads. *J. Polym. Sci. Part A Polym. Chem.* **1992**, *30*, 2443–2449. [CrossRef]
67. Shen, S.; Sudol, E.D.; El-Aasser, M.S. Dispersion polymerization of methyl methacrylate: Mechanism of particle formation. *J. Polym. Sci. Part A Polym. Chem.* **1994**, *32*, 1087–1100. [CrossRef]
68. van Dommelen, R.; Fanzio, P.; Sasso, L. Surface self-assembly of colloidal crystals for micro- and nano-patterning. *Adv. Colloid Interface Sci.* **2018**, *251*, 97–114. [CrossRef]
69. Li, W.; Song, X.; Zhao, X.; Zhang, X.; Chen, R.; Zhang, X.; Jiang, C.; He, J.; Xiao, X. Design of wafer-scale uniform Au nanotip array by ion irradiation for enhanced single conductive filament resistive switching. *Nano Energy* **2020**, *67*, 104213. [CrossRef]
70. Zhang, J.; Li, Y.; Zhang, X.; Yang, B. Colloidal Self-Assembly Meets Nanofabrication: From Two-Dimensional Colloidal Crystals to Nanostructure Arrays. *Adv. Mater.* **2010**, *22*, 4249–4269. [CrossRef]
71. Fischer, U.C.; Zingsheim, H.P. Submicroscopic pattern replication with visible light. *Proc. J. Vac. Sci. Technol.* **1981**, *19*, 881–885. [CrossRef]
72. Deckman, H.W.; Dunsmuir, J.H. Natural lithography. *Appl. Phys. Lett.* **1982**, *41*, 377–379. [CrossRef]
73. Wang, Y.; Zhang, M.; Lai, Y.; Chi, L. Advanced colloidal lithography: From patterning to applications. *Nano Today* **2018**, *22*, 36–61. [CrossRef]
74. Syafiq, A.; Pandey, A.K.; Adzman, N.N.; Rahim, N.A. Advances in approaches and methods for self-cleaning of solar photovoltaic panels. *Sol. Energy* **2018**, *162*, 597–619. [CrossRef]
75. Velev, O.; Jede, T.A.; Lobo, R.F.; Lenhoff, A. Porous silica via colloidal crystallization. *Nat. Cell Biol.* **1997**, *389*, 447–448. [CrossRef]
76. Maskaly, G.R.; Petruska, M.A.; Nanda, J.; Bezel, I.V.; Schaller, R.D.; Htoon, H.; Pietryga, J.M.; Klimov, V.I. Amplified Spontaneous Emission in Semiconductor-Nanocrystal/Synthetic-Opal Composites: Optical-Gain Enhancement via a Photonic Crystal Pseudogap. *Adv. Mater.* **2006**, *18*, 343–347. [CrossRef]
77. Urban, J.J.; Talapin, D.V.; Shevchenko, E.V.; Murray, C.B. Self-Assembly of PbTe Quantum Dots into Nanocrystal Superlattices and Glassy Films. *J. Am. Chem. Soc.* **2006**, *128*, 3248–3255. [CrossRef]
78. Edwards, E.W.; Wang, D.; Möhwald, H. Hierarchical Organization of Colloidal Particles: From Colloidal Crystallization to Supraparticle Chemistry. *Macromol. Chem. Phys.* **2007**, *208*, 439–445. [CrossRef]
79. Zhang, C.; Cvetanovic, S.; Pearce, J.M. Fabricating ordered 2-D nano-structured arrays using nanosphere lithography. *MethodsX* **2017**, *4*, 229–242. [CrossRef]
80. Fang, Y.; Phillips, B.M.; Askar, K.; Choi, B.; Jiang, P.; Jiang, B. Scalable bottom-up fabrication of colloidal photonic crystals and periodic plasmonic nanostructures. *J. Mater. Chem. C* **2013**, *1*, 6031–6047. [CrossRef]
81. Park, B.; Na, S.Y.; Bae, I.-G. Uniform two-dimensional crystals of polystyrene nanospheres fabricated by a surfactant-assisted spin-coating method with polyoxyethylene tridecyl ether. *Sci. Rep.* **2019**, *9*, 1–9. [CrossRef]
82. Birnie, D.P. Spin coating technique. In *Sol-Gel Technologies for Glass Producers and Users*; Aegerter, M.A., Menning, M., Eds.; Springer: New York, NY, USA, 2004; pp. 49–55.
83. Hsieh, C.-H.; Lu, Y.-C.; Yang, H. Self-Assembled Mechanochromic Shape Memory Photonic Crystals by Doctor Blade Coating. *ACS Appl. Mater. Interfaces* **2020**, *12*, 36478–36484. [CrossRef] [PubMed]
84. Prevo, B.G.; Velev, O. Controlled, Rapid Deposition of Structured Coatings from Micro- and Nanoparticle Suspensions. *Langmuir* **2004**, *20*, 2099–2107. [CrossRef] [PubMed]
85. Yang, H.; Jiang, P. Large-Scale Colloidal Self-Assembly by Doctor Blade Coating. *Langmuir* **2010**, *26*, 13173–13182. [CrossRef]
86. Corkery, R.W. Langmuir–Blodgett (L–B) Multilayer Films. *Langmuir* **1997**, *13*, 3591–3594. [CrossRef]
87. Bardosova, M.; Pemble, M.E.; Povey, I.M.; Tredgold, R.H. The langmuir-blodgett approach to making colloidal photonic crystals from silica spheres. *Adv. Mater.* **2010**, *22*, 3104–3124. [CrossRef] [PubMed]
88. Hussain, S.A.; Dey, B.; Bhattacharjee, D.; Mehta, N. Unique supramolecular assembly through Langmuir—Blodgett (LB) technique. *Heliyon* **2018**, *4*, e01038. [CrossRef]
89. Açıkbaş, Y.; Evyapan, M.; Ceyhan, T.; Çapan, R.; Bekaroğlu, Ö. Characterisation of Langmuir–Blodgett films of new multinuclear copper and zinc phthalocyanines and their sensing properties to volatile organic vapours. *Sens. Actuators B Chem.* **2007**, *123*, 1017–1024. [CrossRef]

90. Huang, S.; Tsutsui, G.; Sakaue, H.; Shingubara, S.; Takahagi, T. Experimental conditions for a highly ordered monolayer of gold nanoparticles fabricated by the Langmuir–Blodgett method. *J. Vac. Sci. Technol. B Microelectron. Nanometer Struct.* **2001**, *19*, 2045. [CrossRef]
91. Schwartz, D. Langmuir-Blodgett film structure. *Surf. Sci. Rep.* **1997**, *27*, 245–334. [CrossRef]
92. Kausar, A. Survey on Langmuir–Blodgett Films of Polymer and Polymeric Composite. *Polym. Technol. Eng.* **2016**, *56*, 932–945. [CrossRef]
93. Pechkova, E.; Nicolini, C. Langmuir-Blodgett Protein Multilayer Nanofilms by XFEL. *NanoWorld J.* **2018**, *4*. [CrossRef]
94. Vogel, N.; Goerres, S.; Landfester, K.; Weiss, C. A Convenient Method to Produce Close- and Non-close-Packed Monolayers using Direct Assembly at the Air-Water Interface and Subsequent Plasma-Induced Size Reduction. *Macromol. Chem. Phys.* **2011**, *212*, 1719–1734. [CrossRef]
95. Retsch, M.; Zhou, Z.; Rivera, S.; Kappl, M.; Zhao, X.S.; Jonas, U.; Li, Q. Fabrication of Large-Area, Transferable Colloidal Monolayers Utilizing Self-Assembly at the Air/Water Interface. *Macromol. Chem. Phys.* **2009**, *210*, 230–241. [CrossRef]
96. Hur, J.; Won, Y.-Y. Fabrication of high-quality non-close-packed 2D colloid crystals by template-guided Langmuir–Blodgett particle deposition. *Soft Matter* **2008**, *4*, 1261–1269. [CrossRef] [PubMed]
97. Parchine, M.; McGrath, J.; Bardosova, M.; Pemble, M.E. Large Area 2D and 3D Colloidal Photonic Crystals Fabricated by a Roll-to-Roll Langmuir–Blodgett Method. *Langmuir* **2016**, *32*, 5862–5869. [CrossRef]
98. Reynaert, S.; Moldenaers, P.; Vermant, J. Control over Colloidal Aggregation in Monolayers of Latex Particles at the Oil-Water Interface. *Langmuir* **2006**, *22*, 4936–4945. [CrossRef]
99. Zhavnerko, G.; Marletta, G. Developing Langmuir–Blodgett strategies towards practical devices. *Mater. Sci. Eng. B* **2010**, *169*, 43–48. [CrossRef]
100. Chen, I.-T.; Schappell, E.; Zhang, X.; Chang, C.-H. Continuous roll-to-roll patterning of three-dimensional periodic nanostructures. *Microsyst. Nanoeng.* **2020**, *6*, 1–11. [CrossRef]
101. Li, X.; Gilchrist, J. Large-Area Nanoparticle Films by Continuous Automated Langmuir–Blodgett Assembly and Deposition. *Langmuir* **2016**, *32*, 1220–1226. [CrossRef]
102. Zhang, G.; Wang, D.; Möhwald, H. Decoration of Microspheres with Gold Nanodots—Giving Colloidal Spheres Valences. *Angew. Chem.* **2005**, *117*, 7945–7948. [CrossRef]
103. Zheng, J.; Dai, Z.; Mei, F.; Xiao, X.; Liao, L.; Wu, W.; Zhao, X.; Ying, J.; Ren, F.; Jiang, C. Micro–Nanosized Nontraditional Evaporated Structures Based on Closely Packed Monolayer Binary Colloidal Crystals and Their Fine Structure Enhanced Properties. *J. Phys. Chem. C* **2014**, *118*, 20521–20528. [CrossRef]
104. Li, Z.-W.; Zhou, J.; Zhang, Z.-J.; Dang, H.-X. Self-assembly of carboxyl functionalized polystyrene nano-spheres into close-packed monolayers via chemical adsorption. *Chin. J. Chem.* **2010**, *22*, 1133–1137. [CrossRef]
105. Dickreuter, S.; Gleixner, J.; Kolloch, A.; Boneberg, J.; Scheer, E.; Leiderer, P. Mapping of plasmonic resonances in nanotriangles. *Beilstein J. Nanotechnol.* **2013**, *4*, 588–602. [CrossRef] [PubMed]
106. Haynes, C.L.; Duyne, R.P. Van Nanosphere Lithography: A Versatile Nanofabrication Tool for Studies of Size-Dependent Nanoparticle Optics. *J. Polym. Sci. Part B Polym. Phys.* **2001**, *105*, 5599–5611. [CrossRef]
107. Lee, S.H.; Bantz, K.C.; Lindquist, N.C.; Oh, S.-H.; Haynes, C.L. Self-Assembled Plasmonic Nanohole Arrays. *Langmuir* **2009**, *25*, 13685–13693. [CrossRef] [PubMed]
108. Ai, B.; Yu, Y.; Möhwald, H.; Wang, L.; Zhang, G. Resonant Optical Transmission through Topologically Continuous Films. *ACS Nano* **2014**, *8*, 1566–1575. [CrossRef]
109. Naureen, S.; Sanatinia, R.; Shahid, N.; Anand, S. High Optical Quality InP-Based Nanopillars Fabricated by a Top-Down Approach. *Nano Lett.* **2011**, *11*, 4805–4811. [CrossRef]
110. Li, Y.; Ye, X.; Ma, Y.; Qi, L. Interfacial Nanosphere Lithography toward Ag2S-Ag Heterostructured Nanobowl Arrays with Effective Resistance Switching and Enhanced Photoresponses. *Small* **2014**, *11*, 1183–1188. [CrossRef]
111. Yao, Y.; Yao, J.; Narasimhan, V.K.; Ruan, Z.; Xie, C.; Fan, S.; Cui, Y. Broadband light management using low-Q whispering gallery modes in spherical nanoshells. *Nat. Commun.* **2012**, *3*, 664. [CrossRef]
112. Liu, G.; Li, Y.; Duan, G.; Wang, J.; Changhao, L.; Cai, W. Tunable Surface Plasmon Resonance and Strong SERS Performances of Au Opening-Nanoshell Ordered Arrays. *ACS Appl. Mater. Interfaces* **2012**, *4*, 1–5. [CrossRef] [PubMed]
113. Sun, Z.; Li, Y.; Zhang, J.; Li, Y.; Zhao, Z.; Zhang, K.; Zhang, G.; Guo, J.; Yang, B. A Universal Approach to Fabricate Various Nanoring Arrays Based on a Colloidal-Crystal-Assisted-Lithography Strategy. *Adv. Funct. Mater.* **2008**, *18*, 4036–4042. [CrossRef]
114. Zhang, G.; Wang, D.; Möhwald, H. Patterning Microsphere Surfaces by Templating Colloidal Crystals. *Nano Lett.* **2005**, *5*, 143–146. [CrossRef] [PubMed]
115. Rizzato, S.; Primiceri, E.; Monteduro, A.G.; Colombelli, A.; Leo, A.; Manera, M.G.; Rella, R.; Maruccio, G. Interaction-tailored organization of large-area colloidal assemblies. *Beilstein J. Nanotechnol.* **2018**, *9*, 1582–1593. [CrossRef] [PubMed]
116. Kumar, K.; Kim, Y.-S.; Yang, E.-H. The influence of thermal annealing to remove polymeric residue on the electronic doping and morphological characteristics of graphene. *Carbon* **2013**, *65*, 35–45. [CrossRef]
117. Vlad, A.; Frölich, A.; Zebrowski, T.; Dutu, C.A.; Busch, K.; Melinte, S.; Wegener, M.; Huynen, I. Direct Transcription of Two-Dimensional Colloidal Crystal Arrays into Three-Dimensional Photonic Crystals. *Adv. Funct. Mater.* **2012**, *23*, 1164–1171. [CrossRef]

118. Marlow, F.; Muldarisnur, M.; Sharifi, P.; Brinkmann, R.; Mendive, C.B. Opals: Status and Prospects. *Angew. Chem. Int. Ed.* **2009**, *48*, 6212–6233. [CrossRef]
119. Lee, S.Y.; Gradoń, L.; Janeczko, S.; Iskandar, F.; Okuyama, K. Formation of Highly Ordered Nanostructures by Drying Micrometer Colloidal Droplets. *ACS Nano* **2010**, *4*, 4717–4724. [CrossRef] [PubMed]
120. Lourtioz, J.-M.; Benisty, H.; Berge, V.; Gerard, J.-M.; Maystre, D.; Tchelnokov, A. *Photonic Crystals: Towards Nanoscale Photonic Devices*; Springer: Berlin/Heidelberg, Germany, 2003; ISBN 9783540244318.
121. Armstrong, E.; O'Dwyer, C. Artificial opal photonic crystals and inverse opal structures—Fundamentals and applications from optics to energy storage. *J. Mater. Chem. C* **2015**, *3*, 6109–6143. [CrossRef]
122. Kim, H.; Ge, J.; Kim, J.; Choi, S.-E.; Lee, H.; Lee, H.; Park, W.; Yin, Y.; Kwon, S. Structural colour printing using a magnetically tunable and lithographically fixable photonic crystal. *Nat. Photon.* **2009**, *3*, 534–540. [CrossRef]
123. Kim, S.-H.; Park, H.S.; Choi, J.H.; Shim, J.W.; Yang, S.-M. Integration of Colloidal Photonic Crystals toward Miniaturized Spectrometers. *Adv. Mater.* **2009**, *22*, 946–950. [CrossRef]
124. Xu, H.; Wu, P.; Zhu, C.; Elbaz, A.; Gu, Z.Z. Photonic crystal for gas sensing. *J. Mater. Chem. C* **2013**, *1*, 6087–6098. [CrossRef]
125. Noda, S. Photonic crystals. In *Comprehensive Microsystems*; Elsevier: Amsterdam, The Netherlands, 2007; pp. 101–112. ISBN 9780444521903. [CrossRef]
126. Fudouzi, H.; Xia, Y. Colloidal Crystals with Tunable Colors and Their Use as Photonic Papers. *Langmuir* **2003**, *19*, 9653–9660. [CrossRef]
127. Liu, Y.; Wang, H.; Ho, J.; Ng, R.C.; Ng, R.J.H.; Hall-Chen, V.H.; Koay, E.H.H.; Dong, Z.; Liu, H.; Qiu, C.-W.; et al. Structural color three-dimensional printing by shrinking photonic crystals. *Nat. Commun.* **2019**, *10*, 1–8. [CrossRef]
128. Arsenault, A.C.; Puzzo, D.P.; Manners, I.; Ozin, G.A. Photonic-crystal full-colour displays. *Nat. Photon.* **2007**, *1*, 468–472. [CrossRef]
129. Ding, T.; Song, K.; Clays, K.; Tung, C.-H. Controlled Directionality of Ellipsoids in Monolayer and Multilayer Colloidal Crystals. *Langmuir* **2010**, *26*, 11544–11549. [CrossRef] [PubMed]
130. Hosein, I.D.; Lee, S.; Liddell, C.M. Dimer-Based Three-Dimensional Photonic Crystals. *Adv. Funct. Mater.* **2010**, *20*, 3085–3091. [CrossRef]
131. Hynninen, A.-P.; Thijssen, J.; Vermolen, E.C.M.; Dijkstra, M.; Van Blaaderen, A. Self-assembly route for photonic crystals with a bandgap in the visible region. *Nat. Mater.* **2007**, *6*, 202–205. [CrossRef] [PubMed]
132. Rinne, S.A.; García-Santamaría, F.; Braun, P.V. Embedded cavities and waveguides in three-dimensional silicon photonic crystals. *Nat. Photon.* **2007**, *2*, 52–56. [CrossRef]
133. Blaszczyk-Lezak, I.; Aparicio, F.J.; Borras, A.; Barranco, A.; Alvarez-Herrero, A.; Fernández-Rodríguez, M.; Gonzalez-Elipe, A. Optically Active Luminescent Perylene Thin Films Deposited by Plasma Polymerization. *J. Phys. Chem. C* **2009**, *113*, 431–438. [CrossRef]
134. Qi, M.; Lidorikis, E.; Rakich, P.T.; Johnson, S.G.; Joannopoulos, J.D.; Ippen, E.P.; Smith, H.I. A three-dimensional optical photonic crystal with designed point defects. *Nat. Cell Biol.* **2004**, *429*, 538–542. [CrossRef]
135. Shao, J.; Liu, G.; Zhou, L. *Biomimetic Nanocoatings for Structural Coloration of Textiles*; Elsevier: Amsterdam, The Netherlands, 2016; pp. 269–299.
136. Jiang, C.; Markutsya, S.; Pikus, Y.; Tsukruk, V.V. Freely suspended nanocomposite membranes as highly sensitive sensors. *Nat. Mater.* **2004**, *3*, 721–728. [CrossRef]
137. Feng, D.; Lv, Y.; Wu, Z.; Dou, Y.; Han, L.; Sun, Z.; Xia, Y.; Zheng, G.; Zhao, D. Free-Standing Mesoporous Carbon Thin Films with Highly Ordered Pore Architectures for Nanodevices. *J. Am. Chem. Soc.* **2011**, *133*, 15148–15156. [CrossRef] [PubMed]
138. Jin, J.; Wakayama, Y.; Peng, X.; Ichinose, I. Surfactant-assisted fabrication of free-standing inorganic sheets covering an array of micrometre-sized holes. *Nat. Mater.* **2007**, *6*, 686–691. [CrossRef] [PubMed]
139. Zhang, J.-T.; Chao, X.; Asher, S.A. Asymmetric Free-Standing 2-D Photonic Crystal Films and Their Janus Particles. *J. Am. Chem. Soc.* **2013**, *135*, 11397–11401. [CrossRef] [PubMed]
140. Stein, A.; Li, F.; Denny, N.R. Morphological Control in Colloidal Crystal Templating of Inverse Opals, Hierarchical Structures, and Shaped Particles. *Chem. Mater.* **2008**, *20*, 649–666. [CrossRef]
141. Kim, S.-H.; Lee, S.Y.; Yang, S.-M.; Yi, G.-R. Self-assembled colloidal structures for photonics. *NPG Asia Mater.* **2011**, *3*, 25–33. [CrossRef]
142. Mendes, M.J.; Haque, S.; Sanchez-Sobrado, O.; Araújo, A.; Águas, H.; Fortunato, E.; Martins, R. Optimal-Enhanced Solar Cell Ultra-thinning with Broadband Nanophotonic Light Capture. *iScience* **2018**, *3*, 238–254. [CrossRef]
143. Karmouch, R.; El Hor, H. Solar Cells Performance Reduction under the Effect of Dust in Jazan Region. *J. Fundam. Renew. Energy Appl.* **2017**, *7*. [CrossRef]
144. He, G.; Zhou, C.; Li, Z. Review of Self-Cleaning Method for Solar Cell Array. *Procedia Eng.* **2011**, *16*, 640–645. [CrossRef]
145. Saravanan, V.; Darvekar, S.K. Solar Photovoltaic Panels Cleaning Methods A Review. *Int. J. Pure Appl. Math.* **2018**, *118*, 1–17.
146. Deceglie, M.G.; Ferry, V.E.; Alivisatos, A.P.; Atwater, H.A.; Alivisatos, P. Design of Nanostructured Solar Cells Using Coupled Optical and Electrical Modeling. *Nano Lett.* **2012**, *12*, 2894–2900. [CrossRef]
147. Isabella, O.; Vismara, R.; Linssen, D.; Wang, K.; Fan, S.; Zeman, M. Advanced light trapping scheme in decoupled front and rear textured thin-film silicon solar cells. *Sol. Energy* **2018**, *162*, 344–356. [CrossRef]

148. Righini, G.C.; Enrichi, F. Solar cells' evolution and perspectives: A short review. In *Solar Cells and Light Management*; Elsevier: Amsterdam, The Netherlands, 2020; pp. 1–32. ISBN 978-0-08-102762-2.
149. Mendes, M.J.; Araújo, A.; Vicente, A.; Águas, H.; Ferreira, I.; Fortunato, E.; Martins, R. Design of optimized wave-optical spheroidal nanostructures for photonic-enhanced solar cells. *Nano Energy* **2016**, *26*, 286–296. [CrossRef]
150. Mendes, M.J.; Tobías, I.; Martí, A.; Luque, A. Light concentration in the near-field of dielectric spheroidal particles with mesoscopic sizes. *Opt. Express* **2011**, *19*, 16207–16222. [CrossRef]
151. Li, K.; Haque, S.; Martins, A.; Fortunato, E.; Martins, R.; Mendes, M.J.; Schuster, C. Simple, yet mighty, principles to maximise photon absorption in thin media. *Opt.* **2020**, *7*. [CrossRef]
152. Ariga, K.; Yamauchi, Y.; Mori, T.; Hill, J.P. 25th Anniversary Article: What Can Be Done with the Langmuir-Blodgett Method? Recent Developments and its Critical Role in Materials Science. *Adv. Mater.* **2013**, *25*, 6477–6512. [CrossRef] [PubMed]
153. Kirchartz, T. Photon Management in Perovskite Solar Cells. *J. Phys. Chem. Lett.* **2019**, *10*, 5892–5896. [CrossRef]
154. NREL. *Best Research-Cell Efficiencies Chart*; Rev. 04-06-2020; 2020. Available online: https://www.nrel.gov/pv/cell-efficiency.html (accessed on 4 June 2021).
155. Dudem, B.; Heo, J.H.; Leem, J.W.; Yu, J.S.; Im, S.H. CH3NH3PbI3 planar perovskite solar cells with antireflection and self-cleaning function layers. *J. Mater. Chem. A* **2016**, *4*, 7573–7579. [CrossRef]
156. Thangavel, N.R.; Adhyaksa, G.W.P.; Dewi, H.A.; Tjahjana, L.; Bruno, A.; Birowosuto, M.D.; Wang, H.; Mathews, N.; Mhaisalkar, S. Disordered Polymer Antireflective Coating for Improved Perovskite Photovoltaics. *ACS Photon.* **2020**, *7*, 1971–1977. [CrossRef]
157. Heifetz, A.; Kong, S.-C.; Sahakian, A.V.; Taflove, A.; Backman, V. Photonic Nanojets. *J. Comput. Theor. Nanosci.* **2009**, *6*, 1979–1992. [CrossRef] [PubMed]
158. Chen, J.-D.; Jin, T.-Y.; Li, Y.-Q.; Tang, J.-X. Recent progress of light manipulation strategies in organic and perovskite solar cells. *Nanoscale* **2019**, *11*, 18517–18536. [CrossRef]
159. Erwin, W.R.; Zarick, H.F.; Talbert, E.M.; Bardhan, R. Light trapping in mesoporous solar cells with plasmonic nanostructures. *Energy Environ. Sci.* **2016**, *9*, 1577–1601. [CrossRef]
160. Zhang, H.; Toudert, J. Optical management for efficiency enhancement in hybrid organic-inorganic lead halide perovskite solar cells. *Sci. Technol. Adv. Mater.* **2018**, *19*, 411–424. [CrossRef]
161. Wang, Y.; Wang, P.; Zhou, X.; Li, C.; Li, H.; Hu, X.; Li, F.; Liu, X.; Li, M.; Song, Y. Diffraction-Grated Perovskite Induced Highly Efficient Solar Cells through Nanophotonic Light Trapping. *Adv. Energy Mater.* **2018**, *8*. [CrossRef]
162. Deng, K.; Liu, Z.; Wang, M.; Li, L. Nanoimprinted Grating-Embedded Perovskite Solar Cells with Improved Light Management. *Adv. Funct. Mater.* **2019**, *29*. [CrossRef]
163. Ito, S.; Tanaka, S.; Manabe, K.; Nishino, H. Effects of Surface Blocking Layer of Sb2S3 on Nanocrystalline TiO2 for CH3NH3PbI3 Perovskite Solar Cells. *J. Phys. Chem. C* **2014**, *118*, 16995–17000. [CrossRef]
164. Leijtens, T.; Eperon, G.E.; Pathak, S.; Abate, A.; Lee, M.M.; Snaith, H.J. Overcoming ultraviolet light instability of sensitized TiO2 with meso-superstructured organometal tri-halide perovskite solar cells. *Nat. Commun.* **2013**, *4*, 2885. [CrossRef]
165. Quitsch, W.-A.; Dequilettes, D.W.; Pfingsten, O.; Schmitz, A.; Ognjanovic, S.; Jariwala, S.; Koch, S.; Winterer, M.; Ginger, D.S.; Bacher, G. The Role of Excitation Energy in Photobrightening and Photodegradation of Halide Perovskite Thin Films. *J. Phys. Chem. Lett.* **2018**, *9*, 2062–2069. [CrossRef] [PubMed]
166. Lee, S.-W.; Kim, S.; Bae, S.; Cho, K.; Chung, T.; Mundt, L.E.; Lee, S.; Park, S.; Park, H.; Schubert, M.C.; et al. UV Degradation and Recovery of Perovskite Solar Cells. *Sci. Rep.* **2016**, *6*, 38150. [CrossRef] [PubMed]
167. Farooq, A.; Hossain, I.M.; Moghadamzadeh, S.; Schwenzer, J.A.; Abzieher, T.; Richards, B.S.; Klampaftis, E.; Paetzold, U.W. Spectral Dependence of Degradation under Ultraviolet Light in Perovskite Solar Cells. *ACS Appl. Mater. Interfaces* **2018**, *10*, 21985–21990. [CrossRef]
168. Wang, F.; Zhang, Y.; Yang, M.; Fan, L.; Yang, L.; Sui, Y.; Yang, J.; Zhang, X. Toward ultra-thin and omnidirectional perovskite solar cells: Concurrent improvement in conversion efficiency by employing light-trapping and recrystallizing treatment. *Nano Energy* **2019**, *60*, 198–204. [CrossRef]
169. Aydin, E.; Allen, T.G.; De Bastiani, M.; Xu, L.; Ávila, J.; Salvador, M.; Van Kerschaver, E.; De Wolf, S. Interplay between temperature and bandgap energies on the outdoor performance of perovskite/silicon tandem solar cells. *Nat. Energy* **2020**, *5*, 851–859. [CrossRef]
170. Werner, J.; Sahli, F.; Fu, F.; Leon, J.J.D.; Walter, A.; Kamino, B.A.; Niesen, B.; Nicolay, S.; Jeangros, Q.; Ballif, C. Perovskite/Perovskite/Silicon Monolithic Triple-Junction Solar Cells with a Fully Textured Design. *ACS Energy Lett.* **2018**, *3*, 2052–2058. [CrossRef]
171. SigmaAldrich. Indium Tin Oxide Coated Glass Slide, Square. Available online: https://www.sigmaaldrich.com/catalog/product/aldrich/703192?lang=pt&region=PT (accessed on 23 June 2021).
172. Gao, T.; Wang, B.; Ding, B.; Lee, J.-K.; Leu, P.W. Correction to Uniform and Ordered Copper Nanomeshes by Microsphere Lithography for Transparent Electrodes. *Nano Lett.* **2014**, *14*, 3694. [CrossRef]
173. Lyubchyk, A.; Vicente, A.; Alves, P.U.; Catela, B.; Soule, B.; Mateus, T.; Mendes, M.J.; Águas, H.; Fortunato, E.; Martins, R. Influence of post-deposition annealing on electrical and optical properties of ZnO-based TCOs deposited at room temperature. *Phys. Status Solidi Appl. Mater. Sci.* **2016**, *213*, 2317–2328. [CrossRef]
174. Tuna, O.; Selamet, Y.; Aygun, G.; Ozyuzer, L. High quality ITO thin films grown by dc and RF sputtering without oxygen. *J. Phys. D Appl. Phys.* **2010**, *43*. [CrossRef]

175. Gao, T.; Wang, B.; Ding, B.; Lee, J.-K.; Leu, P.W. Uniform and Ordered Copper Nanomeshes by Microsphere Lithography for Transparent Electrodes. *Nano Lett.* **2014**, *14*, 2105–2110. [CrossRef]
176. Kim, W.-K.; Lee, S.; Lee, D.H.; Park, I.H.; Bae, J.S.; Lee, T.W.; Kim, J.-Y.; Park, J.H.; Cho, Y.C.; Cho, C.R.; et al. Cu mesh for flexible transparent conductive electrodes. *Sci. Rep.* **2015**. [CrossRef]
177. Chapa, M.; Alexandre, M.F.; Mendes, M.J.; Águas, H.; Fortunato, E.; Martins, R. All-Thin-Film Perovskite/C-Si Four-Terminal Tandems: Interlayer and Intermediate Contacts Optimization. *ACS Appl. Energy Mater.* **2019**. [CrossRef]
178. Li, X.-M.; Reinhoudt, D.; Crego-Calama, M. What do we need for a superhydrophobic surface? A review on the recent progress in the preparation of superhydrophobic surfaces. *Chem. Soc. Rev.* **2007**, *36*, 1350–1368. [CrossRef] [PubMed]
179. Liu, Q.; Chen, D.; Kang, Z. One-Step Electrodeposition Process to Fabricate Corrosion-Resistant Superhydrophobic Surface on Magnesium Alloy. *ACS Appl. Mater. Interfaces* **2015**, *7*, 1859–1867. [CrossRef] [PubMed]
180. Zhang, G.; Wang, D.; Gu, Z.-Z.; Möhwald, H. Fabrication of Superhydrophobic Surfaces from Binary Colloidal Assembly. *Langmuir* **2005**, *21*, 9143–9148. [CrossRef]
181. Jindasuwan, S.; Nimittrakoolchai, O.; Sujaridworakun, P.; Jinawath, S.; Supothina, S. Surface characteristics of water-repellent polyelectrolyte multilayer films containing various silica contents. *Thin Solid Films* **2009**, *517*, 5001–5005. [CrossRef]
182. Hikita, M.; Tanaka, K.; Nakamura, T.; Kajiyama, T.; Takahara, A. Super-Liquid-Repellent Surfaces Prepared by Colloidal Silica Nanoparticles Covered with Fluoroalkyl Groups. *Langmuir* **2005**, *21*, 7299–7302. [CrossRef] [PubMed]
183. Meng, E.; Li, P.-Y.; Tai, Y.-C. Plasma removal of Parylene C. *J. Micromech. Microeng.* **2008**, *18*. [CrossRef]
184. Kim, H.; Lee, J.; Kim, B.; Byun, H.R.; Kim, S.H.; Oh, H.M.; Baik, S.; Jeong, M.S. Enhanced Stability of MAPbI3 Perovskite Solar Cells using Poly(p-chloro-xylylene) Encapsulation. *Sci. Rep.* **2019**, *9*, 1–6. [CrossRef] [PubMed]
185. Ortigoza-Diaz, J.; Scholten, K.; Meng, E. Characterization and Modification of Adhesion in Dry and Wet Environments in Thin-Film Parylene Systems. *J. Microelectromech. Syst.* **2018**, *27*, 874–885. [CrossRef]
186. Gołda, M.; Brzychczy-Włoch, M.; Faryna, M.; Engvall, K.; Kotarba, A. Oxygen plasma functionalization of parylene C coating for implants surface: Nanotopography and active sites for drug anchoring. *Mater. Sci. Eng. C* **2013**, *33*, 4221–4227. [CrossRef]
187. Ciurana, J. Designing, prototyping and manufacturing medical devices: An overview. *Int. J. Comput. Integr. Manuf.* **2014**, *27*, 901–918. [CrossRef]
188. Andersson, A.-S.; Brink, J.; Lidberg, U.; Sutherland, D.S. Influence of systematically varied nanoscale topography on the morphology of epithelial cells. *IEEE Trans. Nanobiosci.* **2003**, *2*, 49–57. [CrossRef]
189. Wood, M. Colloidal lithography and current fabrication techniques producing in-plane nanotopography for biological applications. *J. R. Soc. Interface* **2006**, *4*, 1–17. [CrossRef] [PubMed]
190. Dalby, M.J.; O Riehle, M.; Sutherland, D.S.; Agheli, H.; Curtis, A.S.G. Morphological and microarray analysis of human fibroblasts cultured on nanocolumns produced by colloidal lithography. *Eur. Cells Mater.* **2005**, *9*, 1–8. [CrossRef]
191. Wood, M.A.; Wilkinson, C.D.W.; Curtis, A.S.G. The Effects of Colloidal Nanotopography on Initial Fibroblast Adhesion and Morphology. *IEEE Trans. Nanobiosci.* **2006**, *5*, 20–31. [CrossRef] [PubMed]
192. Sutherland, D.S.; Broberg, M.; Nygren, H.; Kasemo, B. Influence of Nanoscale Surface Topography and Chemistry on the Functional Behaviour of an Adsorbed Model Macromolecule. *Macromol. Biosci.* **2001**. [CrossRef]
193. Denis, F.A.; Hanarp, P.; Sutherland, D.S.; Gold, J.; Mustin, C.; Rouxhet, A.P.G.; Dufrêne, Y.F. Protein Adsorption on Model Surfaces with Controlled Nanotopography and Chemistry. *Langmuir* **2002**, *18*, 819–828. [CrossRef]
194. Wang, P.-Y.; Bennetsen, D.T.; Foss, M.; Ameringer, T.; Thissen, H.; Kingshott, P. Modulation of Human Mesenchymal Stem Cell Behavior on Ordered Tantalum Nanotopographies Fabricated Using Colloidal Lithography and Glancing Angle Deposition. *ACS Appl. Mater. Interfaces* **2015**, *7*, 4979–4989. [CrossRef] [PubMed]
195. Dalby, M.J.; Riehle, M.O.; Sutherland, D.S.; Agheli, H.; Curtis, A.S. Fibroblast response to a controlled nanoenvironment produced by colloidal lithography. *J. Biomed. Mater. Res.* **2004**, *69*, 314–322. [CrossRef] [PubMed]

Article

An Effective Method to Accurately Extract the Parameters of Single Diode Model of Solar Cells

Zhaoxu Song, Kun Fang, Xiaofang Sun, Ying Liang, Wei Lin, Chuanzhong Xu, Gongyi Huang and Fei Yu *

College of Information Science and Engineering, Huaqiao University, Xiamen 361021, China; hquszx@163.com (Z.S.); fksaya@126.com (K.F.); xfsun@hqu.edu.cn (X.S.); liangyinghqu@163.com (Y.L.); linwei_0311@126.com (W.L.); xucz@hqu.edu.cn (C.X.); hgy@hqu.edu.cn (G.H.)
* Correspondence: yufei_jnu@126.com

Abstract: A non-iterative method is presented to accurately extract the five parameters of single diode model of solar cells in this paper. This method overcomes the problems of complexity and accuracy by simplifying the calculation process. Key parts of the equation are to be adjusted dynamically so that the desired five parameters can be obtained from the I-V curve. Then, the I-V and P-V characteristic curves of solar cells are used to compare the effectiveness of this method with other methods. Furthermore, the root mean square error analysis shows that this method is more applicable than other methods. Finally, the I-V and P-V characteristics simulated by using the extracted parameters in this method are compared and discussed with the experimental data of solar cells under different conditions. In fact, this extraction process can be regarded as an effective and accurate method to estimate solar cells' single diode model parameters.

Keywords: solar cells; parameter extraction; single diode model; non-iterative

1. Introduction

With the intensification of the greenhouse effect, the demands for clean and sustainable energy resources are sharply increasing worldwide and this has become a public concern [1]. Solar energy is undoubtedly one of the most promising, pollution-free energy sources. Because of solar cells' advantages of energy saving and no pollution, the single diode model of solar cells has become one of the hottest research projects. The main purpose of these studies is to ascertain an analytical solution [2] and parameter extraction [3] to predict the I-V and P-V characteristics of solar cells. At present, analytical solution algorithms have been developed for many years, and the technology tends to be mature and saturated. However, parameter extraction routines still have to face a challenge for a trade-off between accuracy and efficiency. In fact, the accuracy of the single diode model predictions for solar cells' characteristics are fully dependent on the model parameter values being extracted. Although complex extraction [4] procedures can obtain high precision parameter values, it may lead to inefficiency of computational process. Therefore, the single diode model of solar cells urgently needs an accurate and effective method to extract the model parameters.

Up until now, several authors have proposed various methods to determine different parameters in the single diode model. These methods can be divided into two categories [5,6]. One category is non-iterative analysis procedures [7–22], which are reviewed in [23]. They determine the analytical solution by simplifying and replacing the key parts of the equation, and then calculating the parameter values by depending on the information in the datasheet provided by the manufacturer [24–26], which refers to short circuit current, open circuit voltage, maximum power point, or the slope of the intersection of the I-V characteristic curve and the coordinate axes. Although these approaches are relatively simple and the calculation process is fast, the simplification and replacement often lead to a lack of accuracy and to results without physical significance [27]. Additionally, since the parameters are only obtained from the data in the datasheet, the results obtained are also very sensitive to the measurement error. These measurement errors are caused by the different accuracy of the test equipment. Different significant figures will also have a certain impact on the accuracy of the results. The other category is numerical or intelligent algorithm programs [28–32]. These are essentially processes of optimization or fitting, which can minimize the error between the obtained *I-V* or *P-V* characteristic curves and the experimental data, and then obtain high-precision parameter values. However, the inefficiency of the calculation process has always been the biggest problem for this kind of extraction strategy. Briefly, all these methods are almost difficult to have a good trade-off between accuracy and efficiency. Therefore, an efficient and accurate parameter extraction program is still needed to embed the circuit simulator of the model and diagnose the process optimization problem.

In this paper, an effective non-iterative method is proposed to accurately extract five parameters in solar cells' single diode model. The analytical solution of the terminal current-voltage equation of the equivalent circuit model is firstly derived. Subsequently, five basic parameter equations are listed according to data obtained from *I-V* curve. Then, the important parts of the basic circuit equations are simplified and replaced to obtain the five expressions of the parameters. Finally, the five extracted parameters are substituted into the analytical solution to simulate the *I-V* and *P-V* characteristics of solar cells. Simultaneously, five parameter extraction methods described in other works of the literature are compared with the method proposed in this paper. The obtained parameter values and the RMSE are recorded. Furthermore, a comprehensive experimental evaluation is conducted to demonstrate the accuracy and verify the effectiveness of the proposed approach based on different solar cells' photovoltaic technologies, irradiances, and temperatures. The results show that this accurate and efficient strategy can play a good role in single diode model parameter extraction. In fact, the method proposed in this paper is easier to be used to be implement lumped parameter model into simulators in technology. In addition, it also helps to provide an optimization suggestion on solar cells' preparations.

2. Method of Parameter Extraction

The equivalent circuit model of the single diode of solar cells is shown in Figure 1, including a photocurrent source, a single diode, a series resistance, and a shunt resistance. The five parameters of the model are the photocurrent I_{ph}, the diode reverse saturation current I_s, the diode ideal factor n, the parallel resistance R_{sh}, and the series resistance R_s. According to Kirchhoff's current law and Schockley's ideal diode current equation, the terminal current-voltage equation of the circuit model is deduced as follows:

$$I = Iph - \left(\frac{V + IRs}{Rsh}\right) - Is\left(\exp\left(\frac{V + IRs}{nVT}\right) - 1\right) \quad (1)$$

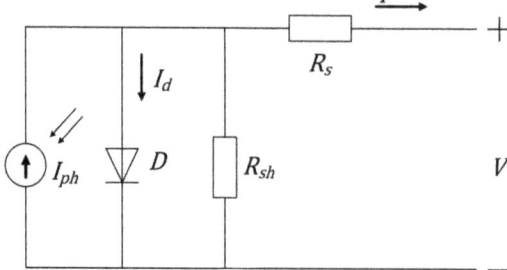

Figure 1. Single diode equivalent circuit model of solar cells.

In Equation (1), V_T is the thermal voltage, which can be calculated by $V_T = kT/q$, where k is the Boltzmann constant, T is the cell temperature, and q is the charge of the electron. According to Equation (1), the main purpose of this research is to adjust these five parameters $I_{ph}, R_s, R_{sh}, I_s, n$ to predict the I-V characteristics, so that they are consistent with the electrostatic performances of solar cells. These adjustments are usually based on the data on the I-V curves measured by experiments or on the datasheet provided by the manufacturer.

The typical I-V curve of solar cells' single diode model is shown in Figure 2. There are three important points, i.e., the short-circuit point, the open-circuit point, and the maximum power point. In fact, the voltage (V) and current (I) values of the three points are the basic template data and always known from solar cells' data sheet, so they are hence used to create relevant equations as follows.

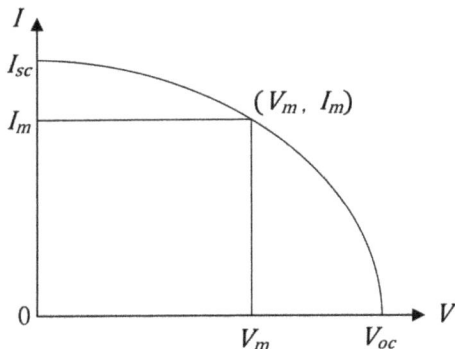

Figure 2. I-V curve of single diode model in solar cells.

At the short-circuit point: $(V = 0, I = I_{sc})$, Equation (1) can be represented as

$$Iph = Isc + Is\left(\exp\left(\frac{IscRs}{nVT}\right) - 1\right) + \frac{IscRs}{Rsh} \qquad (2)$$

At the open-circuit point: $(V = V_{oc}, I = 0)$, Equation (1) can be written by

$$Iph = \frac{Voc}{Rsh} + Is\left(\exp\left(\frac{Voc}{nVT}\right) - 1\right) \qquad (3)$$

At the maximum power point: $(V = V_m, I = I_m)$, these values are substituted into Equation (1), yielding:

$$Iph = \frac{Vm + ImRs}{Rsh} + Im + Is\left(\exp\left(\frac{Vm + ImRs}{nVT}\right) - 1\right) \qquad (4)$$

Under the same irradiance, the left side of Equations (3) and (4) is the same, which means that their right side is equal, yielding:

$$Is\exp\left(\frac{Voc}{nVT}\right) + \frac{Voc - Vm}{Rsh} - Im - \frac{RsIm}{Rsh} - Is\exp\left(\frac{Vm + RsIm}{nVT}\right) = 0 \qquad (5)$$

Generally, these three equations, i.e., Equations (2), (3), and (5), are not enough for extracting the five parameters of the model. Thus, two supplementary equations have to be added to establish an equation set consisting of five equations. In Figure 1, the shunt and series resistances R_{sh} and R_s are estimated as the experimental resistances R_{sho} and R_{so}, which are usually calculated from the slope of the *I-V* curve at short circuit (SC) and open circuit (OC). Therefore, two supplementary equations are written as

$$Rsho = -\frac{dV}{dI}\bigg|_{SC} \tag{6}$$

$$Rso = -\frac{dV}{dI}\bigg|_{OC} \tag{7}$$

Here, R_{sho} and R_{so} can be easily approximated by

$$Rsho \approx -\frac{0.001}{f(0.001) - Isc}(\Omega) \tag{8}$$

$$Rso \approx -\frac{0.001}{0 - f(Voc - 0.001)}(\Omega) \tag{9}$$

It is noted that f is the function of the *I-V* curve in Figure 2. Then, by deriving Equation (1) at the short-circuit point to obtain the expression of $\frac{dV}{dI}\big|_{SC}$ and substituting it into Equation (6), Equation (10) can be obtained as

$$\frac{Rsho - Rs}{Rsh} + \frac{Is}{nVT}(Rsho - Rs)\exp\left(\frac{IscRs}{nVT}\right) - 1 = 0 \tag{10}$$

Similarly, by deriving Equation (1) at the open-circuit point to obtain the expression of $\frac{dV}{dI}\big|_{OC}$ and substituting it into Equation (7), Equation (11) can be expressed as

$$(Rso - Rs)\left(\frac{1}{Rsh} + \frac{Is}{nVT}\exp\left(\frac{Voc}{nVT}\right)\right) - 1 = 0 \tag{11}$$

Above five equations, i.e., Equations (2), (3), (5), (10), and (11), are used to determine the analytical expressions of these five parameters. In order to get more accurate and efficient parameter values, two reasonable approximations need to be considered in these equations. The first one is $I_s\exp(V_{oc}/nV_T) \gg I_s\exp(I_{sc}R_s/nV_T)$ due to $V_{oc} \gg I_{sc}R_s$, compared with the former, the term $I_s\exp(I_{sc}R_s/nV_T)$ can be ignored. In addition, when the value of the term $I_s\exp(I_{sc}R_s/nV_T)$ is too small and has little impact on the whole equation, it can also be replaced by 0. The second one is $R_{sh} \gg R_s$. Thus, $1 + R_s/R_{sh} \approx 1$ and $R_{sho} \approx R_{sh}$ are also valid.

According to these two approximations, the expressions of five parameters can be extracted. However, the analytical equations for determining parameters generally cannot use too many approximate conditions. This may lead to low accuracy of parameters, which makes the calculation results unreliable and unsatisfactory. Therefore, it is necessary to reduce the use of approximation as much as possible and retain important conditions in the calculation. The detailed explanation is presented as follows.

Under the condition of constant irradiance, by taking Equation (2) into Equation (3) and eliminating I_{ph}, and then using the first approximation $I_s\exp(V_{oc}/nV_T) \gg I_s\exp(I_{sc}R_s/nV_T)$, Equation (12) can be obtained as

$$Is\exp\left(\frac{Voc}{nVT}\right) = Isc\left(1 + \frac{Rs}{Rsh}\right) - \frac{Voc}{Rsh} \tag{12}$$

In Equation (10), $(I_s/nV_T)\exp(I_{sc}R_s/nV_T)$ is much smaller than the rest and after simplification, yielding:

$$\frac{Rsho}{Rsh} = 1 + \frac{Rs}{Rsh} \tag{13}$$

According to Equation (13), Equation (12) can be rewritten as

$$Is \exp\left(\frac{Voc}{nVT}\right) = \frac{IscRsho}{Rsh} - \frac{Voc}{Rsh} \qquad (14)$$

Now, Equations (11), (13), and (14) need to be substituted into Equation (5). First, the $I_s\exp(V_{oc}/nV_T)$ of Equation (5) needs to be replaced by the right part of Equation (14). Then, Equation (13) is used to replace R_s/R_{sh} in Equation (5). After replacing these parts above, an intermediate equation can be obtained as

$$\frac{(Isc - Im)Rsho - Vm}{Rsh} = Is \exp\left(\frac{Vm + RsIm}{nVT}\right) \qquad (15)$$

Both sides of the Equation (15) are represented by logarithmic computation, yielding:

$$\ln[(Isc - Im)Rsho - Vm] - \ln Rsh = \ln Is + \frac{Vm + RsIm}{nVT} \qquad (16)$$

Second, Equation (14) is expressed as I_s on the left, the remaining part is on the right, and the right part is used to replace I_s in Equation (16). Another intermediate equation can be obtained as

$$\ln[(Isc - Im)Rsho - Vm] = \ln(IscRsho - Voc) - \frac{Voc}{nVT} + \frac{Vm + RsIm}{nVT} \qquad (17)$$

Finally, Equation (11) needs to be rewritten as R_s on the left and the remainder on the right, and then replace R_s of Equation (17) with the remainder on the right, yielding:

$$\ln[(Isc - Im)Rsho - Vm] - \ln(IscRsho - Voc) = \frac{ImRso + Vm - Voc}{nVT} - \frac{Im}{Isc\frac{Rsho}{Rsh} - \frac{Voc}{Rsh} + \frac{nVT}{Rsh}} \qquad (18)$$

In the above part, we only use the first approximation instead of using the two approximations synchronously as the conventional method. This is mainly because R_s/R_{sh} is much bigger than $I_s\exp(I_{sc}R_s/nV_T)$ and the latter is more complex. This behavior effectively reduces the use of approximation conditions, which is very helpful to improve the accuracy of the parameters. However, in Equation (18), considering $R_{sho} \approx R_{sh}$ has very little effect on the whole equation, Equation (18) can be replaced by Equation (19) as follows.

$$\ln[(Isc - Im)Rsho - Vm] - \ln(IscRsho - Voc) = \frac{ImRso + Vm - Voc}{nVT} - \frac{Im}{Isc - \frac{Voc}{Rsho} + \frac{nVT}{Rsho}} \qquad (19)$$

It is worth noting that we need to solve a quadratic Equation (19) with one unknown parameter n. In order to avoid negative numbers and complex numbers, we choose the negative root as the solution of the parameter n, i.e.,

$$n = \frac{-\sqrt{4ABCRsho + (ACRsho + ImRsho - B)^2} - (ACRsho + ImRsho - B)}{2A \cdot V_T} \qquad (20)$$

Here A, B, and C are symbolled as $A = \ln[(Isc - Im)Rsho - Vm] - \ln(IscRsho - Voc)$, $B = ImRso + Vm - Voc$, $C = Isc - \frac{Voc}{Rsho}$.

According to the order of calculation and considering the second approximation, i.e., $1 + R_s/R_{sh} \approx 1$ and $R_{sho} \approx R_{sh}$ in Equation (12), I_s can be extracted as

$$Is = \left(Isc - \frac{Voc}{Rsho}\right) \exp\left(-\frac{Voc}{nVT}\right) \qquad (21)$$

Similarly, using the approximation $R_{sho} \approx R_{sh}$ in Equation (11), R_s can be extracted as

$$Rs = Rso - \frac{1}{\frac{1}{Rsho} + \frac{Is}{nVT}\exp\left(\frac{Voc}{nVT}\right)} \tag{22}$$

According to Equation (11) and the above three parameters, i.e., n, I_s, and R_s, R_{sh} can be extracted as

$$Rsh = \frac{1}{\frac{1}{Rso-Rs} - \frac{Is}{nVT}\exp\left(\frac{Voc}{nVT}\right)} \tag{23}$$

Finally, by substituting Equations (20)–(23) into Equation (2) and the above four parameters, I_{ph} can be extracted as

$$Iph = Isc\left(1 + \frac{Rs}{Rsh}\right) + Is\left(\exp\left(\frac{IscRs}{nVT}\right) - 1\right) \tag{24}$$

Therefore, the five parameters of the single diode modeling for solar cells can be extracted from Equations (20)–(24) in sequence.

3. Verifications and Discussions

In this part, *I-V*, *P-V*, relative error, absolute error curves, and root-mean-square error (RMSE) are used to verify and compare the accuracy and effectiveness of the proposed parameter extraction strategy. In the verification process, the absolute error represents the absolute difference between the measured value and the real value, and the relative error is calculated by the ratio of the absolute error to the real value, and the result is expressed in the form of percentage. On the one hand, when we extract and compare parameters through a set of initial values, these characteristic curves can clearly show the experimental errors of different methods. In addition, RMSE can also evaluate the quality of all parameter extraction strategies. On the other hand, the performance of the proposed parameter extraction strategy in different cases should also be considered. These points are mainly reflected in PV technologies, irradiance, and temperature. Thus, in these cases, it is important and necessary to evaluate the fitting results of *I-V* and *P-V* curves between the experimental data and the results obtained by using the extracted parameters. Of course, the absolute error curves and RMSE are also obtained to better verify the performance of the method. The detailed verification results and discussion are as follows.

According to the set of initial values, the simulation results are shown in Figures 3 and 4. The comparison results of the parameters are shown in Table 1. First, we fix a set of initial values in Table 1 as reference (Setting). Second, after processing the reference data, we get the five key points mentioned in the second part. Finally, we use these key points to extract parameters so as to compare the proposed method with other methods in the previous literature and draw the corresponding *I-V*, *P-V* and absolute error percentage curves. We can observe from Figures 3 and 4 that only our *I-V* and *P-V* curves agree to the experimental data (scatter points), which is significantly different from other methods. In particular, the part with large gap has been enlarged in Figures 3 and 4 for better observation. All methods are simulated under 1000 W/m^2 and 25 °C, and the obtained parameter results are shown in Table 1.

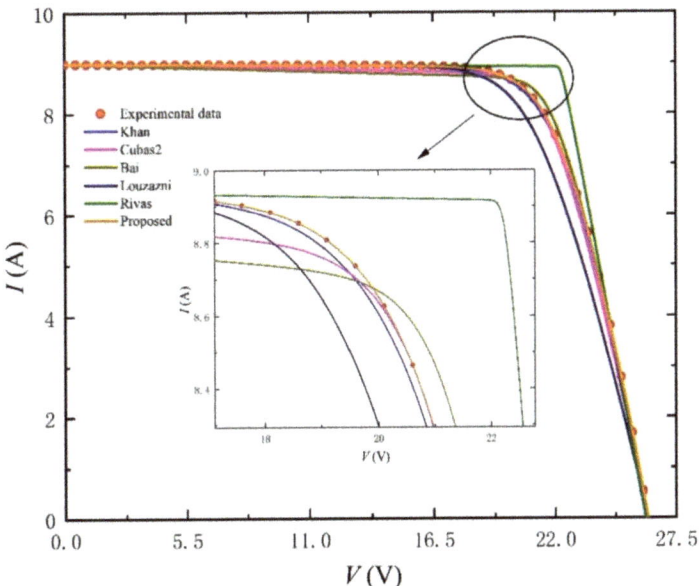

Figure 3. *I-V* curves simulated by using setting initial value and the extracted parameters listed in Table 1 at G = 1000 W/m² and T = 25 °C.

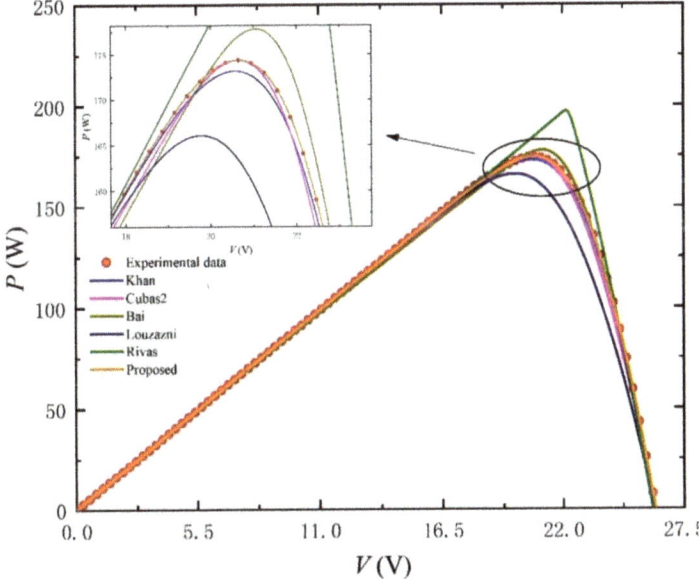

Figure 4. *P-V* curves simulated by using setting initial value and the extracted parameters listed in Table 1 at G = 1000 W/m² and T = 25 °C.

Table 1. Parameter-extraction results and RMSE for I- and P-V Curves in Figures 3 and 4 at G = 1000 W/m^2 and T = 25 °C.

Methods (Year)	n	I_s (A)	R_s (Ω)	R_{sh} (Ω)	I_{ph} (A)	RMSE
Setting	40	1.0×10^{-10}	0.3	300	9	-
Khan (2013)	42.5033	4.4532×10^{-10}	0.2937	294.1	8.9999	0.0301
Cubas2 (2014)	28.0077	2.0138×10^{-15}	0.3807	100.2	9.0251	0.1039
Bai (2014)	21.5822	4.4004×10^{-20}	0.3807	71.5	9.0385	0.1132
Louzazni (2015)	39.9748	9.9279×10^{-11}	0.4166	294.1	9.0037	0.3617
Rivas (2020)	1.7956	1.4471×10^{-242}	0.4114	294.1	9.0035	0.3625
Proposed	39.9672	9.7823×10^{-11}	0.2999	294.1	9.0001	0.0013

It is obvious that the absolute error percentage curve shown in Figure 5 and the RMSE in Table 1 clearly reflecting that the error of the parameter extraction strategy proposed in this paper is the smallest, and the performance of this method is the best. It is worth noting that in Figure 5, it can be observed that the relative error of the proposed method remains almost below 0.1% within the effective range. Compared with other methods, the difference is obvious and the error is smaller, which fully meets the accuracy requirements of the simulations. In addition, the RMSE value of the proposed method in Table 1 is reduced by at least one order of magnitude, compared with other methods, which further reflects the advantages of this method.

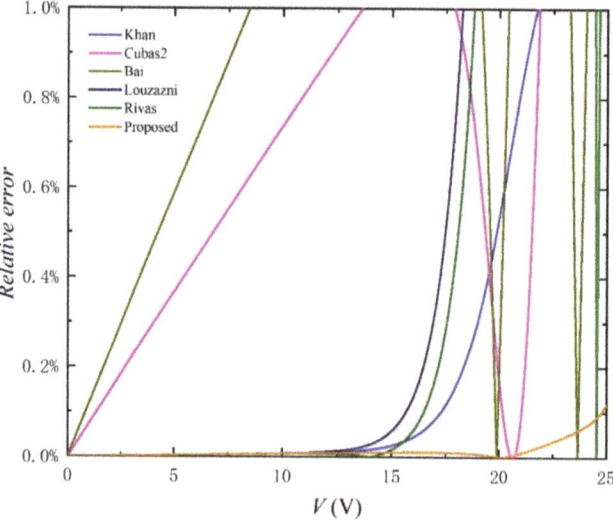

Figure 5. Relative error of I-V curves simulated by using setting initial value and the extracted parameters listed in Table 1.

In order to verify the practicability and effectiveness of this method, the parameter extraction results and calculated RMSE of three different photovoltaic modules (Mono-crystalline, Multi-crystalline, and Thin film) in the literature [33] are recorded in Table 2. Irradiance and temperature are still 1000 W/m^2 and 25 °C, respectively. The corresponding I-V and P-V curves are shown in Figures 6 and 7. We can still observe that the curve obtained by our proposed method is very consistent with the experimental data (scatter points) recorded in the literature. Figure 8 shows the absolute error curves of different photovoltaic technologies, and verifies the high precision and good practicability of our proposed method in photovoltaic technology. It should be noted that there are obvious peaks around V = V_{OC} in Figure 8. This is because in the process of curve fitting, the first approximation exists in the form of exponent. Due to the large range of abscissa and rapid change of exponent, the error will be accumulated and amplified, and there will be an obvious wave crest phenomenon. Of course, relative errors are still smaller than 0.2%.

Table 2. Parameter-extraction results and RMSE for I- and P-V Curves in Figures 6 and 7.

G (W/m^2)	T (°C)	PV Modules	n	I_s (A)	R_s (Ω)	R_{sh} (Ω)	I_{ph} (A)	RMSE
1000	25	Multi-crystalline (S75)	45.0214	9.8695×10^{-9}	0.1995	90.1	4.7103	0.0018
1000	25	Mono-crystalline (SM55)	42.3693	1.8463×10^{-9}	0.4136	140.8	3.4601	0.0017
1000	25	Thin-film (ST40)	55.8424	1.0658×10^{-6}	1.0651	232.6	2.6922	0.0021

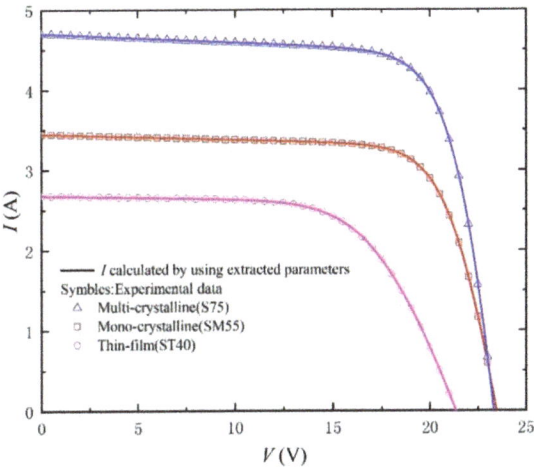

Figure 6. I-V curves measured from experimental data [33] for different PV modules and calculated by using the extracted parameters listed in Table 2.

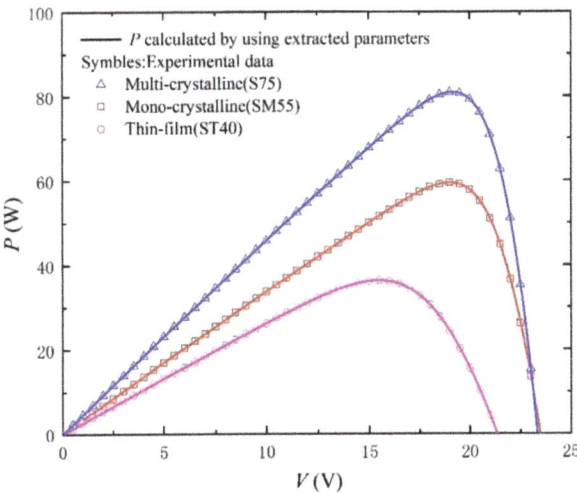

Figure 7. P-V curves measured from experimental data [33] for different PV modules and calculated by using the extracted parameters listed in Table 2.

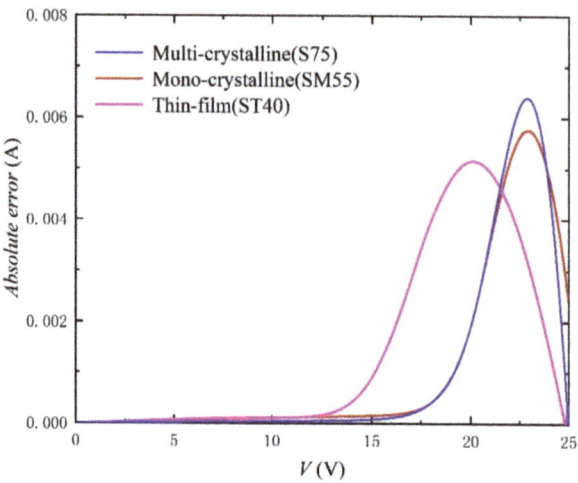

Figure 8. Absolute error curves between experimental data [33] and using the extracted parameters listed in Table 2.

Similarly, in order to better verify the method, we carried out simulation experiments under different irradiances and temperatures. The comparison objects of the experiment are from the literature [33]. The extracted parameters and the calculated RMSE are recorded in Tables 3 and 4. In particular, the values of RMSE are kept below 0.2%, which fully reflects the accuracy of the method. In addition, I-V and P-V curves with different irradiances and temperatures are shown in Figures 9–14. It can be seen that the simulation results are consistent with the experimental data, which highlight the practicability of this method under different irradiances and temperatures. Finally, the absolute error curves of irradiance and temperature are shown in Figures 11 and 14. In short, the above four experiments successfully verify the accuracy and practicability of the proposed parameter extraction strategy.

Table 3. Parameter-extraction results and RMSE for I- and P-V Curves in Figures 9 and 10.

T (°C)	G (W/m²)	n	I_s (A)	R_s (Ω)	R_{sh} (Ω)	I_{ph} (A)	RMSE
25	200	46.6459	2.8645×10^{-8}	0.2978	1111.1	1.6435	0.0002
25	400	46.6232	2.8407×10^{-8}	0.2652	555.6	3.2873	0.0003
25	600	46.6298	2.8482×10^{-8}	0.2458	370.3	4.9309	0.0006
25	800	46.7473	2.9888×10^{-8}	0.2315	285.7	6.5743	0.0013
25	1000	46.6454	2.8656×10^{-8}	0.2212	222.2	8.2182	0.0016

Table 4. Parameter-extraction results and RMSE for I- and P-V Curves in Figures 12 and 13.

G (W/m²)	T (°C)	n	I_s (A)	R_s (Ω)	R_{sh} (Ω)	I_{ph} (A)	RMSE
1000	25	46.6413	2.8607×10^{-8}	0.2212	222.2	8.2182	0.0015
1000	50	43.3215	7.3686×10^{-7}	0.2397	222.2	8.2977	0.0018
1000	75	40.4268	1.2025×10^{-5}	0.2583	222.2	8.3771	0.0017

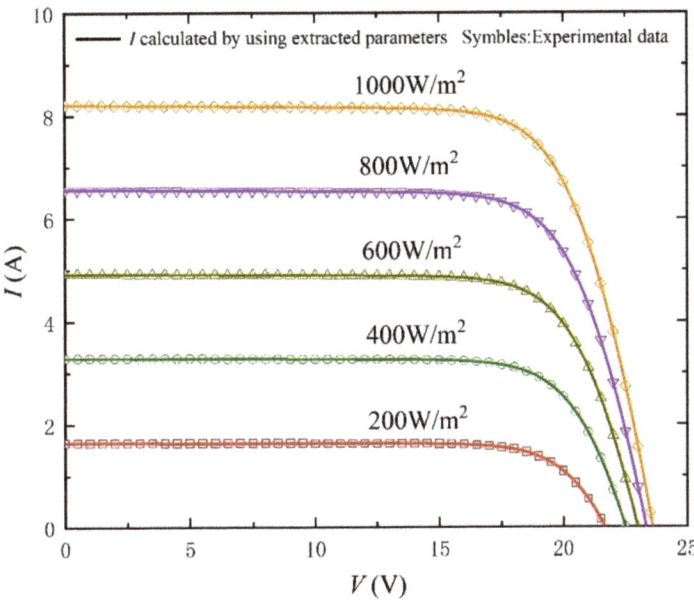

Figure 9. *I-V* curves measured from experimental data [33] at different irradiance levels and T = 25 °C, and calculated by using the extracted parameters listed in Table 3.

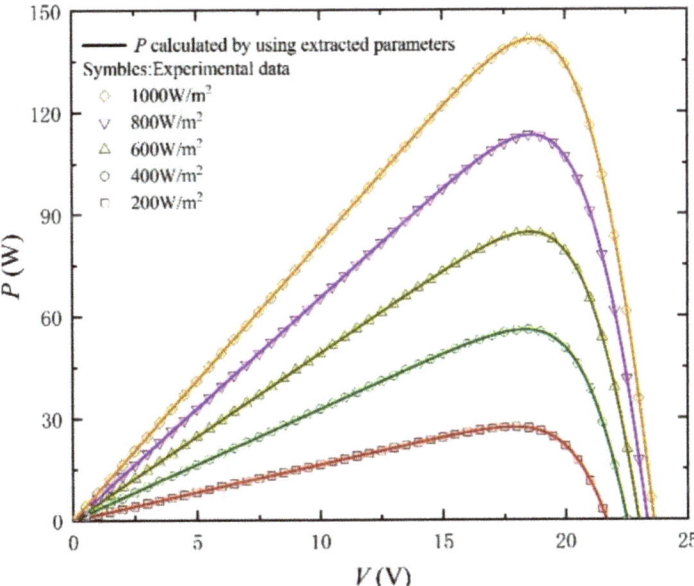

Figure 10. *P-V* curves measured from experimental data [33] at different irradiance levels and T = 25 °C, and calculated by using the extracted parameters listed in Table 3.

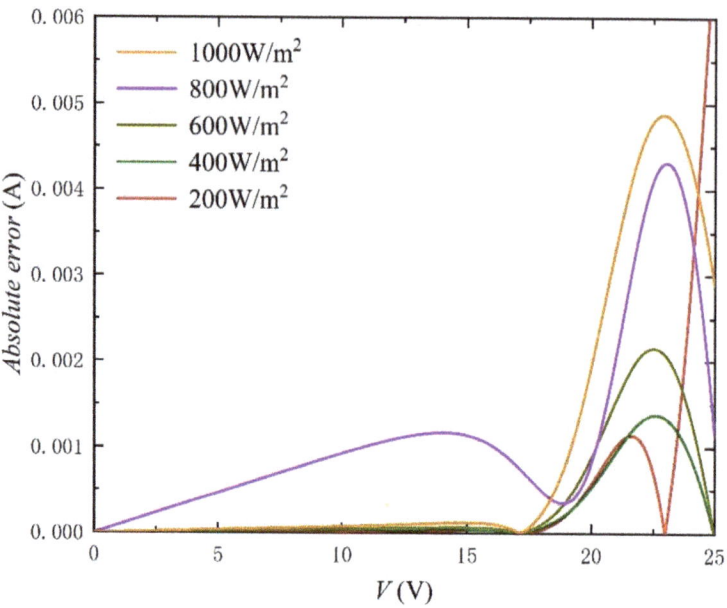

Figure 11. Absolute error curves between experimental data [33] and calculation results by using the extracted parameters listed in Table 3.

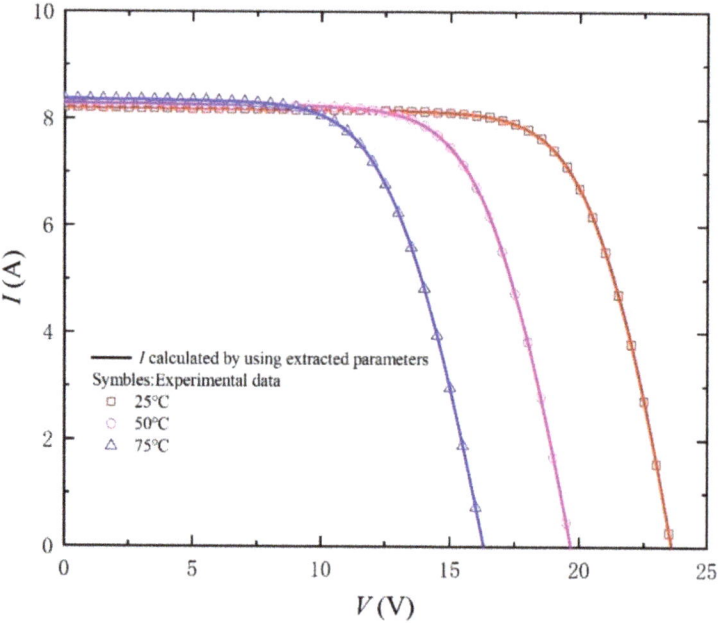

Figure 12. *I-V* curves measured from experimental data [33] at G = 1000 W/m² and different temperatures, and calculated by using the extracted parameters listed in Table 4.

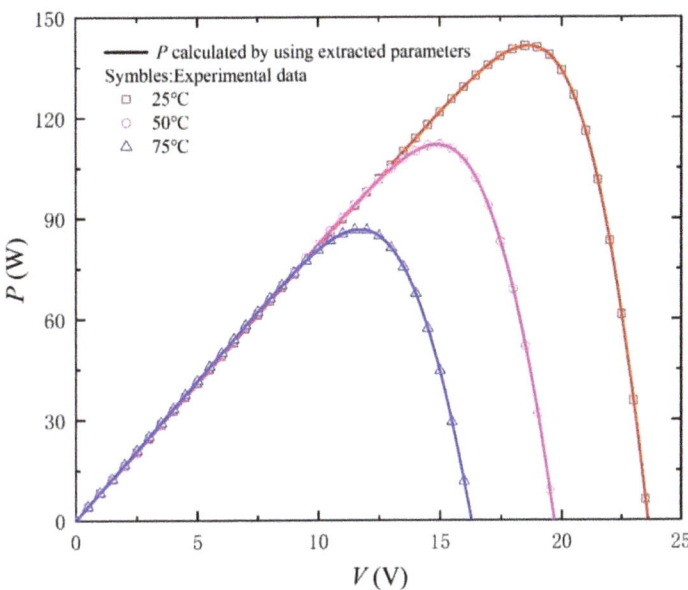

Figure 13. *P-V* curves measured from experimental data [33] at G = 1000 W/m^2 and different temperatures, and calculated by using the extracted parameters listed in Table 4.

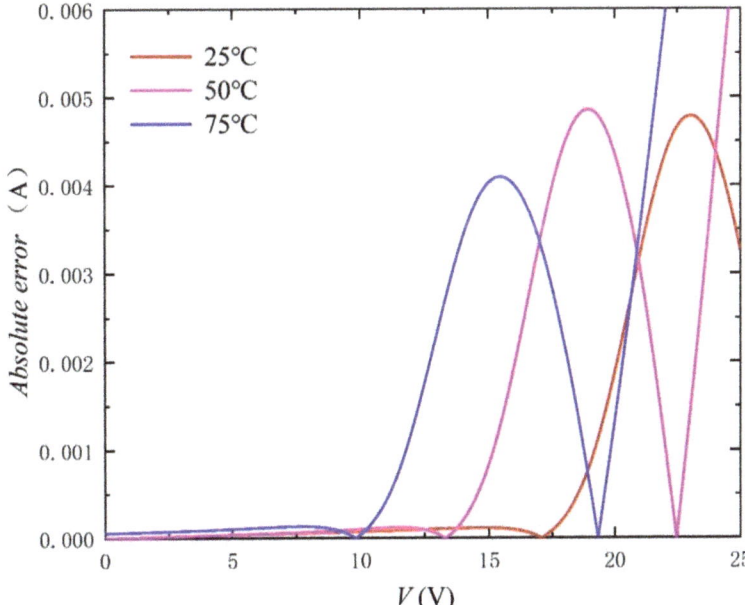

Figure 14. Absolute error curves between experimental data [33] and calculation results by using the extracted parameters listed in Table 4.

4. Conclusions

This paper presents an effective and accurate method to extract the model parameters of solar cells' single diode model. First, the equivalent circuit and I-V curve of single diode model are given to obtain the required circuit equation and key points, including open circuit voltage, short circuit current, and maximum power point. Then, we get the other two constant conditions, R_{sho} and R_{so}, according to the slope of I-V curve at open-circuit point and short-circuit point. In order to overcome the problems of accuracy and complexity, we use the measures of approximations, retain important parts to obtain the simplified five equations, and then derive the five parameter expressions in order. Second, through setting initial values, the I-V and P-V characteristic curves are simulated, and the proposed method is compared with other different methods. According to the simulation results, our proposed method has the best applicability, which not only maintains good accuracy but also simplifies the parameter extraction process. Finally, the fitting and comparison are carried out by I-V, P-V, and absolute error curves under different PV technologies, irradiances, and temperatures. The obtained curves are in good agreement with the experimental data, which also proves the practicability of this method for different preparation conditions and environmental changes. In the future, it may be helpful to prepare solar cells in the face of changeable process conditions.

Author Contributions: Conceptualization, Z.S. and F.Y.; methodology, F.Y.; software, K.F. and X.S.; validation, K.F. and Y.L., W.L. and G.H.; formal analysis, Y.L.; investigation, F.Y.; resources, F.Y.; data curation, F.Y.; writing—original draft preparation, Z.S.; writing—review and editing, F.Y.; visualization, C.X.; supervision, F.Y.; project administration, F.Y.; funding acquisition, F.Y. All authors have read and agreed to the published version of the manuscript.

Funding: This research was funded by National Natural Science Foundation of China and Fundamental Research Funds for the Central Universities, grant number 61904056 and ZQN-809.

Institutional Review Board Statement: Not applicable.

Informed Consent Statement: Informed consent was obtained from all subjects involved in the study.

Data Availability Statement: The data presented in this study are available on a reasonable request from the corresponding author.

Conflicts of Interest: The authors declare no conflict of interest.

References

1. Rivas-Vázquez, J.A.; Loera-Palomo, R.; Álvarez-Macías, C.; Rivero, M.; Sellschopp-Sánchez, S.F. Statistical Method for Single Diode Model Parameters Extraction of a Photovoltaic Module. In Proceedings of the 2020 IEEE International Autumn Meeting on Power, Electronics and Computing (ROPEC), Ixtapa, Mexico, 10–12 November 2020; pp. 1–6. [CrossRef]
2. Ali M, H.; Mojgan, H.; Saad, H.; Hussein, M.H. Solar cell parameters extraction based on single and double-diode models: A review. *Renew. Sustain. Energy Rev.* **2016**, *56*, 494–509.
3. Adelmo, O.C.; Francisco, G.S.; Juan, M.; Andrea, S.G. A review of diode and solar cell equivalent circuit model lumped parameter extraction procedures. Facta universitatis-series. *Electron. Energetics* **2014**, *27*, 57–102.
4. Gomes, M.C.R.; Vitorino, A.M.; Corrêa, R.B.M.; Wang, R.; Fernandes, A.D. Photovoltaic parameter extraction using Shuffled Complex Evolution. In Proceedings of the 2015 IEEE 13th Brazilian Power Electronics Conference and 1st Southern Power Electronics Conference (COBEP/SPEC), Fortaleza, Brazil, 29 November–2 December 2015; pp. 1–6. [CrossRef]
5. Mahmoud, A.Y.; Xiao, W.; Zeineldin, H.H. A Parameterization Approach for Enhancing PV Model Accuracy. *IRE Trans. Ind. Electron.* **2013**, *60*, 5708–5716. [CrossRef]
6. Benahmida, A.; Maouhoub, N.; Sahsah, H. An Efficient Iterative Method for Extracting Parameters of Photovoltaic Panels with Single Diode Model. In Proceedings of the 2020 5th International Conference on Renewable Energies for Developing Countries (REDEC), Marrakech, Morocco, 29–30 June 2020; pp. 1–6. [CrossRef]
7. Phang, J.C.H.; Chan, S.H.; Phillips, J.R. Accurate analytical method for the extraction of solar cell model parameters. *Electron. Lett.* **1984**, *20*, 406–408. [CrossRef]
8. Sera, D.; Teodorescu, R.; Rodriguez, P. Photovoltaic module diagnostics by series resistance monitoring and temperature and rated power estimation. In Proceedings of the 2008 34th Annual Conference of IEEE Industrial Electronics, Orlando, FL, USA, 10–13 November 2008; pp. 2195–2199. [CrossRef]

9. Saleem, H.; Karmalkar, S. An Analytical Method to Extract the Physical Parameters of a Solar Cell From Four Points on the Illuminated $J{-}V$ Curve. *IEEE Electron. Device Lett.* **2009**, *30*, 349–352. [CrossRef]
10. Saloux, E.; Teyssedou, A.; Sorin, M. Explicit model of photovoltaic panels to determine voltages and currents at the maximum power point. *Sol. Energy* **2011**, *85*, 713–722. [CrossRef]
11. Accarino, J.; Petrone, G.; Ramos-Paja, A.C.; Spagnuolo, G. Symbolic algebra for the calculation of the series and parallel resistances in PV module model. In Proceedings of the 2013 International Conference on Clean Electrical Power (ICCEP), Alghero, Italy, 11–13 June 2013; pp. 62–66. [CrossRef]
12. Khan, F.; Baek, S.H.; Park, Y.; Kim, J. Extraction of diode parameters of silicon solar cells under high illumination conditions. *Energy Convers. Manag.* **2013**, *76*, 421–429. [CrossRef]
13. Cubas, J.; Pindado, S.; Victoria, M. On the analytical approach for modeling photovoltaic systems behavior. *J. Power Sources* **2014**, *247*, 467–474. [CrossRef]
14. Cubas, J.; Pindado, S.; De Manuel, C. Explicit Expressions for Solar Panel Equivalent Circuit Parameters Based on Analytical Formulation and the Lambert W-Function. *Energies* **2014**, *7*, 4098–4115. [CrossRef]
15. Bai, J.; Sheng, L.; Hao, Y.; Zhang, Z.; Jiang, M.; Zhang, Y. Development of a new compound method to extract the five parameters of PV modules. *Energy Convers. Manag.* **2014**, *79*, 294–303. [CrossRef]
16. Aldwane, B. Modeling, simulation and parameters estimation for Photovoltaic module. In Proceedings of the 2014 First International Conference on Green Energy ICGE 2014, Sfax, Tunisia, 25–27 March 2014; pp. 101–106. [CrossRef]
17. Cannizzaro, S.; Di Piazza, C.M.; Luna, M.; Vitale, G. PVID: An interactive Matlab application for parameter identification of complete and simplified single-diode PV models. In Proceedings of the 2014 IEEE 15th Workshop on Control and Modeling for Power Electronics (COMPEL), Santander, Spain, 22–25 June 2014; pp. 1–7. [CrossRef]
18. Toledo, F.J.; Blanes, J.M. Geometric properties of the single-diode photovoltaic model and a new very simple method for parameters extraction. *Renew. Energy* **2014**, *72*, 125–133. [CrossRef]
19. Louazni, M.; Aroudam, E.H. An analytical mathematical modeling to extract the parameters of solar cell from implicit equation to explicit form. *Appl. Sol. Energy* **2015**, *51*, 165–171. [CrossRef]
20. Batzelis, I.E.; Papathanassiou, A.S. A Method for the Analytical Extraction of the Single-Diode PV Model Parameters. *IEEE Trans. Sustain. Energy* **2016**, *7*, 504–512. [CrossRef]
21. Hejri, M.; Mokhtari, H.; Azizian, M.R.; Söder, L. An analytical-numerical approach for parameter determination of a five-parameter single-diode model of photovoltaic cells and modules. *Int. J. Sustain. Energy* **2016**, *35*, 396–410. [CrossRef]
22. Senturk, A.; Eke, R. A new method to simulate photovoltaic performance of crystalline silicon photovoltaic modules based on datasheet values. *Renew. Energy* **2017**, *103*, 58–69. [CrossRef]
23. Batzelis, E. Non-Iterative Methods for the Extraction of the Single-Diode Model Parameters of Photovoltaic Modules: A Review and Comparative Assessment. *Energies* **2019**, *12*, 358. [CrossRef]
24. Carrero, C.; Ramirez, D.; Rodriguez, J.; Platero, C.A. Accurate and fast convergence method for parameter estimation of PV generators based on three main points of the I–V curve. *Renew. Energy* **2011**, *36*, 2972–2977. [CrossRef]
25. Chenche, P.; Mendoza, H.; Filho, B. Comparison of four methods for parameter estimation of mono-and multi-junction photovoltaic devices using experimental data. *Renew. Sustain. Energy Rev.* **2018**, *81*, 2823–2838. [CrossRef]
26. Mehta, K.H.; Warke, H.; Kukadiya, K.; Panchal, K.A. Accurate Expressions for Single-Diode-Model Solar Cell Parameterization. *IEEE J. Photovolt.* **2019**, *9*, 803–810. [CrossRef]
27. Kumar, M.; Kumar, A. Power Estimation of Photovoltaic System using 4 and 5-parameter Solar Cell Models under Real Outdoor Conditions. In Proceedings of the 2018 IEEE 7th World Conference on Photovoltaic Energy Conversion (WCPEC) (A Joint Conference of 45th IEEE PVSC, 28th PVSEC & 34th EU PVSEC), Waikoloa, HI, USA, 10–15 June 2018; pp. 0721–0726. [CrossRef]
28. Piliougine, M.; Guejia-Burbano, R.A.; Petrone, G.; Sánchez-Pacheco, F.J.; Mora-López, L.; Sidrach-de-Cardona, M. Parameters extraction of single diode model for degraded photovoltaic modules. *Renew. Energy* **2021**, *164*, 674–686. [CrossRef]
29. ALQahtani, H.A. A simplified and accurate photovoltaic module parameters extraction approach using matlab. In Proceedings of the 2012 IEEE International Symposium on Industrial Electronics, Hangzhou, China, 28–31 May 2012; pp. 1748–1753. [CrossRef]
30. Ulapane, B.N.N.; Dhanapala, H.C.; Wickramasinghe, M.S.; Abeyratne, G.S.; Rathnayake, N.; Binduhewa, J.P. Extraction of parameters for simulating photovoltaic panels. In Proceedings of the 2011 6th International Conference on Industrial and Information Systems, Kandy, Sri Lanka, 16–19 August 2011; pp. 539–544. [CrossRef]
31. Benkercha, R.; Moulahoum, S.; Colak, I.; Taghezouit, B. PV module parameters extraction with maximum power point estimation based on flower pollination algorithm. In Proceedings of the 2016 IEEE International Power Electronics and Motion Control Conference (PEMC), Varna, Bulgaria, 25–28 September 2016; pp. 442–449. [CrossRef]
32. Kumar, M.; Shiva Krishna Rao, K.V.D. Modelling and Parameter Estimation of Solar Cell using Genetic Algorithm. In Proceedings of the 2019 International Conference on Intelligent Computing and Control Systems (ICCS), Madurai, India, 15–17 May 2019; pp. 383–387. [CrossRef]
33. Chennoufi, K.; Ferfra, M. Parameters extraction of photovoltaic modules using a combined analytical-numerical method. In Proceedings of the 2020 5th International Conference on Cloud Computing and Artificial Intelligence: Technologies and Applications (CloudTech), Marrakesh, Morocco, 24–26 November 2020; pp. 1–7. [CrossRef]

Article

Flexible Layered-Graphene Charge Modulation for Highly Stable Triboelectric Nanogenerator

Mamina Sahoo [1], Sz-Nian Lai [2], Jyh-Ming Wu [2,3], Ming-Chung Wu [4] and Chao-Sung Lai [1,5,6,7,*]

1. Department of Electronic Engineering, Chang Gung University, Guishan District, Taoyuan City 33302, Taiwan; msahoo12@gmail.com
2. Department of Materials Science and Engineering, National Tsing Hua University, Hsinchu 30010, Taiwan; snlai712@gapp.nthu.edu.tw (S.-N.L.); jmwuyun@gapp.nthu.edu.tw (J.-M.W.)
3. High Entropy Materials Center, National Tsing Hua University, Hsinchu 30010, Taiwan
4. Department of Chemical and Materials Engineering, Chang Gung University, Taoyuan City 33302, Taiwan; mingchungwu@mail.cgu.edu.tw
5. Artificial Intelligence and Green Technology Research Center, Chang Gung University, Guishan District, Taoyuan City 33302, Taiwan
6. Department of Materials Engineering, Ming Chi University of Technology, Taishan District, New Taipei City 24301, Taiwan
7. Department of Nephrology, Chang Gung Memorial Hospital, Guishan District, Taoyuan City 33305, Taiwan
* Correspondence: cslai@mail.cgu.edu.tw

Abstract: The continuous quest to enhance the output performance of triboelectric nanogenerators (TENGs) based on the surface charge density of the tribolayer has motivated researchers to harvest mechanical energy efficiently. Most of the previous work focused on the enhancement of negative triboelectric charges. The enhancement of charge density over positive tribolayer has been less investigated. In this work, we developed a layer-by-layer assembled multilayer graphene-based TENG to enhance the charge density by creatively introducing a charge trapping layer (CTL) Al_2O_3 in between the positive triboelectric layer and conducting electrode to construct an attractive flexible TENG. Based on the experimental results, the optimized three layers of graphene TENG (3L-Gr-TENG) with CTL showed a 30-fold enhancement in output power compared to its counterpart, 3L-Gr-TENG without CTL. This remarkably enhanced performance can be ascribed to the synergistic effect between the optimized graphene layers with high dielectric CTL. Moreover, the device exhibited outstanding stability after continuous operation of >2000 cycles. Additionally, the device was capable of powering 20 green LEDs and sufficient to power an electronic timer with rectifying circuits. This research provides a new insight to improve the charge density of Gr-TENGs as energy harvesters for next-generation flexible electronics.

Keywords: graphene; triboelectric nanogenerator; charge trapping layer; flexible; stability; energy harvesting

1. Introduction

With the advent of the fourth industrial revolution, the demand for flexible, portable and wearable electronic devices has increased dramatically. However, powering them in a stable manner remains a challenge due to the ongoing energy crisis worldwide [1]. In this aspect, the triboelectric nanogenerator (TENG) is a promising energy harvesting technology owing to its special ability of converting the low-frequency mechanical energy to electrical energy [2]. Basically, the working principle of TENGs depends on the coupling effect of sequential triboelectrification and electrostatic induction, and the fundamental theory lies in Maxwell's displacement current and change in surface polarization [3,4]. Based on this principle TENGs are able to harvest energy from green and renewable sources such as body motions, ocean weaves and wind flows [5–9]. However, despite the rapid advancement in output performance, the triboelectric surface charge decay and poor stability of tribolayer are some of the critical issues of TENGs and could limit certain practical applications. In these aspects, significant research effort has been devoted to enhancing the surface charge density of tribomaterials such as plasma treatment, surface functionalization of triboelectric materials using corona discharge and micro/nanopatterning of tribosurface area [10–12]. Moreover, some researchers are adapting nanostructured surface modification methods, such as templating, appending, etching and crumpling, to achieve a high output performance of TENGs [13,14]. Although the above approaches can be used to fabricate the high-performance TENGs, the complicated fabrication processes and high cost of device design may limit the wide range of practical applications. Hence, there is a pressing need to fill the gap towards potential applications and integrate robust technologies to enhance the output performance of TENGs for future flexible electronic applications.

The triboelectrification process involves charge generation, charge storing and charge decay of triboelectric materials [15]. Therefore, selecting the proper triboelectric materials is the key to improve the output performance of flexible TENGs. In this regard, two-dimensional graphene, a monolayer honeycomb lattice structure of sp^2-bonded carbon atoms, can be a promising material for TENGs owing to its unique properties such as high electrical conductivity, excellent mechanical flexibility, optical transparency and environmental stability [16,17]. The unavoidable wrinkles and ripples of CVD-grown graphene make it more suitable for high output performance due to the enhancement in surface charge during the triboelectrification process. Many studies have reported high output performance of TENG by utilizing graphene as conducting electrode [18,19]. Kim et al. and Liu et al. have demonstrated a flexible TENG, using graphene as a triboelectric material, but unfortunately, the TENG exhibited a low output performance [20,21]. Furthermore, modifying the surface of graphene by plasma treatment, surface modification, micro/nanopatterning is a viable way to enhance the output performance of Gr-TENGs [10–12,22]. However, these methods create defects in the graphene surface, which may deteriorate the graphene and affect device performance. To solve the above problem, many researchers have adopted the interfacial modification of TENG for enhanced output performance [23,24]. Extensive research has been conducted to study the effect of the charge trapping layer on TENG performance. Wu et al. demonstrated a significant enhancement in the surface charge density of TENGs by using reduced graphene oxide (rGO) as a charge trapping layer (CTL) under the friction layer [25]. Furthermore, Cui et al. have shown the improvement in the output performance of TENGs by extending charge decay time and enhancing induced charges with the addition of dielectric layer and charge transport layer in between the triboelectric material and contact electrode [26]. However, most of the previous work has focused on negative triboelectric charge enhancement. Recently, another group has reported high output performance by enhancing the positive charge trap [27]. Based on these previous studies, it has been proved that the enhancement of output performance of TENG presents a positive correlation with the increase in surface charge density and charge trapping sites without degrading the properties of triboelectric material. However, the role played by the charge trapping with multilayer structure has not been investigated. Thus, it is necessary to analyze the multilayer structure with CTL, which influences the output performance of a TENG.

In this work, we report a new approach to achieve high output performance of layer-by-layer assembled multilayer graphene-based TENG by introducing Al_2O_3 CTL in between the positive triboelectric layer and the bottom conducting electrode. The Gr-TENG with Al_2O_3 introduces a mechanism of surface charge enhancement in conduction domains. Relying on the synergistic effect of optimized graphene layers and high dielectric Al_2O_3 CTL, there is a large triboelectric charge yield. The optimized flexible 3L-Gr-TENG with Al_2O_3 exhibits an enhanced output voltage and current of ~55 V and 0.78 µA, respectively. These values were nearly 5-fold higher than those of the counterpart pristine 3L-Gr-TENG (without Al_2O_3 CTL). The output power of the 3L-Gr-TENG is increased from 0.77 to 25 µW (~30 times higher) with the Al_2O_3 CTL under ambient conditions. Importantly, by taking advantage of multilayer graphene as a positive triboelectric layer, the bottom graphene layers can act as a charge transport bridge between Al_2O_3 and the top graphene layer due to its high electrical conductivity, thus accumulating more positive charge on the graphene surface and facilitating the electron flow from graphene to the opposite triboelectric layer. Moreover, the device shows high stability and durability after continuous operation of >2000 cycles. Furthermore, the generated power can light up more than 20 commercial green LEDs and charge various capacitors to power an electronic timer through the rectifying circuits. Most importantly, this work demonstrates a novel and cost-effective method to improve the performance of flexible Gr-TENGs, which can be a power candidate for next-generation flexible energy harvesting systems.

2. Experimental Section
Fabrication of a Gr-TENG with Al_2O_3 as the CTL

Figure 1 shows the schematic diagram for the fabrication process of a flexible Gr-TENG with Al_2O_3 as the CTL over a PET substrate. In this study, the conductor-to-dielectric contact mode was used. To design the vertical contact–separation mode TENG device, the positive friction layer of the TENG was fabricated as follows: (i) First, a 3×3 cm^2, 80 μm thick commercial aluminum (Al) foil was taken as the conducting electrode, which was attached over the polyethylene terephthalate (PET) substrate (~188 μm) with double-sided tape. Then, the Al-foil/PET substrate was cleaned by ethanol and dried in a stream of N_2 gas. (ii) Thereafter, a thin layer of Al_2O_3 (~10 nm), as the CTL, was thermally deposited over the Al-foil/PET substrate by the thermal evaporation method [28]. (iii) Finally, the CVD-grown graphene [29], as a positive friction layer, was transferred over the Al_2O_3/Al-foil/PET substrate using the PMMA transfer method, as shown in Figure S1. In this work, we used monolayer (1L) and multilayer (3L and 5L) graphene as the positive friction layer. Multilayer graphene (3L and 5L) was fabricated by transferring monolayer graphene layer by layer over the Al_2O_3/Al-foil/PET substrate. The positive charge trapping nature and higher relative permittivity of Al_2O_3 helped to enhance the surface charge over graphene for high output performance [30]. In addition, the improvement in the surface roughness of the graphene layer due to Al_2O_3 provided extra support for the enhancement in the electrical output of the Gr-TENG. Subsequently, a commercially available polytetrafluoroethylene (PTFE) film (3×3 cm^2) served as the negative friction layer due to its high electronegativity, which could accept more electrons when rubbed against a positive friction layer [31]. A conducting copper electrode was deposited on the back side of the PTFE film using thermal evaporation. It is noted that the PTFE and graphene surface were placed face-to-face, leaving a small gap between the two contact surfaces. Two thin copper wires were connected to the conducting electrodes (copper and Al foil) to form a complete TENG device. The flexibility of graphene and PTFE is clearly shown in the Figure 1. Such fabrication steps clearly demonstrate an easy and cost-effective fabrication process of the proposed Gr-TENG, which can be suitable for wearable electronic devices.

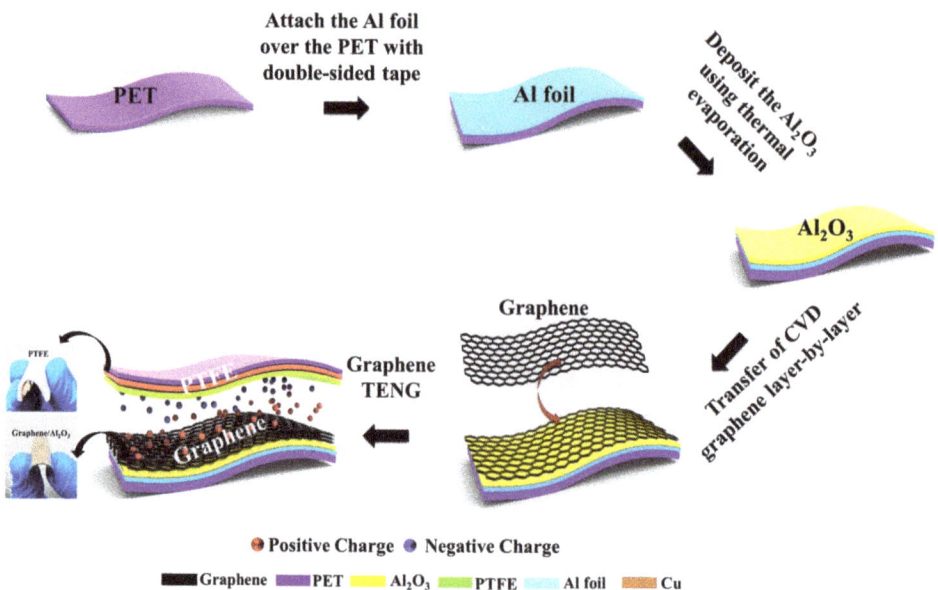

Figure 1. Schematic illustration showing the fabrication process of a flexible Gr-TENG with Al_2O_3 as the CTL.

3. Results and Discussion

3.1. Material Characterization of Graphene Layers/Al_2O_3

Raman spectroscopy is a powerful and nondestructive tool to analyze the quality of graphene. Figure 2a,b shows the Raman spectra of 1L, 3L and 5L graphene before and after their transfer over the Al_2O_3/Al-foil/PET substrate. Figure 2a shows the graphene layers (1L, 3L and 5L) over the Al-foil/PET substrate, where the G peak (at ~1582 cm^{-1}) and the 2D peak (at ~2700 cm^{-1}) are the characteristics of the sp^2-hybridized C–C bonds in graphene. Basically, these two bands are used to determine the number of layers in graphene. In addition, a negligible D peak (at ~1359 cm^{-1}) corresponds to atomic-scale defects or lattice disorder in graphene [32]. Here, the negligible intensity of the D peak indicates the low density of defects and a highly crystalline phase of graphene. However, with an increase in the number of graphene layers (3L and 5L), the G band becomes more intense, and the 2D peaks become broader and slightly upshifted with respect to monolayer graphene [33], as shown in Figure 2a. The slight upshift may be due to the unintentional strain originating from the growth of the copper substrate and the unavoidable formation of wrinkles during the graphene transfer process [34]. Moreover, the I_{2D}/I_G of ~2.18 of CVD-grown graphene indicates high-quality monolayer graphene and gradually decreases with an increase in the number of layers, as shown in Figure S2a. The intensity of the D peak slightly increases after transferring the graphene layers (1L, 3L and 5L) over the Al_2O_3/Al-foil/PET substrate, as shown in Figure 2b and Figure S2b. This result indicates an increase in substrate roughness due to the deposition of thin-layer Al_2O_3 over the Al-foil/PET substrate, which is beneficial for enhancing the output performance because the effective contact increases during the triboelectrification process.

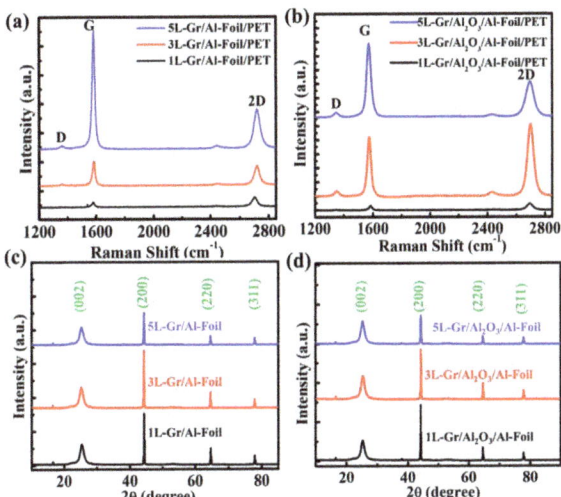

Figure 2. The Raman spectra of (**a**) graphene/Al-foil/PET and (**b**) graphene/Al$_2$O$_3$/Al-foil/PET. XRD patterns of (**c**) graphene/Al-foil/PET and (**d**) graphene/Al$_2$O$_3$/Al-foil/PET.

To further confirm the quality of graphene (1L, 3L and 5L) on the Al-foil/PET and Al$_2$O$_3$/Al-foil/PET substrates, we performed XRD, which is shown in Figure 2c,d. Both figures show that the diffraction peak at a 2θ of 26.4° corresponds to the (002) lattice orientation of hexagonal graphitic carbon, which indicates the successful fabrication of high-quality graphene [35]. Additionally, the other peaks at 44.7, 65.1 and 78.3° correspond to the (200), (220) and (311) lattice orientations, respectively, which are attributed to Al (JCPDS card No. 04-0787) as the graphene is transferred over the Al-foil substrate. It is clearly visible that the positions of different graphene layers over the Al-foil/PET and Al$_2$O$_3$/Al-foil/PET substrates retain the same peak position, indicating that the crystalline structure of graphene is restored after transfer over the Al$_2$O$_3$/Al-foil/PET substrates, which is helpful for the enhancement of the electrical output of the TENG.

Further, to investigate the surface morphology of graphene after the insertions of Al$_2$O$_3$ CTL, FESEM analysis was conducted, as illustrated in Figure 3a,b. Regarding the FESEM analysis, a sample size of 1 cm^2 was used. As clearly seen from Figure 3b, the graphene over Al$_2$O$_3$/Al-foil/PET substrate exhibits a rougher wrinkled surface morphology in comparison with the graphene over Al-foil/PET substrate (Figure 3a) due to the underneath nanostructure roughness of Al$_2$O$_3$, which supports the output enhancement of TENGs due to the enhancement of effective contact area during triboelectrification process. Additionally, we performed energy-dispersive X-ray spectroscopy (EDS) to quantify the atomic composition of the graphene sample. The graphene samples contain carbon (C), oxygen (O) and aluminum (Al), as depicted in Figure S3. Furthermore, the EDS elemental mapping (Figure 3c and Table 1) confirms the uniform distribution of C, O and Al. According to the elemental analysis, the occurrence of aluminum (44.51%) and the low atomic percentage of oxygen (1.10%) are due to the transfer of graphene over conducting electrode (Al foil) and the unavoidable oxidation of graphene. However, the atomic percentage of oxygen gradually increases to 6.08 % (Table 1) when the graphene is transferred on the Al$_2$O$_3$/Al-foil/PET substrate. This increase in oxygen will help to enhance the output performance because oxygen has excellent electron-donating ability due to its high Lewis basicity, which makes the graphene layer more tribopositive [36].

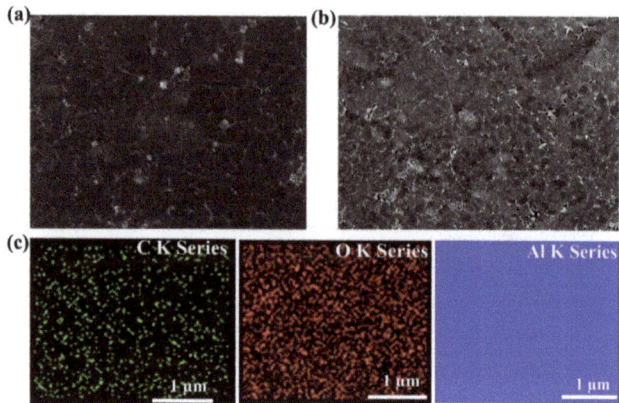

Figure 3. FESEM image of the graphene surface on (**a**) Al-foil/PET and (**b**) Al$_2$O$_3$/Al-foil/PET. (**c**) EDS elemental mapping of the graphene/Al$_2$O$_3$/Al-foil/PET presenting C K series, O K series and Al K series.

Table 1. EDS elemental analysis of graphene over Al-foil/PET and Al$_2$O$_3$/Al-foil/PET.

Elements		Samples	
		Graphene/Al-foil	Graphene/Al$_2$O$_3$/Al-foil
C K	Weight %	34.90	39.39
	Atomic %	54.39	57.88
O K	Weight %	0.94	5.51
	Atomic %	1.10	6.08
Al K	Weight %	64.16	55.10
	Atomic %	44.51	36.04

Typically, the surface roughness of triboelectric materials plays a crucial role in the enhancement of TENG output performance. Thus, to examine the surface roughness of the triboelectric layer, we performed AFM analysis. Figure 4a,b shows the 3D AFM image of the three-layer graphene on the Al-foil/PET and Al$_2$O$_3$/Al-foil/PET substrates. The surface roughness values of graphene (1L, 3L and 5L) on the Al-foil/PET substrate were 7.78, 9.57 and 11.2 nm, respectively (Figure S4a,b). However, the surface roughness of graphene (1L, 3L and 5L) was further enhanced to 11.8, 16.0 and 14.2 nm after the fusion of Al$_2$O$_3$ CTL underneath the graphene layers (Figure S4c,d). This result indicates that the random layer-by-layer transfer of graphene and the nanostructure surface roughness of Al$_2$O$_3$ enhance the surface roughness of graphene layers [37]. However, after three layers of graphene transfer, the subsequent layers are not much more affected by the surface roughness of Al$_2$O$_3$ due to the increase in thickness of the multiple layers of graphene. Therefore, the surface roughness of 5L-Gr is less than that of 3L-Gr over the Al$_2$O$_3$/Al-foil/PET substrate. Regarding TENGs, the increase in the surface roughness of graphene is beneficial for enhancing the electrical output because the surface roughness increases the effective contact area of the graphene friction layer. Regarding TENGs, the increase in the surface roughness of graphene is beneficial for enhancing the electrical output because the surface roughness increases the effective contact area of the graphene friction layer.

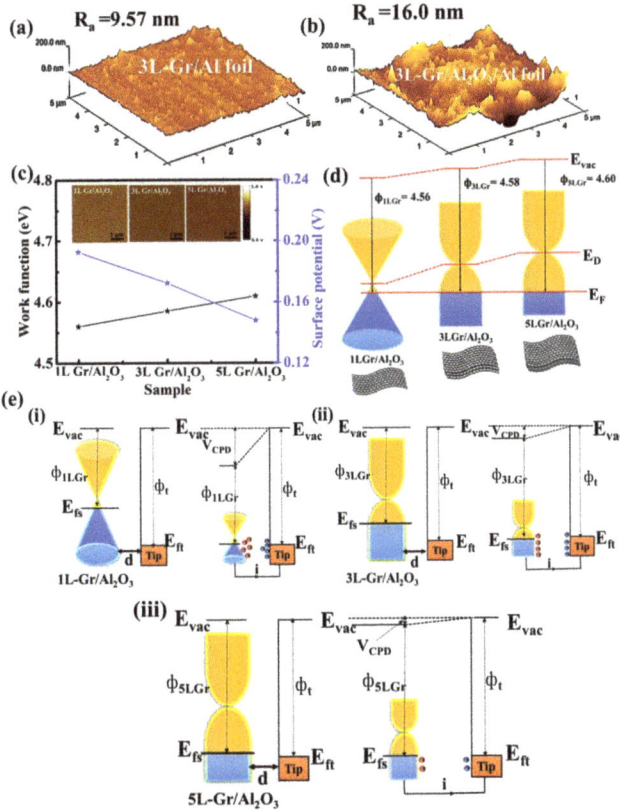

Figure 4. 3D AFM images of (**a**) 3L-Gr/Al-foil/PET substrate and (**b**) 3L-Gr/Al$_2$O$_3$/Al-foil/PET substrate. (**c**) Work function measurements of 1L-, 3L- and 5L-Gr on the Al$_2$O$_3$/Al-foil substrate by KPFM. Inset showing the surface potential of graphene layers (1L, 3L and 5L) over Al$_2$O$_3$. Schematic illustration of (**d**) energy band diagrams for 1L-Gr, 3L-Gr and 5L-Gr over Al$_2$O$_3$. (**e**) Electronic energy levels of graphene samples and AFM tip without and with electrical contact for three cases: (**i**) tip and the 1L-Gr over Al$_2$O$_3$/Al-foil/PET, (**ii**) tip and the 3L-Gr over Al$_2$O$_3$/Al-foil/PET and (**iii**) tip and the 5L-Gr over Al$_2$O$_3$/Al-foil/PET. E$_{vac}$ is the vacuum energy level. E$_{fs}$ and E$_{ft}$ are Fermi energy levels of the graphene sample and tip, respectively. V$_{CPD}$ is the contact potential difference, and Φ$_{nGr}$ and φ$_t$ are work functions of the graphene layer (n = 1, 3 and 5) and tip, respectively.

In addition, we have investigated the work function of graphene layers using KPFM to comprehensively understand the physical mechanism of triboelectric graphene layers on the output performance of TENGs. Figure 4c shows the work function (WF) and surface potential of the graphene layer (1L, 3L and 5L) on the Al$_2$O$_3$/Al-foil/PET substrate. The corresponding work functions were measured to be Φ$_{1LGr}$ = 4.56 eV, Φ$_{3LGr}$ = 4.58 eV and Φ$_{5LG}$ = 4.60 eV, respectively. The measured WF of graphene follows an increasing trend with an increase in the number of layers, which agrees well with the previously reported literature [20,38]. A similar trend occurs when graphene is transferred on the Al-foil/PET substrate as shown in Figure S5. However, the surface potential of graphene gradually decreases with an increase in the number of graphene layers (see the detail in Figure S6). The key factors for the gradual increase in the WF of graphene can be explained as follows: (i) the shift in the Fermi level of graphene with respect to the number of layers (Figure 4d) affects the charge transfer at the substrate interface, and the charge distribution within the graphene layers affects the WF of graphene; (ii) the increase in graphene thickness with respect to the number of layers may affect the WF of graphene; and (iii) the underlying substrate is also an important factor that affects the WF of graphene [39]. Moreover, the change in WF affects the surface potential of graphene layers [40], as illustrated in Figure 4e. When the 1L-Gr/Al$_2$O$_3$ sample and tip are brought close to each other for electron tunneling, the equilibrium forces align the fermi level of the 1L-Gr sample with the tip. Upon electrical contact, the 1L-Gr/Al$_2$O$_3$ sample and tip are charged, and V$_{CPD}$ is formed between the tip and 1L-Gr/Al$_2$O$_3$ sample (Figure 4e(i)). With the increase in the number of layers (3L and 5L), the characteristic energy level of graphene surface further drops, resulting in a smaller contact potential difference, as shown in Figure 4e(ii–iii). Therefore, the proper optimization of the graphene layer is very important for enhancing the output performance of TENGs because the larger potential difference between the triboelectric layer leads to a better output performance.

3.2. Working Mechanism of Gr-TENG with Al$_2$O$_3$ as CTL

Prior to performing the electrical measurement and examining the relationship between the output performance of the flexible pristine Gr-TENG and Gr-TENG with Al$_2$O$_3$, the basic operating mechanism of the flexible Gr-TENG with Al$_2$O$_3$ is clearly elaborated and illustrated in Figure 5. In general, the working mechanism of a TENG is based on the coupling effect, triboelectrification and electrostatic induction effect between triboelectric layers. In the initial position when the triboelectric graphene layers (as the tribopositive bottom layer) and PTFE (as the tribonegative top layer) are separated by a certain distance, there is no charge generation on the surface of graphene and PTFE. Therefore, no electric potential between the electrodes is observed, and no signal is observed (Figure 5a). Once the external impact is applied to the electrode, both triboelectric layers (graphene and PTFE) are brought into contact with each other. According to the triboelectric series, PTFE is much more triboelectrically negative than the graphene layer. Hence, due to the triboelectrification phenomenon, electrons (negative charges) are injected into the PTFE film from the graphene layer while leaving positive charges on its surface, as shown in Figure 5b. Since the high-k dielectric and positive charge trapping nature of Al$_2$O$_3$ [28] as the CTL exists under the graphene layer, the positive charge density over the graphene surface will be enhanced. This increases the flow of free electrons from the graphene surface to the PTFE film, causing the accumulation of more positive and negative surface charges on the graphene and PTFE surfaces, respectively. When the external force is withdrawn, both triboelectric layers separate, resulting in a potential difference between the electrodes. The difference in electric potential leads to current flow from the positively charged graphene to negatively charged PTFE via an external load (Figure 5c). This current flow is due to the electrostatic induction effect. Afterward, when both triboelectric layers (graphene and PTFE) are completely separated, a new equilibrium state occurs, in which no current conduction takes place, as shown in Figure 5d. After complete separation, if the device is pressed again, a reverse current will flow back through the external load, as depicted in Figure 5e. The repetition of this working mechanism leads to the generation of a periodic alternating current (AC). One typical signal of Gr-TENG with Al$_2$O$_3$ upon pressing and releasing is shown in Figure 5f. It is important to note that the charge generation and accumulation over the triboelectric layer are strongly related to the optimized graphene layers with the Al$_2$O$_3$ CTL. Moreover, the tribopositive charge is not only generated on the top surface of multilayer graphene but also generated on the surface of Al$_2$O$_3$, which continuously supplies the positive charge to the top graphene layers, which will increase the potential difference when the device is released. The high dielectric constant of Al$_2$O$_3$ CTL enhanced the total capacitance of Gr-TENG, which will support the output enhancement. By considering the contact–separation mode, when the two triboelectric layers contact each other, the open-circuit voltage (V$_{OC}$) at zero transferred charge can be expressed according to the following equation [41]:

$$V_{OC} = \frac{\sigma_0 \times X(t)}{\varepsilon_0} \quad (1)$$

where ε_0, σ_0 and $x(t)$ represent the vacuum permittivity, surface charge density and distance between graphene and PTFE triboelectric layers, respectively. According to Equation (1), the V_{OC} is strongly related to the σ_0. However, the surface σ_0 depends on the device capacitance because the TENG acts as both energy storage and energy generation device. Therefore, the device capacitance can be calculated as follows [42]:

$$C_{max} = \frac{\varepsilon_0 \varepsilon_r A}{d} \qquad (2)$$

where ε_r is defined as relative permittivity, A is the surface area and d is the thickness of the triboelectric layer. According to Equation (2), a thin film with high relative permittivity can exhibit high output performance. Therefore, for the enhanced output performance, the optimized number of graphene layers needs to be investigated.

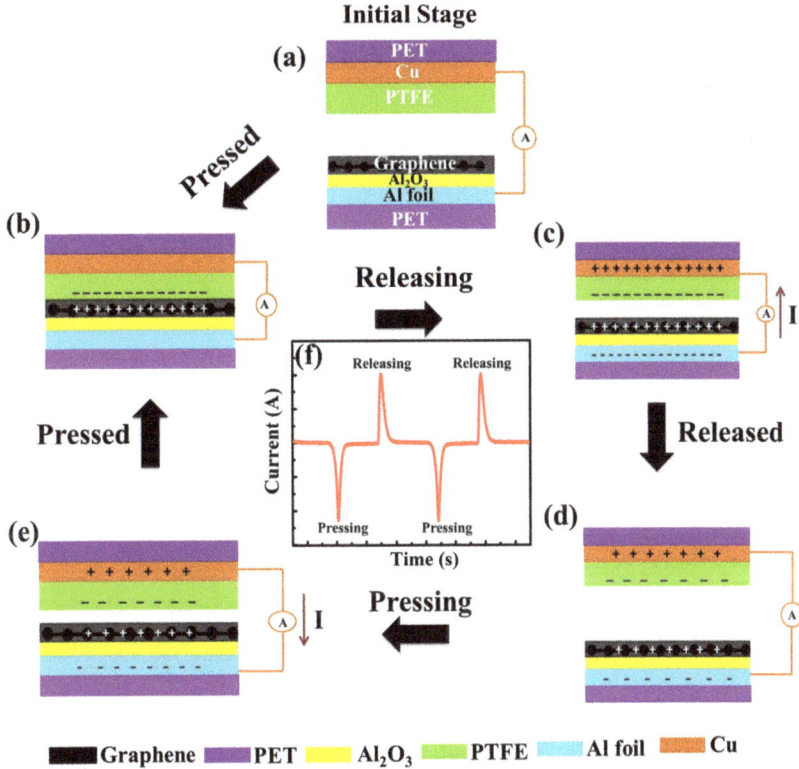

Figure 5. Schematic diagram showing the working mechanism of Gr-TENG with Al_2O_3. (**a**) Initial stage of two different triboelectric layers when no external impact is applied. (**b**) Triboelectric charges are generated on the surfaces of graphene and PTFE when they are in contact with each other. (**c**) Separation of graphene and the PTFE triboelectric layer begins. The current flows from the bottom electrode (graphene) to the top electrode (PTFE) to maintain electrical equilibrium. (**d**) Complete separation of the graphene and PTFE triboelectric layer and reaching electrical equilibrium causes no electron flow. (**e**) Pressing the graphene and PTFE triboelectric layer into contact again causes the current to flow from the top electrode to the bottom electrode. (**f**) One typical signal of Gr-TENG with Al_2O_3 upon pressing and releasing.

3.3. Electrical Characterization of Gr-TENG with Al_2O_3 CTL

The electrical output current and voltage of the Gr-TENG (1L, 3L, and 5L) with and without Al_2O_3 CTL were systematically investigated, as shown in Figure 6. These measurements were performed at a frequency of 1 Hz. A comparison of the electrical output of Gr-TENG (1L, 3L, and 5L) without the Al_2O_3 CTL is depicted in Figure 6a,b. Initially, the 3L-GR-TENG over the Al-foil/PET substrate exhibits the maximum I_{SC} of ~155.9 nA and V_{OC} of ~12 V compared to those of 1L-Gr-based (I_{SC} ~43.1 nA and V_{OC} ~7.5 V), 2L-Gr-based (I_{SC} ~66.1 nA and V_{OC} ~9 V (Figure S7)), 4L-Gr-based (I_{SC} ~113.9 nA and V_{OC} ~11 V (Figure S7)) and 5L-Gr-based (I_{SC} ~108 nA and V_{OC} ~9.5 V) TENGs. The experimental results indicate that with the increase in the number of graphene layers, the conductivity gradually increases due to the decrease in sheet resistance, as shown in Figure S8, which enhances the output I_{SC} and V_{OC} of the 3L-Gr-TENG. However, there is a decrease in electrical output for the 4L- and 5L-Gr-based TENGs compared with the 3L-Gr-based TENG. It should be noted that the increase in output I_{SC} and V_{OC} not only depends on the conductivity but also has a relationship with the work function of the triboelectric layer (i.e., the graphene in this work). The work function of graphene increases with an increase in the number of graphene layers, as discussed in Figure 4d. A higher work function means a large amount of energy is needed to extract electrons from the surface. Therefore, there should be a balance between the conductivity and work function to identify the number of graphene layers (here, 3L-Gr) that effectively enhances the output performance of the Gr-TENG. Although the performance of the 3L-Gr-based TENG is better than that of the 1L-, 2L-, 4L- and 5L-Gr-based TENGs, the electrical output of the Gr-TENG is not sufficient for many practical applications.

Furthermore, we anticipate that applying a CTL (Al_2O_3) under the graphene will be an effective method to further enhance the output performance of Gr-TENGs because the electrical output of TENG strongly depends on the surface charge density [42]. Therefore, following the above experimental demonstration of Gr-TENG with respect to the number of graphene layers as a proof of concept, Gr-TENG with Al_2O_3 as the CTL structure was systematically investigated. Apparently, the output I_{SC} and V_{OC} of the Gr-TENG with Al_2O_3 as the CTL are remarkably enhanced compared with those of the pristine Gr-TENG, as shown in Figure 6c,d. The 1L-, 2L-, 3L-, 4L- and 5L-Gr-TENGs with Al_2O_3 possess output I_{SC} and V_{OC} of 0.33 µA and 32 V, 0.43 µA and 37 V (Figure S9), 0.78 µA and 55 V, 0.63 µA and 45 V (Figure S9) and 0.55 µA and 40.3 V, respectively. The enhancement in the output I_{SC} and V_{OC} is mainly attributed to the following: (i) The increase in surface charge density over the graphene surface is due to the high charge storage capacity of Al_2O_3 [43]. (ii) The small intrinsic carrier density of Al_2O_3 compared with graphene is also one of the dominant factors [44]. Thus, more induced electrostatic charges resulting from triboelectrification may provide an enhanced electrical output of Gr-TENGs. (iii) The relative increase in the surface roughness of graphene due to Al_2O_3 increases the contact area between PTFE and the graphene layer during the triboelectrification process, providing additional support for the enhancement in output performance compared with the flexible Gr-TENG without CTL. Furthermore, Figure 6e,f clearly and intuitively reveals the strong dependency of output I_{SC} and V_{OC} of Gr-TENG on the Al_2O_3 CTL. Although the outputs increase with the CTL and lower work function of triboelectric material, the 3L-Gr-TENG with Al_2O_3 exhibits a maximum electrical output compared to the 1L-, 2L-, 4L- and 5L-Gr-TENGs with Al_2O_3. This result indicates that not only the work function but also the higher surface roughness of 3L-Gr-TENG with Al_2O_3 (Figure 4b) plays a crucial role in the enhancement of output performance of TENG. Moreover, the high conductivity of multilayer graphene further enhances the output performance of TENG compared to the 1L-Gr-based TENG. Therefore, there should be a balance between the work function, conductivity and surface roughness to identify the number of graphene layers that effectively enhances the output performance of the Gr-TENG with Al_2O_3 CTL. Based on the above results and discussion, the 3L-Gr-TENG with Al_2O_3 as CTL exhibits the optimal output performance; therefore, we considered this sample for subsequent experiments in this study.

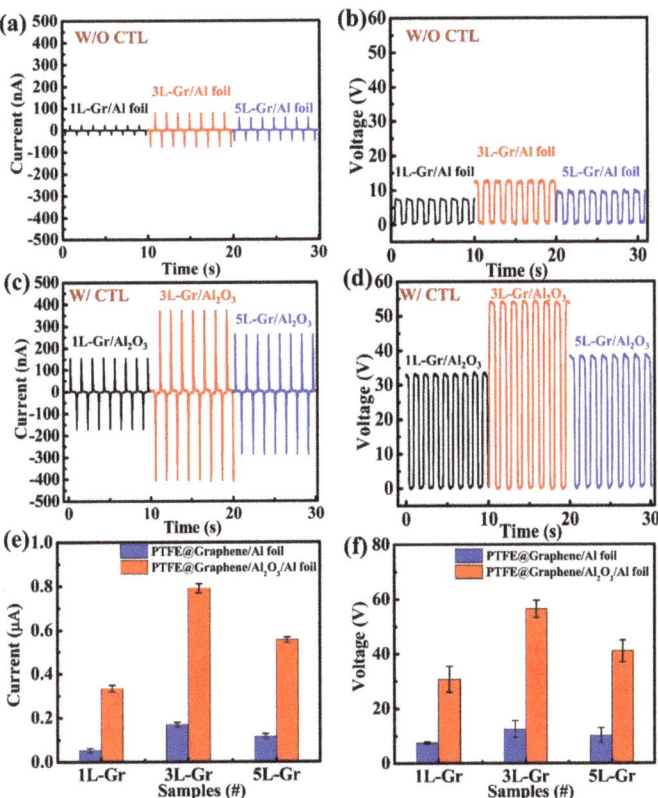

Figure 6. Electrical output of the Gr-TENG: (**a**) Short-circuit current (I_{SC}) and (**b**) open-circuit voltage (V_{OC}) of 1L-, 3L- and 5L-Gr-TENGs without Al_2O_3 CTL. (**c**) I_{SC} and (**d**) V_{OC} of 1L-, 3L- and 5L-Gr-TENGs with Al_2O_3 CTL. Average mean (**e**) current and (**f**) voltage generated by pristine Gr-TENGs (1L, 3L and 5L) and Gr-TENGs (1L, 3L and 5L) with Al_2O_3 CTL. Error bars indicate standard deviations for 4 sets of data points.

To experimentally investigate the effect of the charge trapping layer in Gr-TENGs, we fabricated a metal–insulator–metal (MIM) device structure in which 3L graphene over Al_2O_3 was sandwiched between the metal electrodes. CV analysis was performed to detect the charge trapping phenomenon. Figure 7a shows the CV curve of the Al/3L-Gr/Al_2O_3/Al device at two different frequencies. The capacitance values at 100 kHz and 1 MHz are ~1.25 nF and ~340 pF, respectively. The decrease in capacitance value with increasing frequency is due to the reduced space charge polarization effect [45]. Furthermore, the significant increase in the hysteresis window with increasing sweep voltage, as shown in Figure S10, indicates the increased trapping of charge carriers in the Al_2O_3 CTL. This charge trapping capability of graphene over Al_2O_3 promotes the triboelectric charge storage and accumulation for the enhanced TENG output performance. Moreover, the surface charge densities (σ) of Gr-TENG (1L-, 3L- and 5L-Gr) with Al_2O_3 are comparatively higher than that of Gr-TENG without CTL, as shown in Figure 7b. This indicates that surface charge density is influenced by not only the number of layers but also the existence of Al_2O_3 CTL. However, with the increase in the number of layers (from 1L-Gr to 3L-Gr) with Al_2O_3 CTL, the surface charge density increases, but the further increase in the number of layers (5L-Gr) does not contribute to the increase in surface charge density of Gr-TENG, which supports the electrical analysis as demonstrated in Figure 6e,f. For practical applications, Figure 7c shows the effect of the output voltage and current of the optimized flexible 3L-Gr-TENG with Al_2O_3 as a function of external load resistance ranging from 10 MΩ to 1 GΩ. The TENG output voltage gradually increases with increasing load resistance, while the output current value follows the opposite trend due to ohmic loss. Figure 7d shows the effective electrical output power of the graphene TENG with Al_2O_3 as charge trapping layer as a function of external load resistance. This output power was calculated by using Equation (3):

$$P = V \times I \qquad (3)$$

where V and I correspond to the output peak voltage and current value at various load resistances. The maximum value of the output power reaches 25 µW at a loading resistance of 300 MΩ, which is 30 times larger than that of the pristine 3L-Gr-TENG, as shown in Figure S11a,b. In addition, the 3L-Gr-TENG with Al_2O_3 device shows a maximum power density of 6.25 µW/cm^2 at a load resistance of ~300 MΩ, which is higher than that of a previously reported graphene-based TENG [20,43,46]. This indicates the high potential of the proposed device to support portable electronic devices.

To further explore the stability and durability of the flexible 3L-Gr-TENG with Al_2O_3 as the CTL, we continuously applied >2000 cycles (Figure 7e). Notably, there are no significant changes in the output voltage in the initial and final stages after ~2000 cycles, as shown in the Figure 7e inset, confirming the high stability and durability of the TENG device. This demonstrates the outstanding mechanical stability of graphene material the and continuous supply of positive charge from Al_2O_3 CTL to the graphene, which leads to a largely enhanced surface charge density and thus the ability to harvest mechanical energy for a long period of time. Furthermore, Figure 7f shows a schematic illustration of the charge trapping mechanism of the 3L-Gr-TENG with Al_2O_3. In the case of pristine 3L-Gr-TENG (Figure 7f(i)), under external impact, when both the triboelectric layers, PTFE (top layer) and 3L graphene (bottom layer), come into contact, triboelectric charges (positive charge over graphene and negative charge over PTFE) are generated on their surfaces according to the triboelectric series, as discussed above. The surface charges can be shifted by the electric field and combined with the induced opposite charges. This charge combination can result in a sharp deterioration in electrical output [47], as observed in Figure 6a,b. In contrast, the insertion of the Al_2O_3 CTL in between the triboelectric material (3L-Gr) and conducting electrode (Al-foil) contributes to an enhancement in the surface charge density. The remarkable key mechanism for this enhancement can be explained as follows: (i) The high positive charge storage capacity of high-k Al_2O_3 [42] underneath the graphene layers promotes charge retention at the surface, resulting in an enhancement in surface charge density that enhances the output performance of TENG. (ii) The multilayered graphene (3L-Gr) plays the dual role of the triboelectric layer and the charge transport layer between the Al_2O_3 CTL and top graphene layer due to its high electrical conductivity, which helps to improve the surface charge density (Figure 7b) since it facilitates the charge accumulation process [26]. (iii) Last but not least, the nanomorphology structure of Al_2O_3 CTL underneath the graphene plays an important role in the enhancement of surface charge density due to its enlarged effective contact area compared to the flat surface of pristine 3L-Gr-TENG, which in turn enhances the output performance of the Gr-TENG device. Consequently, all the above demonstrations indicate that the proposed optimized flexible 3L-Gr-TENG with Al_2O_3 as the CTL possesses promising practical applications for portable electronic devices.

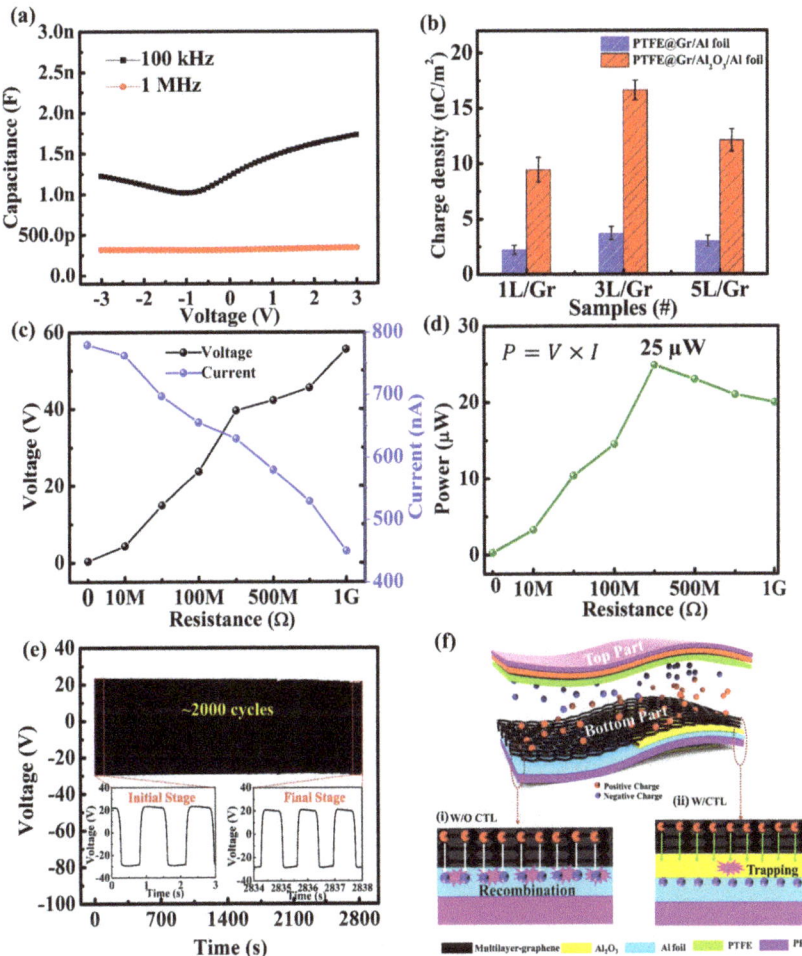

Figure 7. Performances of the 3L-Gr-TENG with the CTL: (**a**) CV characteristics of Al/Al$_2$O$_3$/3L-Gr/Al at frequencies of 100 kHz and 1 MHz. (**b**) Surface charge density of graphene (1L, 3L and 5L)-based TENG with and without Al$_2$O$_3$ as CTL. (**c**) Dependence of the output voltage and current outputs as a function of different resistors as external loads. (**d**) Relationship between electrical output power and external loading resistance. (**e**) Mechanical stability and durability test of the TENG with the continuous application of ~2000 cycles. The inset shows the output voltage at the initial stage and the final stage after ~2000 cycles. (**f**) Schematic illustrations showing the charge-trapping mechanism of 3L-Gr-TENG without and with Al$_2$O$_3$ charge trapping layer.

3.4. Applications of the Gr-TENG with Al$_2$O$_3$ as CTL

To further demonstrate the application of a Gr-TENG with Al$_2$O$_3$ as the CTL as a power source for portable electronics, we successfully lit 20 commercial green light-emitting diodes (LEDs), as shown in Figure 8a (Video S1 in the Supplementary Materials). However, the electrical output of the TENGs is an alternating current (AC) signal that is not suitable to operate portable electronic devices. Therefore, to supply a continuous current to electronic devices, we used a full-wave bridge rectifier circuit to convert the AC signal into a direct current (DC) signal, which was further utilized to charge a commercial capacitor. Finally, the stored energy in the capacitor can be used to operate the electronic device, such as a portable timer. Figure 8b shows the schematic diagram of the bridge rectifier circuit to charge the capacitor in which the capacitor is serially connected to the circuit for energy storage. Figure 8c shows the charging of 0.1, 1 and 2.2 μF capacitors. The capacitor with low capacitance, i.e., the 0.1 μF capacitor, was charged to 19 V in 60 s with the continuous contact–separation process. However, capacitors with high capacitance, namely the 1 and 2.2 μF capacitors, were charged only to 13.6 and 6.3 V in 60 s, respectively. Furthermore, an electronic timer was directly powered by the stored charge, as shown in Figure 8d (Figure S12 and Video S2 in the Supplementary Materials). Although the harvested power may seem low, charging the capacitor allows for the timer to be turned on and requires no battery. Thus, one could envision that by increasing the working area of the Gr-TENG with Al_2O_3, it can be used to drive the electronic timer for a long time.

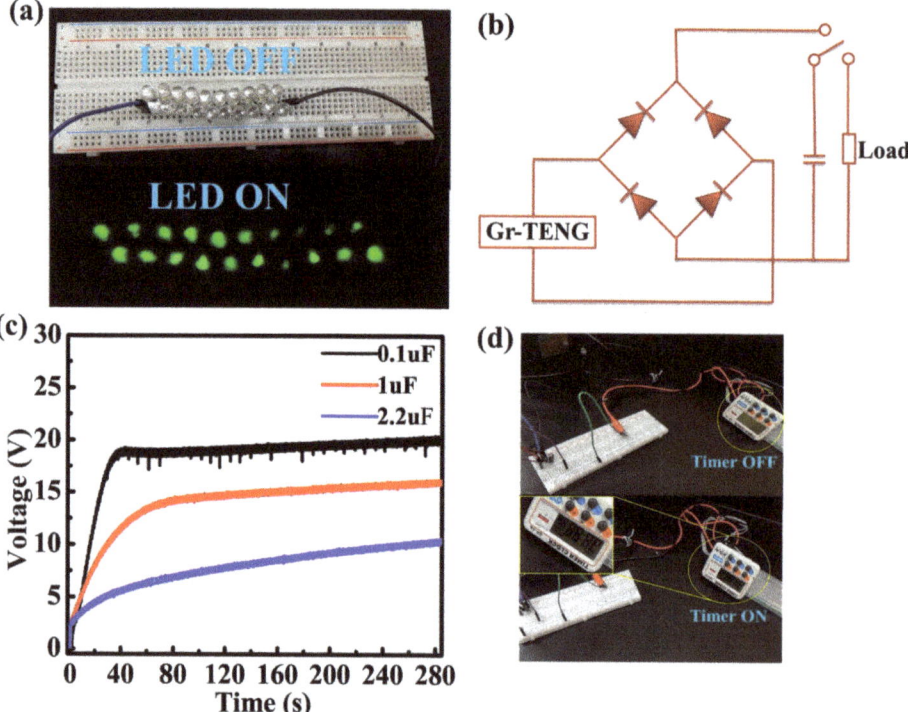

Figure 8. Applications of the Gr-TENG with Al_2O_3 as the CTL as a power supply: (**a**) Photograph showing 20 green light-emitting diodes (LEDs) being powered. (**b**) Circuit diagram of the bridge-rectifier for charging a capacitor and turning on a timer. (**c**) Charging curves of capacitors with various capacitances (0.1, 1 and 2.2 μF). (**d**) Photograph of powering a timer.

According to the aforementioned experimental results, we hereby conclude that the optimized structure, proper selection of material and surface modification of the tribolayer without degrading its inherent property such as the addition of a CTL to Gr-TENGs promotes the ability to maintain a high surface charge density, resulting in enhanced output power. A brief comparison of the electrical output performance of Gr-based TENGs with and without Al_2O_3 CTL is summarized in Table S1. It can be seen that the Gr-TENG with Al_2O_3 as CTL shows higher output performance than the Gr-TENG without CTL. These results indicate that the CTL in between the friction layer and conducting electrode is an effective path to improve the triboelectric property of Gr-TENGs. Thus, it is an effective strategy to fabricate the high-performance flexible Gr-TENG with Al_2O_3 CTL as an energy harvester.

4. Conclusions

In summary, we demonstrated a novel and simple fabrication methodology for enhanced triboelectric performance by introducing Al_2O_3 as a CTL between a positive triboelectric material (graphene) and a bottom contact electrode (Al foil). The strong tendency to repel electrons and the positive charge trapping nature of Al_2O_3 help to enhance the charge density on the graphene layer. By varying the number of graphene layers (1L, 3L, and 5L) and evaluating the electrical performance, we found the optimal layered structure (3L-Gr) of a flexible Gr-TENG. Maximum V_{OC} and I_{SC} values of ~55 V and 0.78 μA were achieved by the 3L-Gr-TENG with an Al_2O_3 CTL. Additionally, this TENG exhibited a maximum power of ~25 μW at a load resistance of ~300 MΩ, which was 30 times higher than that of the pristine 3L-Gr-TENG. Finally, the generated output power was capable of driving 20 commercially available green LEDs connected in series and able to turn on an electronic timer by using a rectifier circuit. Therefore, based on the above results, we believe that our proposed structure holds high promise for enhancing the surface charge density of Gr-TENGs by fusion of CTL, which possesses promising applications for future flexible and portable energy harvesting systems.

Supplementary Materials: The following are available online at https://www.mdpi.com/article/10.3390/nano11092276/s1, Materials and method, Figure S1: Schematic illustration of graphene transfers over Al_2O_3/Al-foil/PET substrate by PMMA technique, Figure S2: Raman analysis the I_{2D}/I_G of graphene layers (1L, 3L and 5L) over Al-foil/PET substrate and I_D/I_G of graphene layers (1L, 3L and 5L) over Al_2O_3/Al-foil/PET substrate, Figure S3: EDS spectra of graphene over Al-foil/PET and Al_2O_3/Al-foil/PET substrate, Figure S4: 3D AFM images of the transferred graphene layers (1L and 5L) on the Al-foil/PET and Al_2O_3/Al-foil/PET substrate, Figure S5: Work function measurements of graphene (1L, 3L and 5L) on the Al-foil substrate, Figure S6: KPFM surface potential measurement of graphene layers (1L, 3L and 5L) over Al-foil/PET and Al_2O_3/Al-foil/PET substrate, Figure S7. Electrical output of the Gr-TENG:(a) Short-circuit current (I_{SC}) and (b) open-circuit voltage (V_{OC}) of 2L- and 4L-Gr-TENGs without Al_2O_3 CTL, Figure S8: Sheet resistance of graphene layers (1L, 3L and 5L), Figure S9. Electrical output of the Gr-TENG:(a) Short-circuit current (I_{SC}) and (b) open-circuit voltage (V_{OC}) of 2L- and 4L-Gr-TENGs with Al_2O_3 CTL. Figure S10: CV hysteresis characteristics of 3L-Gr-TENG with Al_2O_3 as CTL with different sweeping voltages, Figure S11: Electrical performance of flexible 3L-Gr-TENG without Al_2O_3 CTL at various external load resistances, Figure S12: Electrical capacitive load characteristics of Gr-TENG with Al_2O_3 as CTL showing the charging and discharging curve for 1 μF capacitor with respect to time; inset shows the powering of a portable electronic timer by the charged capacitor, Table S1. Comparison of electrical output performance of Gr-TENGs with and without Al_2O_3 CTL samples used in this study, Video S1: Green LEDs were directly lit up by the Gr-TENG with Al_2O_3 CTL, Video S2: An electronic timer was powered using a capacitor charged by the Gr-TENG with CTL.

Author Contributions: Conceptualization, M.S., and C.-S.L.; methodology, M.S. and S.-N.L.; validation, M.S. and J.-M.W.; formal analysis, M.S.; investigation, M.S. and S.-N.L.; data curation, M.S., M.-C.W. and C.-S.L.; writing—original draft preparation, M.S.; writing—review and editing, M.S., J.-M.W. and C.-S.L.; visualization, M.S. and M.-C.W.; supervision, M.-C.W. and C.-S.L.; project administration, C.-S.L.; funding acquisition, C.-S.L. All authors have read and agreed to the published version of the manuscript.

Funding: This research was funded by the Ministry of Science and Technology, Taiwan, under grant MOST (109-2221-E-182-013-MY3, 110-2622-8-182-001-TS1, 110-2119-M-492-002-MBK). This work was supported by Chang Gung Memorial Hospital (CMRPD2K0171 and CORPD2J0072).

Institutional Review Board Statement: Not applicable.

Informed Consent Statement: Not applicable.

Data Availability Statement: Data are contained within the article or Supplementary Material.

Acknowledgments: The authors would like to thank Ministry of Science and Technology and Chang Gung Memorial Hospital for the financial support.

Conflicts of Interest: The authors declare no conflict of interest.

References

1. Zhu, G.; Su, Y.; Bai, P.; Chen, J.; Jing, Q.; Yang, W.; Wang, Z.L. Harvesting Water Wave Energy by Asymmetric Screening of Electrostatic Charges on a Nanostructured Hydrophobic Thin-Film Surface. *ACS Nano* **2014**, *8*, 6031–6037. [CrossRef]
2. Wu, J.M.; Chang, C.K.; Chang, Y.T. High-output current density of the triboelectric nanogenerator made from recycling rice husks. *Nano Energy* **2016**, *19*, 39–47. [CrossRef]
3. Fan, F.-R.; Tian, Z.-Q.; Wang, Z.L. Flexible triboelectric generator. *Nano Energy* **2012**, *1*, 328–334. [CrossRef]
4. Zhang, C.; Zhou, L.; Cheng, P.; Yin, X.; Liu, D.; Li, X.; Guo, H.; Wang, Z.L.; Wang, J. Surface charge density of triboelectric nanogenerators: Theoretical boundary and optimization methodology. *Appl. Mater. Today* **2019**, *18*, 100496. [CrossRef]
5. Yang, J.; Chen, J.; Yang, Y.; Zhang, H.; Yang, W.; Bai, P.; Su, Y.; Wang, Z.L. Broadband Vibrational Energy Harvesting Based on a Triboelectric Nanogenerator. *Adv. Energy Mater.* **2013**, *4*. [CrossRef]
6. Wu, C.; Kim, T.W.; Li, F.; Guo, T. Wearable Electricity Generators Fabricated Utilizing Transparent Electronic Textiles Based on Polyester/Ag Nanowires/Graphene Core–Shell Nanocomposites. *ACS Nano* **2016**, *10*, 6449–6457. [CrossRef] [PubMed]
7. Wang, S.; Wang, X.; Wang, Z.L.; Yang, Y. Efficient Scavenging of Solar and Wind Energies in a Smart City. *ACS Nano* **2016**, *10*, 5696–5700. [CrossRef] [PubMed]
8. Mule, A.R.; Dudem, B.; Yu, J.S. High-performance and cost-effective triboelectric nanogenerators by sandpaper-assisted micropatterned polytetrafluoroethylene. *Energy* **2018**, *165*, 677–684. [CrossRef]
9. Gupta, A.; Kumar, A.; Khatod, D.K. Optimized scheduling of hydropower with increase in solar and wind installations. *Energy* **2019**, *183*, 716–732. [CrossRef]
10. Shao, J.; Tang, W.; Jiang, T.; Chen, X.; Xu, L.; Chen, B.; Zhou, T.; Deng, C.R.; Wang, Z.L. A multi-dielectric-layered triboelectric nanogenerator as energized by corona discharge. *Nanoscale* **2017**, *9*, 9668–9675. [CrossRef]
11. Juárez-Moreno, J.; Ávila-Ortega, A.; Oliva, A.; Avilés, F.; Cauich-Rodríguez, J. Effect of wettability and surface roughness on the adhesion properties of collagen on PDMS films treated by capacitively coupled oxygen plasma. *Appl. Surf. Sci.* **2015**, *349*, 763–773. [CrossRef]
12. Fan, F.-R.; Lin, L.; Zhu, G.; Wu, W.; Zhang, R.; Wang, Z.L. Transparent Triboelectric Nanogenerators and Self-Powered Pressure Sensors Based on Micropatterned Plastic Films. *Nano Lett.* **2012**, *12*, 3109–3114. [CrossRef]
13. Zou, Y.; Xu, J.; Chen, K.; Chen, J. Advances in Nanostructures for High-Performance Triboelectric Nanogenerators. *Adv. Mater. Technol.* **2021**, *6*, 2000916. [CrossRef]
14. Yu, A.; Zhu, Y.; Wang, W.; Zhai, J. Progress in Triboelectric Materials: Toward High Performance and Widespread Applications. *Adv. Funct. Mater.* **2019**, *29*. [CrossRef]
15. Lv, S.; Zhang, X.; Huang, T.; Yu, H.; Zhang, Q.; Zhu, M. Trap Distribution and Conductivity Synergic Optimization of High-Performance Triboelectric Nanogenerators for Self-Powered Devices. *ACS Appl. Mater. Interfaces* **2021**, *13*, 2566–2575. [CrossRef]
16. Singh, V.; Joung, D.; Zhai, L.; Das, S.; Khondaker, S.I.; Seal, S. Graphene based materials: Past, present and future. *Prog. Mater. Sci.* **2011**, *56*, 1178–1271. [CrossRef]
17. Jin, Y.; Ka, D.; Jang, S.; Heo, D.; Seo, J.; Jung, H.; Jeong, K.; Lee, S. Fabrication of Graphene Based Durable Intelligent Personal Protective Clothing for Conventional and Non-Conventional Chemical Threats. *Nanomaterials* **2021**, *11*, 940. [CrossRef] [PubMed]
18. Pace, G.; Ansaldo, A.; Serri, M.; Lauciello, S.; Bonaccorso, F. Electrode selection rules for enhancing the performance of triboelectric nanogenerators and the role of few-layers graphene. *Nano Energy* **2020**, *76*, 104989. [CrossRef]
19. Zhao, P.; Bhattacharya, G.; Fishlock, S.J.; Guy, J.G.; Kumar, A.; Tsonos, C.; Yu, Z.; Raj, S.; McLaughlin, J.A.; Luo, J.; et al. Replacing the metal electrodes in triboelectric nanogenerators: High-performance laser-induced graphene electrodes. *Nano Energy* **2020**, *75*, 104958. [CrossRef]
20. Kim, S.; Gupta, M.K.; Lee, K.Y.; Sohn, A.; Kim, T.Y.; Shin, K.S.; Kim, D.; Kim, S.K.; Lee, K.H.; Shin, H.J.; et al. Transparent flexible graphene triboelectric nanogenerators. *Adv. Mater.* **2014**, *26*, 3918–3925. [CrossRef] [PubMed]
21. Chandrashekar, B.N.; Deng, B.; Smitha, A.S.; Chen, Y.; Tan, C.; Zhang, H.; Peng, H.; Liu, Z. Roll-to-Roll Green Transfer of CVD Graphene onto Plastic for a Transparent and Flexible Triboelectric Nanogenerator. *Adv. Mater.* **2015**, *27*, 5210–5216. [CrossRef]
22. Chen, H.; Xu, Y.; Zhang, J.; Wu, W.; Song, G. Enhanced stretchable graphene-based triboelectric nanogenerator via control of surface nanostructure. *Nano Energy* **2019**, *58*, 304–311. [CrossRef]
23. Feng, Y.; Zheng, Y.; Zhang, G.; Wang, D.; Zhou, F.; Liu, W. A new protocol toward high output TENG with polyimide as charge storage layer. *Nano Energy* **2017**, *38*, 467–476. [CrossRef]
24. Park, H.-W.; Huynh, N.D.; Kim, W.; Lee, C.; Nam, Y.; Lee, S.; Chung, K.-B.; Choi, D. Electron blocking layer-based interfacial design for highly-enhanced triboelectric nanogenerators. *Nano Energy* **2018**, *50*, 9–15. [CrossRef]
25. Wu, C.; Kim, T.W.; Choi, H.Y. Reduced graphene-oxide acting as electron-trapping sites in the friction layer for giant triboelectric enhancement. *Nano Energy* **2017**, *32*, 542–550. [CrossRef]
26. Cui, N.; Gu, L.; Lei, Y.; Liu, J.; Qin, Y.; Ma, X.-H.; Hao, Y.; Wang, Z.L. Dynamic Behavior of the Triboelectric Charges and Structural Optimization of the Friction Layer for a Triboelectric Nanogenerator. *ACS Nano* **2016**, *10*, 6131–6138. [CrossRef] [PubMed]
27. Gao, L.; Hu, D.; Qi, M.; Gong, J.; Zhou, H.; Chen, X.; Chen, J.; Cai, J.; Wu, L.; Hu, N.; et al. A double-helix-structured triboelectric nanogenerator enhanced with positive charge traps for self-powered temperature sensing and smart-home control systems. *Nanoscale* **2018**, *10*, 19781–19790. [CrossRef] [PubMed]
28. Winters, M.; Sveinbjörnsson, E.Ö.; Melios, C.; Kazakova, O.; Strupiński, W.; Rorsman, N. Characterization and physical modeling of MOS capacitors in epitaxial graphene monolayers and bilayers on 6H-SiC. *AIP Adv.* **2016**, *6*, 085010. [CrossRef]

29. Sahoo, M.; Wang, J.-C.; Nishina, Y.; Liu, Z.; Bow, J.-S.; Lai, C.-S. Robust sandwiched fluorinated graphene for highly reliable flexible electronics. *Appl. Surf. Sci.* **2019**, *499*, 143839. [CrossRef]
30. Kim, Y.J.; Lee, J.; Park, S.; Park, C.; Park, C.; Choi, H.-J. Effect of the relative permittivity of oxides on the performance of triboelectric nanogenerators. *RSC Adv.* **2017**, *7*, 49368–49373. [CrossRef]
31. Diaz, A.; Felix-Navarro, R. A semi-quantitative tribo-electric series for polymeric materials: The influence of chemical structure and properties. *J. Electrost.* **2004**, *62*, 277–290. [CrossRef]
32. Wong, F.R.; Ali, A.A.; Yasui, K.; Hashim, A.M. Seed/Catalyst-Free Growth of Gallium-Based Compound Materials on Graphene on Insulator by Electrochemical Deposition at Room Temperature. *Nanoscale Res. Lett.* **2015**, *10*, 1–10. [CrossRef]
33. Ferrari, A.C.; Meyer, J.; Scardaci, V.; Casiraghi, C.; Lazzeri, M.; Mauri, F.; Piscanec, S.; Jiang, D.; Novoselov, K.; Roth, S.; et al. Raman Spectrum of Graphene and Graphene Layers. *Phys. Rev. Lett.* **2006**, *97*, 187401. [CrossRef]
34. Huang, M.; Yan, H.; Heinz, T.F.; Hone, J. Probing Strain-Induced Electronic Structure Change in Graphene by Raman Spectroscopy. *Nano Lett.* **2010**, *10*, 4074–4079. [CrossRef] [PubMed]
35. Badri, M.A.S.; Salleh, M.M.; Noor, N.F.M.; Rahman, M.Y.A.; Umar, A.A. Green synthesis of few-layered graphene from aqueous processed graphite exfoliation for graphene thin film preparation. *Mater. Chem. Phys.* **2017**, *193*, 212–219. [CrossRef]
36. Ding, P.; Chen, J.; Farooq, U.; Zhao, P.; Soin, N.; Yu, L.; Jin, H.; Wang, X.; Dong, S.; Luo, J. Realizing the potential of polyethylene oxide as new positive tribo-material: Over 40 W/m2 high power flat surface triboelectric nanogenerators. *Nano Energy* **2018**, *46*, 63–72. [CrossRef]
37. Ye, Z.; Balkanci, A.; Martini, A.; Baykara, M.Z. Effect of roughness on the layer-dependent friction of few-layer graphene. *Phys. Rev. B* **2017**, *96*, 115401. [CrossRef]
38. Naghdi, S.; Sanchez-Arriaga, G.; Rhee, K.Y. Tuning the work function of graphene toward application as anode and cathode. *J. Alloys Compd.* **2019**, *805*, 1117–1134. [CrossRef]
39. Seo, J.-T.; Bong, J.; Cha, J.; Lim, T.; Son, J.; Park, S.H.; Hwang, J.; Hong, S.; Ju, S. Manipulation of graphene work function using a self-assembled monolayer. *J. Appl. Phys.* **2014**, *116*, 084312. [CrossRef]
40. Jiang, H.; Lei, H.; Wen, Z.; Shi, J.; Bao, D.; Chen, C.; Jiang, J.; Guan, Q.; Sun, X.; Lee, S.-T. Charge-trapping-blocking layer for enhanced triboelectric nanogenerators. *Nano Energy* **2020**, *75*, 105011. [CrossRef]
41. Harnchana, V.; Van Ngoc, H.; He, W.; Rasheed, A.; Park, H.; Amornkitbamrung, V.; Kang, D.J. Enhanced Power Output of a Triboelectric Nanogenerator using Poly(dimethylsiloxane) Modified with Graphene Oxide and Sodium Dodecyl Sulfate. *ACS Appl. Mater. Interfaces* **2018**, *10*, 25263–25272. [CrossRef]
42. Niu, S.; Wang, Z.L. Theoretical systems of triboelectric nanogenerators. *Nano Energy* **2015**, *14*, 161–192. [CrossRef]
43. Han, S.A.; Lee, K.H.; Kim, T.-H.; Seung, W.; Lee, S.K.; Choi, S.; Kumar, B.; Bhatia, R.; Shin, H.-J.; Lee, W.-J.; et al. Hexagonal boron nitride assisted growth of stoichiometric Al_2O_3 dielectric on graphene for triboelectric nanogenerators. *Nano Energy* **2015**, *12*, 556–566. [CrossRef]
44. Yu, Y.; Li, Z.; Wang, Y.; Gong, S.; Wang, X. Sequential Infiltration Synthesis of Doped Polymer Films with Tunable Electrical Properties for Efficient Triboelectric Nanogenerator Development. *Adv. Mater.* **2015**, *27*, 4938–4944. [CrossRef]
45. Patnam, H.; Dudem, B.; Graham, S.A.; Yu, J.S. High-performance and robust triboelectric nanogenerators based on optimal microstructured poly(vinyl alcohol) and poly(vinylidene fluoride) polymers for self-powered electronic applications. *Energy* **2021**, *223*, 120031. [CrossRef]
46. Chen, H.; Xu, Y.; Bai, L.; Jiang, Y.; Zhang, J.; Zhao, C.; Li, T.; Yu, H.; Song, G.; Zhang, N.; et al. Crumpled Graphene Triboelectric Nanogenerators: Smaller Devices with Higher Output Performance. *Adv. Mater. Technol.* **2017**, *2*, 1700044. [CrossRef]
47. Kim, D.W.; Lee, J.H.; Kim, J.K.; Jeong, U. Material aspects of triboelectric energy generation and sensors. *NPG Asia Mater.* **2020**, *12*, 1–17. [CrossRef]

Review

Cellulose Nanocrystals (CNC)-Based Functional Materials for Supercapacitor Applications

Arulppan Durairaj [1], Moorthy Maruthapandi [1], Arumugam Saravanan [1], John H. T. Luong [2] and Aharon Gedanken [1,*]

1. Department of Chemistry, Bar-Ilan Institute for Nanotechnology and Advanced Materials, Bar-Ilan University, Ramat-Gan 52900, Israel; chemdraj@gmail.com (A.D.); lewismartin.jesus@gmail.com (M.M.); saran.bc94@gmail.com (A.S.)
2. School of Chemistry, University College Cork, T12 YN60 Cork, Ireland; luongprof@gmail.com
* Correspondence: gedanken@biu.ac.il or gedanken@mail.biu.ac.il; Tel.: +972-3-5318315; Fax: +972-3-7384053

Abstract: The growth of industrialization and the population has increased the usage of fossil fuels, resulting in the emission of large amounts of CO_2. This serious environmental issue can be abated by using sustainable and environmentally friendly materials with promising novel and superior performance as an alternative to petroleum-based plastics. Emerging nanomaterials derived from abundant natural resources have received considerable attention as candidates to replace petroleum-based synthetic polymers. As renewable materials from biomass, cellulose nanocrystals (CNCs) nanomaterials exhibit unique physicochemical properties, low cost, biocompatibility and biodegradability. Among a plethora of applications, CNCs have become proven nanomaterials for energy applications encompassing energy storage devices and supercapacitors. This review highlights the recent research contribution on novel CNC-conductive materials and CNCs-based nanocomposites, focusing on their synthesis, surface functionalization and potential applications as supercapacitors (SCs). The synthesis of CNCs encompasses various pretreatment steps including acid hydrolysis, mechanical exfoliation and enzymatic and combination processes from renewable carbon sources. For the widespread applications of CNCs, their derivatives such as carboxylated CNCs, aldehyde-CNCs, hydride-CNCs and sulfonated CNC-based materials are more pertinent. The potential applications of CNCs-conductive hybrid composites as SCs, critical technical issues and the future feasibility of this endeavor are highlighted. Discussion is also extended to the transformation of renewable and low-attractive CNCs to conductive nanocomposites using green approaches. This review also addresses the key scientific achievements and industrial uses of nanoscale materials and composites for energy conversion and storage applications.

Keywords: cellulose nanocrystal; surface functionalization; conductive electrodes; energy storage; supercapacitors

1. Introduction

Environmental safety and renewable sources are two prerequisites for energy conversion and storage fields. However, the extensive usage of petroleum-based synthetic polymers for energy conversion applications results in severe environmental issues [1,2]. For the last two decades, the widespread utilization of synthetic polymers has instigated critical issues through an accumulation of plastic wastes, the depletion of fossil fuels and global climate change. Owing to diversified industrialization, the global CO_2 emissions from petroleum-based fossil fuels increase daily, and this trend will continue in the future [3]. To mitigate these environmental issues, sustainable and renewable biopolymers can be an alternative source for the development of efficient viable products [4]. Natural biomass and biowaste including lignocellulose are abundant and renewable sources for the production of value-added products without any harmful effect on the ecosystem [5]. The most abundant cellulose has many attractive features such as nontoxicity, renewability, biodegradability, low cost and colloidal stability [6]. The cellulosic natural biomass contains various other components including waxes, lignin, pectin and inorganic nitrogenous salts [7]. A pretreatment process is necessary to fractionate nanocellulose from cellulose resources such as plants, woods, algae, tunicate and bacteria with a polysaccharide structure [8–10]. It comes up as a promising, eco-friendly and sustainable candidate with a unique structure and remarkable properties for further structural modifications. Nanocellulose has been successfully investigated for its various applications in energy, packaging, paper making, sensors, cosmetics, coating and environmental remediation areas [11,12]. Nanocellulose [13] is termed to depict two emerging dominant materials: cellulose nanocrystals (CNCs) or nanocrystalline cellulose (NCCs) [14] and cellulose nanofiber (CNF) [15]. Nanocellulose can be prepared from cellulose, a linear chain of β-D-glucopyranose units, via β-1, 4 glycosidic bonds with intensive intra/intermolecular hydrogen bonding. The preparation of CNCs is obtained by traditional acid hydrolysis [16], mechanical disintegration [17], chemical oxidation and hydrolysis by ammonium persulfate [14].

A cellulose nanocrystal resembles a needle-shaped structure; it is 100–500 nm in length and 1–50 nm in diameter [14]. CNCs are the most ideal material, with attractive features including good mechanical strength, colloidal stability, biodegradability and very low cytotoxicity [18,19]. CNCs have gained considerable attention because of their attractive cost and diversified industrial applications and bioapplications [20–23]. In this context, CNCs have been advocated for supercapacitor applications and lithium-ion batteries [24–26]. One of the three energy storage technologies [26], along with mechanical energy storage and biological energy storage, electrochemical energy storage systems comprise lithium batteries, lithium-sulfur batteries and supercapacitors [27,28]. SCs have gained extensive consideration due to their charge-discharge ability, high cycle stability, high energy conversion and environmentally friendly technologies [29–31]. SCs have a high specific capacity and high energy density in contrast with traditional capacitors. Compared with secondary batteries, they are generating less pollution and fewer chemicals. They also have an extended life cycle, a broad operating temperature and a fast charge-discharging capability [32,33]. Currently, SCs are widely used for energy vehicles, wearable electronic products, etc. [34,35].

SC devices consist of electrodes, electrolytes, current collectors and a diaphragm. As the backbone of the supercapacitor, an electrode is an important tool for the effective conversion and storage of electrochemical energy [36,37]. Supercapacitors are energy-stored energy devices, which are based on three major principles: EDLC (electrical double layer capacitance), pseudocapacitance and asymmetric supercapacitance. EDLC is attained by the separation of charges between the Helmholtz double-layer and the diffusion layer at the electrolyte interface [29]. Graphene oxide, activated carbon, reduced graphene oxide and heteroatom doped carbon have a high power density and recyclability. Unlike EDLC, pseudocapacitors are formed through the reduction–oxidation reactions resulting from chemical transformations [30]. The major principle involved in pseudocapacitors is the transfer of electron charge between the electrode and electrolyte through reduction–oxidation reactions, electrosorption and intercalation. Pseudocapacitor materials are generally made up of metal oxides such as RuO_2, NiO, MnO_2, Co_3O_4, conducting polymers and metal sulfides. Pseudocapacitors exhibit a high specific capacitance, a high energy density and a high power density; however, a limited lifetime and reduced cell voltage are their two major disadvantages [31]. Asymmetric supercapacitors (ASCs) or hybrid supercapacitors are fabricated by combining two different electrode materials with special properties for increasing the cell voltage and power density of supercapacitor materials. Supercapacitors exhibit fast charge-discharge, an extended cycle lifetime, a high efficiency for the charge and discharge cycle, eco-friendliness, etc. [37]. Therefore, the selection of novel electrode materials is one of the main tasks for the fabrication of high-performance SCs. To date, carbon-based materials, transition metals, graphene-based materials and conductive materials are commonly used electrode materials for SCs [38–40]. These nanomaterials were successfully applied as supercapacitor electrodes for many energy applications including for automobiles and electronic gadgets such as watches, laptops, etc. [41,42]. However, the existing SC electrode system often fails to construct efficient electrochemical energy charge-discharge behavior [43,44]. Thus, research endeavors have focused on the development of green carbon-based nanomaterials with highly efficient capacitor electrodes. This strategy is considered a good pathway to reduce the agro-industrial waste and generate revenue for supply chains—a significant shift toward the circular economy [45]. Porous green-based carbon materials with a high surface area and bounteous heteroatoms can serve as electrode materials with boosted electrochemical activities [46,47]. In particular, CNCs can be the ideal material for SCs due to their high crystalline nature, surface functionalization ability and abundant nature [48,49]. However, critical findings on the collective evaluation of CNC-based electrode materials in the field of SCs remain a subject of future endeavors.

This comprehensive review highlights the research findings of novel CNCs designed for energy storage applications. It elaborates on the significance of renewable and innocuous CNCs as low-cost natural resources and essential features in supercapacitor applications. Different synthesized methods of CNCs will be discussed including the pretreatment of cellulose sources, processing parameters and surface functionalization. The pretreatment process plays a vital role in eliminating the lignin and other unwanted impurities from the cellulose and in increasing the crystallinity of the CNCs. For the enhanced dispersion of CNC in hydrophilic and hydrophobic matrixes, selective surface modification techniques including chemical modification and enzymatic modification are discussed. CNC electrodes in SCs consist of a conductive polymer/CNC, porous carbon derived from CNCs, hybrid CNC electrodes as well as corresponding doped materials (Figure 1). This in-depth review highlights the key challenges and prospects of CNC-based SC electrodes.

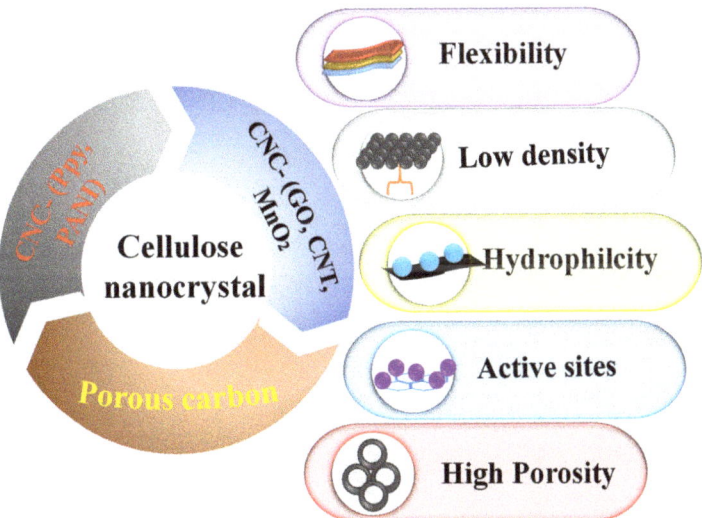

Figure 1. Schematic illustration of the hybrid cellulose nanocrystal and its properties.

2. Cellulose Nanocrystals (CNCs): Structure and Properties

CNCs have gained enormous attraction due to their novel nanostructure and highly crystalline nature [14,21]. These emerging materials enable the development of highly efficient nanocomposites with unique properties. Besides renewable and abundant plants and woods, [50,51], CNCs are often derived from cotton, bacterial tunicate, microcrystalline cellulose (MCC) and biowaste materials [14,21,52,53]. CNCs are prepared from cellulose under controlled hydrolysis conditions, resulting in a stable CNC suspension. Cellulose has two distinct components: amorphous and crystalline [14,21]. When it is subjected to conventional acid hydrolysis [54] or chemical oxidation and hydrolysis [14], the amorphous part is removed from parental cellulose, resulting in a shorter CNC with high crystallinity. Among the various acids and oxidants, the amorphous domain of cellulose can be easily removed by concentrated sulfuric acid [54]. Compared to cellulose, CNCs are shorter β(1-4) connected chains of anhydrous-glucopyranose units (AGU)—a short-rod-like shape or whisker shape (diameter: 2–20 nm and length: 100–500 nm) with higher crystallinity [14,55,56]. Like cellulose, all the hydroxyls are placed in every equatorial position in the chair conformation, which allows for the stable nature of CNCs. The equatorial hydroxyl groups are also stabilized by intramolecular hydrogen bonding, providing enhanced mechanical strength. The CNC properties can be altered by surface modification with suitable materials [14,57] and gold nanoparticles for specific applications [58]. Some other important properties of CNCs are (i) their high mechanical strength, (ii) the creation of 3D nanostructured nanomaterials through their intermolecular interactions and (iii) their tunable surface modification [14,56,59]. The schematic preparation of CNCs is portrayed in Figure 2.

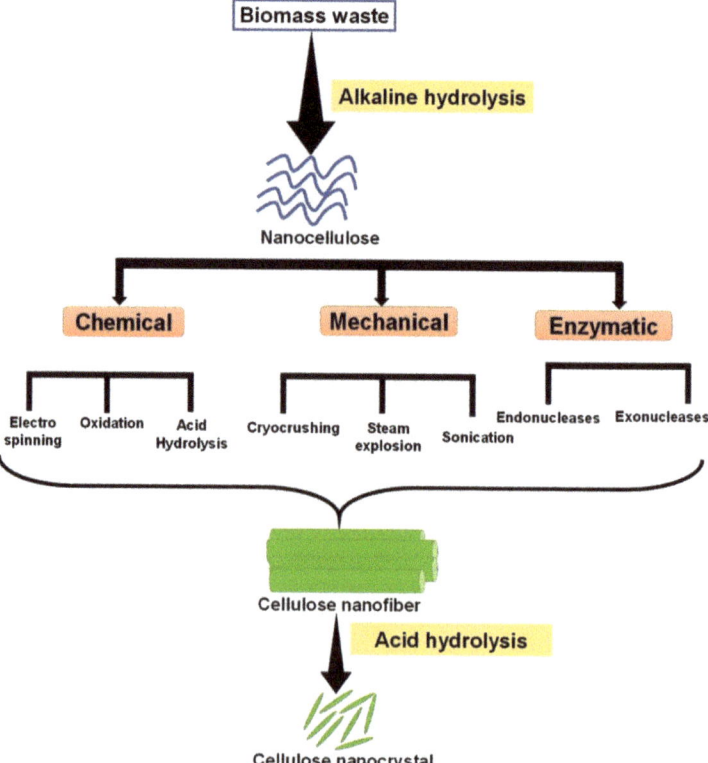

Figure 2. Different schemes for making cellulose nanocrystals (CNCs).

The development of surface-modified CNCs offers a plethora of potential applications encompassing energy, packaging, special papers, paints, building materials, aerospace, biomedical materials, pharma industries, cosmetics, the electronic and automotive industries, etc. [60–62].

3. Pretreatment Process

As an attractive source of the production of nanocrystalline cellulose, natural biomass contains various components such as cellulose, lignin, hemicelluloses, etc. [63]. Various types of pretreatment processes, shown in Table 1, must be carried out for the removal of noncellulosic components [64]. The traditional pretreatment process includes alkaline treatment and acid chloride treatment to remove both hemicellulose and lignin from renewable biomass. These methods are often referred to as "delignification", as they effectively eliminate lignin and other components from the biomass, resulting in pure cellulose. The acid hydrolysis of cellulose from different sources, e.g., cotton and wood, has been frequently used to prepare CNCs [65]. Besides two strong acids, sulfuric and hydrochloric acids, phosphoric acid and hydrobromic acids are mostly used to prepare rod-like CNCs from cellulose feedstocks. Strong acids are polluted, corrosive and adversely affect the heat resistant nature of CNCs [14,56].

Table 1. Comparison of various pretreatment processes of CNCs.

Methods	Size (nm)	Advantages	Disadvantages
Acid hydrolysis	Diameter: 3–15 Length: 100–300	✓ Uniform ✓ High crystalline	• High pollution
Enzymolysis	Diameter: 3–50 Length: 100–1800	✓ Fewer reagents ✓ Less pollution	• Hard preparation • Not uniform
Physical (mechanical, ultrasonic)	Diameter: 3–50 Length: 100–2000	✓ Large quantity ✓ Simple synthesis	• Very large size • Large energy consumption
Physical + green chemistry	Diameter: <20 Length: 100–500	✓ Maximum yield ✓ Less pollution	• The preparation process is complex

Greener methods have emerged to minimize the usage of sulfuric acid for the synthesis of CNCs en masse. Of note is the preparation of CNCs from palm oil through the TCF (total chlorine-free) method; however, the resulting CNCs are subjected to notable degradation during this treatment [66]. Two ionic liquids, 1-propyl-3-methylimidazolium chloride and 1-ethyl-3-methylimidazolium chloride, are used to prepare CNCs (average diameter of 20 nm) from Avicel [67], the commercial Sigma-Aldrich microcrystalline cellulose (MCC). Avicel is prepared by the acid hydrolysis of specialty wood pulp and purified and partially depolymerized alpha-cellulose. The yield of CNC from biomass can be improved through various strategies such as solution plasma processing technology [68], ultrasound-assisted enzymatic hydrolysis [69], microwave-assisted acid hydrolysis [70] and sonication-assisted TEMPO ($C_9H_{18}NO$, 2,2,6,6-tetramethylpiperidine 1-oxyl, 2,2,6,6-tetramethyl-1-piperidinyloxy) oxidation [71]. One-pot TEMPO-periodate oxidation reactions form highly-carboxylated CNCs [72]; however, TEMPO is very expensive, which is a deterrent factor for the mass preparation of CNCs. In this context, the use of ammonium persulfate (APS) [14,56], a patented technology [73], offers a low-cost chemical for the synthesis of carboxylated CNCs.

CNCs can also be prepared from bamboo pulp fibers using a weak acid, e.g., maleic acid, together with a ball mill pretreatment process. The ball milling mechanical force decomposes the bamboo fiber and promotes acid hydrolysis effectively. The yield of ball-milled pretreated CNC is 10.55–24.50% higher than that of the normal acid hydrolysis without ball milling mechanical forces [74]. Ultrasonication can be used to minimize the amount of sulfuric acid in the preparation of CNCs from MCC [75]. Ultrasonicated CNCs exhibit higher thermal stability compared to their counterparts obtained by concentrated acid hydrolysis. Perhaps ultrasonicated pretreatment is one of the facile techniques for making CNCs from biomass feedstocks [75]. Microwave pretreatment assisting with alkali treatment is another green approach for the synthesis of CNCs from seaweed. This technique eliminates the wax from seaweed fibers and reduces the alkali effect with a short heating time [76]. Of interest is the design of a deep eutectic solvent method to prepare high crystalline CNCs without any chemical functionalization. The deep eutectic solvent method may be an environmentally friendly, renewable, biodegradable and non-toxic technique to prepare CNCs in the future [77]. The environmentally friendly pretreated steps are critical in the large-scale production of CNCs for industrial applications.

4. Surface Modification of CNCs

With high crystallinity, CNCs are not reactive, and the hydrophilic behavior of CNCs, particularly sulfonated and carboxylated CNCs, impedes their dispersion in hydrophobic matrices. Therefore, the incorporation of pertinent groups on CNCs is required to increase their physiochemical properties and endow innovative applications. The surface functionalization of CNCs fulfills the recent challenging needs in developing applications such as wastewater treatment, polymer composites, barrier films, textiles, energy and biomedical applications [78,79]. In brief, the CNC surface can be modified with covalent methods (acetylation, amidation, benzoylation, silanization, esterification, isocyanation and polymer grafting) and non-covalent methods (the introduction of polymers, surfactants and compatibilizing agents). Among the various routes available for the surface functionalization of CNCs [80], the reaction must be performed under strictly controlled conditions to preserve the distinct crystalline behavior of CNCs. Functionalized CNCs from innovative techniques (Figures 3 and 4) will foster advanced and novel applications of CNCs.

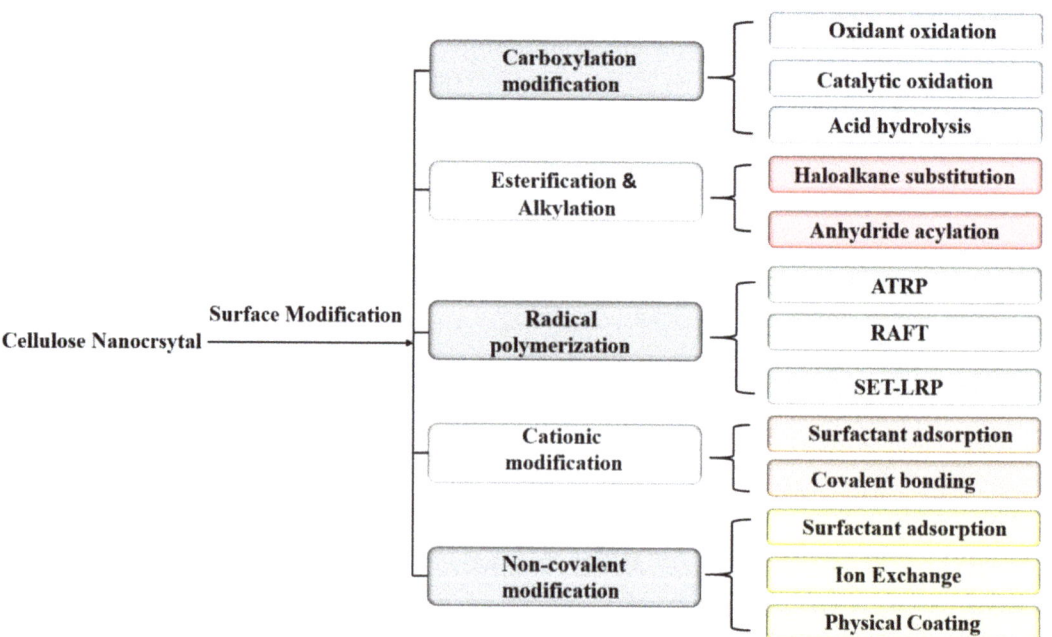

Figure 3. Surface functionalization of CNCs by different routes.

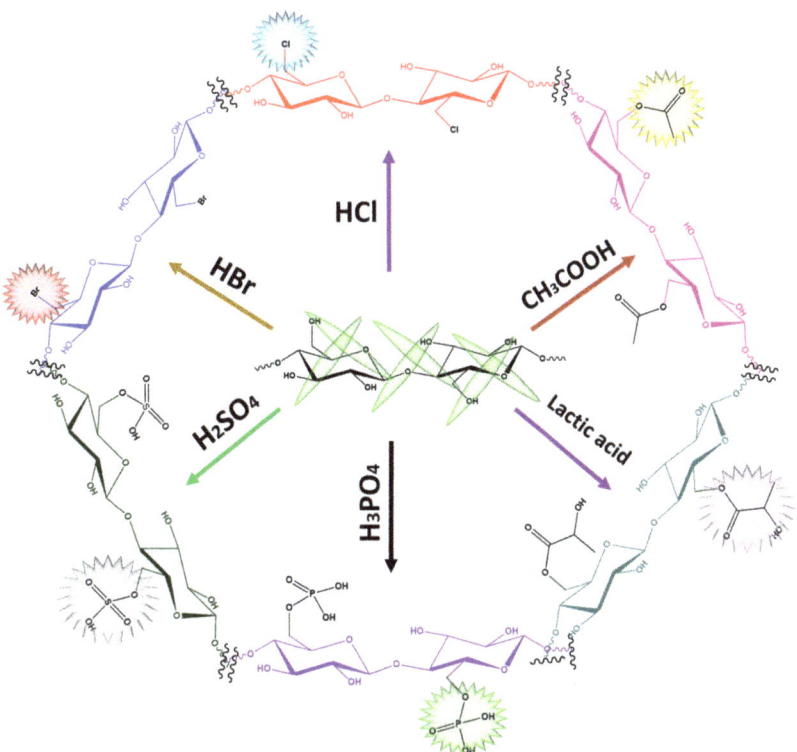

Figure 4. Common surface functionalization techniques of CNCs.

5. Oxidation of Cellulose Nanocrystals (CNCs)

A commercial oxidizing reagent (TEMPO) selectively converts some hydroxyl functional groups to carboxyl functional groups on the cellulose moiety at the C-6 position. The oxidized CNCs with carboxyl groups from the HCl hydrolysis pretreated cellulose fibers show a better dispersion in the aqueous medium [14,81]. The TEMPO-oxidized CNCs exhibit the same morphology and excellent dispersion in water after the incorporation of the carboxyl group [82,83]. Other oxidizing reagents such as ammonium persulfate [14,56] and chlorite-periodate are successfully used for CNC surface modification through the oxidation process, as mentioned earlier. APS can defibrillate and remove the amorphous domain of cellulose effectively, resulting in uniform high-crystalline carboxylated CNCs [14,84].

6. Acetylation of CNCs

Acetylated CNCs can be obtained by reacting CNCs with acetic anhydride under the dimethylformamide (DMF) solvent medium [85]. After heating at 105 °C for 24 h and thorough washing with water/methanol/acetone to remove unreacted CNCs, the resulting CNCs are easily dispersed on the nonpolar polymeric matrix by the reduction of hydrogen bonds. The combined acid hydrolysis and acetylation reduce the crystallinity and adversely affect the morphology of the acetylated CNCs. CNCs extracted from cotton fibers can also be acetylated by vinyl acetate, and this one-step pathway is carried out in dimethylformamide (DMF) at 95 °C [86].

7. Sulfonation of CNC

The acid hydrolysis of cellulose by sulfuric acid produces CNCs with some sulfate moieties [87], which increase their dispersion in aqueous media. Even with the optimized procedure, the resulting sulfate functional groups on the CNCs are still limited, and this is a subject of future endeavors for the preparation of CNCs with sufficient sulfate groups [9].

8. Silylation of CNCs

Hydrophobic alkyl groups such as chlorosilanes and alkoxysilanes can be introduced to CNCs by silylation with silane coupling agents. Normally, silane coupling agents are used in composites for adhesive applications. They contain several functional groups, which easily couple with different nanomaterials through covalent linkage or van der Waals interaction [88]. Silanized CNCs become more hydrophobic, with improved dispersion in organic solvents compared to pristine CNCs. The coupling of the silane functional groups on CNCs starts with hydrolysis, which introduces alkoxyl groups in acidic, basic and/or neutral pH conditions to form the silanol (Si–O–Si). The formation of silanol is evidence of the elimination of hydroxyl or water from the CNCs [89]. Silane coupling agents can be used to develop silylated CNCs. During the silanization, hydrolysis and condensation reactions occur on the CNC surface. The combined esterification and depolymerization process leads to the creation of Si-O-C bonds [90].

9. Carbamation

CNCs subjected to isocyanates such as 2,4-diisocyanatotoluene, 3,5-dimethylphenyl isocyanate, n-octadecyl isocyanates (OI), hexamethylene diisocyanate and methylene diphenyl diisocyanates form carbamated CNCs with enhanced hydrophobicity [91,92]. For example, isocyanate modified CNCs can be prepared using isophorone diisocyanate under dimethylsulfoxide (DMSO) at 60 °C [93]. Modified CNCs can be dispersed in a polyurethane matrix after the incorporation of IDPI (isophorone diisocyanate). Carbamated CNCs are prepared by the treatment of pristine CNCs with 2,4-toluene diisocyanate in the presence of triethylamine. The *p*-NCO and *o*-NCO functional groups replace the OH groups of CNCs, which have a higher reactivity than pristine CNCs [94].

10. CNCs: Key to Developing Sustainable Electrochemical Energy Material

Upcoming sustainable development focuses on utilizing renewable materials for the development of value-added, environmentally friendly products. CNCs derived from renewable biomass have been advocated for diverse applications in the paper industries, energy storage, remediation, optoelectronic fields, etc. Of interest are the potential uses of CNCs in SCs, Li-ion batteries, solar cells and electrochemical energy storage devices [95,96]. However, CNCs still require enormous surface conductivity to become ideal materials for important energy applications. In this context, the CNC surface can be modified with conductive polymers, MWCNT (multiwalled carbon nanotube), graphene nanosheets and some inorganic nanomaterials, as shown in Figure 5. This endeavor aims to develop highly efficient energy storage systems containing CNCs. In this context, low-cost and flexible CNCs have shown great promise as effective electrode materials [97]. Two major approaches have been applied to develop CNC-based SCs such as coating the conductive material on the surface of CNCs and mixing conductive materials with CNCs via in situ polymerization and blending. The in situ polymerization strategy is followed to develop a conductive polymer on the CNC matrix from the repetitive building block of monomers to build CNC-conductive polymer composites. The blending approach is used to combine different active nanomaterials with CNCs to make effective hybrid composites via synergistic effects [98]. In addition, co-precipitation and doping methods have been applied to prepare CNC-conductive electrodes. The facile preparation of conductive cellulose nanocrystal electrodes is depicted in Figure 6.

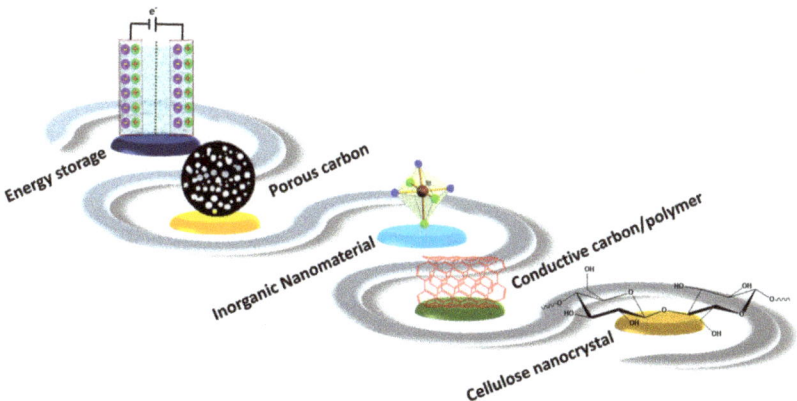

Figure 5. Schematic representation of CNC building blocks and active materials for supercapacitor devices.

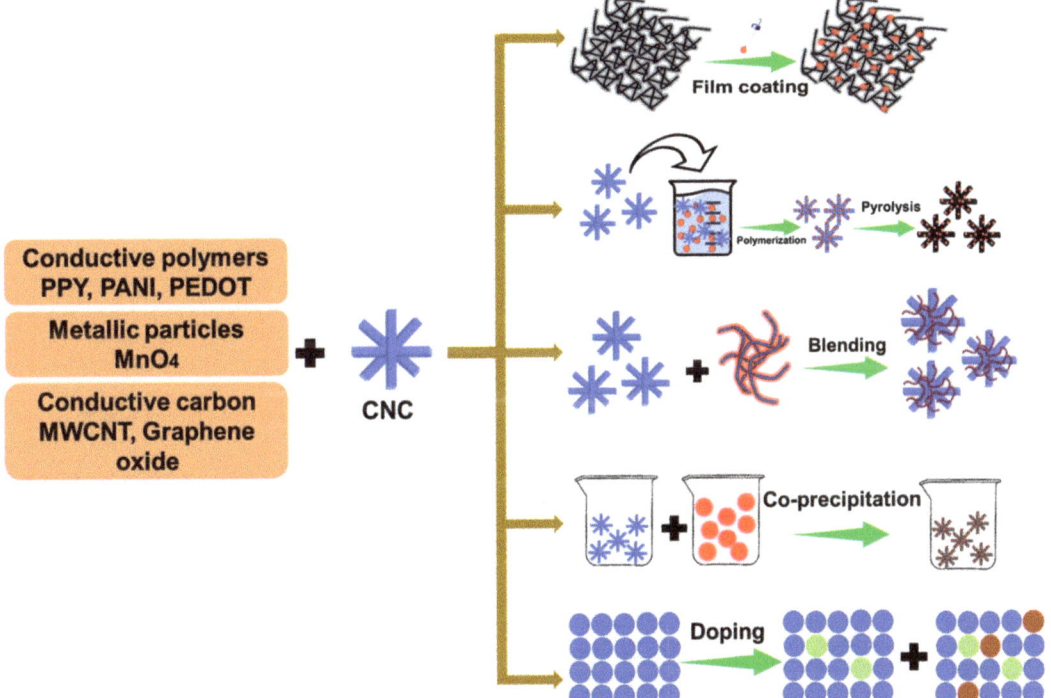

Figure 6. Various methodologies to develop novel conductive cellulose nanocrystals (CNCs).

11. Conductive Polymer CNCs

Two traditional and popular conjugated polymers, e.g., polypyrrole (PPy) and polyaniline (PANI), are conducting macromolecules with their backbone comprising alternating single and multiple bonds [99]. With their unique properties, conjugated polymers are widely studied as electrochemical capacitors, solar cells, OLED, fuel cells and sensors. However, the solubility of conjugated polymers is very limited [100], and this setback can be overcome by the preparation of nanocomposites between conducting polymers and CNCs. In principle, the conjugated polymers such as PPy, PANI, polyacetylene, polythiophene, poly(p-phenylene vinylene) and their derivatives are used to make CNC-conductive polymer electrodes. As an example, the fabrication of a poly(3,4-ethylenedioxythiophene) (PEDOT)/CNC conductive electrode can be realized by a simple electrochemical polymerization technique [101]. The formation of PEDOT on CNC depends on the effect of the electrochemical polymerization potential, including the applied potential, deposition time and concentration of the precursors. The PEDOT/CNC electrode exhibits a specific capacity capacitance (C_s) of 117.02 Fg^{-1} at 100 mVs^{-1}, with an E_s (energy density) and P_s (power density) of 11.44 $Whkg^{-1}$ and 99.85 Wkg^{-1}, respectively, at a 0.2 Ag^{-1} current density. The combination of PEDOT and CNCs shows a lower charge transfer resistance (R_{ct} = 0.53 Ω), with a retention capacitance of 86% after 1000 cycles. The presence of SO_3^- groups being counter ions to the positively charged PEDOT on the CNC surface serves as a supportive framework in the formation of PEDOT/CNC. The functional groups on modified CNCs also prevent the severe structural changes of the electrode during electrochemical storage applications [101].

A novel, simple and scalable conductive PPy/CNC electrode for supercapacitor applications is attained via the in situ chemical polymerization method [**102**]. CNCs treated with TEMPO possess carboxylic acids on their CNC surface (Figure 7). The pyrrole monomer is then deposited and polymerized on the surface of carboxylated CNCs to construct a PPy/CNC electrode with 248 Fg^{-1} (C_s) at 10 mVs^{-1}, compared to the 90 Fg^{-1} of a pristine PPy electrode. Hence, carboxylated CNCs play a vital role in the charge-discharge mechanism during electrochemical energy storage applications. The extraordinary specific capacitance behavior of PPy/CNCs can be attributed to the favorable polymerization of PPy on carboxylated CNCs, the strong interaction between CNCs and PPy, the porous nature of PPy/CNC and the low density of CNCs [102]. Another attempt is the fabrication of a porous CNC/PPy electrode using the electrochemical co-deposition method [103]. Again, CNCs prepared from cotton by acid hydrolysis are subjected to TEMPO to form carboxylated CNCs. The CNC/PPy electrode is developed by the electrodeposition of pyrrole and carboxylated CNCs. The Cl^- ions from the PPy are incorporated to balance the positively charged CNC skeleton. The CNC/PPy and PPy electrodes exhibit 336 Fg^{-1} and 258 Fg^{-1}, respectively, and the former's high C_s is attributed to the high porous morphology in the composite. In addition, the PPy favors ion transfer between the CNCs and electrolyte interface. The CNC/PPy electrodes are also compared with the PPy/carbon nanotube electrode prepared by similar electrodeposition. The CNC/PPy nanocomposites reveal comparable specific capacitance and good stability compared to the CNT/PPy electrode [103].

Similarly, the CNC/PANI and CNC/PEDOT composites have been proven as electrochemical supercapacitors [104]. CNCs are prepared from a cotton source and are oxidized by TEMPO. The CNC/PANI and CNC/PEDOT electrodes are then prepared by an electrodeposition process consisting of CNCs, HCl, aniline and 3,4-ethylenedioxythiophene. The electrodepositing polymer films exhibit a porous morphology, which facilitates the movement of ions between the electrode and electrolyte interface. The C_s of the CNC/PANI, PANI, CNC/PEDOT and PEDOT is 488 Fg^{-1}, 358 Fg^{-1}, 69 Fg^{-1} and 58 Fg^{-1}, respectively. The CNC/PANI electrodes show higher C_s than the CNC/PEDOT due to the strong deposition of PANI on CNCs. At a high charge density, the PEDOT is not deposited on the CNC surface but forms a gel-like film. The CNC/PEDOT displays reduced volumetric results during the charge-discharge reaction. In the case of CNC/PANI, PANI is strongly deposited on the CNC surface, and it forms a porous morphology at a higher deposition charge to support ion and electrolyte interface movement during the supercapacitor process [104].

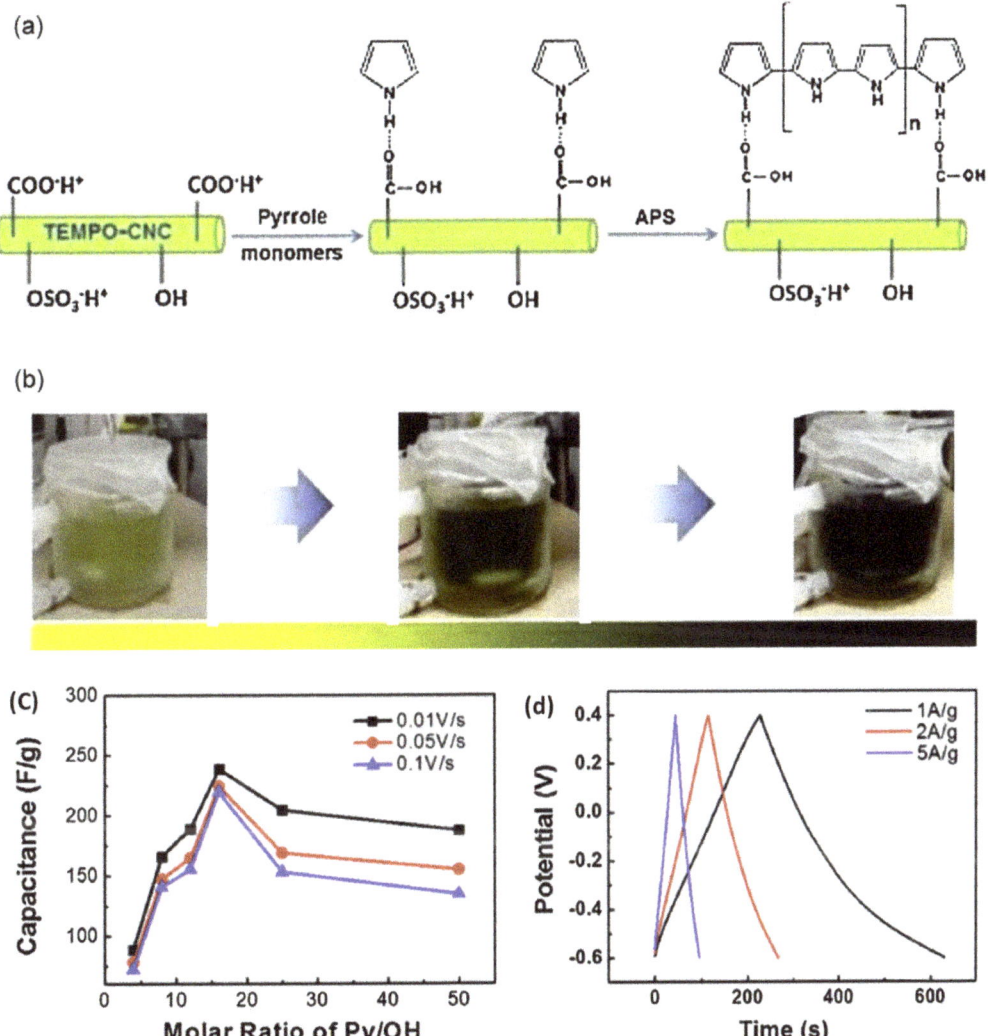

Figure 7. (**a,b**) In situ polymerization synthesis of CNC/PPy, (**c**) Effect of PPy on the capacitance of CNC/PPy, (**d**) Charge and discharge test of CNC/PPy at 1 Ag^{-1}, 2 Ag^{-1} and 5 Ag^{-1} (charge current). Adapted with permission from [102]. Copyright 2014 Elsevier.

A different approach is the fabrication of layered CNC/PPy/PVP electrodes as superior electrochemical supercapacitors [105]. The hydrophilic CNC surface is modified by the adsorption of polyvinylpyrrolidone (PVP), enabling the growth of the PPy polymer on the CNC surface. The conductive CNC/PPy/PVP composites display a smooth, uniform PPY coating and a higher capacitance than the CNC/PPy electrode. The CNC/PPy/PVP composite exhibits a C_s of 338.6 Fg^{-1} at 2 A g^{-1}, with a retention rate of 87.3% after 2000 cycles. The higher capacitance and superior cyclic stability are attributed to the smooth and uniform deposition of PPy on the CNC surfaces, facilitating the charge transfer and diffusion between the electrode/electrolyte interface. The hydrogen bonding and the hydrophobic interaction of PPy and PVP provide a stable PPy coating layer on the CNC surface. A pseudo-3-layer structure of the CNC/PPy/PVP also plays a role to enhance the super capacitance [105].

12. Cellulose Nanocrystal-Conductive Hybrid Electrode

High conductive inorganic nanomaterials/CNC hybrids became very attractive for various potential applications owing to their unique size dependence and optical/catalytic properties [106]. Graphene oxide, transition metals, noble nanoparticles and metal oxides with high electrical conductivity are ideal materials for energy storage applications. The cellulose nanocrystals produce a strong interaction between the CNC surface and the inorganic nanomaterials via aggregation behavior. The dispersion of various nanoparticles on the CNC surface can develop new versatile nanomaterials for energy storage and environmental applications [107]. Carbon-based materials (graphene oxide, carbon nanotube (CNT), graphene, etc.) with outstanding conductive features are also chemically stable even under harsh environments. The deposition of carbon-based materials on the CNC surface results in lower conductivity due to the weak interaction and layer fall-off. The blending method is a suitable technique to trap the carbon-based nanomaterial inside the CNCs, which generates good electrical conductivity. Metal oxides exhibit higher electrical conductivity than graphitic particles; therefore, outstanding supercapacitor materials can be fabricated by combining them with CNCs. The metal oxide/CNC hybrid composites are prepared by coating, doping and co-precipitation methods [108]. Of note is the preparation of an Al-CNC transparent hydrogel electrolyte for supercapacitor applications [109]. CNCs are derived from brewery residues and are subjected to TEMPO (Figure 8). To obtain the transparent Al-CNC hydrogel, carboxylated CNCs are combined with high valence Al^{3+} via physical linking interaction. The Al-CNC hydrogel exhibits high electric conductivity, mechanical properties and optical properties. For comparison, Al^{3+} ion liquid electrolytes (LE) are prepared without CNCs. The Al-CNC electrolyte exhibits good ionic conductivity, high optical transmittance, excellent compression strength and tolerating properties to different angles (0° and 90°). The animal bone is carbonized at 900 °C to prepare porous carbon (PC) with a large surface area. An Al-CNC hydrogel-based system shows good electrochemical super capacitance and mechanical durability compared to the Al-LE system. The PC/Al-CNC electrode exhibits a specific capacitance of 804 Fg^{-1} at 1 Ag^{-1} compared to the 737 Fg^{-1} of the PC/Al-LE electrode. This result evinces that the 3D network of the transparent Al-CNC hydrogel electrolyte is attributed to enhanced specific capacitance [109].

A high-performance and flexible PANI@CNT−CNC/PVA− PAA electrode (Figure 9) can be prepared by electrospinning and thermal treatment followed by a polymerization coating process for supercapacitor applications [110]. Poly(vinyl alcohol, PVA) and poly(acrylic acid, PAA) are utilized to prepare an electrospun nanofibrous membrane, and CNCs stabilized CNTs are used to enhance the mechanical and electrochemical functions. CNCs are obtained using sulfuric acid hydrolysis and ultrasonic treatment. During this step, negatively charged ions are formed to stabilize the CNTs on the CNC surface. To obtain the CNT−CNC/PVA−PAA membrane, a lab-scale electrospinning method is applied. The thermal treatment process triggers the esterification crosslinking reaction between the OH groups of CNCs and the –COOH groups of PAA. To enhance the electrochemical behavior of the CNT−CNC/PVA−PAA membrane, aniline is deposited and polymerized on the membrane to generate PANI@CNT−CNC/PVA−PAA. The specific capacitance of the PANI@ CNT−CNC/PVA−PAA is 164.6 Fg^{-1}, which can be attributed to the synergetic effect between the CNC-CNT hybrid and PANI. The high porosity, the large surface area and the existence of PANI can enhance the capacitance of the prepared membrane. Particularly, a large surface area constructs enormous active sites for the faradaic reaction and increases the interaction between the electrode/electrolyte interface. The PVA–PAA polymeric matrix also promotes the electrolyte interface area to enlarge the redox reactions on the PANI.

The developed electrode is considered a symmetrical supercapacitor by sandwiching the PVA–PAA–KCl electrolyte between the PANI@CNT–CNC/PVA–PAA electrodes. The symmetrical supercapacitor exhibited superior capacitance values of 155 F g^{-1}, with a retention rate of 92% after 2000 cycles. The calculated specific energy (E_s) and specific power (P_s) values are 13.8 Wh kg^{-1} and 200.3 W kg^{-1} for the PANI@CNT–CNC/PVA–PAA electrode [110].

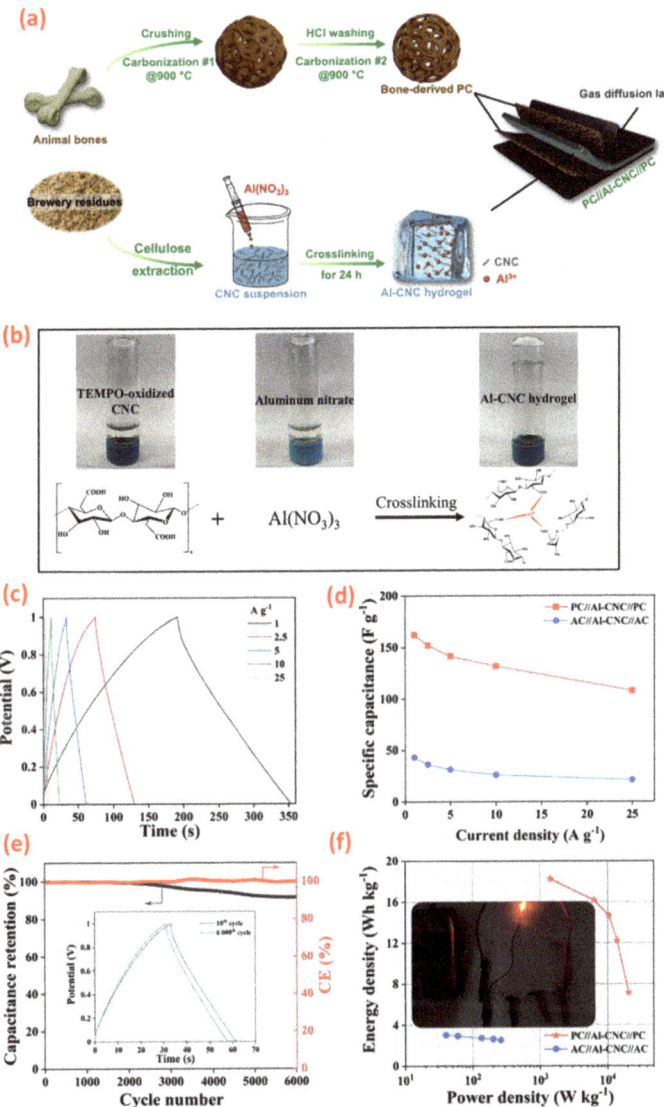

Figure 8. (**a**) The preparation of PC//Al–CNC//PC as a flexible supercapacitor, (**b**) The preparation of TEMPO mediated Al–CNC hydrogel, (**c–f**) The electrochemical properties, such as the GCD curve, specific capacitance, stability analysis and Ragone plot, of the PC//Al–CNC//PC electrode. Adapted with permission from [109]. Copyright 2022 Elsevier.

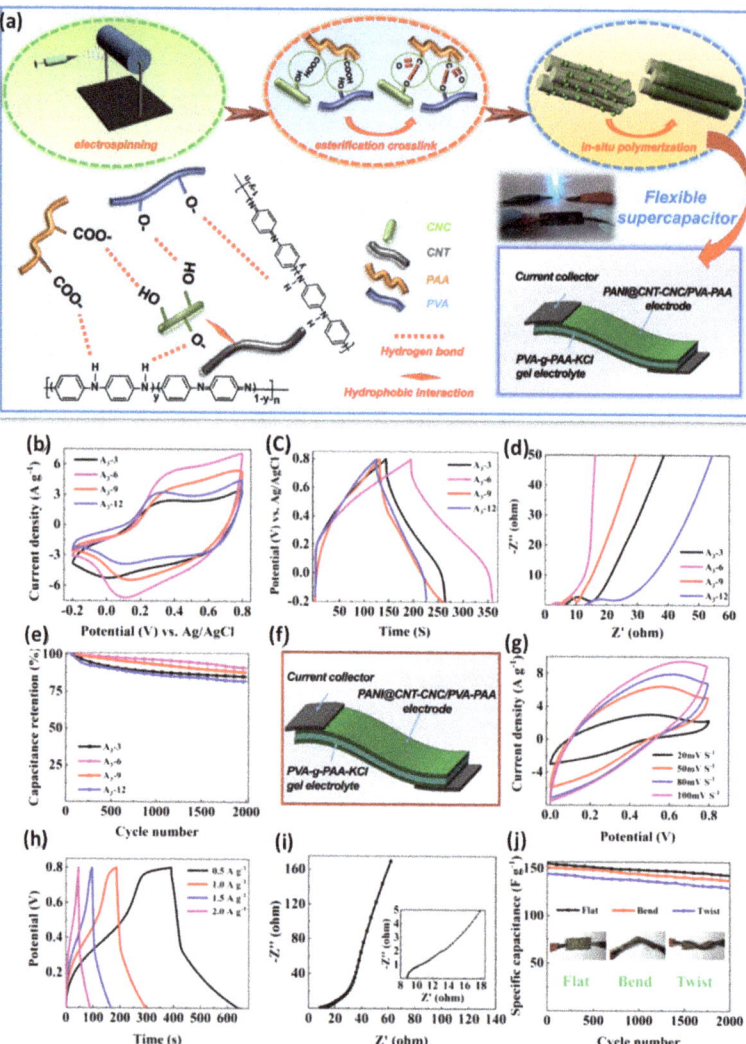

Figure 9. (a) The preparation PANI@ CNT−CNC/PVA−PAA; the electrochemical performance of the core-shell-structured PANI@CNT−CNC/PVA−PAA nanofibrous electrodes. (b) CV curves, (c) GCD curves, (d) EIS analysis and (e) cycling life of PANI@CNT−CNC/PVA−PAA. (f) The symmetrical flexible supercapacitor. (g) CV curves, (h) GCD curves, (i) Nyquist plots and (j) cycling life of the symmetrical PANI@ CNT−CNC/PVA−PAA flexible supercapacitor. Adapted with permission from [110]. Copyright 2019 American Chemical Society (ACS).

A lightweight CNC–MWCNT–PPy electrode has been proven to be an efficient supercapacitor. CNCs are modified with NaIO$_4$ and adipic acid dihydrazide to make aldehyde-modified CNCs (CNC–CHO) and hydrazide-modified CNCs (CNC–NHNH$_2$), respectively. Sol-gel cross-linking enables the preparation of CNC aerogels from the CNC–CHO and CNC–NHNH$_2$ with enhanced mechanical properties due to the chemical bonding between the amine and aldehyde groups. To facilitate the dispersion of MWCNTs on the CNC surface, the MWCNTs are suspended in a mixture consisting of sodium dodecyl sulfate (SDS) and taurocholic acid sodium salt [111]. The incorporation of these surfactants with MWC-

NTs enhances the reinforcing ability and high conductivity. The SDS and TCH (taurocholic acid sodium salt) surfactants help the dispersion of MWCNTs on the surface-functionalized CNC aerogels. The PPy conductive polymer is then deposited on the MWCNT/CNC aerogels by in situ polymerization of pyrrole using APS as an initiator. The lightweight CNC–MWCNT–PPy electrodes exhibited high super capacitance and a flexible nature along with a compressible nature. The dispersion and deposition of MWCNTs and PPy on the CNC surface result in 2.1 Fcm^{-2}.

An effective one-step hydrothermal preparation of CNC-MnO_2 nanocomposites as solid-state electrochemical supercapacitors using commercial microcrystalline cellulose and $KmnO_4$ has been reported [112]. A solid-state supercapacitor is assembled with sandwich typed CNC–MnO_2 electrodes and the PVA/KOH electrolyte (Figure 10). The fabricated electrode exhibits a specific capacitance of 306.3 Fg^{-1}, with good cyclic stability. The enhanced capacitance of the CNC–MnO_2 reflects the large surface area of the MnO_2 and CNCs with abundant hydrophilic hydroxyl groups. The CNC–MnO_2 composite enhances the electrolyte uptake and transport through the pores and voids of the CNCs and MnO_2, providing an excess ion diffusion pathway for the charge-discharge process. The energy density of CNC–MnO_2 (42.59 Wh kg^{-1}) is higher than that of pristine MnO_2. The CNC–MnO_2 is combined with the textile fabrics and conductive PPy to fabricate flexible electrodes for smart supercapacitor applications [112].

The synthesis of ternary PEDOT/CNC/MnO_2 can be performed by one-step electropolymerization [113]. MnO_2 is considered a good electrode material, but its conductivity is very modest, whereas PEDOT has mechanical flexibility and good electrical conductivity. However, the PEDOT/MnO_2 composite still exhibits poor charge-discharge stability due to the effect of the continuous influx and outflow of electrolyte ions. To overcome this setback, CNCs are incorporated with PEDOT/MnO_2 to improve super capacitance properties. A ternary PEDOT/CNC/MnO_2 hybrid electrode exhibits a specific capacitance of 144.69 Fg^{-1} at 25 mVs^{-1} compared to PEDOT/CNC (63.57 Fg^{-1}). The specific power (P_s) and specific energy (E_s) of PEDOT/CNC/MnO_2 are 494.9 Wkg^{-1} and 10.3 Wg^{-1}h, respectively. The PEDOT/MnO_2 builds up the redox-active surface area, resulting in enhanced electrochemical capacitance [113].

13. Carbonized CNC Electrodes

High surface area porous carbon has often been synthesized from renewable and non-renewable resources such as waste-derived biomass, petroleum and coal through chemical activation and templating methods [114,115]. The fabrication of porous carbon from non-renewable resources is non-sustainable and very expensive. Highly efficient 3D porous carbon can be prepared from CNC using carbonization. During the carbonization process, non-conductive CNC is converted into conductive porous carbon materials. It is important to develop high-performance electrochemical supercapacitor materials without the influence of conductive polymers [116,117]. CNC-based carbon materials are prepared from CNCs with silica precursors by pyrolysis and etching. The fabricated carbon with a high surface area of 1400 m^2/g possesses a C_s of 170 Fg^{-1} at 230 mAg^{-1} under the acidic electrolyte. The silica moiety triggers the generation of mesoporous carbon with a high surface area. The CNC-silica composite exhibits excellent super capacitance and good electrochemical retention stability [118].

A high surface area freestanding carbon film from CNCs and CNFs is created by the atomic layer deposition method [119]. Alumina is deposited on the CNC/CNF film at a low temperature and is carbonized at 900 °C for 2 h under the inner atmosphere. The Al_2O_3 layer helps to prevent CNC/CNF aggregation and preserve the fine structures of porous carbon during the carbonization process. The CNC/CNF film shows outstanding electrical conductivity and a surface area of 1200 m^{2}g^{-1} compared to the traditionally prepared activated carbons. The specific capacitance of the carbonized CNF and CNC/CNF is 50 Fg^{-1} and 152 Fg^{-1}, respectively. The interaction between the CNCs and CNFs augments the ion-transport efficiency, resulting in the higher performance of super capacitance activity in the CNC/CNF film. However, the specific capacitance is only 152 F g^{-1}, as the carbonized CNC/CNF film is interconnected with strong chemical interaction without any active materials with heteroatoms or functional groups [119].

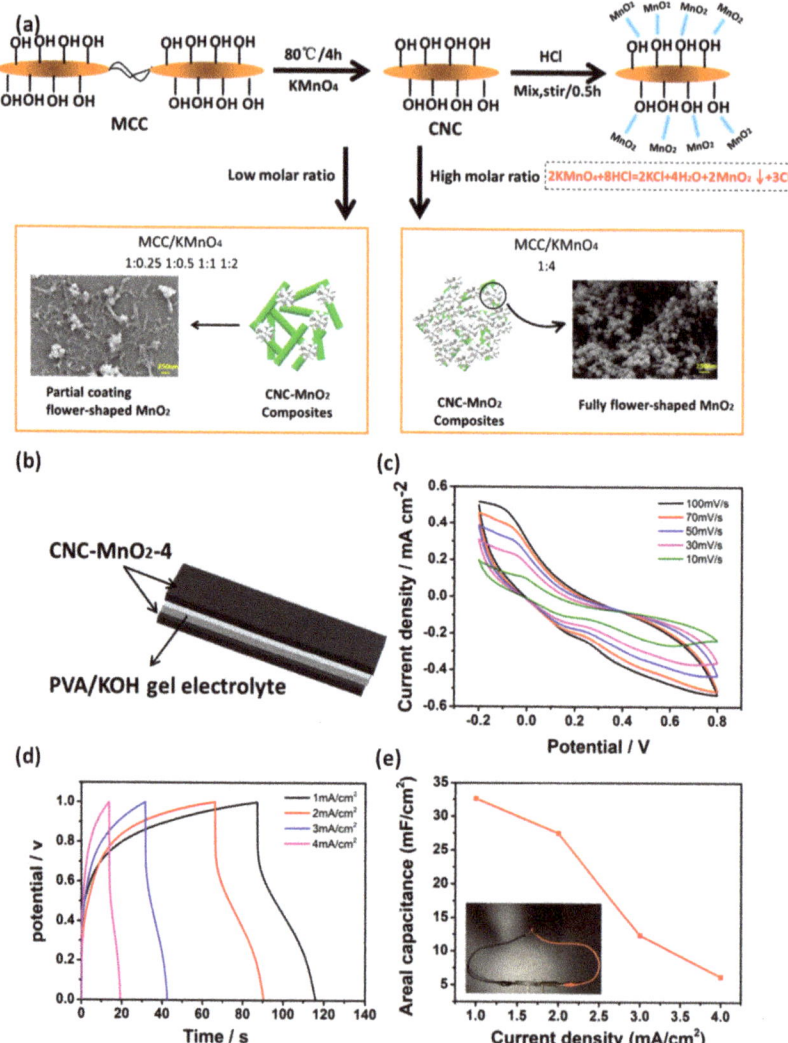

Figure 10. (a,b) Schematic synthetic route of the solid-state supercapacitor using the CNC–MnO$_2$ electrode. (c–e) CV curves, GCD curves and the areal capacitance of CNC–MnO$_2$. Adapted with permission from [112]. Copyright American Chemical Society (ACS).

The incorporation of heteroatoms (nitrogen) can be an effective approach to generating more active sites on the carbonized CNCs, which leads to the altering of the microstructure and properties of the CNCs [120]. An N-MFCNC nanocomposite can be prepared from melamine-formaldehyde functionalized CNC by one-step pyrolysis. The N-MFCNC electrode displayed a specific capacitance of 328 F g^{-1} at 10 mVs^{-1} under acidic electrolyte, with cyclic stability of 95.4%. The existence of numerous nitrogen-active sites enhances the super capacitance activity of carbonized N-MFCNC [120].

A brief discussion is extended to the preparation of an N-doped highly porous carbon material [121] with a rod-like arrangement by the self-templated strategy from the in situ growth of ZIF-8 and CNCs followed by pyrolysis (Figure 11). The CNC structure is covered by micro and mesoporous ZIF-8-derived hollow carbon moiety without disturbing the chiral nematic phases of the CNCs. In addition, ZIF-8 helps to form a hierarchical porous structure and helically rod-like nanoparticles, enhancing the active sites and shortening the mass diffusion distance. Interestingly, CNCs are also involved in augmenting the surface area and pore size of the electrode materials. The CNC-ZIF-8-derived porous carbon exhibits 172 Fg^{-1} at 0.1 Ag^{-1}, with long cyclic stability and a retention of 95% after 5000 cycles. The energy density (E_s) and power density (P_s) of N-doped carbon are 23.75 Whkg^{-1} and 50 W/kg^{-1}, respectively. Such superior features of the N-doped carbon can be attributed to the conductive carbon rod-like helical structure and its porous nature, which facilitate fast electron migration in the electrode. The presence of nitrogen atoms increases the active sites and wettable nature of the electrode. In addition, the existence of macro, meso and micropores offers the spaces for storing electrolytes and rapid ion transportation and enhances the EDLC capacitance [121].

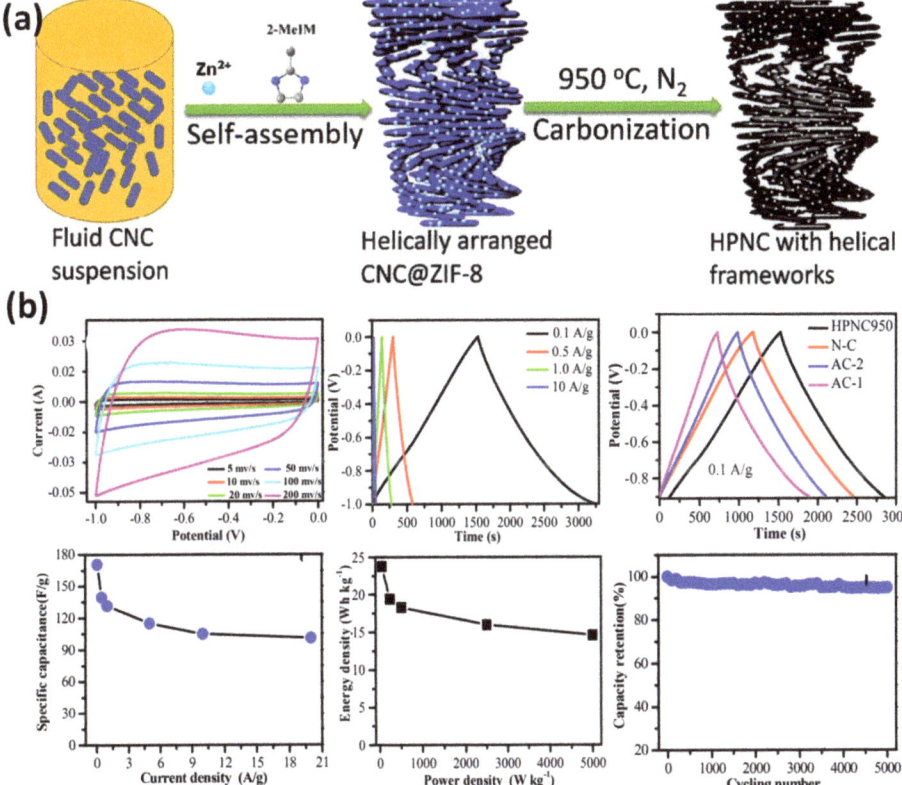

Figure 11. (**a**) Facile synthetic route of hierarchically porous N-doping carbon (HPNC), (**b**) Electrochemical characteristics of an HPNC composite in an aqueous KOH electrolyte. Adapted with permission from [121]. Copyright 2018 American Chemical Society (ACS).

Of note is also the synthesis of N-doped porous carbon from a CNC suspension and water-soluble urea by carbonization for supercapacitor applications (Figure 12). Typically, the CNC surface is functionalized by acid hydrolysis to generate surface negative charges with sulfate groups. The synthesized N-doped carbon displays a surface area of 366.5 m^2/g, a porous microstructure and improved electrochemical properties, with a capacitance retention of 91.2% after 1000 cycles. The N-doped carbon composite shows an electrochemical super capacitance of 570.6 Fg^{-1} at 1 Ag under the basic electrolyte. The result of the higher super capacitance and double layer capacitance characteristics of the N-doped carbon is due to the presence of nitrogen heteroatom in the composite. A symmetrical supercapacitor is fabricated with N-doped carbon materials for investigating practical super capacitance performance. The N-doped carbon electrode exhibits a specific capacity of 119 Fg^{-1}, with an outstanding stability retention of 99.8% after 5000 cycles [122].

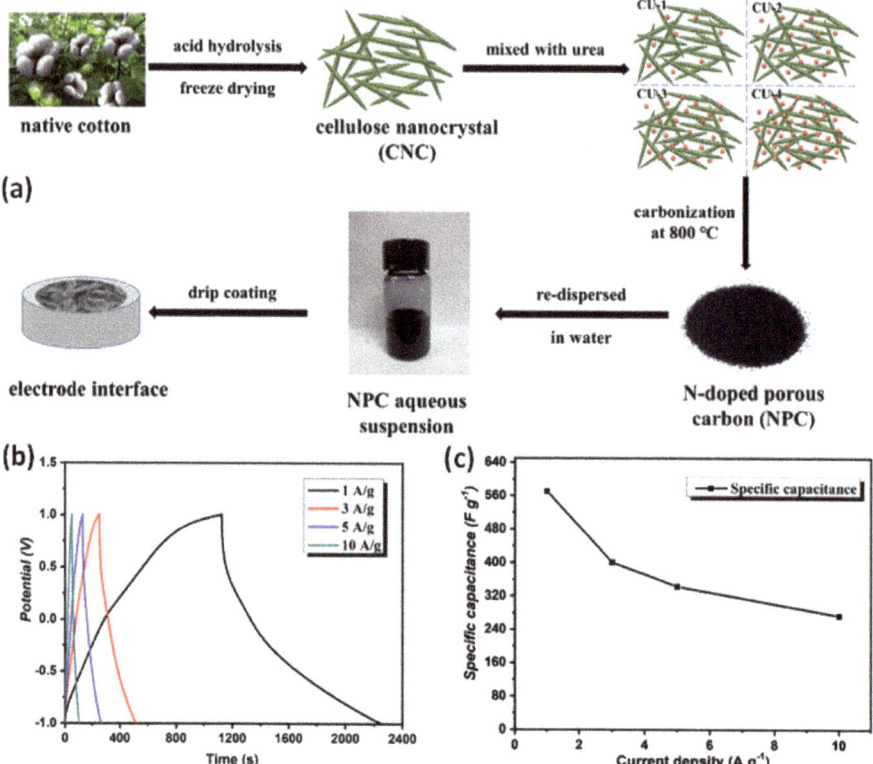

Figure 12. (a) N−doped cellulose nanocrystals (N−CNCs) are prepared from cotton and urea. (b,c) Electrochemical GCD curve and specific capacitance of N−CNCs under different current densities. Adapted with permission from [122]. Copyright 2020 Elsevier.

CNCs with high crystallinity due to the specific chemical composition and intermolecular interactions are considered advanced functional materials for various energy applications including supercapacitors and Li-ion batteries. The conversion of CNCs to porous carbon materials further enhances the supercapacitance properties. Renewable and low-cost CNCs are holding an incredible significance and are ideal candidates with an enormous potential for the development of supercapacitor electrodes. A comparison of the capacitance, power density and energy density of various CNC-based electrodes is depicted in Table 2.

Table 2. Comparison of CNC-based supercapacitor electrodes.

Composite	Fabrication	Electrolyte	Capacitance (F/g)	Power Density (Wh/kg)	Energy Density (Wkg^{-1})	Ref
PEDOT/NCC	Electrochemical deposition	1.0 M KCl	117.02	11.44	99.85	[101]
PPY/CNC	TEMPO/polymerization	0.5 M KCl	248	-	-	[102]
PPY/CNC	Electrochemical deposition	0.1 M KCl	256	-	-	[103]
PANI/CNC	Electrochemical deposition	1.0 M HCl	488	-	-	[104]
PEDOT/CNC	Electrochemical deposition	1.0 M HCl	69	-	-	[104]
PPY/PVP/CNC	Polymerization	0.5 M KCl	338.6	-	-	[105]
Biowaste-derived carbon	Carbonization	Al-CNC hydrogel	804	425	18.2	[109]
PANI@ CNT−CNC/PVA−PAA	Electrospinning/Polymerization	1 M H$_2$SO$_4$	164.6	13.8	200.3	[110]
CNC-MWCNT-PPY	Polymerization	0.5 M Na$_2$SO$_4$	2.1 F cm^{-2}	-	-	[111]
CNC-MnO$_2$	Thermal synthesis	PVA/KOH	306.3	42.59	-	[112]
PEDOT/CNC /MnO$_2$	Electro polymerization	1 M KCl	144.69	10.3	494.9	[113]
Porous carbon from CNC	Silica etching	1 M H$_2$SO$_4$	170	-	-	[118]
CNC/CNF	ALD	2 M KOH	152	-	-	[119]
N-CNC	Pyrolysis	1 M H$_2$SO$_4$	352	48.8	39.85	[120]
N-CNC	Pyrolysis	6 M KOH	172	23.75	70	[121]
N-CNC	pyrolysis	6 M KOH	570.6	23.75	50	[122]

14. Conclusions and Outlook

In recent years, the continuous research interest regarding the fabrication of biodegradable and renewable materials owes to the necessity of a sustainable and productive environment as an alternative to fossil fuels. Renewable and innocuous cellulose nanocrystals have been widely utilized for advanced functional and environmentally friendly materials in supercapacitors, batteries and fuel cells due to the merits of their chemical composition, their strong intermolecular interaction and their high crystallinity. The focus of our review is on explaining the importance of CNCs as a sustainable natural resource and their applications as electrochemical supercapacitors. This review discusses the extraction, surface functionalization and morphological characteristics of CNCs to ensure the ability of CNCs to function in the energy storage field. The pretreatment step is always important and needed for the removal of unwanted chemical components such as lignin, pectin and other impurities. The surface functionalization process improves the dispersion ability of CNCs within both hydrophilic and hydrophobic polymer matrices. Promising CNCs are modified with various conductive polymers, conductive carbons and inorganic hybrids for developing new biodegradable green nanomaterials for the most significant energy systems. In addition, the key features for fabricating advanced electrode materials by converting non-conductive CNCs to conductive CNCs have been discussed. The optimization of the heteroatom doping strategy with sustainable and low-cost CNCs will achieve the market commercialization of conductive electronic and storage devices. Undoubtedly, some challenges are existing, and significant research challenges are required to solve the following issues for a better understanding of CNC characteristics and sustainable energy applications.

i. The need to fabricate hierarchical CNCs with an existing millimeter thickness and multi-scale porous microstructure. Hierarchical CNCs would become a great potential candidate for electrochemical systems.

ii. The control and optimization of the surface functionalization on the CNC will be crucial to making novel CNCs for both hydrophobic and hydrophilic matrices.

iii. Biosynthetic alternatives allow for the development of advanced high-performance CNC materials. The incorporation of biological molecules on the CNC surface by bacterial cellulose can make a better impact on their superior properties.
iv. There is still a large research gap in improving the performance of super capacitance, energy and power densities.
v. The construction of metal-free porous and heteroatom-doped CNC using a low cost and sustainable approach from abundant biomass is still needed.

This review fosters research endeavors to develop advanced manufacturing processes and characteristics of CNCs that reflect the ultimate value of their potential as supercapacitors. It also promotes the sustainable utilization of renewable biomass and biopolymer resources. Besides CNCs, carboxylated nanocrystalline chitin can be prepared by the hydrolysis and oxidation of chitin by ammonium persulfate [123]. Chitin is another abundant fibrous macromolecule that forms the major constituent in the exoskeleton of arthropods and fungal cell walls. Like CNCs, supercapacitors can be prepared from carboxylated nanocrystalline chitin, a subject of future endeavors. Like CNCs [124,125], the cytotoxicity of chitin nanocrystals [123] is very minimal, and this feature is a prerequisite for the production and application of nanomaterials en masse.

Conducting PPY has been used with CNCs and other polymers toward the development of supercapacitors. However, this conducting polymer can be decorated with zinc vanadium oxide and used as a supercapacitor electrode material [126]. Other competitive techniques include using graphitic carbon nitride-doped copper–manganese alloy as an electrode material for energy storage [127]. Of interest is the fabrication of a lightweight, flexible self-charging power pack by the prudent integration of two paper-based high-performance triboelectric nanogenerators [128]. A novel two-dimensional donor-acceptor conjugated copolymer has been used as single-junction polymer solar cells with over 10% efficiency [129]. A review of the reliability of supercapacitors in energy storage applications is available elsewhere [130,131], and the discussion topics include the failure mechanisms, lifetime modeling and reliability-oriented design of SCs.

Author Contributions: Conceptualization, M.M.; methodology, A.S.; writing—original draft preparation, A.D.; writing—review and editing, J.H.T.L.; supervision, A.G. All authors have read and agreed to the published version of the manuscript.

Funding: This research received no external funding.

Institutional Review Board Statement: Not applicable.

Informed Consent Statement: Not applicable.

Data Availability Statement: No new data were created or analyzed in this study. Data sharing is not applicable to this article.

Conflicts of Interest: The authors declare no conflict of interest.

References

1. Philippot, M.; Alvarez, G.; Ayerbe, E.; Van Mierlo, J.; Messagie, M. Eco-Efficiency of a Lithium-Ion Battery for Electric Vehicles: Influence of Manufacturing Country and Commodity Prices on GHG Emissions and Costs. *Batteries* **2019**, *5*, 23. [CrossRef]
2. Dwivedi, P.; Mishra, P.K.; Mondal, M.K.; Srivastava, N. Non-Biodegradable Polymeric Waste Pyrolysis for Energy Recovery. *Heliyon* **2019**, *5*, e02198. [CrossRef] [PubMed]
3. Dilshad, E.; Waheed, H.; Ali, U.; Amin, A.; Ahmed, I. General Structure and Classification of Bioplastics and Biodegradable Plastics. In *Bioplastics Sustainable Development*; Kuddus, M., Roohi, Eds.; Springer Nature: Singapore, 2021; pp. 61–82. [CrossRef]
4. Guo, A.F.; Li, J.F.; Li, F.Y.; Xu, J.; Zhang, C.W.; Chen, S. Compression Behavior of Biodegradable Thermoplastic Plasticizer-Containing Composites. *Strength Mater.* **2019**, *51*, 18–25. [CrossRef]
5. Klemm, D.; Kramer, F.; Moritz, S.; Lindström, T.; Ankerfors, M.; Gray, D.; Dorris, A. Nanocelluloses: A New Family of Nature-Based Materials. *Angew. Chemie Int. Ed.* **2011**, *50*, 5438–5466. [CrossRef] [PubMed]
6. Bhaladhare, S.; Das, D. Cellulose: A Fascinating Biopolymer for Hydrogel Synthesis. *J. Mater. Chem. B* **2022**, *10*, 1923–1945. [CrossRef]

7. Karimi, K.; Taherzadeh, M.J. A Critical Review of Analytical Methods in Pretreatment of Lignocelluloses: Composition, Imaging, and Crystallinity. *Bioresour. Technol.* **2016**, *200*, 1008–1018. [CrossRef]
8. Mishra, R.K.; Sabu, A.; Tiwari, S.K. Materials Chemistry and the Futurist Eco-Friendly Applications of Nanocellulose: Status and Prospect. *J. Saudi Chem. Soc.* **2018**, *22*, 949–978. [CrossRef]
9. Trache, D.; Tarchoun, A.F.; Derradji, M.; Hamidon, T.S.; Masruchin, N.; Brosse, N.; Hussin, M.H. Nanocellulose: From Fundamentals to Advanced Applications. *Front. Chem.* **2020**, *8*, 392. [CrossRef]
10. Pui, K.; Shak, Y.; Pang, Y.L.; Keat, S.; Review, M.; Rahman, A.; Long, J.S.; Long, S.; Kajang, D.; Ehsan, M.; et al. Nanocellulose: Recent Advances and Its Prospects in Environmental Remediation. *Beilstein J. Nanotechnol.* 9232 **2018**, *9*, 2479–2498. [CrossRef]
11. Dhali, K.; Ghasemlou, M.; Daver, F.; Cass, P.; Adhikari, B. A Review of Nanocellulose as a New Material towards Environmental Sustainability. *Sci. Total Environ.* **2021**, *775*, 145871. [CrossRef]
12. Jaffar, S.S.; Saallah, S.; Misson, M.; Siddiquee, S.; Roslan, J.; Saalah, S.; Lenggoro, W. Recent Development and Environmental Applications of Nanocellulose-Based Membranes. *Membranes* **2022**, *12*, 287. [CrossRef]
13. Lin, N.; Dufresne, A. Nanocellulose in Biomedicine: Current Status and Future Prospect. *Eur. Polym. J.* **2014**, *59*, 302–325. [CrossRef]
14. Leung, A.C.W.; Hrapovic, S.; Lam, E.; Liu, Y.; Male, K.B.; Mahmoud, K.A.; Luong, J.H.T. Characteristics and Properties of Carboxylated Cellulose Nanocrystals Prepared from a Novel One-step Procedure. *Small* **2011**, *7*, 302–305. [CrossRef]
15. Kargarzadeh, H.; Mariano, M.; Huang, J.; Lin, N.; Ahmad, I.; Dufresne, A.; Thomas, S. Recent Developments on Nanocellulose Reinforced Polymer Nanocomposites: A Review. *Polymer* **2017**, *132*, 368–393. [CrossRef]
16. Salas, C.; Nypelö, T.; Rodriguez-Abreu, C.; Carrillo, C.; Rojas, O.J. Nanocellulose Properties and Applications in Colloids and Interfaces. *Curr. Opin. Colloid Interface Sci.* **2014**, *19*, 383–396. [CrossRef]
17. Mondal, S. Preparation, Properties and Applications of Nanocellulosic Materials. *Carbohydr. Polym.* **2017**, *163*, 301–316. [CrossRef]
18. Abitbol, T.; Rivkin, A.; Cao, Y.; Nevo, Y.; Abraham, E.; Ben-Shalom, T.; Lapidot, S.; Shoseyov, O. Nanocellulose, a Tiny Fiber with Huge Applications. *Curr. Opin. Biotechnol.* **2016**, *39*, 76–88. [CrossRef]
19. George, J.; Sabapathi, S.N. Cellulose Nanocrystals: Synthesis, Functional Properties, and Applications. *Nanotechnol. Sci. Appl.* **2015**, *8*, 45. [CrossRef]
20. Tavakolian, M.; Jafari, S.M.; van de Ven, T.G.M. A Review on Surface-Functionalized Cellulosic Nanostructures as Biocompatible Antibacterial Materials. *Nano-Micro Lett.* **2020**, *12*, 73. [CrossRef]
21. Lam, E.; Male, K.B.; Chong, J.H.; Leung, A.C.W.; Luong, J.H.T. Applications of Functionalized and Nanoparticle-Modified Nanocrystalline Cellulose. *Trends Biotech.* **2012**, *30*, 283–290. [CrossRef]
22. He, X.; Male, K.B.; Nesterenko, P.N.; Brabazon, D.; Paull, B.; Luong, J.H.T. Adsorption and Desorption of Methylene Blue on Porous Carbon Monoliths and Nanocrystalline Cellulose. *ACS Appl. Mater. Interf.* **2013**, *5*, 8796–8804. [CrossRef] [PubMed]
23. Mahmoud, K.A.; Male, K.B.; Hrapovic, S.; Luong, J.H.T. Cellulose Nanocrystal/Gold Nanoparticle Composite as a Matrix for Enzyme Immobilization. *ACS Appl. Mater. Interf.* **2009**, *1*, 1383–1386. [CrossRef] [PubMed]
24. Tafete, G.A.; Abera, M.K.; Thothadri, G. Review on Nanocellulose-Based Materials for Supercapacitors Applications. *J. Energy Storage* **2022**, *48*, 103938. [CrossRef]
25. Afif, A.; Rahman, S.M.; Tasfiah Azad, A.; Zaini, J.; Islam, M.A.; Azad, A.K. Advanced Materials and Technologies for Hybrid Supercapacitors for Energy Storage—A Review. *J. Energy Storage* **2019**, *25*, 100852. [CrossRef]
26. Trahey, L.; Brushett, F.R.; Balsara, N.P.; Ceder, G.; Cheng, L.; Chiang, Y.M.; Hahn, N.T.; Ingram, B.J.; Minteer, S.D.; Moore, J.S.; et al. Energy Storage Emerging: A Perspective from the Joint Center for Energy Storage Research. *Proc. Natl. Acad. Sci. USA* **2020**, *117*, 12550–12557. [CrossRef] [PubMed]
27. Bashir, T.; Irfan, R.M. A Review of the Energy Storage Aspects of Chemical Elements for Lithium-Ion Based Batteries. *Energy Mater.* **2021**, *1*, 100019. [CrossRef]
28. Kebede, A.A.; Kalogiannis, T.; Van Mierlo, J.; Berecibar, M. A Comprehensive Review of Stationary Energy Storage Devices for Large Scale Renewable Energy Sources Grid Integration. *Renew. Sustain. Energy Rev.* **2022**, *159*, 112213. [CrossRef]
29. Sam, D.K.; Sam, E.K.; Durairaj, A.; Lv, X.; Zhou, Z.; Liu, J. Synthesis of Biomass-Based Carbon Aerogels in Energy and Sustainability. *Carbohydr. Res.* **2020**, *491*, 107986. [CrossRef]
30. Durairaj, A.; Sakthivel, T.; Ramanathan, S.; Obadiah, A.; Vasanthkumar, S. Conversion of Laboratory Paper Waste into Useful Activated Carbon: A Potential Supercapacitor Material and a Good Adsorbent for Organic Pollutant and Heavy Metals. *Cellulose* **2019**, *26*, 3313–3324. [CrossRef]
31. Durairaj, A.; Liu, J.; Lv, X.; Vasanthkumar, S.; Sakthivel, T. Facile Synthesis of Waste-Derived Carbon/MoS_2 Composite for Energy Storage and Water Purification Applications. *Biomass Convers. Biorefinery* **2021**, *11*, 1–12. [CrossRef]
32. Li, B.; Dai, F.; Xiao, Q.; Yang, L.; Shen, J.; Zhang, C.; Cai, M. Nitrogen-Doped Activated Carbon for a High Energy Hybrid Supercapacitor. *Energy Environ. Sci.* **2016**, *9*, 102–106. [CrossRef]
33. Zhang, M.; Fan, H.; Zhao, N.; Peng, H.; Ren, X.; Wang, W.; Li, H.; Chen, G.; Zhu, Y.; Jiang, X.; et al. 3D Hierarchical $CoWO_4/Co_3O_4$ Nanowire Arrays for Asymmetric Supercapacitors with High Energy Density. *Chem. Eng. J.* **2018**, *347*, 291–300. [CrossRef]
34. Wang, Y.; Zhong, W.H. Development of Electrolytes towards Achieving Safe and High-Performance Energy-Storage Devices: A Review. *ChemElectroChem* **2015**, *2*, 22–36. [CrossRef]
35. Kim, B.K.; Sy, S.; Yu, A.; Zhang, J. Electrochemical Supercapacitors for Energy Storage and Conversion. In *Handbook of Clean Energy Systems*; John Wiley & Sons, Ltd.: Chichester, UK, 2015; pp. 1–25.

36. Tang, Z.; Tang, C.H.; Gong, H. A High Energy Density Asymmetric Supercapacitor from Nano-Architectured Ni(OH)$_2$/Carbon Nanotube Electrodes. *Adv. Funct. Mater.* **2012**, *22*, 1272–1278. [CrossRef]
37. Han, S.; Wu, D.; Li, S.; Zhang, F.; Feng, X. Porous Graphene Materials for Advanced Electrochemical Energy Storage and Conversion Devices. *Adv. Mater.* **2014**, *26*, 849–864. [CrossRef]
38. Wesley, R.J.; Durairaj, A.; Ramanathan, S.; Obadiah, A.; Justinabraham, R.; Lv, X.; Vasanthkumar, S. Potato Peels Biochar Composite with Copper Phthalocyanine for Energy Storage Application. *Diam. Relat. Mater.* **2021**, *115*, 108360. [CrossRef]
39. Zhang, Q.Z.; Zhang, D.; Miao, Z.C.; Zhang, X.L.; Chou, S.L. Research Progress in MnO$_2$–Carbon Based Supercapacitor Electrode Materials. *Small* **2018**, *14*, 1702883. [CrossRef]
40. Hussain, S.Z.; Ihrar, M.; Hussain, S.B.; Oh, W.C.; Ullah, K. A Review on Graphene Based Transition Metal Oxide Composites and Its Application towards Supercapacitor Electrodes. *SN Appl. Sci.* **2020**, *2*, 764. [CrossRef]
41. Zhang, L.; Zhao, X.S. Carbon-Based Materials as Supercapacitor Electrodes. *Chem. Soc. Rev.* **2009**, *38*, 2520–2531. [CrossRef]
42. Zhou, Y.; Qi, H.; Yang, J.; Bo, Z.; Huang, F.; Islam, M.S.; Lu, X.; Dai, L.; Amal, R.; Wang, C.H.; et al. Two-Birds-One-Stone: Multifunctional Supercapacitors beyond Traditional Energy Storage. *Energy Environ. Sci.* **2021**, *14*, 1854–1896. [CrossRef]
43. Hou, J.; Shao, Y.; Ellis, M.W.; Moore, R.B.; Yi, B. Graphene-Based Electrochemical Energy Conversion and Storage: Fuel Cells, Supercapacitors and Lithium Ion Batteries. *Phys. Chem. Chem. Phys.* **2011**, *13*, 15384–15402. [CrossRef] [PubMed]
44. Chen, G.Z. Supercapacitor and Supercapattery as Emerging Electrochemical Energy Stores. *Int. Mater. Rev.* **2016**, *62*, 173–202. [CrossRef]
45. Jiang, H.; Lee, P.S.; Li, C. 3D Carbon Based Nanostructures for Advanced Supercapacitors. *Energy Environ. Sci.* **2012**, *6*, 41–53. [CrossRef]
46. Wu, Z.Y.; Liang, H.W.; Chen, L.F.; Hu, B.C.; Yu, S.H. Bacterial Cellulose: A Robust Platform for Design of Three Dimensional Carbon-Based Functional Nanomaterials. *Acc. Chem. Res.* **2016**, *49*, 96–105. [CrossRef]
47. Wang, X.; Yao, C.; Wang, F.; Li, Z. Cellulose-Based Nanomaterials for Energy Applications. *Small* **2017**, *13*, 1702240. [CrossRef]
48. Cheng, H.; Lijie, L.; Wang, B.; Feng, X.; Mao, Z.; Vancso, G.J.; Sui, X. Multifaceted Applications of Cellulosic Porous Materials in Environment, Energy, and Health. *Prog. Polym. Sci.* **2020**, *106*, 101253. [CrossRef]
49. Du, X.; Zhang, Z.; Liu, W.; Deng, Y. Nanocellulose-Based Conductive Materials and Their Emerging Applications in Energy Devices—A Review. *Nano Energy* **2017**, *35*, 299–320. [CrossRef]
50. Habibi, Y.; Lucia, L.A.; Rojas, O.J. Cellulose Nanocrystals: Chemistry, Self-Assembly, and Applications. *Chem. Rev.* **2010**, *110*, 3479–3500. [CrossRef]
51. Hua, K.; Carlsson, D.O.; Ålander, E.; Lindström, T.; Strømme, M.; Mihranyan, A.; Ferraz, N. Translational Study between Structure and Biological Response of Nanocellulose from Wood and Green Algae. *RSC Adv.* **2013**, *4*, 2892–2903. [CrossRef]
52. Mokhena, T.C.; John, M.J. Cellulose Nanomaterials: New Generation Materials for Solving Global Issues. *Cellulose* **2019**, *27*, 1149–1194. [CrossRef]
53. Farooq, A.; Patoary, M.K.; Zhang, M.; Mussana, H.; Li, M.; Naeem, M.A.; Mushtaq, M.; Farooq, A.; Liu, L. Cellulose from Sources to Nanocellulose and an Overview of Synthesis and Properties of Nanocellulose/Zinc Oxide Nanocomposite Materials. *Int. J. Biol. Macromol.* **2020**, *154*, 1050–1073. [CrossRef] [PubMed]
54. Klemm, D.; Heublein, B.; Fink, H.P.; Bohn, A. Cellulose: Fascinating Biopolymer and Sustainable Raw Material. *Angew. Chemie Int. Ed.* **2005**, *44*, 3358–3393. [CrossRef]
55. De France, K.J.; Hoare, T.; Cranston, E.D. Review of Hydrogels and Aerogels Containing Nanocellulose. *Chem. Mater.* **2017**, *29*, 4609–4631. [CrossRef]
56. Lam, E.; Leung, A.C.W.; Liu, Y.; Majid, E.; Hrapovic, S.; Male, K.B.; Luong, J.H.T. Green Strategy Guided by Raman Spectroscopy for the Synthesis of Ammonium Carboxylated Nanocrystalline Cellulose and the Recovery of Byproducts. *ACS Sustain. Chem. Eng.* **2013**, *1*, 278–283. [CrossRef]
57. Yang, X.; Cranston, E.D. Chemically Cross-Linked Cellulose Nanocrystal Aerogels with Shape Recovery and Superabsorbent Properties. *Chem. Mater.* **2014**, *26*, 6016–6025. [CrossRef]
58. Lam, E.; Hrapovic, S.; Majid, E.; Chong, J.H.; Luong, J.H.T. Catalysis using Gold Nanoparticles Decorated on Nanocrystalline Cellulose. *Nanoscale* **2012**, *4*, 997–1002. [CrossRef] [PubMed]
59. Wang, Z.; Tammela, P.; Strømme, M.; Nyholm, L. Cellulose-Based Supercapacitors: Material and Performance Considerations. *Adv. Energy Mater.* **2017**, *7*, 1700130. [CrossRef]
60. Blanco, A.; Monte, M.C.; Campano, C.; Balea, A.; Merayo, N.; Negro, C. Nanocellulose for Industrial Use: Cellulose Nanofibers (CNF), Cellulose Nanocrystals (CNC), and Bacterial Cellulose (BC). *Handb. Nanomater. Ind. Appl.* **2018**, 74–126. [CrossRef]
61. Shankaran, D.R. Cellulose Nanocrystals for Health Care Applications. *Appl. Nanomater.* **2018**, 415–459. [CrossRef]
62. Leung, A.C.W.; Lam, E.; Chong, J.H.; Hrapovic, S.; Luong, J.H.T. Reinforced Plastics and Aerogels by Nanocrystalline Cellulose. *J. Nanoparticle Res.* **2013**, *15*, 1636. [CrossRef]
63. Brinchi, L.; Cotana, F.; Fortunati, E.; Kenny, J.M. Production of Nanocrystalline Cellulose from Lignocellulosic Biomass: Technology and Applications. *Carbohydr. Polym.* **2013**, *94*, 154–169. [CrossRef] [PubMed]
64. Julie Chandra, C.S.; George, N.; Narayanankutty, S.K. Isolation and Characterization of Cellulose Nanofibrils from Arecanut Husk Fibre. *Carbohydr. Polym.* **2016**, *142*, 158–166. [CrossRef]
65. Nickerson, R.F.; Harree, J.A. Cellulose Intercrystalline Structure. *Ind. Eng. Chem.* **1947**, *39*, 1507–1512. [CrossRef]

66. Lamaming, J.; Hashim, R.; Leh, C.P.; Sulaiman, O. Properties of Cellulose Nanocrystals from Oil Palm Trunk Isolated by Total Chlorine Free Method. *Carbohydr. Polym.* **2017**, *156*, 409–416. [CrossRef]
67. Babicka, M.; Woźniak, M.; Dwiecki, K.; Borysiak, S.; Ratajczak, I. Preparation of Nanocellulose Using Ionic Liquids: 1-Propyl-3-Methylimidazolium Chloride and 1-Ethyl-3-Methylimidazolium Chloride. *Molecules* **2020**, *25*, 1544. [CrossRef]
68. Surov, O.V.; Voronova, M.I.; Rubleva, N.V.; Kuzmicheva, L.A. A Novel Effective Approach of Nanocrystalline Cellulose Production: Oxidation–Hydrolysis Strategy. *Cellulose* **2018**, *25*, 5035–5048. [CrossRef]
69. Cui, S.; Zhang, S.; Ge, S.; Xiong, L.; Sun, Q. Green Preparation and Characterization of Size-Controlled Nanocrystalline Cellulose via Ultrasonic-Assisted Enzymatic Hydrolysis. *Ind. Crops Prod.* **2016**, *83*, 346–352. [CrossRef]
70. Chowdhury, Z.Z.; Hamid, S.B.A. Preparation and Characterization of Nanocrystalline Cellulose Using Ultrasonication Combined with a Microwave-Assisted Pretreatment Process. *BioResources* **2016**, *11*, 3397–3415. [CrossRef]
71. Rohaizu, R.; Wanrosli, W.D. Sono-Assisted TEMPO Oxidation of Oil Palm Lignocellulosic Biomass for Isolation of Nanocrystalline Cellulose. *Ultrason. Sonochem.* **2017**, *34*, 631–639. [CrossRef]
72. Mendoza, D.J.; Browne, C.; Raghuwanshi, C.S.; Simon, G.P.; Garnier, G. One-shot TEMPO-Periodate Oxidation of Native Cellulose. *Carbohydr. Polym.* **2019**, *226*, 115292. [CrossRef]
73. Leung, C.W.; Luong, J.H.T.; Hrapovic, S.; Lam, E.; Liu, Y.; Male, K.B.; Mahmoud, K.; Rho, D. Cellulose Nanocrystals from Renewable Biomass. U.S. Patent No. 8,900,706 B2, 2 December 2014.
74. Seta, F.T.; An, X.; Liu, L.; Zhang, H.; Yang, J.; Zhang, W.; Nie, S.; Yao, S.; Cao, H.; Xu, Q.; et al. Preparation and Characterization of High Yield Cellulose Nanocrystals (CNC) Derived from Ball Mill Pretreatment and Maleic Acid Hydrolysis. *Carbohydr. Polym.* **2020**, *234*, 115942. [CrossRef] [PubMed]
75. Pang, Z.; Wang, P.; Dong, C. Ultrasonic Pretreatment of Cellulose in Ionic Liquid for Efficient Preparation of Cellulose Nanocrystals. *Cellulose* **2018**, *25*, 7053–7064. [CrossRef]
76. Singh, S.; Gaikwad, K.K.; Lee, Y.S. Antimicrobial and Antioxidant Properties of Polyvinyl Alcohol Bio Composite Films Containing Seaweed Extracted Cellulose Nano-Crystal and Basil Leaves Extract. *Int. J. Biol. Macromol.* **2018**, *107*, 1879–1887. [CrossRef]
77. Smirnov, M.A.; Sokolova, M.P.; Tolmachev, D.A.; Vorobiov, V.K.; Kasatkin, I.A.; Smirnov, N.N.; Klaving, A.V.; Bobrova, N.V.; Lukasheva, N.V.; Yakimansky, A.V. Green Method for Preparation of Cellulose Nanocrystals Using Deep Eutectic Solvent. *Cellulose* **2020**, *27*, 4305–4317. [CrossRef]
78. Calvino, C.; Macke, N.; Kato, R.; Rowan, S.J. Development, Processing and Applications of Bio-Sourced Cellulose Nanocrystal Composites. *Prog. Polym. Sci.* **2020**, *103*, 101221. [CrossRef]
79. Trache, D.; Thakur, V.K.; Boukherroub, R. Cellulose Nanocrystals/Graphene Hybrids—A Promising New Class of Materials for Advanced Applications. *Nanomaterials* **2020**, *10*, 1523. [CrossRef]
80. Eyley, S.; Thielemans, W. Surface Modification of Cellulose Nanocrystals. *Nanoscale* **2014**, *6*, 7764–7779. [CrossRef]
81. Nechyporchuk, O.; Belgacem, M.N.; Bras, J. Production of Cellulose Nanofibrils: A Review of Recent Advances. *Ind. Crops Prod.* **2016**, *93*, 2–25. [CrossRef]
82. Araki, J.; Wada, M.; Kuga, S. Steric Stabilization of a Cellulose Microcrystal Suspension by Poly(Ethylene Glycol) Grafting. *Langmuir* **2000**, *17*, 21–27. [CrossRef]
83. Habibi, Y. Key Advances in the Chemical Modification of Nanocelluloses. *Chem. Soc. Rev.* **2014**, *43*, 1519–1542. [CrossRef]
84. Cheng, M.; Qin, Z.; Liu, Y.; Qin, Y.; Li, T.; Chen, L.; Zhu, M. Efficient Extraction of Carboxylated Spherical Cellulose Nanocrystals with Narrow Distribution through Hydrolysis of Lyocell Fibers by Using Ammonium Persulfate as an Oxidant. *J. Mater. Chem. A* **2013**, *2*, 251–258. [CrossRef]
85. Daud, J.B.; Lee, K.-Y. Surface Modification of Nanocellulose. *Handb. Nanocellulose Cellul. Nanocomposites* **2017**, 101–122. [CrossRef]
86. Lee, J.H.; Park, S.H.; Kim, S.H. Surface Alkylation of Cellulose Nanocrystals to Enhance Their Compatibility with Polylactide. *Polymers* **2020**, *12*, 178. [CrossRef]
87. Çetin, N.S.; Tingaut, P.; Özmen, N.; Henry, N.; Harper, D.; Dadmun, M.; Sèbe, G. Acetylation of Cellulose Nanowhiskers with Vinyl Acetate under Moderate Conditions. *Macromol. Biosci.* **2009**, *9*, 997–1003. [CrossRef] [PubMed]
88. Mariano, M.; El Kissi, N.; Dufresne, A. Cellulose Nanomaterials: Size and Surface Influence on the Thermal and Rheological Behavior. *Polímeros* **2018**, *28*, 93–102. [CrossRef]
89. Thakur, V.K.; Thakur, M.K. Processing and Characterization of Natural Cellulose Fibers/Thermoset Polymer Composites. *Carbohydr. Polym.* **2014**, *109*, 102–117. [CrossRef]
90. Sharma, A.; Thakur, M.; Bhattacharya, M.; Mandal, T.; Goswami, S. Commercial Application of Cellulose Nano-Composites—A Review. *Biotechnol. Rep.* **2019**, *21*, e00316. [CrossRef]
91. Issa, A.A.; Luyt, A.S. Kinetics of Alkoxysilanes and Organoalkoxysilanes Polymerization: A Review. *Polymers* **2019**, *11*, 537. [CrossRef]
92. Zhang, X.; Wang, L.; Dong, S.; Zhang, X.; Wu, Q.; Zhao, L.; Shi, Y. Nanocellulose 3,5-Dimethylphenylcarbamate Derivative Coated Chiral Stationary Phase: Preparation and Enantioseparation Performance. *Chirality* **2016**, *28*, 376–381. [CrossRef]
93. Guo, J.; Du, W.; Gao, Y.; Cao, Y.; Yin, Y. Cellulose Nanocrystals as Water-in-Oil Pickering Emulsifiers via Intercalative Modification. *Colloids Surf. A Physicochem. Eng. Asp.* **2017**, *529*, 634–642. [CrossRef]
94. Girouard, N.M.; Xu, S.; Schueneman, G.T.; Shofner, M.L.; Meredith, J.C. Site-Selective Modification of Cellulose Nanocrystals with Isophorone Diisocyanate and Formation of Polyurethane-CNC Composites. *ACS Appl. Mater. Interfaces* **2016**, *8*, 1458–1467. [CrossRef] [PubMed]

95. Abushammala, H. On the Para/Ortho Reactivity of Isocyanate Groups during the Carbamation of Cellulose Nanocrystals Using 2,4-Toluene Diisocyanate. *Polymers* **2019**, *11*, 1164. [CrossRef] [PubMed]
96. Wu, X.; Lu, C.; Han, Y.; Zhou, Z.; Yuan, G.; Zhang, X. Cellulose Nanowhisker Modulated 3D Hierarchical Conductive Structure of Carbon Black/Natural Rubber Nanocomposites for Liquid and Strain Sensing Application. *Compos. Sci. Technol.* **2016**, *124*, 44–51. [CrossRef]
97. Zhuo, H.; Hu, Y.; Tong, X.; Zhong, L.; Peng, X.; Sun, R. Sustainable Hierarchical Porous Carbon Aerogel from Cellulose for High-Performance Supercapacitor and CO_2 Capture. *Ind. Crops Prod.* **2016**, *87*, 229–235. [CrossRef]
98. Bhanvase, B.A.; Sonawane, S.H. Ultrasound Assisted in Situ Emulsion Polymerization for Polymer Nanocomposite: A Review. *Chem. Eng. Process. Process Intensif.* **2014**, *85*, 86–107. [CrossRef]
99. Liang, H.W.; Guan, Q.F.; Zhu, Z.Z.; Song, L.T.; Yao, H.B.; Lei, X.; Yu, S.H. Highly Conductive and Stretchable Conductors Fabricated from Bacterial Cellulose. *NPG Asia Mater.* **2012**, *4*, e19. [CrossRef]
100. Luong, J.H.T.; Narayan, T.; Solanki, S.; Malhotra, B.D. Recent Advances of Conducting Polymers and Their Composites for ElectrochemicalBiosensing Applications. *J. Funct. Biomat.* **2021**, *11*, 71. [CrossRef]
101. Shi, Y.; Peng, L.; Ding, Y.; Zhao, Y.; Yu, G. Nanostructured Conductive Polymers for Advanced Energy Storage. *Chem. Soc. Rev.* **2015**, *44*, 6684–6696. [CrossRef]
102. Ravit, R.; Abdullah, J.; Ahmad, I.; Sulaiman, Y. Electrochemical Performance of Poly(3, 4-Ethylenedioxythipohene)/Nanocrystalline Cellulose (PEDOT/NCC) Film for Supercapacitor. *Carbohydr. Polym.* **2019**, *203*, 128–138. [CrossRef]
103. Wu, X.; Chabot, V.L.; Kim, B.K.; Yu, A.; Berry, R.M.; Tam, K.C. Cost-Effective and Scalable Chemical Synthesis of Conductive Cellulose Nanocrystals for High-Performance Supercapacitors. *Electrochim. Acta* **2014**, *138*, 139–147. [CrossRef]
104. Liew, S.Y.; Thielemans, W.; Walsh, D.A. Electrochemical Capacitance of Nanocomposite Polypyrrole/Cellulose Films. *J. Phys. Chem. C* **2010**, *114*, 17926–17933. [CrossRef]
105. Liew, S.Y.; Thielemans, W.; Walsh, D.A. Polyaniline- and Poly(Ethylenedioxythiophene)-Cellulose Nanocomposite Electrodes for Supercapacitors. *J. Solid State Electrochem.* **2014**, *18*, 3307–3315. [CrossRef]
106. Wu, X.; Tang, J.; Duan, Y.; Yu, A.; Berry, R.M.; Tam, K.C. Conductive Cellulose Nanocrystals with High Cycling Stability for Supercapacitor Applications. *J. Mater. Chem. A* **2014**, *2*, 19268–19274. [CrossRef]
107. Forner-Cuenca, A.; Brushett, F.R. Engineering Porous Electrodes for Next-Generation Redox Flow Batteries: Recent Progress and Opportunities. *Curr. Opin. Electrochem.* **2019**, *18*, 113–122. [CrossRef]
108. Valentini, L.; Cardinali, M.; Fortunati, E.; Torre, L.; Kenny, J.M. A Novel Method to Prepare Conductive Nanocrystalline Cellulose/Graphene Oxide Composite Films. *Mater. Lett.* **2013**, *105*, 4–7. [CrossRef]
109. Imai, M.; Akiyama, K.; Tanaka, T.; Sano, E. Highly Strong and Conductive Carbon Nanotube/Cellulose Composite Paper. *Compos. Sci. Technol.* **2010**, *70*, 1564–1570. [CrossRef]
110. Al Haj, Y.; Mousavihashemi, S.; Robertson, D.; Borghei, M.; Pääkkönen, T.; Rojas, O.J.; Kontturi, E.; Kallio, T.; Vapaavuori, J. Biowaste-Derived Electrode and Electrolyte Materials for Flexible Supercapacitors. *Chem. Eng. J.* **2022**, *435*, 135058. [CrossRef]
111. Han, J.; Wang, S.; Zhu, S.; Huang, C.; Yue, Y.; Mei, C.; Xu, X.; Xia, C. Electrospun Core-Shell Nanofibrous Membranes with Nanocellulose-Stabilized Carbon Nanotubes for Use as High-Performance Flexible Supercapacitor Electrodes with Enhanced Water Resistance, Thermal Stability, and Mechanical Toughness. *ACS Appl. Mater. Interfaces* **2019**, *11*, 44624–44635. [CrossRef]
112. Shi, K.; Yang, X.; Cranston, E.D.; Zhitomirsky, I. Efficient Lightweight Supercapacitor with Compression Stability. *Adv. Funct. Mater.* **2016**, *26*, 6437–6445. [CrossRef]
113. Chen, L.M.; Yu, H.Y.; Wang, D.C.; Yang, T.; Yao, J.M.; Tam, K.C. Simple Synthesis of Flower-like Manganese Dioxide Nanostructures on Cellulose Nanocrystals for High-Performance Supercapacitors and Wearable Electrodes. *ACS Sustain. Chem. Eng.* **2019**, *7*, 11823–11831. [CrossRef]
114. Ravit, R.; Azman, N.H.N.; Kulandaivalu, S.; Abdullah, J.; Ahmad, I.; Sulaiman, Y. Cauliflower-like Poly(3,4-Ethylenedioxythipohene)/Nanocrystalline Cellulose/Manganese Oxide Ternary Nanocomposite for Supercapacitor. *J. Appl. Polym. Sci.* **2020**, *137*, 49162. [CrossRef]
115. Johnsen, D.L.; Zhang, Z.; Emamipour, H.; Yan, Z.; Rood, M.J. Effect of Isobutane Adsorption on the Electrical Resistivity of Activated Carbon Fiber Cloth with Select Physical and Chemical Properties. *Carbon* **2014**, *76*, 435–445. [CrossRef]
116. Lakhi, K.S.; Park, D.H.; Al-Bahily, K.; Cha, W.; Viswanathan, B.; Choy, J.H.; Vinu, A. Mesoporous Carbon Nitrides: Synthesis, Functionalization, and Applications. *Chem. Soc. Rev.* **2017**, *46*, 72–101. [CrossRef] [PubMed]
117. Li, M.; Zhang, Y.; Wang, X.; Ahn, W.; Jiang, G.; Feng, K.; Lui, G.; Chen, Z. Gas Pickering Emulsion Templated Hollow Carbon for High Rate Performance Lithium Sulfur Batteries. *Adv. Funct. Mater.* **2016**, *26*, 8408–8417. [CrossRef]
118. Lu, W.; Wang, T.; He, X.; Sun, K.; Huang, Z.; Tan, G.; Eddings, E.G.; Adidharma, H.; Fan, M. A New Method for Preparing Excellent Electrical Conductivity Carbon Nanofibers from Coal Extraction Residual. *Clean. Eng. Technol.* **2021**, *4*, 100109. [CrossRef]
119. Shopsowitz, K.E.; Hamad, W.Y.; MacLachlan, M.J. Chiral Nematic Mesoporous Carbon Derived From Nanocrystalline Cellulose. *Angew. Chem. Int. Ed.* **2011**, *50*, 10991–10995. [CrossRef]
120. Li, Z.; Ahadi, K.; Jiang, K.; Ahvazi, B.; Li, P.; Anyia, A.O.; Cadien, K.; Thundat, T. Freestanding Hierarchical Porous Carbon Film Derived from Hybrid Nanocellulose for High-Power Supercapacitors. *Nano Res.* **2017**, *10*, 1847–1860. [CrossRef]

121. Wu, X.; Shi, Z.; Tjandra, R.; Cousins, A.J.; Sy, S.; Yu, A.; Berry, R.M.; Tam, K.C. Nitrogen-Enriched Porous Carbon Nanorods Templated by Cellulose Nanocrystals as High Performance Supercapacitor Electrodes. *J. Mater. Chem. A* **2015**, *3*, 23768–23777. [CrossRef]
122. Wang, Y.; Liu, T.; Lin, X.; Chen, H.; Chen, S.; Jiang, Z.; Chen, Y.; Liu, J.; Huang, J.; Liu, M. Self-Templated Synthesis of Hierarchically Porous N-Doped Carbon Derived from Biomass for Supercapacitors. *ACS Sustain. Chem. Eng.* **2018**, *6*, 13932–13939. [CrossRef]
123. Wang, S.; Dong, L.; Li, Z.; Lin, N.; Xu, H.; Gao, S. Sustainable Supercapacitors of Nitrogen-Doping Porous Carbon Based on Cellulose Nanocrystals and Urea. *Int. J. Biol. Macromol.* **2020**, *164*, 4095–4103. [CrossRef]
124. Luong, J.H.T.; Lam, E.; Leung, C.W.; Hrapovic, S.; Male, K.B. Chitin Nanocrystals and Process for Preparation Thereof. US Patent No. US10214596B2, 2019.
125. Mahmoud, K.A.; Mena, J.A.; Male, K.B.; Hrapovic, S.; Kamen, A.; Luong, J.H.T. Effect of Surface Charge on the Cellular Uptake and Cytotoxicity of Fluorescent Labeled Cellulose Nanocrystals. *ACS Appl. Mater. Interf.* **2010**, *2*, 2924–2932. [CrossRef] [PubMed]
126. Male, K.B.; Leung, A.C.W.; Montes, J.; Kamen, A.; Luong, J.H.T. Probing Inhibitory Effects of Nanocrystalline Cellulose: Inhibition versus Surface Charge. *Nanoscale* **2012**, *4*, 1373–1379. [CrossRef] [PubMed]
127. Halder, L.; Das, A.K.; Maitra, A.; Bera, A.; Paria, S.; Karan, S.K.; Si, S.K.; Ojha, S.; De, A.; Bhanu Bhusan Khatua, B.B. A Polypyrrole-adorned, Self-supported, Pseudocapacitive Zinc Vanadium Oxide Nanoflower and Nitrogen-doped Reduced Graphene Oxide-based Asymmetric Supercapacitor Device for Power Density Applications. *New J. Chem.* **2020**, *44*, 1063–1075. [CrossRef]
128. Siwa, S.S.; Zhang, Q.; Sun, C.; Thakur, V.T. Graphitic Carbon Nitride Doped Copper–Manganese Alloy as High–Performance Electrode Material in Supercapacitor for Energy Storage. *Nanomaterials* **2020**, *10*, 2.
129. Maitra, A.; Bera, R.; Halder, L.; Bera, A.; Paria, S.; Karan, S.K.; Si, S.K.; De, A.; Ojha, S.; Khatua, B.B. Photovoltaic and triboelectrification empowered light-weight flexible self-charging asymmetric supercapacitor cell for self-powered multifunctional electronics. *Renew. Sustain. Energy Rev.* **2021**, *151*, 111595. [CrossRef]
130. Liu, C.; Yi, C.; Wang, K.; Yang, Y.; Bhatta, R.S.; Tsige, M.; Xiao, S.; Gong, X. Single-Junction Polymer Solar Cells with Over 10% Efficiency by a Novel Two-Dimensional Donor–Acceptor Conjugated Copolymer. *ACS Appl. Mater. Interfaces* **2015**, *7*, 4928–4935. [CrossRef]
131. Liu, S.; Wei, L.; Wang, H. Review on Reliability of Supercapacitors in Energy Storage Applications. *Appl. Energy* **2020**, *278*, 115436. [CrossRef]

Article

The Key Role of Tin (Sn) in Microstructure and Mechanical Properties of Ti₂SnC (M₂AX) Thin Nanocrystalline Films and Powdered Polycrystalline Samples

Snejana Bakardjieva [1,2,*], Jiří Plocek [1], Bauyrzhan Ismagulov [1,3], Jaroslav Kupčík [1], Jiří Vacík [4], Giovanni Ceccio [4], Vasily Lavrentiev [4], Jiří Němeček [5], Štefan Michna [2] and Robert Klie [6]

1. Institute of Inorganic Chemistry of the Czech Academy of Sciences, 250 68 Husinec-Rez, Czech Republic; plocek@iic.cas.cz (J.P.); ismagulov@iic.cas.cz (B.I.); kupcik@iic.cas.cz (J.K.)
2. Faculty of Mechanical Engineering, JE Purkyně University, Pasteurova 1, 400 96 Ústí nad Labem, Czech Republic; stefan.michna@ujep.cz
3. Department of Inorganic Chemistry, Faculty of Science, Charles University in Prague, Albertov 6, 128 43 Prague, Czech Republic
4. Nuclear Physics Institute, Czech Academy of Sciences, 250 68 Husinec-Rez, Czech Republic; vacik@ujf.cas.cz (J.V.); ceccio@ujf.cas.cz (G.C.); lavrentiev@ujf.cas.cz (V.L.)
5. Faculty of Civil Engineering, Czech Technical University in Prague, Thakurova 7, 166 29 Prague, Czech Republic; jiri.nemecek@fsv.cvut.cz
6. Department of Physics, The University of Illinois at Chicago, Chicago, IL 60607, USA; rfklie@uic.edu
* Correspondence: snejana@iic.cas.cz

Citation: Bakardjieva, S.; Plocek, J.; Ismagulov, B.; Kupčík, J.; Vacík, J.; Ceccio, G.; Lavrentiev, V.; Němeček, J.; Michna, Š.; Klie, R. The Key Role of Tin (Sn) in Microstructure and Mechanical Properties of Ti₂SnC (M₂AX) Thin Nanocrystalline Films and Powdered Polycrystalline Samples. *Nanomaterials* **2022**, *12*, 307. https://doi.org/10.3390/nano12030307

Academic Editors: Federico Cesano, Simas Rackauskas and Mohammed Jasim Uddin

Received: 16 December 2021
Accepted: 12 January 2022
Published: 18 January 2022

Publisher's Note: MDPI stays neutral with regard to jurisdictional claims in published maps and institutional affiliations.

Copyright: © 2022 by the authors. Licensee MDPI, Basel, Switzerland. This article is an open access article distributed under the terms and conditions of the Creative Commons Attribution (CC BY) license (https://creativecommons.org/licenses/by/4.0/).

Abstract: Layered ternary Ti₂SnC carbides have attracted significant attention because of their advantage as a M₂AX phase to bridge the gap between properties of metals and ceramics. In this study, Ti₂SnC materials were synthesized by two different methods—an unconventional low-energy ion facility (LEIF) based on Ar+ ion beam sputtering of the Ti, Sn, and C targets and sintering of a compressed mixture consisting of Ti, Sn, and C elemental powders up to 1250 °C. The Ti₂SnC nanocrystalline thin films obtained by LEIF were irradiated by Ar⁺ ions with an energy of 30 keV to the fluence of 1.10^{15} cm^{-2} in order to examine their irradiation-induced resistivity. Quantitative structural analysis obtained by Cs-corrected high-angle annular dark-field scanning transmission electron microscopy (HAADF-STEM) confirmed transition from ternary Ti₂SnC to binary Ti$_{0.98}$C carbide due to irradiation-induced β-Sn surface segregation. The nanoindentation of Ti₂SnC thin nanocrystalline films and Ti₂SnC polycrystalline powders shows that irradiation did not affect significantly their mechanical properties when concerning their hardness (H) and Young's modulus (E). We highlighted the importance of the HAADF-STEM techniques to track atomic pathways clarifying the behavior of Sn atoms at the proximity of irradiation-induced nanoscale defects in Ti₂SnC thin films.

Keywords: Ti₂SnC; M₂AX; powders; thin films; STEM; nanoindentation

1. Introduction

The MAX phases are a family of about 90+ carbides (or nitrides) synthesized up to now with a basic stoichiometry nomenclature $M_{n+1}AX_n$, where M is an early transition d metal (i.e., Sc, Ti, V, Cr, Zr, Nb, Mo, Hf, and Ta), A represents an element (mainly) from the IIIA or IVA groups of the periodic table (i.e., Al, Si, P, S, Ga, Ge, As, Cd, In, Sn, Tl, and Pb), and X is carbon or nitrogen. The index n can be 1, 2, or 3, so as the stoichiometry can vary, phases AMXMA (211), AMXMXMA (312), or AMXMXMXMA (413) can be formed [1]. The MAX phases have a hexagonal crystal structure (a space group D^4_{6h}, P63/mmc) with two units per cell. The cell consists of the M₆X octahedra that alternate with a single layer of the A elements [2].

The idea of the ternary transition metal carbides referred to as the H phases, where H denotes hexagonal close-packed structures, was developed by Nowotny and Jeitschko almost 60 years ago [3]. Later, the ternary transition metal carbides were classified as the M_{n+1}

AX_n phases [1–4] due to their unique hybrid structure with mixed covalent/ionic/metallic properties. This unusual conjunction influences their thermodynamic stability and mechanical properties and predicts that MAX phases could be highly regarded candidates for applications in extreme conditions. Some of the MAX compounds have already driven interest in nuclear engineering as materials with high potential for future fission and fusion reactors [5]. For instance, high radiation resistance was shown for the first time on titanium aluminum and titanium silicon carbides irradiated with high fluence heavy ions (Xe^+, 6.25×10^{15} ions cm^{-2}) [6]. The received experimental data suggested that after irradiation, the structure of the MAX compounds keeps well-ordered. This observation was acknowledged in new experiments [7] with other members of the MAX group. Titanium tin carbide (e.g., Ti_2SnC M_2AX), discovered already in 1963 [3], demonstrates unusual material characteristics, such as high tolerance to mechanical damage, high modulus elasticity, and good integral stability at high temperature [8–11]. Surprisingly, there is a lack of relevant data on the ion beam and/or neutron radiation tolerance of Ti_2SnC. Perhaps this is due to the rather high cross sections for neutron-induced γ–(gamma) activation of the Sn isotopes (~0.6 b for natural Sn) with relatively long lifetimes (e.g., ~129 d for $^{122}Sn + n$), which means that the Ti_2SnC may be unattractive for nuclear engineering technology. However, as a promising coating material, Ti_2SnC may still be interesting, and it is worth studying its irradiation resistance. The Ti_2SnC is especially synthesized using a finely dispersed powder of the Ti, Sn, and C phases, mixed in stoichiometric ratios, grounded, pressed, and sintered at high temperatures [12,13]. Other techniques, such as spark plasma sintering [14], have been invented and used for Ti_2SnC fabrication [15]. The applied methods, however, documented that together with the Ti_2SnC composite, some precipitates (such as TiC, Ti_6Sn_6, or Sn) are also detected. It turns out that to fully transform the correct stoichiometric ratio of the Ti-Sn-C to the acceptable ternary M_2AX phase is still a challenge.

In this research, the Ti_2SnC thin nanocrystalline films (TNCFs) were synthesized using an unconventional low-energy ion facility (LEIF) based on ion beam sputtering combined with further a low- temperature thermal processing up to 150 °C. The Ti_2SnC powdered polycrystalline samples (PPS) were fabricated by simplified sintering of a compressed mixture consisting of (Ti, Sn, C) elemental powders. The goal of this study is to compare the morphological and nanomechanical features of Ti_2SnC M_2AX materials prepared by using different synthetic methods. In addition, the irradiation tolerance of the Ti_2SnC TNCFs was examined. The irradiation was carried out by a heavy Ar^+ ion with an energy of 30 keV to the fluence of 1.10^{15} cm^{-2}. We provide an experiment in the understanding of Sn atoms surface segregation and highlight the importance of aberration-corrected STEM techniques including high-angle annular dark-field detector (HAADF) to track atomic pathway clarifying the behavior of Sn atoms at the proximity of irradiation-induced nanoscale defects in Ti_2SnC TNCFs.

2. Materials and Methods

2.1. Synthesis of Ti_2SnC M_2AX PPSs

The Ti_2SnC M_2AX PPSs were fabricated using a simplified method of sintering raw elemental powders. The experimental setup is presented in Figure 1a.

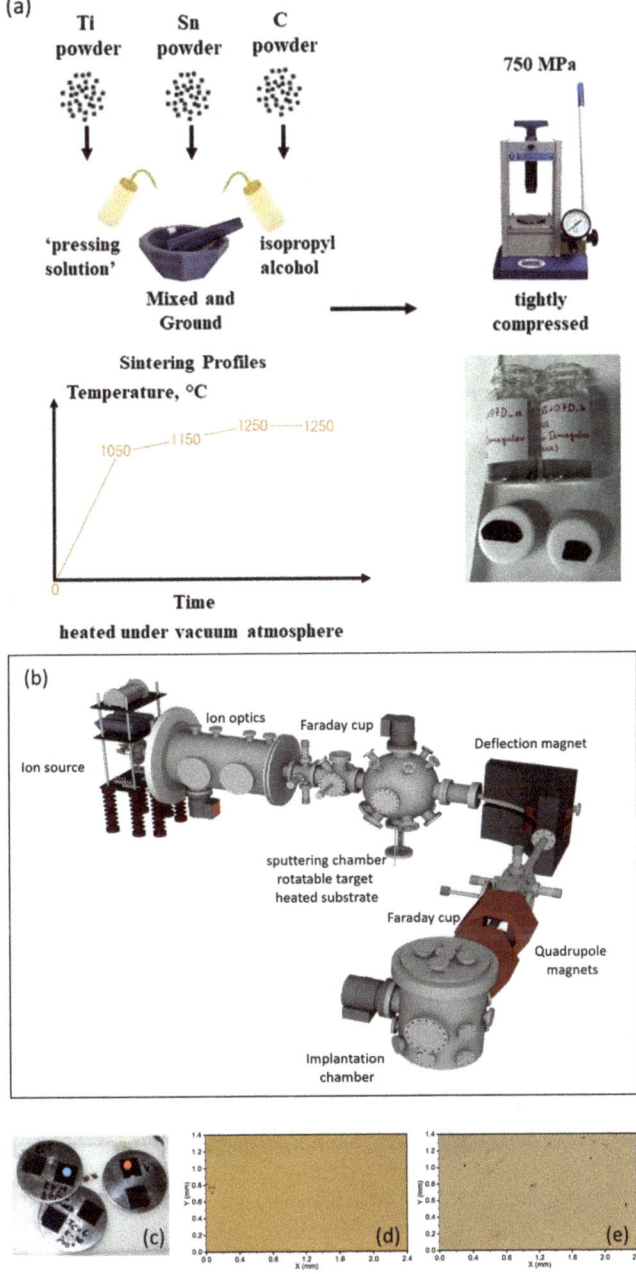

Figure 1. Schematic drawing of the Ti$_2$SnC synthesis (**a**) Ti$_2$SnC PPS by sintering of raw elemental powders with optical micrographs of Ti$_2$SnC PPS at low and high magnification. (**b**) Ion beam sputter deposition setup by LEIF for the synthesis of Ti$_2$SnC TNCFs. (**c**) Optical micrographs Ti$_2$SnC TNCFs by LEIF. (**d**) High magnification from the blue point marks the area where an optical micrograph of a single area for unirradiated Ti$_2$SnC_AGTNCFs was acquired. (**e**) High magnification from the red point mark the area where an optical micrograph of a single area for irradiated Ti$_2$SnC_Ar$^+$ TNCFs was acquired.

Stoichiometric amounts of Ti (99.7%, Aldrich, powder), Sn (≥93%, Aldrich, powder), and graphite (diamond powder) as raw materials were mixed with isopropyl alcohol at a 1/0.8/0.9 molar ratio and ground in an agate mortar. After thorough grinding, 0.1 mL of a 'pressing solution' (ethanol solution of polyethylene glycol 400, 1% w/w) was added to the fine ground suspension and mixed thoroughly again. After evaporation of the alcohol component, the solid mixture of reagents was tightly compressed (at a specified pressure of 750 MPa) to form a pellet with a diameter of 1.3 cm. The pellet was then placed into a corundum tube of a laboratory furnace and heated at a temperature up to 1250 °C under a vacuum. The optimal annealing regime to receive the Ti_2SnC M_2AX PPS was determined to be following (1. step 0–1050 C–20 C/min, 2. step 1050–1150C–10C/min, 3. step 1150–1250C–5C/min, 4. step 1250C-delay 2 h). The as-obtained pellet was milled and heated again under the same regime. The optical micrographs of Ti_2SnC PPS are presented in Figure 1a.

2.2. Synthesis of Ti_2SnC TNCFs

A set of Ti_2SnC TNCFs was prepared by ion beam sputtering (IBS), as well as controlled thermal processing. The ions were generated in the high-current ion source (duoplasmatron) placed in the LEIF (lab-made Low Energy Ion Facility assembled by NPI) of the CANAM (Center of Accelerators and Nuclear Analytical Methods) research infrastructure in the NPI Rez [16]. The LEIF facility lets us utilize different gaseous ions with energy in the range of 100 eV to 35 keV, and an ion current up to 500 µA. The beam spot size of the Gaussian-shaped ion beam was about 20 mm. In this report, the Ar^+ ions-beam has been used (apart from singly charged Ar^+, double-charged Ar^{2+} ions were also present, though only a fraction of a few %). The current and energy of the Ar^+ ion have been varied to measure the optimal values for the manufacture of the titanium tin carbide. For the fabrication of the Ti_2SnC TNCFs, highly purified targets of Ti (99.995%), Sn (99.999%), and C (99.999%), all of the MaTeck (MaTeck Material Technologie & Kristalle GmbH, Juelich, Germany) materials have been used. The specimens were placed on a metallic (Cu) stocker (a frame with an equilateral triangle shape), each on a different side of the frame, that was mounted in the sputtering chamber of the LEIF system. The schema of the LEIF sputtering is presented in Figure 1b and described in detail elsewhere [17].

The targets with a size of about 5 cm in diameter covered the dimensions of the frame, so the sputtering from the target holder itself was stopped. The holder was connected to a metallic (Cu) axis (controlled from the outside of the chamber) and cooled down forcefully (by distilled water of an external cooling system) when specimens were overheated. The holder was revolving in 3 shifts with a speed of 1 rotation per 100 s. The revolution was performed automatically using a stepper motor (Accu-Glass Products. Inc., Valencia, CA, USA) operated by a PC. In each shift of 60°, a corresponding composite (Ti, Sn, or C) was sputtered for a definite period t_{phase} to deposit a necessary amount Δ (~10^{15} cm^{-2}) of the specimens' material on the substrate. The sputtering times were defined by the deposition rates DR_{phase} of the particular composites and by the stoichiometric ratio of the Ti_2SnC phase (2:1:1): $t_{Ti} = 2\Delta/DR_{Ti}$, $t_{Sn} = \Delta/DR_{Sn}$, $t_C = \Delta/DR_C$. An approximately 1 nm thick layer with 2Ti+1Sn+1C atomic mixture was deposited during each rotation. The deposition was conducted on Si wafers or Mo TEM grids with ion energy of 25 keV and an ion current of 400 micro A. The deposition rates were held permanent and they were defined for Ti—0.85×10^{15} cm^{-2}, Sn—9.80×10^{15} cm^{-2} and C—0.76×10^{15} cm^{-2} per min using RBS. A set of the samples was annealed at 150 °C for 24 h in a vacuum to induce interphase chemical interaction and complete formation of the stoichiometrically correct Ti_2SnC M_2AX phase. Samples were labeled Ti_2SnC_AGTNCF, where AG was denoted "as-growth" thin film. The Ti_2SnC_AGTNCF was further irradiated by 35 keV Ar^+ ions in order to examine their irradiation-induced resistivity. Optical micrographs Ti_2SnC TNCFs obtained by LEIF are presented in Figure 1c–e.

2.3. Ion Beam Irradiation

In order to get pieces of knowledge about the radiation tolerance of the $Ti_2SnC\_AGNGTFs$, a heavy Ar^+ ion with an energy of 30 keV to the fluence of 1.10^{15} cm^{-2} was applied. It is assumed that Ar^+ ions with an energy of 30 keV generate an irradiated (damaged) area of 30 nm deep from the film surface. Using the SRIM-2013 code, the dpa value for this fluency was evaluated at 9.49 dpa (in the calculation, the density of Ti_2SnC—6.36 g cm^{-3}, displacement energy—25 keV, and 'energy to recoil'—81 eVÅ^{-1} were considered). The $Ti_2SnC\_AGTNCFs$ were tested by several nuclear analytical methods. The thickness was examined with a sub-nanometer precision profilometer KLA-Tencor Alpha-Step IQ Surface Profiler/, as well as by Rutherford backscattering spectrometry (RBS; lab-made, assembled in NPI).

2.4. Methods of Characterization

The powder diffraction patterns of the Ti_2SnC PPS were obtained with a PANalytical X'PertPRO MPD diffractometer (Malvern, United Kingdom) equipped with the Cu Kα tube (λ = 0.15406 nm). The diffractometer was operated at 40 kV and 30 mA with a 0.5° divergent slit coupled with a 0.1 mm receiving slit. Room temperature diffractograms were recorded in the transmission regime in the range from 5° to 85° at a 2θ step size of 0.01°. The phase composition of the measured powdered sample was calculated by the Rietveld analysis in an automatic mode of HighScore software 5.0 (Malvern, United Kingdom).

Scanning electron microscopy (SEM) was used for the characterization and imaging of the fine surface structure of the prepared Ti_2SnC_TNCFs and Ti_2SnC PPS. For measurement, a JSM 6510LV system (low vacuum JEOL microscope, Jeol Ltd., Tokyo, Japan) with an acceleration voltage of 0.5–30 kV was used. For analysis, the secondary electron imaging mode (SE) was applied.

A detailed microstructural analysis of electron diffraction, and also elemental mapping, were carried out on a high-resolution transmission electron microscope (HRTEM) JEOL JEM 3010 Jeol Ltd., Tokyo, Japan). The microscope was operated at 300 kV (using a LaB$_6$ cathode; the point resolution was 1.7 Å), and it was equipped with an energy-dispersive X-ray (EDX) detector (Oxford Instruments, High Wycombe, UK) for elemental analysis, and a Gatan CCD camera (1024 × 1024 pixels) for image recording. The obtained images were analyzed using the Digital Micrograph software 3.5 package Gatan, California, USA), the EDX analysis was processed with the INCA software package (High Wycombe, UK). Electron diffraction patterns were evaluated using the ICDD PDF-2 database, Newtown Square, PA, USA [18] and ProcessDiffraction V_8.7.1. Q software 7 package (Budapest, Hungary) [19]. For the TEM analysis, a small bit of the pellet sample was crushed, dispersed in ethanol, and the obtained suspension was sonicated for 2 min. A drop of the very dilute suspension was then placed on a holey-carbon coated Cu-grid and allowed to dry by evaporation at ambient temperature.

The atomic resolution Z-contrast images of $Ti_2SnC\_AGTNCFs$ and $Ti_2SnC\_Ar^+TNCFs$ were collected using the JEOL ARM200CF (Jeol Ltd., Tokyo, Japan) aberration-corrected STEM with a cold-field emission gun operated at an acceleration voltage of 80 kV. The high-angle annular dark-field (HAADF) images were acquired using an annular dark-field detector with a collection angle ranging from 90 to 175 mrad. The probe convergence semi-angle was set to 29 mrad, which yields a probe size of 1 Å at 80 kV and a probe current of 62 pA [20].

The surface topography of the Ti_2SnC_TNCFs fabricated by LEIF was studied by atomic force microscopy (AFM) using the NTEGRA scanning probe microscope (NT-MDT Spectrum Instruments, Moscow, Russia). The AFM experiments were performed under ambient conditions using tapping mode for the acquisition of the sample surface images (AFM topography).

The nanomechanical properties of the $Ti_2SnC\_AGTNCFs$, $Ti_2SnC\_Ar^+TNCFs$, and Ti_2SnC PPS were inspected by nanoindentation with a Hysitron Tribolab TI-700 Nanoindenter (Bruker Nano GmbH, Berlin, Germany) equipped with a Berkovich tip. Indentations were made at 4 distant locations (6 indentations at each), and each measurement consisted of 10 cycles with penetration depths between 20–150 nm (with a contact depth of 10–130 nm) to analyze changes induced by irradiation in the material's properties. The Hysitron TI-700 used with Berkovich tip is capable of quantitative measurements for depths larger than 10 nm, for which the tip calibration done on fused silica standard was performed. In the calibration procedure, the tip radius is not explicitly assessed, but the contact area is evaluated before the measurements. Measurements in larger depths can be considered accurate with the accuracy of the polynomial contact area calibration function ($R^2 = 0.999$).

3. Results and Discussion

3.1. Elemental Detection by Nuclear Analytical Methods

In Figure 2, both RBS and non-Rutherford spectra of $Ti_2SnC\_AGTNCFs$ are presented together with the results obtained by simulations with SIMNRA code (performed on the Tandetron 4230 MC). It was registered that *$Ti_2SnC\_ADTNCFs$* were contaminated by oxygen up to a level of about 35%. However, the ratio of the Ti, Sn, and C elements keeps up the stoichiometric ratio (Ti/Sn/C~2/1/1).

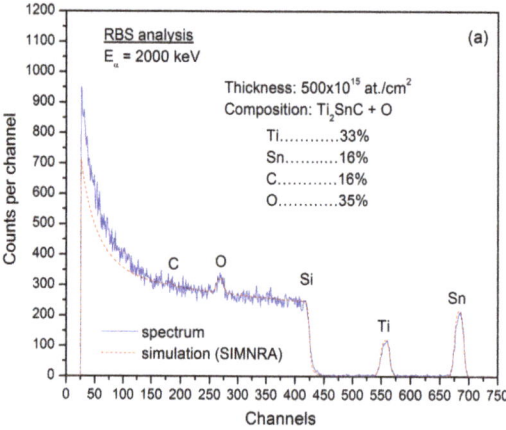

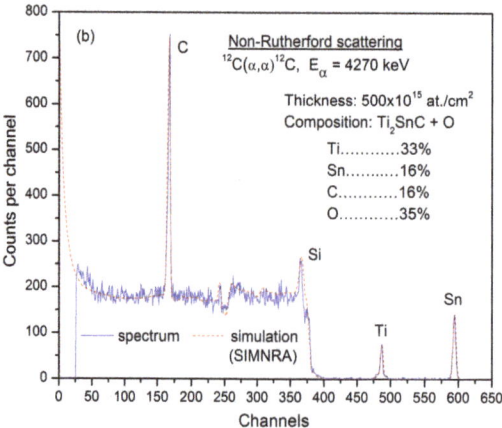

Figure 2. (a) RBS spectra of the $Ti_2SnC\_AGTNCFs$. (b) non-Rutherford scattering spectra of $Ti_2SnC\_AGTNCFs$ by simulation of the energy spectra using the SIMNRA code.

We can suggest that Ti$_2$SnC_AGTNCFs were (partially) oxidized either during the deposition process (with the residual oxygen in the sputtering chamber), or through the annealing in a relatively low-level vacuum of 10^{-4} Pa [21]. Therefore, the synthetic challenge is to avoid oxidation contamination which occurs commonly when LEIF deposition is used [22].

The thickness of the Ti$_2$SnC_AGTNCFs on the polished Si wafers with a size of about 1 cm^2 was detected to be in the diapason 460–920 × 10^{15} cm^{-2}.

3.2. Morphology of Ti$_2$SnC_AD, Ti$_2$SnC_Ar$^+$TNCFs, and Ti$_2$SnC PPS Imaged by AFM and SEM

Figure 3 shows the AFM results obtained from analysis of the Ti$_2$SnC_AGTNCFs and Ti$_2$SnC_Ar$^+$TNCFs taken from the sample surface area of (1 × 1) µm^2.

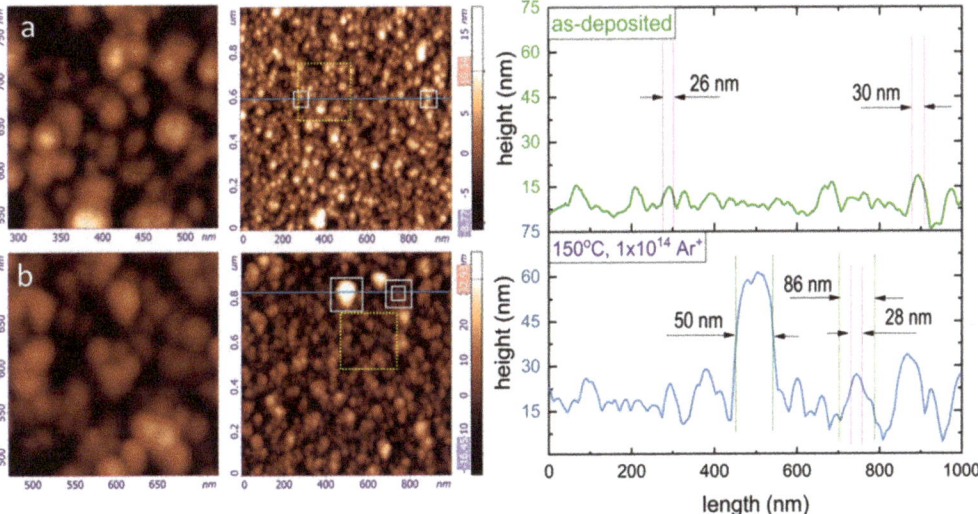

Figure 3. AFM characterization of the Ti$_2$SnC_AGTNCF and Ti$_2$SnC_Ar$^+$TNCF. (**a**) AFM topography of Ti$_2$SnC_AGTNCF and image magnified part from the yellow dash square (right and left images, respectively) with its corresponding surface profile. (**b**) AFM topography of the surface profile of Ti$_2$SnC_Ar$^+$TNCF and the image magnified part from the yellow dashed square (right and left images, respectively) with its corresponding surface profile.

Surface profile plots were prepared along the horizontal blue lines in the AFM images (see the plots from the right panel in Figure 3a,b). It is seen that the Ti$_2$SnC_AGTNCFs surface (Figure 3a) consists of the nanoparticles (NPs), which are mostly separated each from other. Different color of the NPs reflects their different height, suggesting the formation of non-uniform agglomerates during the film deposition. The latter effect results in the relatively high surface roughness, which was found to be SR_{rms} = 3.275 nm (SR_{rms} denotes the root mean square roughness of the surface). The details of the NPs agglomerations are seen in the magnified image of the (250 × 250) nm^2-sized surface area (see the AFM image from the left panel), selected by the dotted-line square in the original AFM image. Analysis of the magnified image and the surface profile reveals the size distribution of the NPs, with a lateral size of 20–30 nm. Annealing at 150 °C (T_a = 150 °C) and irradiation with the Ar$^+$ ion beam modified the morphology of the Ti$_2$SnC_Ar$^+$TNCFs. Surface roughness was found to be significantly higher (SR_{rms} = 7.057 nm) than that of Ti$_2$SnC_AGTCNF. According to the magnified image (the left AFM image in Figure 3b) and the surface profile, bigger agglomerates with a lateral size of 50–90 nm appear on the Ti$_2$SnC_Ar$^+$TNCFs surface. The spatial density of the agglomerations is also higher, and the size distribution shifting to the larger NP size was observed. It was rather hard to obtain AFM vital data for Ti_2SnC PPS. The polishing technique and mechanical surface treatment during sample preparation were the main obstacles for representative AFM imaging. Despite that, Figure S1 shows the SEM surface morphology of the Ti_2SnC PPS. It is found that misaligned grains and pores are formed on the Ti_2SnC PPS surface during hydrothermal synthesis. A smooth surface and fully dense microstructure without defects can be seen in SEM micrographs of $Ti_2SnC\_AGTNCFs$ (Figure S1a). When irradiation with an Ar^+ ion beam was applied, the small round-shaped features on the surface of the $Ti_2SnC\_Ar^+TNCFs$ appeared (Figure S1b), suggesting that Ar^+ ion-beam irradiation-induced blistering and/or surface defects. In order for the structure of $Ti_2SnC\_AGTNCFs$ and Ti_2SnC Ar^+TNCFs to be solved, atomic-resolution bright-filed (BF) and HAADF imaging in Cs-corrected STEM was further performed and models for atomic ordering of $Ti_2SnC\_$ AGTNCFs and Ti_2SnC Ar^+TNCFs structures were proposed.

3.3. Structural Analysis of the Ti_2SnC_AGTNCF with Atomic-Resolution STEM

Figure 4 shows the aberration-corrected BF and HAADF STEM micrographs of Ti_2SnC_AGTNCF acquired from different magnification. Two regions can be distinguished in the HAADF-STEM images in Figure 4a,b; matrix with lower intensity (labeled 1) and well-crystallized spherical NPs (labeled 2), which present higher intensity (see a yellow boxed region in Figure 4b). This contrast can be associated with the atomic weight dependence of constituted elements. It should be noted that the contrast of Sn appears brighter than the Ti since the atomic number of Sn ($Z = 50$) is larger than that of Ti ($Z = 22$) in the HAADF electron scattering regime. The corresponding SAED (Figure 4c) confirms a mixture of two phases: tetragonal SnO (JCPDS 06–0395 space group P4/nmm) and hexagonal Ti_2InC (JCPDS PDF 01-089-5590, space group $P6_3/mmc$). A zoomed STEM image of the matrix is displayed in Figure 4f with an incident electron beam along the [001] direction. The measured lattice fringe spacings $d_{(100)} = 0.27$ nm correspond to hexagonal Ti_2SnC. HAADF STEM analysis (Figure 4g–i) further provides additional evidence of the formation of SnO (SnO-Sn^{2+}). In Figure 4i, (101) atomic planes with interplanar spacing $d = 0.29$ nm for tetragonal SnO with growth in the c parameter direction on the surface of Ti_2SnC can be seen. Therefore, a protective Ar_2 atmosphere could avoid the oxidation of Sn to the thermodynamically more stable SnO_2 (SnO_2-Sn^{+4}) [23,24]. A similar mechanism of SnO formation/SnO_2 avoiding was reported upon using a protective N_2 atmosphere for the synthesis of uniform nanocrystalline SnO layers [25,26].

Simultaneously performed EDS-STEM quantitative analysis (Figure S2a) confirmed the same elements (Ti, Sn, C, O) as detected by RBS and non-Rutherford scattering (see Figure 2). It can be seen from the elemental distribution maps (Figure S2b) that the mapping image of Sn is interconnected with the O mapping image. Such a correspondence reveals that Sn appears in oxygen-enriched regions, for example, spherical shaped NPs on the surface, and indicates that Sn can exist as an Sn-oxide rather than metal Sn [27].

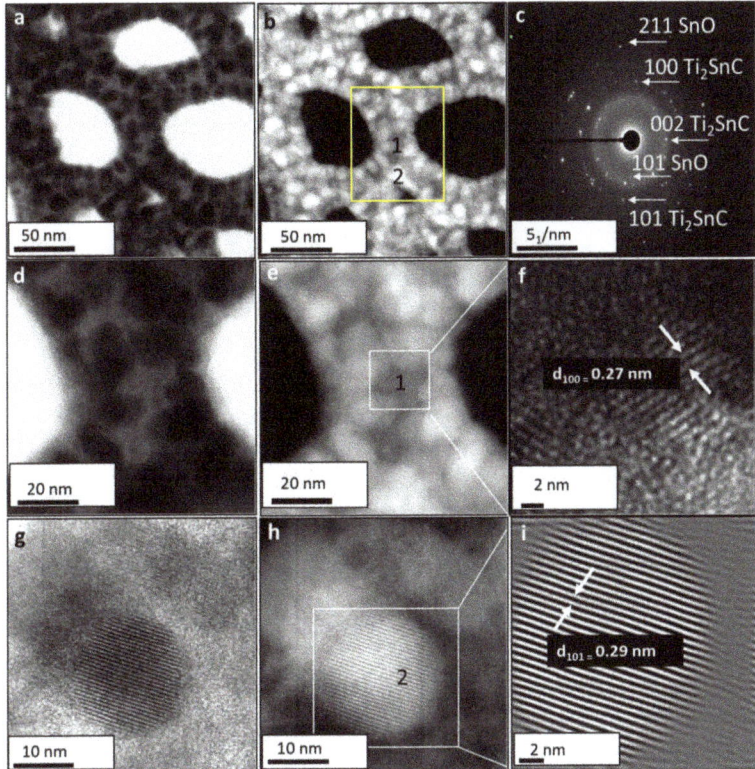

Figure 4. Aberration-corrected STEM images of unirradiated Ti$_2$SnC_AGTNCF. (**a**) Low magnification BF-STEM image and (**b**) HAADF-STEM image where yellow boxed area outlines matrix with lower intensity (labeled 1) and brighter grains (labeled 2). (**c**) Corresponding SAED pattern showing atomic planes of SnO and Ti$_2$SnC. (**d**) High magnification BF-STEM image and (**e**) HAADF_STEM image showing zoomed matrix. (**f**) High magnification from the white marked area in (**e**) where lattice fringe spacing d$_{(100)}$ = 0.27 nm corresponds to hexagonal Ti$_2$SnC was indexed. (**g**) high magnification BF-STEM image and (**h**) HAADF_STEM image showing zoomed grain with higher intensity. (**i**) High magnification from the white marked area in (**h**) where lattice fringe spacing d$_{(101)}$ = 0.29 nm corresponds to tetragonal SnO was indexed.

3.4. Structural Analysis of the Ti$_2$SnC_Ar$^+$TNCF with Atomic-Resolution STEM

A structural anomaly in Ti$_2$SnC_Ar$^+$NCTF was found to be driven during Ar$^+$ ion beam irradiation. Low magnification BF (Figure 5a) and HAADF-STEM (Figure 5d) micrographs revealed well-crystallized nanograins. Figure 5b,e report magnified BF and HAADF STEM images, where interconnected disc-like NPs (labeled 1) and spherical NPs (labeled 2) were observed (Figure 4b,e).

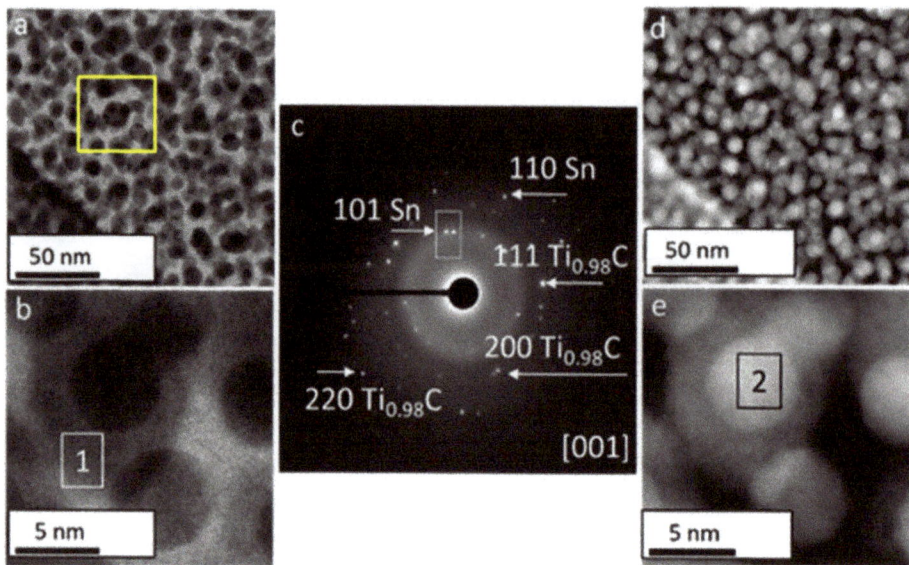

Figure 5. Aberration-corrected STEM images of irradiated Ti$_2$SnC_Ar$^+$TNCF. BF and HAADF-STEM images of Ti$_2$SnC_Ar$^+$TNCF (**a**) low magnification BF-STEM image and (**b**) BF-STEM high magnification of single Sn particle. (**c**) SAED from yellow boxed region in (**a**) showing atomic planes of metallic Sn and Ti$_{0.98}$C. (**d**) HAADF-STEM image at low magnification and (**e**) HAADF-STEM image at high magnification of Sn particle appeared with bright contrast because of the contribution of scattered electrons to the Sn image.

Besides the distinct morphology, the SAED pattern (Figure 5b) taken from the yellow boxed area in Figure 5a did not match those of Ti$_2$SnC. It was found that the lattice spacing and the angle of hexagonal Ti$_2$SnC lattice structure undergo a fundamental transformation. For identified [001] zone axis, the main (111), (220), and (200) lattice planes for cubic Ti$_{0.98}$C phase (i.e., space group *Fm-3m*, JCPDS PDF No. 04-004-2862) were detected. Furthermore, brighter reflections on the ED patterns were found to be extremely close to the (101) and (110) planes, matching well with an β-Sn with tetragonal symmetry (i.e., space group *I4/mmm*, JCPDS PDF No. 00-018-1380). We noticed that some diffraction spots (white boxed in SAED pattern) do become dimmers. High contrast variation in the HAADF STEM image (Figure 5e) suggests that the surface of disc-like NPs (labeled 1) is covered with smaller spherical NPs with bright contrast (labeled 2). Concerning the difference in atomic numbers of constituent elements and approached distinct Z-contrast, we could infer that Ar$^+$ ion beam irradiation promoted the growth of fine spherical β-Sn NPs running along disc-like Ti$_{0.98}$C NPs [28–30]. The proposed dual heterostructure that evolved under the Ar$^+$ ion-beam irradiation process was further confirmed by aberration-corrected HAADF-STEM imaging and simulation of experimental SAED patterns. We discuss first the structure of disc-like NPs in Figure 6a. The FFT in the [110] orientation (inset in Figure 6a) reports the major lattice planes (111), (220), and (200) for the cubic Ti$_{0.98}$C phase. Figure 6b shows an aberration-corrected HAADF-STEM image viewed along with the [001] zone axis. The structure is maintained to the surface on the (001) planes as obtained from the yellow marked area in Figure 6a. At this surface, there are only bright atomic columns, suggesting an atomic arrangement expected for the lattice with the space group Fm-3m, where the Ti^{2+} ions occupy the tetrahedral sites. The structure model of Ti$_{0.98}$C is superimposed on the HAADF-STEM image. For irradiated Ti$_2$SnC_Ar$^+$TNCF the atomic stacking transforms from ABABA/TiCSnCTi to ABBA//TiCCTi (Figure 6c) [21]. We could expect that the atomic arrangements change to ABBA because all Sn atoms are segregated from the Ti$_2$SnC lattice and transition from M$_2$AX to M$_{0.98}$X structure has occurred. The simulated ED pattern of cubic Ti$_{0.98}$C agrees with the experimental FFT. Simulated [001] HRTEM image at a focus value f = −440 Å and a thickness t = 19 Å in Figure 6d is in line with the experimental contrast. The Ti columns can be seen in the simulated [110] HRTEM image that the Ti position appears as bright dots. The interlayer spacing of 0.24 nm is between the (111) cubic planes in the Ti$_{0.98}$C structure. The unit cell for the cubic Ti$_{0.98}$C arrangement is overlaid on the simulated STEM image.

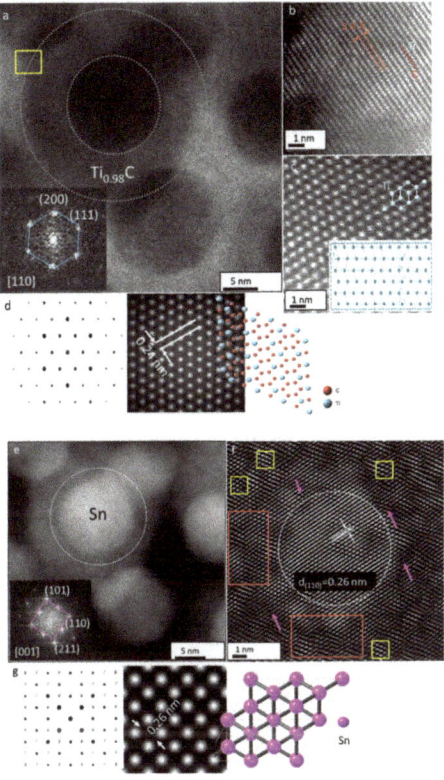

Figure 6. Crystalline structures of Ti$_2$SnC_Ar$^+$TNCF. (**a**) Aberration-corrected BF-STEM image with gray dashed lines outlines the disk-like grain as it degrades to Ti$_{0.98}$C. Inset: FFT phase diagrams of the corresponding STEM image. (**b**) HAADF-STEM image of the yellow boxed area in disk-like grain in (**a**) with clearly resolved atomic columns. The lattice spacing is measured as 2.0 Å. The structure model of Ti$_{0.98}$C is superimposed on the STEM image. (**c**) Atomic stacking transformation from ABABA in Ti$_2$SnC to ABBA in Ti$_{0.98}$C. (**d**) Simulated SAED patterns of Ti$_2$SnC_Ar$^+$TNCF structure. Transformation to the Ti$_{0.98}$C configuration with an interlayer spacing of 0.24 nm between the (111) planes due to the escape of Sn has occurred. The unit cell for Ti$_{0.98}$C composition with the cubic arrangement is overlaid on the simulated SAED image. Blue and red atoms refer to Ti and C, respectively. (**e**) aberration-corrected HAADF-STEM image with a white dashed line to outline the spherical NP as it segregates to β-Sn. Inset: FFT phase diagrams of the corresponding STEM image. (**f**) The corresponding HAADF-STEM image with clearly resolved atomic columns of tetragonal β-Sn. The lattice spacing is measured as 0.26 nm. (**g**) Simulated SAED patterns of the tetragonal β-Sn structure. Single Sn grain with an interlayer spacing of 0.26 nm between the (110) planes is observed. The unit cell for β-Sn composition with the tetragonal arrangement is overlaid on simulated SAED image: magenta atom refers to Sn.

The white contrast in the HAADF-STEM images in Figure 6e suggested that spherical NP with corresponding FFT (inset in Figure 6e) resemble tetragonal β-Sn. We can index the set of lattice planes (001), (101), and (200) and identify the [001] axis (the c axis) as the orientation of segregated Sn. The original STEM image was filtered by applying a Fourier mask to remove the noise and obtain clearer lattice periodicity (Figure 6f). As a result, lattice fringes with a $d_{(110)}$ spacing of 0.26 nm, consistent with the tetragonal crystalline structure of Sn (JCPDS PDF No. 00-018-1380) was obtained. The weaker lattice fringes in the background appeared due to the adjacent cubic structure of the bottom disk-like $Ti_{0.98}C$. Here, we were able to identify the cubic $Ti_{0.98}C$ (selected in yellow regions). Alongside the ordered structure, certain line defects–dislocation lines (marked with magenta-colored arrows) and collision cascades (selected in red regions) into the $Ti_{0.98}C$ lattice were observed. These data appear to suggest that Sn segregation into small atomic clusters prefers to grow near the dislocation lines. It could be proposed that Sn clusters and irradiation-induced defect cores (dislocations and cascades) nucleated and grew together [31,32]. Simulated ED patterns of Sn along with the [110] zone axis (Figure 6g) show that $d_{(110)}$ = 0.26 nm crystal plane in tetragonal Sn is present. The atomic structure model of Sn along (010) in $Ti_2SnC\_Ar^+NCTF$ is overlaid with a simulated ED pattern.

Figure S3a shows the EDS-STEM spectrum, atomic % of elements (Table inset in Figure S3a), and EDS elemental mapping (Figure S3b) of $Ti_2SnC\_Ar^+NCTF$. We found the presence of Ti, Sn, and C elements. The Ti/Sn atomic ratio in $Ti_2SnC\_Ar^+NCTF$ decreases from the initial 1.17 in Ti_2SnC_AGNCTF to the final 0.85 in $Ti_2SnC\_Ar^+TNCF$. It can be seen from the mapping images of Ti and C (Figure S3) that both elements become woven together and distributed all over the film, whereas the Sn mapping image suggests that Sn is not interconnected with Ti. Such an observation can indicate that Sn separated rather than being interwoven with Ti_2SnC structures, which was proved true from the HAADF-STEM and SAED observation. It is worthy to mention that the EDS spectra and the quantitative EDS mapping did not indicate the presence of oxygen. The absence of oxygen in $Ti_2SnC\_Ar^+NCTF$ may be explained by the irradiation-induced ionization effect, which is reported to be pronounced not only in weakly bound Van der Waals elements but could be also occurred in systems with stronger bonds [33–35]. This ionization process can change both the equilibrium state and geometry of the overall system, including the Ti_2SnC matrix and surface SnO "etched out" the Ti_2SnC. Subsequently, the charge distribution in SnO, as well as its binding energy, may be altered. Therefore, we can assume that the 30 keV Ar^+ with a fluence of 10^{15} cm^{-2} could provide enough energy to overcome the binding energy of 485.57 eV of the Sn $3d_{5/2}$ core level of SnO, and further contribute to the fragmentation of the SnO [36]. These structural fragments could be Sn atoms and oxygen [37]. In the presence of Ar_2, chemical rearrangement of elements into Ar^+-O_2 mixture and argon-oxygen ions as $Ar(O_2)_n^+$, $Ar_2(O_2)_n^+$ maybe generated [38]. Moreover, the presence of Ar peak in the EDS spectrum is very hard to evidence (Figure S3) since its only characteristic diffraction peak at 3.0 keV [39] is overlapped with Sn shoulder or probably because the Ar peak intensity is close to the noise level of the EDS spectrum. Additionally, the EDS technique could not be applicable for the detection of $Ar_n(O_2)_m^+$ traces.

3.5. Mechanism of Irradiation-Induced Structural Transformation in Ti_2SnC TNCFs

Although all of Sn containing ternary Ti_2SnC M_2AX phase is very studied, the role of Sn element is still under debate. It is well known that Sn belongs to the Carbon family, group 14 (IVA) of the periodic table. Unlike other elements in the group, the Sn exists in two different allotropes, metallic β-Sn (malleable) and nonmetallic α-Sn (brittle). Despite that β-Sn is the more common stable form, back transition process, from α-Sn to β-Sn at low temperatures of −50 °C is also well documented. This transition is called tin pest and hints at the different properties of Sn-based compounds. Apart from the recently reported ability of Sn to segregate in the early stages of crystallization and to act as heterogeneous nucleation sites for the secondary precipitated phase [40], other competing mechanisms claimed out that the Sn atoms could be activated to excessive secondary segregation into facets along with the site of the yet crystallized matrix [41].

The segregation of β-Sn and formation of dual β-Sn/Ti$_{0.98}$C heterogeneous structure in irradiated Ti$_2$SnC_Ar$^+$TNCF was unraveled using atomic level direct experimental STEM observation (see Section 3.4). Close inspection of structures is present in the aberration-corrected HAADF images in Figure 7. The original ADF images are filtered using the annular mask tool in Digital Micrograph to remove high-frequency noise, and presented at the same magnification. Figure 7a reveals that Ti$_2$SnC_AGTNCF was dislocation-free, but lattice parameters obtained by Single Crystal software (Oxford, England) based on SEAD patterns were estimated to be larger than proposed for hexagonal Ti$_2$SnC (JCPDS PDF 01-089-5590, space group P6$_3$/mmc). This observation suggests that residual strain in Ti$_2$SnC_AGTNCF has remained during the preparation.

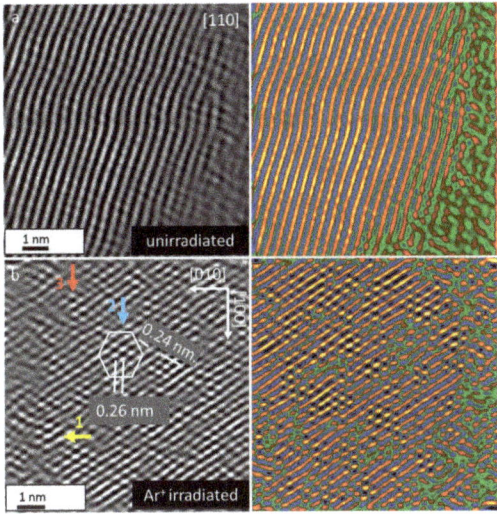

Figure 7. Aberration-corrected HAADF image of (**a**) Ti$_2$SnC_AGTNCF along the [110] direction and (**b**) Ti$_2$SnC_Ar$^+$TNCF. The regions of distorted structure are indicated by yellow, blue and red arrows. The atomic arrangements of β-Sn with spacing d$_{(110)}$ of 0.26 nm interacted with Ti$_{0.98}$C with spacing d$_{(111)}$ of 0.24 nm are well recognized. Corresponding filtered images to (**a**) and (**b**) by annular mask tool in Digital Micrograph are presented at the same magnification.

The Ti$_2$SnC_Ar$^+$TNCF (Figure 7b) was found to have a highly distorted structure. The regions of distorted structure are indicated with yellow, blue, and red arrows. Suffering from Ar$^+$ ion beam irradiation, the Sn atomic displacement phenomenon, as the first consequence of irradiation, can result in the formation of point defects in Ti$_2$SnC lattice [42]. Additionally, irradiation-induced dislocations can provide channels for very fast Sn mass transport. Coming back to the thickness of the Ti$_2$SnC_AGTNCF estimated to be in the diapason 460–920 × 10^{15} cm^{-2} (Section 3.1), an onset of the metamorphic layer due to melting of Sn (231.9 °C) could take place during the Ar$^+$ ion-beam irradiation. For Ti$_2$SnC, the migration energy barrier of Sn determined by ab initio calculation was found to be low enough (0.66 eV) to allow the self-diffusion of Sn atoms. These irradiation-induced defects can generate atomic transport in Ti$_2$SnC lattice and as a sequence, an extreme case, when all Sn atoms are extracted from Ti$_2$SnC could have occurred [43,44]. The STEM analysis reveals that in the region of the distorted structure indicated with a blue arrow, an isolated particle having a spacing of 0.26 nm that could be attributed to the β-Sn, is in contact with the particle, having space of 0.24 nm, that corresponded to Ti$_{0.98}$C. Even though staring

Ti$_2$SnC_AGTNCF has M$_2$AX stochiometric, Ar$^+$ ion-beam irradiation lowered stability of the Ti$_2$SnC and promoted its decomposition into Sn and Ti$_{0.98}$C. Our observations are in line with the attempts to correlate lattice parameters' c/a ratio with the stability of the 211 MAX phases as a function of Sn concentration. In addition, Ti$_2$SnC combines a small M-atom with a large A-atom and the distortions due to the steric effect in both building blocks of MAX, octahedral and trigonal prisms, should be considered [45]. Therefore, radiation-induced dislocations and point defects in Ti$_2$SnC can trigger diffusion and segregation of Sn atoms in irradiation-induced metastable Ti$_2$SnC structure [46]. The rate at which the Sn concentration increases depends probably on the Ti$_2$SnC NCs orientation, i.e., the precipitation of β-Sn has a specific crystallographic orientation relationship with the Ti$_2$SnC matrix. In our case, β-Sn nucleated on Ti$_{0.98}$C along the (110) planes as a result of the lattice stress, which induced a ⟨110⟩-oriented β-Sn pattern on the Ti$_{0.98}$C surface (see Figure 7b) [47–49]. Concerning the above-discussed results, a model for describing the irradiation-induced behavior of Ti$_2$SnC could be proposed to follow the steps: introducing of metastable Ti$_2$SnC phase → spontaneously growth of Sn core at the initial stage of irradiation → interaction between Sn core and irradiation-induced defects → remove of metal β-Sn and restoring of Ti$_2$SnC to equilibrium Ti$_{0.98}$C concentration [21,50–52].

3.6. Structural Analysis of the Ti$_2$SnC_PPS with HRTEM/SAED

Microstructure and phase composition of Ti$_2$SnC PPS were investigated by HRTEM/SAED and XRD. HRTEM micrograph in Figure 8a demonstrates another method of preparation graded material with various particle shapes. High magnification from the red boxed region in Figure 8a confirmed plates with an average size of 200 nm each, decorated with small nanograins located on the top edge of each plate (see Figure 8b). The corresponding SAED pattern (inset of Figure 8b) depicts the well-crystallized hexagonal Ti$_2$SnC with resolved (100) and (101) lattice plane (JCPDS PDF 01-089-5590, space group P6$_3$/mmc) along the [110] zone axis. An amorphous layer with a thickness of 10 nm (marked with red arrows in Figure 8b) was also well recognized. As expected, the HRTEM image (Figure 8b$_1$) of single nanograin confirmed lattice fringes with spacing 0.23 nm observed for Ti$_2$SnC. An HRTEM micrograph in Figure 8c corresponding to the blue framed area in Figure 8a revealed particles with different morphology as compared to the plate-shaped particles. Single NC with a length of about 50 nm and anisotropic 1D growth achieved through the preferred [111] orientation can be observed. The corresponding SAED pattern (inset in Figure 8c) confirmed the well-crystallized cubic TiC$_{0.55}$ with resolved (111) and (200) lattice plane and highly ordered lattice fringes with d$_{(111)}$ spacing of 0.24 nm (Figure 8c$_1$) observed for TiC$_{0.55}$ with JCPDS PDF No. 04-018-5143.

Our HRTEM/SAED observations are in line with the XRD analysis of the Ti$_2$SnC_PPS. As one can see in Figure S4, the XRD pattern exhibits 70.4% of a single Ti$_2$SnC phase following JCPDS PDF 04-005-0037. The diffraction peaks are sharp and confirmed a sample with high crystallinity. The dominance of the highest peaks, i.e., (103) for the in-plane pattern of hexagonal Ti$_2$SnC phase, is well recognized. The calculated value for NPs size by Scherrer's formula was 120 nm [53]. In addition, there is a presence of 11.2% TiC$_{0.55}$ JCPDS PDF 04-018-5143 phase and 9% of Sn JCPDS PDF04-004-7745. Our XRD results are consistent with those of Li et al. [13] as well as with M.W. Barsoum, [54] which confirmed that the content of Ti$_2$SnC increases with increasing the temperature. Therefore, when the reaction temperature increases up to 1250 °C, the Ti$_2$SnC becomes the prevailing phase. Additionally, our results confirmed that when Sn presents in the composition range lower than 10%, no stable intermetallic impurities such as Ti$_3$Sn, Ti$_6$Sn$_5$, Ti$_2$Sn, and Ti$_5$Sn$_3$ will be formed, and Ti$_2$SnC can retain a single phase up to 70% homogeneity [55,56].

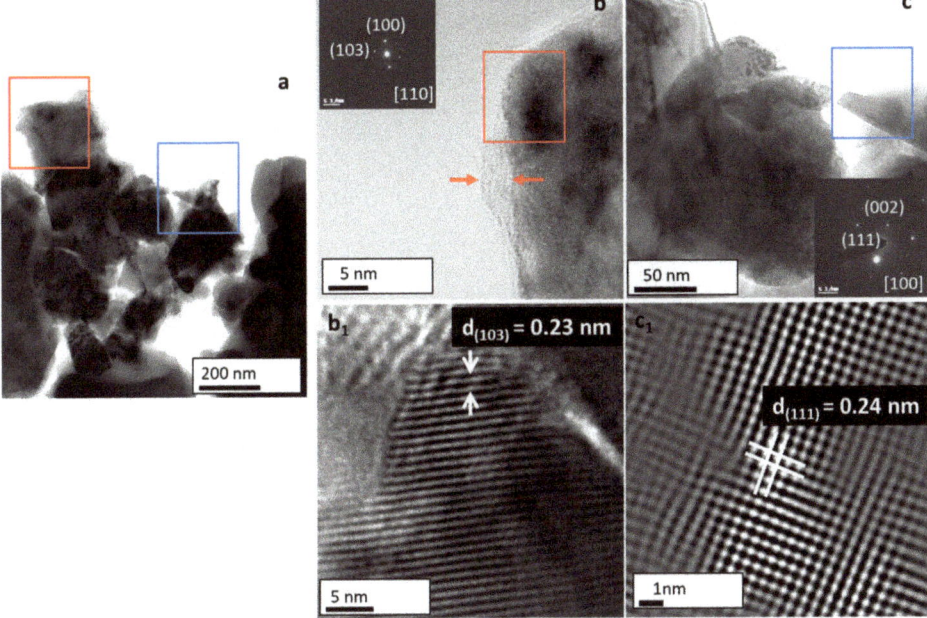

Figure 8. HRTEM study of Ti$_2$SnC PPS. (**a**) Representative low magnification micrograph of Ti$_2$SnC. (**b**) High magnification of red boxed region in (**a**) and SAED pattern with the (110) and (211) planes observed for cubic Sn. (**c**) High magnification of blue boxed region in (**a**) and SAED pattern with the (002) and (100) planes observed for hexagonal Ti$_2$SnC. (**b$_1$**) The magnified HRTEM image of the selected grain in red boxed area in (**b**) with lattice spacing matching these indexed in SAED pattern and (**c$_1$**) high magnification image of the selected grain in blue boxed area of Ti$_2$SnC in (**c**) with lattice spacing matching these indexed in SAED pattern.

3.7. Nanomechanical Properties Ti$_2$SnC_AG, Ti$_2$SnC_Ar$^+$TNCFs and Ti$_2$SnC_PPS

In this section, we consider the mechanical properties of Ti$_2$SnC_AGTNCF, Ti$_2$SnC_Ar$^+$TNCF, and Ti$_2$SnC_PPS. Table 1 shows values of Young's modulus (E) and hardness (H) calculated from contact depths of 10–40 nm using linear extrapolation to zero depth (from 10–80 nm for Ti$_2$SnC_PPS samples, respectively). The reduced modulus (Er) and hardness (H) were evaluated for each loading step by the Oliver and Pharr method [57]. Young's modulus (E) was calculated from the reduced modulus based on the assumption of the sample Poisson's ratio of 0.24 [58]. Figures 9 and 10 show results of elastic moduli and hardness in individual points represented by dots in the figures while red lines in Figures 9 and 10 represent a linear fit from the 10–40 nm or 10–80 nm region on respective samples. The non-constant trend indicates an influence of the harder substrate for larger penetration depths meaning the true surface properties in the sub-10 nm region can be even lower. Extrapolated values to zero depth are theoretical surface characteristics not influenced by the substrate effects on Ti$_2$SnC_AG/Ar$^+$TNCF samples or structural effects in the case of the bulk samples Ti$_2$SnC_PPS pristine (non-irradiated) (Figure 11) and Ti$_2$SnC_PPS irradiated (Figure 12). Nevertheless, any comparison of pristine and irradiated samples made from the results holds. The values of the substrate Si/(001) wafer as reference material are also included [17].

Table 1. Young's modulus (E) and hardness (H) (values of linear fit at zero depth) of Ti$_2$SnC_AGTNCF (film), Ti$_2$SnC_Ar$^+$TNCF (film), Ti$_2$SnC_PPS pristine (bulk), Ti$_2$SnC_PPS irradiated (bulk), and Si substrate.

Sample	E(GPa)	H(GPa)
Ti$_2$SnC_AGTNCF (film)	87.7	2.48
Ti$_2$SnC_Ar$^+$TNCF (film)	97.6	2.44
Ti$_2$SnC_PPS pristine (bulk)	104.9	2.56
Ti$_2$SnC_PPS irradiated (bulk)	100.8	2.08
Si substrate, ref. [17]	166.6	15.3

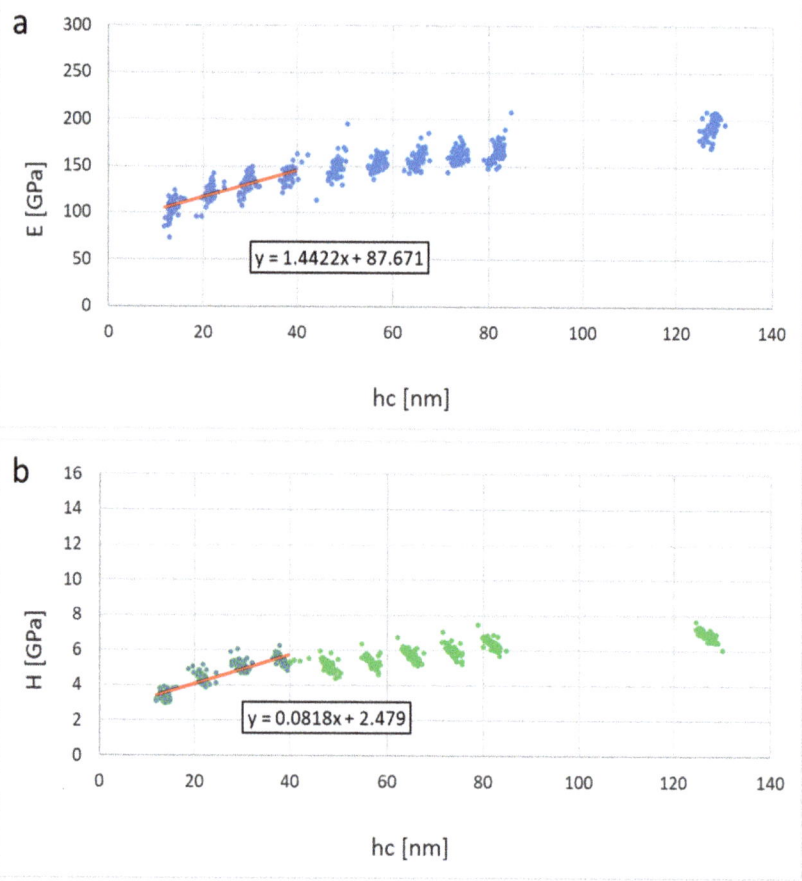

Figure 9. Results for Ti$_2$SnC_AGTNCF (**a**) Young's modulus vs. contact depth, (**b**) hardness vs. contact depth.

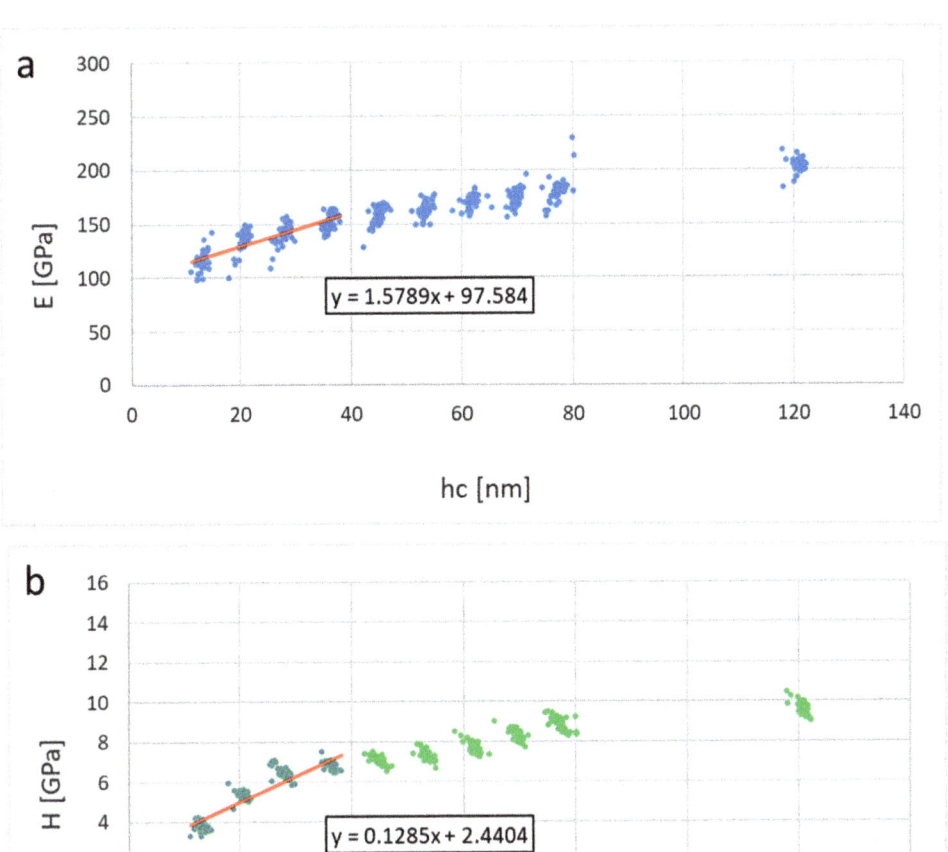

Figure 10. Results for Ti$_2$SnC_Ar$^+$TNCF (**a**) Young's modulus vs. contact depth, (**b**) hardness vs. contact depth.

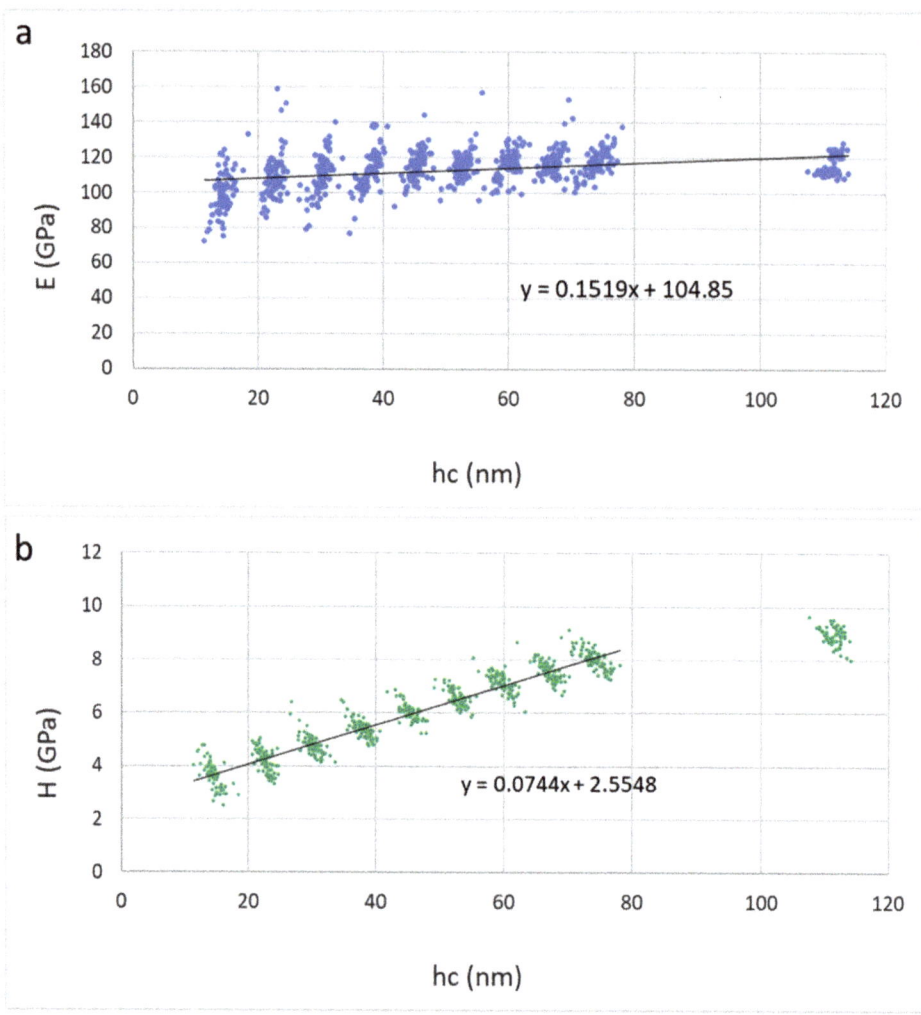

Figure 11. Results for Ti_2SnC_PPS pristine (**a**) Young's modulus vs. contact depth, (**b**) hardness vs. contact depth.

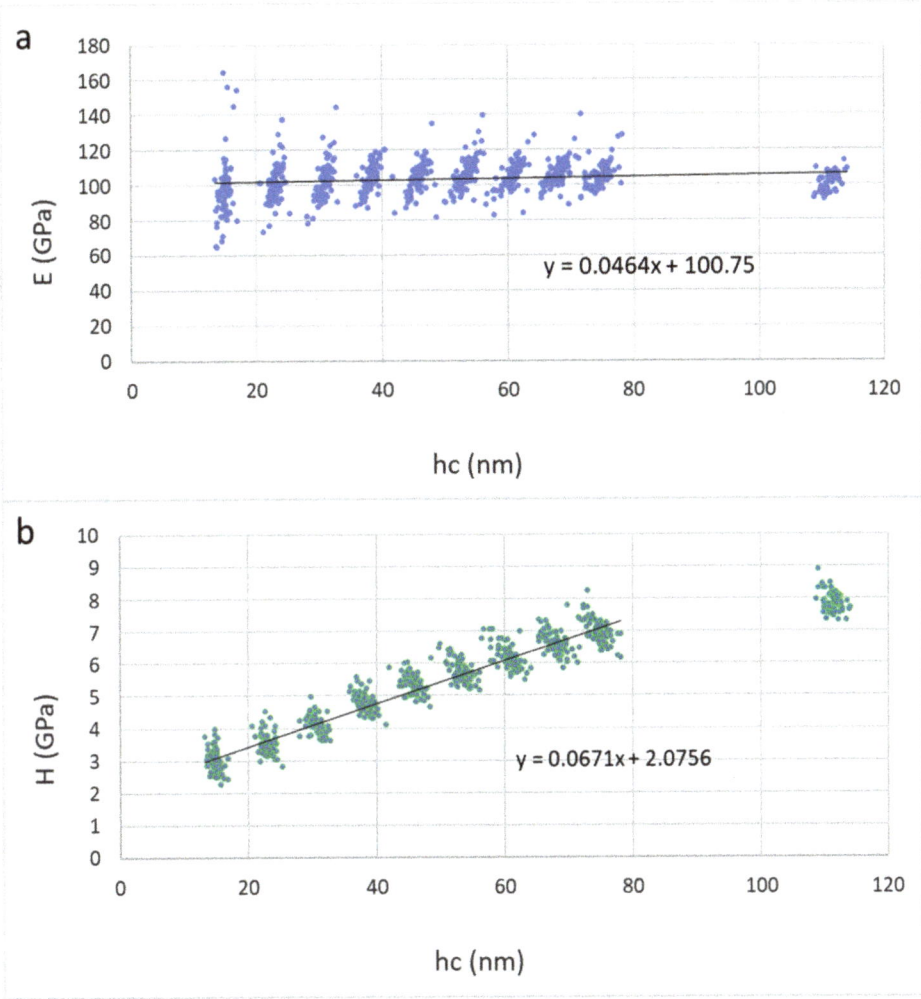

Figure 12. Results for Ti$_2$SnC_PPS irradiated (**a**) Young's modulus vs. contact depth, (**b**) hardness vs. contact depth.

The data for the irradiated Ti$_2$SnC_Ar$^+$TNCF shows commensurable hardness with Ti$_2$SnC_AGTNCF, which means that even irradiated with an Ar$^+$ ion beam, this material may have a high resistance to plastic straining. It seems logical to predict the role of Sn in affecting the mechanical properties of the irradiated film. It was recently reported [58] that a negligible amount of 0.1 at.% of Sn in binary Al/Cu alloys could enhance the hardening of the resultant Al/Cu/Sn material even in low temperatures (100–200 °C). Our EDS analysis finds out ~20 at.% β-Sn in Ti$_2$SnC_Ar$^+$TNCF. Therefore, our results are in line with the statement that the concentration of Sn is the most liable concerning the hardness properties of the materials. On the other hand, more significant elastic modulus increases than those for the Ti$_2$SnC_AGTNCF were observed. Based on the HAADF-STEM results, we could assume that the microsegregation of Sn atoms on the Ti$_{0.98}$C surface could involve local shear strain [59,60]. Local modulation of the Ti$_2$SnC_Ar$^+$TNCF structure with point defects/voids (marked with red arrows in Figure S5a–d) is well established [61]. Probably due to a low degree of surface defects and low concentration of microsegregation conducted by Sn, the elastic modulus of the Ti$_2$SnC_Ar$^+$TNCF is not seriously affected. The nanoindentation results for Ti$_2$SnC_AGTNCF and Ti$_2$SnC_Ar$^+$TNCFs are graphically presented in Figures 9 and 10. The hardness value and Young's modulus for Ti$_2$SnC_PPB are higher when compared to those obtained forTi$_2$SnC_AGTNCF and Ti$_2$SnC_Ar$^+$TNCFs (Table 1). Additionally, Young's modulus for Ti$_2$SnC_PPB irradiated is higher than those of Ti$_2$SnC_AGTNCF and Ti$_2$SnC_Ar$^+$TNCFs. The Young's modulus and hardness for Ti$_2$SnC_PPB (pristine) and Ti$_2$SnC_PPB irradiated are graphically presented in Figures 11 and 12, respectively. Although it is hard to generalize, the reason for such a difference between mechanical properties of bulk and thin films may be a dependency on material properties. For instance: (i) the Ti$_2$SnC_AGTNCF and Ti$_2$SnC_Ar$^+$TNCF are thin enough, which can lead to their higher defect density when correlated with bulk Ti$_2$SnC PPS, (ii) preferred orientation of grains in a sputter-deposited c-axis oriented Ti$_2$SnC films. We can assume that irradiation could maximize the coupling between grains that can reflect the degree of texture and mechanical properties as well [62,63], (iii) differences in grain size between Ti$_2$SnC_AGTNCF (~13–14 nm) and Ti$_2$SnC_Ar$^+$TNCF (~15–16 nm), and Ti$_2$SnC_PPB (grains size $\geq$100 nm) (see the size distribution for Ti$_2$SnC_AG and Ti$_2$SnC_Ar$^+$TNCF in Figure S6 obtained by ImageJ software (Madison, WI, USA)) [64], and Section 3.6. Surprisingly, the particle size distribution of the irradiated Ti$_2$SnC_Ar$^+$NCTF was found to follow the same order as before irradiation (see Figure S6), probably due to variation of the phase composition in Ti$_2$SnC_Ar$^+$TNCF. This result suggested that non-only thin nanocrystalline films with an average size below 20 nm could be considered as effective materials with enhanced radiation damage tolerance, but powdered polycrystalline Ti$_2$SnC could be considered as a promising resistant material when irradiated with Ar$^+$ ion beam, the fluence of which does not go over 10^{15} cm^{-2} [65,66].

4. Conclusions

- In conclusion, the microstructure and mechanical properties of Ti$_2$SnC TNCFs, synthesized by an unconventional low-energy ion facility (LEIF) based on Ar$^+$ ion beam sputtering of the Ti, Sn, and C targets have been investigated. Combining high-resolution HAADF-STEM analysis with simulations of SAED patterns, observed that Ti$_2$SnC_AGTNCFs coexist with SnO due to oxidation of Sn during the preparation process. A significant microstructural instability was observed after irradiation of the Ti$_2$SnC_AGTNCFs with Ar$^+$ ion beam having an energy of 30 keV and fluence of 1.10^{15} cm^{-2}. The results from simulated SEAD patterns are compatible with experimental HAADF-STEM analysis and have suggested the existence of a heterostructure composed of binary Ti$_{0.98}$C carbide and metallic β-Sn, which could be attributed to the irradiation-induced instability of the ultrathin Ti$_2$SnC film. In addition, Ar$^+$ ion-beam irradiation-induced dislocation and point defects can provide channels for very fast Sn mass transport. The analysis by nanoindentation showed that the irradiated Ti$_2$SnC TNCFs and irradiated Ti$_2$SnC_PPS exhibited promising Young's modulus and hardness even for the locally disordered structure in Ti$_2$SnC TNCFs. This fact opens the possibility of exploiting the β-Sn/Ti$_{0.98}$C structure as a composite where a harsh radiation environment could have occurred.
- HRTEM/SAED observations and XRD analyses of the Ti$_2$SnC_PPS documented 70.4% of a single Ti$_2$SnC phase. The calculated value for NPs size by Scherrer's formula was estimated to be 120 nm. The presence of 9% of Sn avoided the formation of stable intermetallic impurities (Ti$_3$Sn, Ti$_6$Sn$_5$, Ti$_2$Sn, and Ti$_5$Sn$_3$) in Ti$_2$SnC_PPS. The Ti$_2$SnC_PPS irradiated yield the lowest hardness (H) when compared with Ti$_2$SnC_PPS and unirradiated and irradiated Ti$_2$SnC TNCFs. Probably, the low degree of nano crystallinity and tendency to agglomeration upon irradiation can contribute to the surface hardness of polycrystalline bulk materials.
- The approach here presented may be extendable to other M$_2$AX nanostructured materials, and can keep attention for material science applications ranging from protective

nanocoating films, ion-beam irradiation resistant parts for nuclear applications and nanoceramics, to their utilization as a precursor for MX phases.

Supplementary Materials: The following are available online at https://www.mdpi.com/article/10.3390/nano12030307/s1, Figure S1: Morphology of Ti_2SnC M_2AX samples (a) Ti_2SnC_AGTNCF, (b) $Ti_2SnC\_Ar^+TNCF$ and (c) Ti_2SnC_PPS obtained by SEM with SE detector; Figure S2: (a) EDS-STEM analysis and (b) EDS-STEM mapping Ti_2SnC_AGTNCF. HAADF-STEM image of Ti_2SnC_AGTNCF at low magnification and Table with at. % of elements are present as an inset in Figure S2a; Figure S3: (a) EDS-STEM analysis and (b) EDS-STEM mapping $Ti_2SnC\_Ar^+TNCF$. HAADF-STEM image of $Ti_2SnC\_Ar^+TNCF$ at low magnification and Table with at. % of elements are present as an inset in Figure S3a; Figure S4: XRD pattern of Ti_2SnC PPS: Titanium Tin Carbide Ti_2SnC–70.4% Reference code: 04-005-0037, Tin Sn–9.1% Reference code: 04-004-7745, Titanium Carbide $TiC0.55$–11.2% Reference code: 04-018-5143, Titanium Tin Carbide Ti_2SnC–9.3% Reference code: 04-005-0049; Figure S5: (a–d) HRTEM images of irradiation-induced defect cores in $Ti_2SnC\_Ar^+TNCF$. The yellow marked regions indicated defect cores in a different region of interest, where the interaction of dislocation lines, voids, and atoms segregated on the surface (marked with red arrows) are well visible; Figure S6. STEM images and corresponding particle size distribution estimated by ImageJ software [64]. (a) Ti_2SnC_AGTNCF and (b) $Ti_2SnC\_Ar^+TNCF$.

Author Contributions: Conceptualization, S.B. and J.V.; methodology, J.P.; software, G.C. and B.I.; validation, B.I., V.L. and J.K.; formal analysis, G.C.; investigation, J.N.; resources, S.B.; data curation, J.N.; writing—original draft preparation, J.V. and B.I.; writing—review and editing, S.B.; visualization, G.C.; supervision, S.B., Š.M. and R.K.; project administration, J.V.; funding acquisition, S.B. and J.V. All authors have read and agreed to the published version of the manuscript.

Funding: This research is funded by the Czech Science Foundation (project No. 18-21677S). This work was supported by OP VVV Project Development of new nano and micro coatings on the surface of selected metallic materials–NANOTECH ITI II. Reg. No. CZ.02.1.01/0.0/0.0/18_069/0010045. The aberration-corrected JEOL ARM200CF at UIC Chicago was acquired using two grants from the National Science Foundation (DMR-0959470, DMR-1626065).

Institutional Review Board Statement: Not applicable.

Informed Consent Statement: Informed consent was obtained from all subjects involved in the study.

Data Availability Statement: MDPI Research Data Policies at https://www.mdpi.com/ethics (accessed on 12 January 2022).

Acknowledgments: We would like to thank the reviewers for their thoughtful comments and efforts towards improving our manuscript.

Conflicts of Interest: The authors declare no conflict of interest.

References

1. Barsoum, M.W. The $M_{n+1}AX_n$ Phases: A new class of solids: Thermodynamically stable nanolaminates. *Prog. Solid State Chem.* **2000**, *28*, 201–281. [CrossRef]
2. Eklund, P.; Beckers, M.; Jansson, U.; Hogberg, H.; Hultman, L. The $M_{n+1}AX_n$ Phases: Materials science and thin-film processing. *Thin. Solid Film.* **2010**, *518*, 1851–1878. [CrossRef]
3. Jeitschko, W.; Nowotny, H.; Benesovsky, F. Kohlenstoffhaltige ternäre verbindungen (H-phase). *Mon. Chem. Verwandte Teile And. Wiss.* **1963**, *94*, 672–676. [CrossRef]
4. Barsoum, M.W.; El-Raghy, T. The Max Phases: Unique new carbide and nitride materials: Ternary ceramics turn out to be surprisingly soft and machinable, yet also heat-tolerant, strong and lightweight. *Am. Sci.* **2001**, *89*, 334–343. [CrossRef]
5. Tallman, D.J. On the Potential of Max Phases for Nuclear Applications. Ph.D. Thesis, Drexel University, Philadelphia, PA, USA, June 2015.
6. Whittle, K.R.; Blackford, M.G.; Aughterson, R.D.; Moricca, S.; Lumpkin, G.R.; Riley, D.P.; Zaluzec, N.J. Radiation tolerance of $M_{n+1}AX_n$ Phases, Ti_3AlC_2 and Ti_3SiC_2. *Acta Mater.* **2010**, *58*, 4362–4368. [CrossRef]
7. Liu, C.; Shi, L.; Qi, Q.; O'Connor, D.J.; King, B.V.; Kisi, E.H.; Qing, X.B.; Wang, B.Y. Surface damage of Ti_3SiC_2 By Mev Iodine bombardment. *Nucl. Instrum. Methods Phys. Res. Sect. B: Beam Interact. Mater. At.* **2013**, *307*, 536–540. [CrossRef]
8. Wu, J.Y.; Zhou, Y.C.; Wang, J.Y. Tribological behavior of Ti_2SnC particulate reinforced copper matrix composites. *Mater. Sci. Eng.* **2006**, *422*, 266–271. [CrossRef]

9. Dong, H.Y.; Yan, C.K.; Chen, S.Q.; Zhou, Y.C. Solid–Liquid reaction synthesis and thermal stability of Ti$_2$SnC powders. *J. Mater. Chem.* **2001**, *11*, 1402–1407. [CrossRef]
10. Li, Y.; Bai, P. The microstructural evolution of Ti$_2$SnC from Sn–Ti–C system by Self-propagating High-temperature Synthesis (SHS). *Int. J. Refract. Met. Hard Mater.* **2011**, *29*, 751–754. [CrossRef]
11. Li, S.; Bei, G.-P.; Zhai, H.-X.; Zhou, Y. Bimodal microstructure and reaction mechanism of Ti$_2$SnC synthesized by a high—temperature reaction using Ti/Sn/C and Ti/Sn/TiC powder compacts. *J. Am. Ceram. Soc.* **2006**, *89*, 3617–3623. [CrossRef]
12. Yeh, C.L.; Kuo, C.W. Effects of TiC addition on formation of Ti$_2$SnC by self-propagating combustion of Ti–Sn–C–TiC powder compacts. *J. Alloy. Compd.* **2010**, *502*, 461–465. [CrossRef]
13. Li, S.; Bei, G.-P.; Zhai, H.-X.; Zhou, Y. Synthesis of Ti$_2$SnC from Ti/Sn/TiC powder mixtures by pressureless sintering technique. *Mater. Lett.* **2006**, *60*, 3530–3532. [CrossRef]
14. Guillon, O.; Gonzales-Julian, J.; Dargatz, B.; Kessel, T.; Schierning, G.; Rathel, J.; Herrmann, M. Field-assisted sintering technology/spark plasma sintering: Mechanisms, materials, and technology developments. *Adv. Eng. Mater.* **2014**, *16*, 830–849. [CrossRef]
15. Lu, C.; Wang, Y.; Wang, X.; Zhang, J. Synthesis of Ti$_2$SnC under optimized experimental parameters of pressureless spark plasma sintering assisted by al addition. *Adv. Mater. Sci. Eng.* **2018**, *2018*, 9861894. [CrossRef]
16. Canam Project. Available online: Http://Www.Ujf.Cas.Cz/En/Research-Development/Large-Research-Infrastructures-And-Centres/Canam/About-The-Project/ (accessed on 12 December 2021).
17. Bakardjieva, S.; Ceccio, G.; Vacik, J.; Calcagno, L.; Cannavo, A.; Horak, P.; Lavrentiev, V.; Nemecek, J.; Michalcova, A.; Klie, R. Surface morphology and mechanical properties changes induced In Ti$_3$InC$_2$ (M$_3$AX$_2$) thin nanocrystalline films by irradiation of 100 Kev Ne$^+$ ions. *Surf. Coat. Technol.* **2021**, *426*, 127775. [CrossRef]
18. Gates-Rector, S.D.; Blanton, T.N. The Powder Diffraction File: A Quality Materials Characterization Database. *Powder Diffr.* **2019**, *34*, 352–360. [CrossRef]
19. Lábár, J.L. Consistent Index. A (Set) SAED Pattern(S) ProcessDiffraction Program. *Ultramicroscopy* **2005**, *103*, 237–249.
20. Klie, R.F.; Zheng, J.C.; Zhu, Y.; Varela, M.; Wu, J.; Leighton, C. Direct measurement of the low-temperature spin-state transition in LaCoO$_3$. *Phys. Rev. Lett.* **2007**, *99*, 047203. [CrossRef]
21. Zhang, J.; Liu, B.; Wang, J.Y.; Zhou, Y.C. Low-temperature instability of Ti$_2$SnC: A combined transmission electron microscopy, differential scanning calorimetry, and X-ray diffraction investigations. *J. Mater. Res.* **2009**, *24*, 39–49. [CrossRef]
22. Cannavò, A.; Vacik, J.; Bakardjieva, S.; Kupcik, J.; Lavrentiev, V.; Ceccio, G.; Horak, P.; Nemecek, J.; Calcagno, L. Effect of medium energy He$^+$, Ne$^+$ And Ar$^+$ ion irradiation on the Hf-In-C thin film composites. *Thin. Solid Films* **2021**, *743*, 139025. [CrossRef]
23. Dias, J.S.; Batista, F.R.M.; Bacani, R.; Triboni, E.R. Structural characterization of SnO nanoparticles synthesized by the hydrothermal and microwave routes. *Sci. Rep.* **2020**, *10*, 9446. [CrossRef]
24. Vázquez-López, A.; Maestre, D.; Ramirez-Castellanos, J.; Cremades, A. In situ local oxidation of SnO induced by laser irradiation: A stability study. *Nanomaterials* **2021**, *11*, 976. [CrossRef]
25. Jaśkaniec, S.; Kavanagh, S.R.; Coelho, J.; Ryan, S.; Hobbs, C.; Walsh, A.; Scanlon, D.O.; Nicolosi, V. Solvent engineered synthesis of layered SnO for high-performance anodes. *npj 2D Mater. Appl.* **2021**, *5*, 27. [CrossRef]
26. Campo, C.M.; Rodríguez, J.E.; Ramírez, A.E. Thermal behaviour of romarchite phase SnO in different atmospheres: A hypothesis about the phase transformation. *Heliyon* **2016**, *2*, E00112. [CrossRef] [PubMed]
27. Moreno, M.S.; Egerton, R.F.; Midgley, P.A. Differentiation of Tin Oxides using electron energy-loss spectroscopy. *Phys. Rev. B* **2004**, *69*, 233304. [CrossRef]
28. Adelmann, C.; Brault, J.; Rouviere, J.-R.; Mariette, H.; Mula, G.; Daudin, B. Atomic-layer epitaxy of GaN quantum wells and quantum dots on (0001) AlN. *J. Appl. Phys.* **2002**, *91*, 5498–5500. [CrossRef]
29. Voyles, P.M.; Muller, D.A.; Grazul, J.L.; Citrin, P.H.; Gossmann, H.J. Atomic-scale imaging of individual dopant atoms and clusters in highly N-Type bulk Si. *Nature* **2002**, *416*, 826–829. [CrossRef]
30. Rafferty, B.; Nellist, D.; Pennycook, J. On the origin of transverse incoherence in z-contrast stem. *Microscopy* **2001**, *50*, 227–233. [CrossRef]
31. Williams, R.K.; Wiffen, F.W.; Bentley, J.; Stiegler, J.O. Irradiation induced precipitation in tungsten based, W-Re alloys. *Metall. Trans.* **1983**, *14*, 655–666. [CrossRef]
32. Nordlund, K.; Averback, R.S. Point defect movement and annealing in collision cascades. *Phys. Rev.* **1997**, *56*, 2421. [CrossRef]
33. Haberland, H. A model for the processes happening in a rare-gas cluster after ionization. *Surf. Sci.* **1985**, *156*, 305–312. [CrossRef]
34. Ding, A.; Futrell, J.H.; Cassidy, R.A.; Cordis, L.; Hesslich, J. Mass spectrometric and photoionisation investigations of the structure of heterogeneous clusters. *Surf. Sci.* **1985**, *156*, 282–291. [CrossRef]
35. Begemann, W.; Dreihöfer, S.; Meiwes-Broer, K.H.; Lutz, H.O. Sputtered metal cluster ions: Unimolecular decomposition and collision induced fragmentation. *Z. Phys. D At. Mol. Clust.* **1986**, *3*, 183–188. [CrossRef]
36. Fondell, M.; Gorgoi, M.; Boman, M.; Lindblad, A. An HAXPES study of sn, SnS, SnO and SnO$_2$. *J. Electron. Spectrosc. Relat. Phenom.* **2014**, *195*, 195–199. [CrossRef]
37. Hayashi, Y.; Matsumoto, K. Fragmentation of Sputtered Neutrals by SNMS. *J. Mass Spectrom. Soc. Jpn.* **1991**, *39*, 93–96. [CrossRef]
38. Chapon, C.; Gillet, M.F.; Henry, C.R. (Eds.) *Small Particles and Inorganic Clusters: Proceedings of the Fourth International Meeting on Small Particles and Inorganic Clusters University Aix-Marseille III Aix-en-Provence*; Springer Science & Business Media: Marseille, France, 2012.

39. Bras, P.; Sterner, J.; Platzer-Björkman, C. Investigation of blister formation in sputtered Cu_2ZnSnS_4 absorbers for thin film solar cells. *J. Vac. Sci. Technol. A: Vac. Surf. Film.* **2015**, *33*, 061201. [CrossRef]
40. Hardy, H.K.; Silcock, J.M. The Phase Sections at 500 and 350 °C of Al Rich Al-Cu-Li Alloys. *J. Inst. Met.* **1955**, *84*, 423–428.
41. Parish, C.M.; Field, K.G.; Certain, A.G.; Wharry, J.P. Application of STEM Characterization for Investigating Radiation Effects in BCC Fe-Based. *J. Mater. Res.* **2015**, *30*, 1275–1289. [CrossRef]
42. Schäublin, R. Effect of helium on irradiation-induced hardening of iron: A simulation point of view. *J. Nucl. Mater.* **2007**, *362*, 152–160. [CrossRef]
43. Riviere, J.P. Radiation Induced Point Defects and Diffusion. In *Application of Particle and Laser Beams in Materials Technology*; Misaelides, P., Ed.; NATO ASI Series (Applied Sciences); Springer: Dordrecht, The Netherlands, 1995; Volume 283. [CrossRef]
44. Lecterc, S.J.; Li, X.; Lescoat, M.L.; Fortuna, F.; Gentils, A. Microstructure of Au-ion irradiated 316L and FeNiCr austenitic stainless steels. *J. Nucl. Mater.* **2016**, *480*, 436–446.
45. Lapauw, T.; Tunca, B.; Potashnikov, D.; Pesach, A.; Ozeri, O.; Vleugels, J.; Lambrinou, K. The double solid solution (Zr, Nb)$_2$(Al, Sn)C MAX phase: A steric stability approach. *Sci. Rep.* **2018**, *8*, 12801. [CrossRef] [PubMed]
46. Titus, M.S.; Rhein, R.K.; Wells, P.B.; Dodge, P.C.; Viswanathan, G.B.; Mills, M.J.; Van derVan, A.; Pollock, T.M. Solute segregation and deviation from bulk thermodynamics at nanoscale crystalline defects. *Sci. Adv.* **2016**, *2*, e1601796. [CrossRef]
47. Mukherjee, S.; Assali, S.; Oussama Moutanabbir, O. Atomic Pathways of Solute Segregation in the Vicinity of Nanoscale Defects. *Nano Lett.* **2021**, *23*, 9882–9888. [CrossRef] [PubMed]
48. Rathika, R.; Kovendhan, M.; Paul Joseph, D.; Vijayarangamuthu, K.; Sendil Kamur, A.; Venkateswaran, C.; Asokan, K.; Johnson Jeyakumar, S. 200 MeV Ag^{15+} ion beam irradiation induced modifications in spray deposited MoO_3 thin films by fluence variation. *Nucl. Eng. Technol.* **2019**, *51*, 1983–1990. [CrossRef]
49. Wang, J.Y.; Zhou, Y.C.; Liao, T.; Zhang, J.; Lin, Z.J. A first principles investigation of the phase stability of Ti_2AlC with Al vacancies. *Scr. Mater.* **2008**, *58*, 227–231. [CrossRef]
50. Herring, C.; Galt, J.K. Elastic and plastic properties of very small metal specimens. *Phys. Rev.* **1952**, *85*, 1060. [CrossRef]
51. Sears, G.W. A mechanism of whisker growth. *Acta Metall.* **1955**, *3*, 367. [CrossRef]
52. Tu, K.N.; Thompson, R.D. Kinetics of interfacial reaction in bimetallic Cu-Sn thin films. *Acta Metall.* **1982**, *30*, 947–952. [CrossRef]
53. Zenou, V.Y.; Bakardjieva, S. Microstructural analysis of undoped and moderately Sc-doped TiO_2 anatase nanoparticles using Scherrer equation and Debye function analysis. *Mater. Charact.* **2018**, *144*, 287–296. [CrossRef]
54. Barsoum, M.W.; Yaroschuk, G.; Tyagi, S. Fabrication and characterization of M_2SnC (M= Ti, Zr, Hf and Nb). *Scr. Mater.* **1997**, *37*, 1583–1591. [CrossRef]
55. Yin, F.; Tedenac, J.C.; Gascoin, F. Thermodynamic modelling of the Ti–Sn system and calculation of the Co–Ti–Sn system. *Comput. Coupling Phase Diagr. Thermochem.* **2007**, *31*, 370–379. [CrossRef]
56. Yang, J.Y.; Kim, W.J. The effect of addition of Sn to copper on hot compressive deformation mechanisms, microstructural evolution and processing maps. *J. Mater. Res. Technol.* **2020**, *9*, 749–761. [CrossRef]
57. Oliver, W.C.; Pharr, G.M. An improved technique for determining hardness and elastic modulus using load and displacement sensing indentation experiments. *J. Mater. Res.* **1992**, *7*, 1564–1583. [CrossRef]
58. Lapauw, T.; Vanmeensel, K.; Lambrinou, K.; Vleugels, J. Rapid synthesis and elastic properties of fine-grained Ti_2SnC produced by spark plasma sintering. *J. Alloy. Compd.* **2015**, *631*, 72–76. [CrossRef]
59. Nie, J.F. Applications of atomic-resolution HAADF-STEM and EDS-STEM characterization of light alloys. In *IOP Conference Series: Materials Science and Engineering*; IOP Publishing: Riso, Denmark, 2017; p. 012005. [CrossRef]
60. Nie, J.F.; Aaronson, H.I.; Muddle, B.C. *Proceedings of an International Conference on Solid-Solid Phase Transform*; Koiwa, M., Otsuka, K., Miyazaki, T., Eds.; The Japan Institute of Metals: Tokyo, Japan, 1999; p. 157.
61. Nie, J.F.; Muddle, B.C. The lattice correspondence and diffusional-displacive phase transformations. *Mater. Forum* **1999**, *23*, 23–40.
62. Bakardjieva, S.; Horak, P.; Vacik, J.; Cannavo, A.; Lavrentiev, V.; Torrisi, A.; Michalcova, A.; Klie, R.; Rui, X.; Calcagno, L. Effect of Ar^+ irradiation of Ti_3InC_2 at different ion beam fluences. *Surf. Coat. Technol.* **2020**, *394*, 125834. [CrossRef]
63. Högberg, H.; Emmerlich, J.; Eklund, P.; Wilhelmsson, O.; Palmquist, J.P.; Jansson, U.; Hultman, L. Growth and property characterization of epitaxial MAX-phase thin films from the Ti$_{n+1}$ (Si, Ge, Sn) C$_n$ systems. In *Advances in Science and Technology*; Trans Tech Publications Ltd.: Acireale, Italy, 2006; Volume 45, pp. 2648–2655.
64. Rueden, C.T.; Schindelin, J.; Hiner, M.C.; DeZonia, B.E.; Walter, A.E.; Arena, E.T.; Eliceiri, K.W. ImageJ2: ImageJ for the next generation of scientific image data. *BMC Bioinform.* **2017**, *18*, 529. [CrossRef] [PubMed]
65. Kalita, P.; Ghosh, S.; Gutierrez, G.; Rajput, P.; Grover, V.; Sattonnay, G.; Avasthi, D.K. Grain size effect on the radiation damage tolerance of cubic zirconia against simultaneous low and high energy heavy ions: Nano triumphs bulk. *Sci Rep.* **2021**, *11*, 10886. [CrossRef]
66. Liew, P.J.; Yap, C.Y.; Wang, J. Surface modification and functionalization by electrical discharge coating: A comprehensive review. *Int. J. Extrem. Manuf.* **2020**, *2*, 012004. [CrossRef]

Article

Investigating the Interface between Ceramic Particles and Polymer Matrix in Hybrid Electrolytes by Electrochemical Strain Microscopy

Philipp M. Veelken [1,2], Maike Wirtz [1,2], Roland Schierholz [1], Hermann Tempel [1], Hans Kungl [1], Rüdiger-A. Eichel [1,2,3] and Florian Hausen [1,2,3,*]

1. Institute of Energy and Climate Research, IEK-9, Forschungszentrum Jülich, 52425 Jülich, Germany; ph.veelken@fz-juelich.de (P.M.V.); maike.wirtz@rwth-aachen.de (M.W.); r.schierholz@fz-juelich.de (R.S.); h.tempel@fz-juelich.de (H.T.); h.kungl@fz-juelich.de (H.K.); r.eichel@fz-juelich.de (R.-A.E.)
2. Institute of Physical Chemistry, RWTH Aachen University, Landoltweg 2, 52074 Aachen, Germany
3. Jülich-Aachen Research Alliance, Section JARA-Energy, 52425 Jülich, Germany
* Correspondence: f.hausen@fz-juelich.de; Tel.: +49-2461-61-4412

Abstract: The interface between ceramic particles and a polymer matrix in a hybrid electrolyte is studied with high spatial resolution by means of Electrochemical Strain Microscopy (ESM), an Atomic Force Microscope (AFM)-based technique. The electrolyte consists of polyethylene oxide with lithium bis(trifluoromethanesulfonyl)imide (PEO_6–LiTFSI) and $Li_{6.5}La_3Zr_{1.5}Ta_{0.5}O_{12}$ (LLZO:Ta). The individual components are differentiated by their respective contact resonance, ESM amplitude and friction signals. The ESM signal shows increased amplitudes and higher contact resonance frequencies on the ceramic particles, while lower amplitudes and lower contact resonance frequencies are present on the bulk polymer phase. The amplitude distribution of the hybrid electrolyte shows a broader distribution in comparison to pure PEO_6–LiTFSI. In the direct vicinity of the particles, an interfacial area with enhanced amplitude signals is found. These results are an important contribution to elucidate the influence of the ceramic–polymer interaction on the conductivity of hybrid electrolytes.

Keywords: Atomic Force Microscopy; Electrochemical Strain Microscopy; hybrid electrolyte; Energy Storage; lithium transport; lithium distribution; all-solid-state electrolytes

1. Introduction

Over the last few years, the interest in all-solid-state batteries (ASSBs) has increased due to their enhanced safety and theoretical capacity compared to conventional organic, liquid electrolyte batteries [1–3]. Polymers, ceramics and polymer/ceramic hybrid materials are under development for application in ASSBs. Polymer electrolytes allow for improved electrolyte–electrode interfaces compared to ceramic-based electrolytes due to their higher mechanical flexibility [4–6]. However, polymer electrolytes display comparably low ionic conductivities [5,7]. In contrast to polymer electrolytes, ceramic electrolytes demonstrate superior conductivities [8–10].

Ceramic electrolytes are brittle and inherently exhibit a high rigidity. Furthermore, their interface towards electrodes is hindered due to their rough surface structure. The current research focuses on overcoming these limitations by employing hybrid electrolytes based on a polymer matrix and added ceramic particles. Hybrid electrolytes exhibit important advantages of multiple solid-state electrolyte types, such as superior electrode–electrolyte contact and flexibility [11–14].

Keller et al. gave an excellent overview of recent developments and problems regarding different hybrid electrolyte types [15]. With the addition of ceramic particles into the polymer electrolyte, the goal is to increase the global conductivity of the polymer electrolyte while retaining the flexibility.

The literature points to different conductivity tendencies in polymer electrolytes with added ceramic electrolyte particles. There are reports showing an increase in the ionic conductivity with added ceramic particles [16–18]. However, there are also studies showing the complete opposite—a decrease in the ionic conductivity when the polymer electrolyte is filled with ceramic particles [19,20]. Hence, hybrid electrolytes require careful optimization of the lithium salt concentration in order to achieve high ionic conductivities [21].

The lithium ion conductivities of the individual ceramic and polymer components in hybrid electrolytes are only two of the factors that determine the overall conductivity of a hybrid electrolyte. A pronounced influence on the conductivity of the hybrids results from the modification of the material in the vicinity of ceramic particles and from the transition resistance between polymer and ceramic components in the hybrid. The presence of ceramic particles prevents local poly(ethylene oxide) (PEO) chain organization and leads to a high degree of disorder in the polymer neighboring the ceramic particles [22–24].

Moreover, in the interface region between the polymer and ceramic particles, Lewis acid–base interactions with the electrolyte ionic species form and promote lithium salt dissociation [24]. Dixit et al. found, by simulations, that, inside composite materials, an interfacial area forms between bulk polymer phases and single ceramic particles [25,26].

Furthermore, they modeled the interfacial conductivity by Effective Mean Field Theory and stated that the interfacial conductivity depends on the composition of the hybrid electrolyte, as, for 25 wt.% $Li_7La_3Zr_2O_{12}$ (LLZO) in PEO the interfacial conductivity was lower than in a hybrid film with 75 wt.% LLZO in PEO. The transition of the lithium ions across the interface between the polymer and the ceramic determines the possible pathways for the lithium transport within the hybrid electrolyte. The activation energies for the ion transfer across the polymer/ceramic (PEO/LLZO) interface were up to 96 kJ mol^{-1} (0.9 eV) [27], and interface resistances within the hybrid material were high.

Information from typically applied methods, such as electrochemical impedance spectroscopy (EIS) and cyclic voltammetry can be collected as an average from the entire sample, i.e., globally [21,28]. On the other hand, information with a high spatial resolution, i.e., on a local scale, are required to identify the transport path for the lithium ions and to improve the materials and cell designs. In particular, the interactions between different types of materials, such as polymers and ceramics in hybrid electrolytes have to be understood in detail.

Zheng et al. employed isotopically labeled ^{6}Li NMR in a LLZO-PEO hybrid electrolyte to observe the lithium diffusion inside cycled symmetrical battery cells [29]. After cycling, they found that, at a high ceramic content, the lithium attempted to primarily move through the ceramic phase. ^{6}Li NMR was also applied by Li et al., showing comparable results in $Li_{10}GeP_2S_{12}$-PEO [30]. For a lower ceramic content (10 wt.%), the main conduction pathway was through the polymer phase, while, with increasing ceramic content (>50 wt.%), the conduction pathway was mostly inside the ceramic phase.

Recently, based on NMR experiments, Ranque et al. suggested that the ion transport between the polymer and ceramic phase is possible also for low (10%) $Li_{6.55}Ga_{0.15}La_3Zr_2O_{12}$ content while being, however, comparably slow [31].

In this study, Electrochemical Strain Microscopy (ESM) was employed to investigate the local ionic conductivity in a hybrid electrolyte. Typically, ESM is used on electrode materials [32,33]. Generally, in Electrochemical Strain Microscopy (ESM) an alternating voltage with the same frequency as the Contact Resonance Frequency (CRF) between the conductive tip and sample is applied in contact mode. The electrical field at the tip forces mobile lithium ions inside the material towards or away from the tip [34,35]. The signal origin of ESM in electrolytes was discussed recently.

Schön et al. showed on $Li_{1.3}Al_{0.3}Ti_{1.7}(PO_4)_3$ (LATP) that the dominant contribution to the resulting ESM signal is caused by electrostatic forces [36]. A link between the chemical composition and local tip-sample interaction was found. The aim of this work is to investigate the interfacial area between ceramic particles and a surrounding polymer matrix, as the ion transport through this interfacial area is still under debate. Therefore, applying ESM on a hybrid all-solid-state electrolyte offers insights into the local ionic mobility and transport between different electrolyte materials.

2. Experimental Section

The preparation method of $Li_{6.5}La_3Zr_{1.5}Ta_{0.5}O_{12}$ (LLZO:Ta) is described in detail in [21]. It is important to note that OH groups are present on the ceramic particles. The synthesis of the hybrid electrolyte was performed in inert gas atmosphere inside a glovebox (MBraun, Stratham, NH, USA) to ensure the exclusion of water and oxygen (H_2O < 0.1 ppm, O_2 < 0.1 ppm). The hybrid electrolyte film was synthesized using the solution-casting method with anhydrous acetonitrile inside the glovebox. The ratio of ethylene oxide monomer groups to lithium ions (Li:EO = 1:x) was used to define the LiTFSI (Sigma-Aldrich, St. Louis, MO, USA, 99.95%) concentration.

In this case, a film with a ratio of Li:EO = 1:6 was synthesized. We dispersed 50 wt.% LLZO:Ta powder with respect to PEO (MW = 1,000,000 g mol^{-1}, Alfa Aesar, Ward Hill, MA, USA) in acetonitrile and added to the polyethylene oxide with lithium bis(trifluoromethanesulfonyl)imide (PEO_6–LiTFSI) solution. For the PEO_6–LiTFSI with 50 wt.% LLZO:Ta, as investigated here, no increase of the ionic conductivity in comparison to the pure polymer was observed by EIS [21]. For the Atomic Force Microscope (AFM) and ESM measurements, a Bruker Dimension Icon (Bruker, Santa Barbara, CA, USA) operating inside a glovebox (H_2O < 0.1 ppm, O_2 < 0.1 ppm, Argon filled) was used.

In the ESM mode, we tracked the contact resonance frequency and amplitude with a phase-locked loop (HF2LI, Zurich Instruments, Zurich, Switzerland). For the measurements, a cantilever with a free resonance peak around 75 kHz and a conductive platinum iridium (Pt/Ir) coating (PPP-EFM, nominal spring constant 2.8 N m^{-1}, Nano World AG, Neuchatel, Switzerland) was utilized. A baseline correction was performed with the respective contact resonance frequencies. For the amplitude distribution, all peaks were fitted as Gaussian according to Equation (1):

$$y = y_0 + \frac{A e^{\frac{-4ln(2)(x-x_c)^2}{w^2}}}{w \sqrt{\frac{\pi}{4ln(2)}}} \quad (1)$$

where y_0 is the baseline, A is the area under the peak, w is the width of the peak and x_c is the center of the peak. The Focused Ion Beam (FIB) polishing process was conducted with a NanoLab 460FI (FEI, Waltham, MA, USA). The sample transfer between the FIB and AFM was realized in an inert gas shuttle. The Scanning Electron Microscopy (SEM) image was recorded on the same material but from a different batch. The sample was broken in liquid nitrogen and measured with a FEI Quanta FEG 650 with an accelerating voltage of 10 kV utilizing a Back Scattered Electron (BSE) detector.

3. Results and Discussion

ESM measurements require a specifically prepared surface of the PEO_6–LiTFSI film containing 50 wt.% LLZO:Ta to avoid cross-talk between topographical features and the ESM amplitude signal. Therefore, the solution-cast hybrid electrolyte film was carefully polished using a focused ion beam (FIB). For ESM experiments, it is of the utmost importance to clearly distinguish between isolated ceramic particles and the surrounding polymer matrix to identify the interfacial region between both materials. Individual particles within the polymer matrix were identified based on their respective frictional response.

For the interpretation of ESM signals, another important prerequisite is to always record the amplitude signal at the respective Contact Resonance Frequency (CRF) of the cantilever. However, variations in the mechanical properties strongly influence the resulting frequencies and ESM signals. Figure 1 shows a typical SEM-BSE image of a hybrid electrolyte film, while the AFM topography and friction images display the resulting smooth hybrid electrolyte film of the polished surface.

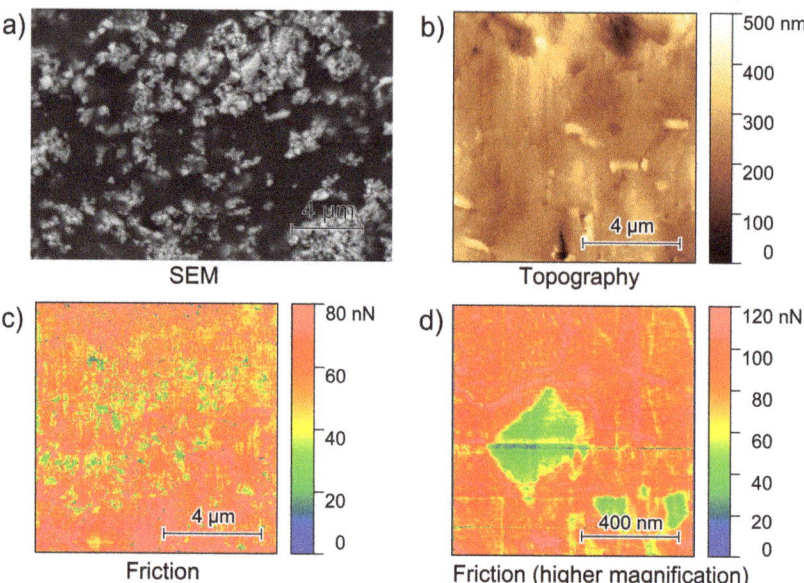

Figure 1. (**a**) SEM-BSE image of a PEO$_6$–LiTFSI with 50 wt.% LLZO:Ta film, (**b**) 10 µm × 10 µm topography, (**c**) 10 µm × 10 µm friction image of a PEO$_6$–LiTFSI with 50 wt.% LLZO:Ta film polished with a focused ion beam and (**d**) a higher magnification of a friction image showing a single ceramic particle within the polymer matrix.

The SEM-BSE image depicted in Figure 1a exhibits multiple bright spots on an area of 20 µm × 14 µm, representing the high Z LLZO:Ta-particles. The dark area in the SEM image in Figure 1a reveals the PEO$_6$–LiTFSI, as this material represents the matrix in which LLZO:Ta particles are embedded. Inside the polymer matrix, the LLZO:Ta particles tend to form agglomerates. In addition to the SEM image, a typical 10 µm × 10 µm AFM topography image is shown in Figure 1b. Vertical lines are observed from the ion milling process.

These effects are typical artifacts, known as curtaining, from the focused ion beam polishing process as this was executed from the top of the image to the bottom. The image Figure 1c illustrates a 10 µm × 10 µm friction image that was recorded simultaneously with Figure 1b. Within a homogeneous matrix exhibiting a friction value of 60–80 nN, several spots showing lower friction values are observed. A higher magnification of an isolated square-shaped particle exhibiting low friction values is shown in Figure 1d.

The friction images demonstrate the materials' differences, similar to the SEM image of a different area. Lower friction values are associated with the ceramic particles, while higher friction values are assigned to the bulk polymer phase. Due to the softness of the PEO$_6$–LiTFSI polymer and the stiffness of the ceramic, they show different tip–surface interactions. The friction image reveals some agglomerates of ceramic particles, i.e., lower friction values inside the bulk polymer phase as also detected by SEM with a particle size between 0.4 and 1 µm and a mean particle size of <1 µm [21].

Figure 2 shows CRF curves on individual spots on the polymer matrix PEO$_6$–LiTFSI (red) and on LLZO:Ta ceramic particles (green). The CRF between the AFM tip and the LLZO:Ta particle is observed between 297 and 304 kHz. In comparison, the frequency of the identical tip in contact with the polymer matrix is found in the region of 266 to 279 kHz. Additionally, we observed that the polymer peaks are broader and exhibit lower amplitude values compared with the peaks obtained on the ceramic particle. The difference between the polymer peaks and ceramic peaks verifies the highly individual properties of both materials within the hybrid electrolyte film. As the polymer is a softer material than the ceramic particle, the damping of the cantilever causes the range to shift towards lower frequencies and become broader in comparison to the ceramic peaks [37].

This verifies that polymer matrix and ceramic particles can be differentiated by AFM-based techniques employing a sophisticated sample preparation method. This is especially relevant as the topography image does not show the materials' differences.

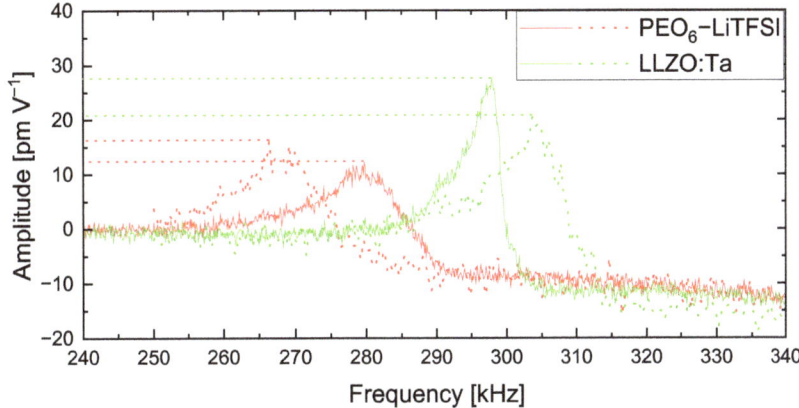

Figure 2. Contact resonance frequency curves on PEO_6–LiTFSI (red) and LLZO:Ta (green).

3.1. Electrochemical Strain Signal Distribution on the Individual Polymer and Ceramic Electrolyte

The amplitude distribution is studied on the single materials to differentiate the ESM amplitude signal on the pure PEO_6–LiTFSI and a LLZO:Ta pellet. The lower amplitude of the polymer peak might be attributed to the lower lithium ion conductivity of the polymer compared to the ceramic electrolyte. A closer inspection of the amplitude signal on a 10 µm × 10 µm area of pure PEO_6–LiTFSI reveals that it follows a bimodal distribution as depicted in the graph in Figure 3a. Two peaks exhibiting amplitude values of 5.9 pm V^{-1} and of 10.4 pm V^{-1} are found.

The first peak shows a Full-Width-Half-Maximum (FWHM) value of 1.9 pm V^{-1}, while the second broader peak has a FWHM of 7.0 pm V^{-1}. Higher amplitude values indicate a higher lithium mobility in this particular area. This observation illustrates heterogeneous lithium mobility in a polymer matrix in agreement with earlier observations and is discussed in literature [38,39]. The variation in lithium ion conductivity is typically caused by non-conducting crystalline regions and ion-conducting amorphous regions within the polymer. Therefore, the first peak is assigned to the crystalline phase, thus, resulting in a narrow FWHM. In comparison, the amorphous conducting phase is critically influenced by the inhomogeneity of the microstructure and, hence, shows a broader distribution in the second peak.

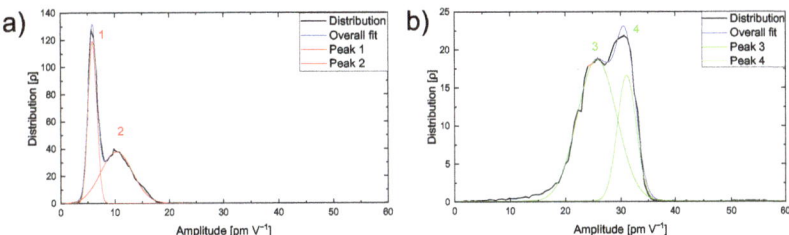

Figure 3. (**a**) Graph with the amplitude distribution of a 10 µm × 10 µm area on a pure PEO_6–LiTFSI electrolyte film and (**b**) graph with the amplitude distribution of a 500 nm × 500 nm area on a LLZO:Ta pellet.

The amplitude distribution on a 500 nm × 500 nm area on a LLZO:Ta pellet is displayed in Figure 3b. The graph shows two distinguishable peaks, one at an amplitude value of 25.7 pm V^{-1} and the other at 31.1 pm V^{-1}. The first peak is broader with a FWHM of 8.3 pm V^{-1}, as the second peak is narrower with a FWHM of 4.0 pm V^{-1}.

In comparison to the amplitude peaks on PEO$_6$–LiTFSI, the LLZO:Ta peaks are found at higher amplitude values. This correlates with the different ionic conductivities of the PEO$_6$–LiTFSI and LLZO:Ta electrolytes. The PEO$_6$–LiTFSI shows lower amplitude values with a lower ionic conductivity, whereas the LLZO:Ta shows higher amplitude values with a higher ionic conductivity.

The two distinct peaks in the distribution of the LLZO:Ta ceramic particles might originate from microstructural or morphological effects influencing the local ionic conductivity within the material. To further understand the ESM amplitude signal response of the hybrid electrolyte film, a similar approach was conducted on the polished sample within an area exhibiting particles as well as the polymer matrix.

3.2. Electrochemical Strain Signal Distribution Inside the Hybrid Electrolyte

Figure 4 shows the topography, ESM amplitude and ESM frequency images, as well as the amplitude distribution of a 10 μm × 10 μm area on the polished PEO$_6$–LiTFSI with 50 wt.% LLZO:Ta film. The topography image in Figure 4a illustrates the surface of the polished area, exhibiting a height difference on the order of 300 nm. A few cracks on the surface, together with vertical lines from the top to the bottom that can be attributed to the curtaining effect, are observed.

The simultaneous recording of the ESM amplitude, the CRF and the friction force signals is very important to gain more information about materials differences than from topographical measurements alone. Therefore, in Figure 4b, the ESM amplitude signal image is shown. At the top part of the image, multiple green and blue spots with higher amplitude values are seen. Additionally, in the bottom part of the image, more green areas are observed.

Orange regions of lower amplitude values are visible in the middle of the image, as well as a few orange regions on the right side. Furthermore, there are areas with lower amplitude values in between and around the green and blue spots. The amplitude distribution of the PEO$_6$–LiTFSI electrolyte in Figure 3a shows amplitude values up to 20 pm V^{-1}, verifying that the orange areas within the hybrid electrolyte film in Figure 4b reflect the polymer part of the electrolyte.

As the ceramic particles show significantly higher amplitude values, as indicated by the amplitude distribution in Figure 3b, the green and blue spots with amplitude values up to 45 pm V^{-1} are attributed to the ceramic particles. As in Figure 1a, the ceramic particles form agglomerates.

Figure 4c shows the variations of the contact resonance frequency. This is simultaneously recorded with the topography and amplitude image. From the top right and middle to the lower left area decreased CRFs are observed. This is in agreement with the amplitude image in Figure 4b and the friction image in Figure 4d displaying higher friction in the same regions although there is no exact correlation between the individual channels. A possible reason for the weak correlation might be an overlapping thin polymer film that influences the various signals differently.

The graph in Figure 4e illustrates the corresponding amplitude distribution of Figure 4b. Two peaks are found in the amplitude distribution: The narrower peak at 12.8 pm V^{-1} exhibits a FWHM of 2.8 pm V^{-1}, while the broader second peak with a maximum at 27.0 pm V^{-1} possesses a FWHM of 21.5 pm V^{-1}.

Hence, the first peak is rather similar to the second peak in Figure 3a, i.e., reflecting the higher conductive amorphous polymer phase. This important observation confirms the strategy of incorporating particles into a polymer matrix to prevent crystallization and increase the amorphous fraction of the polymer phase. The second peak 3' of the hybrid sample (Figure 4e) is rather broad and exhibits significantly larger amplitude values, comparable to peaks 3 and 4 in the distribution on the LLZO:Ta pellet in Figure 3b. As this peak is absent in the pure PEO_6–LiTFSI phase and has a similar position to the peaks on the LLZO:Ta pellet, it is expected to originate from the added ceramic particles.

The obtained ESM amplitudes in the hybrid electrolyte reflect the identical ESM amplitudes as obtained for the individual materials. However, an additional higher amplitude was found corresponding to the interfacial area. This increased lithium ion mobility has not been previously observed by EIS [21] as the interfacial areas are not connected. Thus, the interfacial area around te particles is of particular interest, and we focus on this in the next chapter.

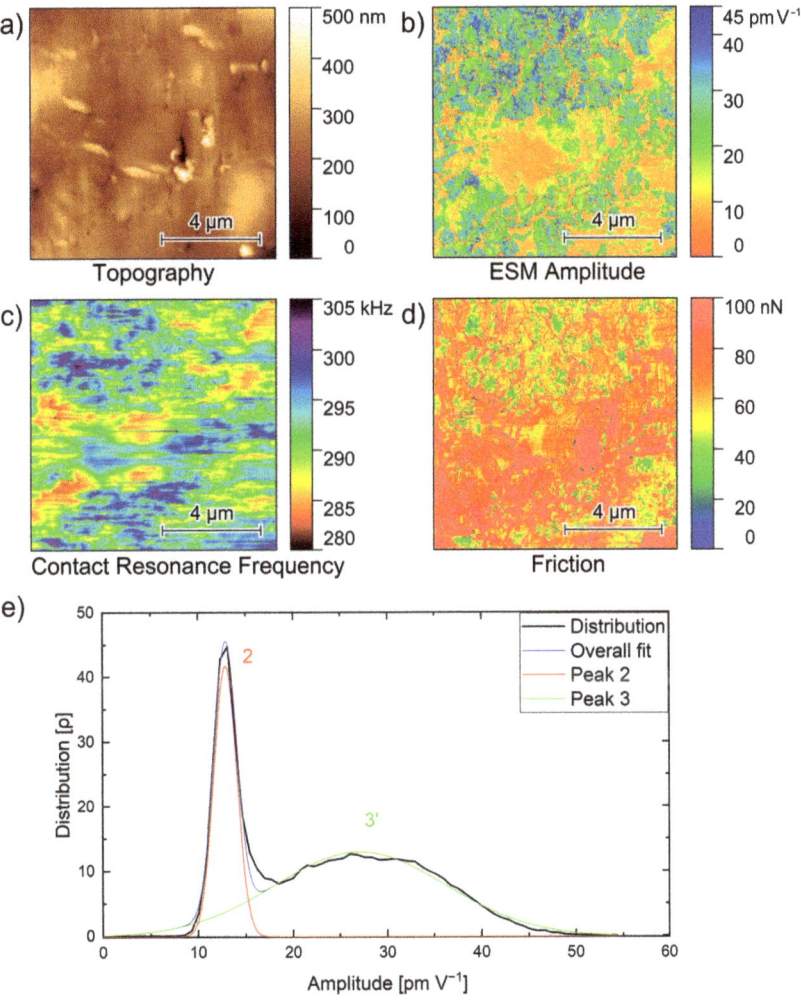

Figure 4. (a) Topography, (b) amplitude, (c) contact resonance frequency, (d) friction images and (e) amplitude distribution graph of Figure 4b, of a 10 μm × 10 μm area of a PEO_6–LiTFSI film with 50 wt.% LLZO:Ta film polished with a focused ion beam. Peak 3' reflects the overlapping peaks 3 and 4.

3.3. Interfacial Analysis between Ceramic Particles and Bulk Polymer

The interfacial area between a single LLZO:Ta particle and the surrounding PEO_6–LiTFSI is analyzed in greater detail. Figure 5 shows the amplitude signal, Contact Resonance

Frequency and friction force images of a 200 nm × 200 nm area of the hybrid electrolyte. With the focus on the interfacial area, Figure 5a displays lower amplitude values on the lower left, while the right hand side exhibits higher amplitude values. The upper left exhibits higher values as well.

The CRF map in Figure 5b illustrates a similar pattern with mostly low CRFs on the left and high CRFs on the right, thus, allowing a clear differentiation between the polymer and ceramic phase. Though subsurface effects on the top left area might influence the amplitude image. The friction force image supports the assignment of the polymer and ceramic phase, with high friction forces on the left and low friction forces on the right-hand side of the map. Therefore, we concluded that the left side of the image shows a bulk polymer phase, while the right side shows a single ceramic particle. In this case, the boundary between both materials is clearly identifiable.

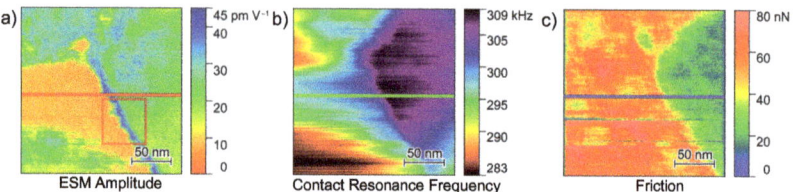

Figure 5. (a) Amplitude, (b) contact resonance frequency and (c) friction force images of a 200 nm × 200 nm area between the bulk PEO_6–LiTFSI and a single LLZO:Ta particle. The line sections indicated in the middle of the images are shown in Figure 6.

It is important to note that the amplitude image shows an area with significantly increased amplitude values at the transition between the bulk polymer phase and the ceramic particle. To assign this region to the polymer or ceramic material, line sections of the amplitude, CRF and friction force at identical positions, are visualized in Figure 6.

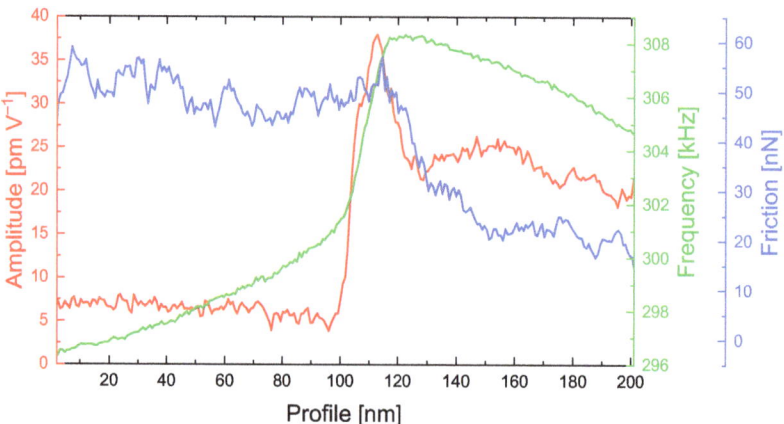

Figure 6. Corresponding line sections as highlighted in Figure 5 of the ESM amplitude, contact resonance frequency and friction force on a PEO_6–LiTFSI with 50 wt.% LLZO:Ta film. The line sections are averaged over five lines.

As in the images shown in Figure 5, the left side of the graph shows the polymer phase, while the right side shows the ceramic particle. The transition between both materials can easily be recognized between 100 and 120 nm. Interestingly, the amplitude signal exhibits a spike at 110 nm, i.e., exactly in the transition zone. The CRF increases drastically at the same position. However, the friction force decreases at 120 nm and, thus, indicates where the ceramic particle is located.

Combining this information, high friction and a low CRF on the polymer phase and lower friction but higher CRF on the ceramic particle, we verified that the interfacial area with high ESM amplitudes was mostly present on the polymer phase. In agreement with reports from the literature [25,26,31], these findings lead to the major conclusion that, indeed, the transition zone plays an important role in lithium ion transport in PEO_6–LiTFSI with LLZO:Ta electrolytes. However, the increased ESM amplitude might also reflect the accumulation of lithium ions within this transition zone with a dimension of approximately 20 nm in front of the ceramic particles.

The effect of the transition zone is also reflected in the amplitude distribution of the region from Figure 5a as shown in Figure 7.

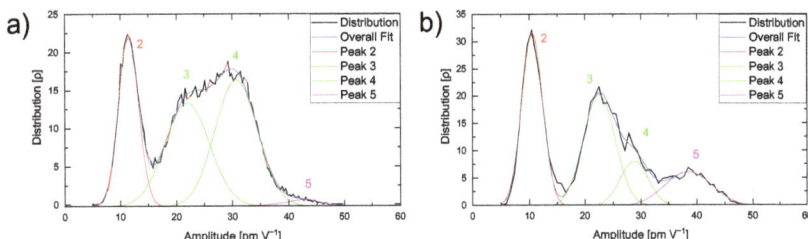

Figure 7. (**a**) Amplitude distribution of 200 nm × 200 nm of a PEO_6–LiTFSI with 50 wt.% LLZO:Ta film and (**b**) of a smaller region as indicated by the red square shown in Figure 5a.

The distribution in Figure 7a displays four distinguishable peaks. While the first peak 2 appears at around 11.6 pm V^{-1} and reflects the amorphous polymer phase as discussed above, the second peak 3 at 22.0 pm V^{-1} and the third peak 4 at 30.7 pm V^{-1} correspond to the ceramic particle, comparable to Figure 3b.

In comparison to the single peak observed in Figure 4e, the high spatial resolution reveals even small variations in the ESM amplitude signal within the hybrid electrolyte. The rather small fourth peak 5 at 42.9 pm V^{-1} is only obtained at sufficiently high spatial resolution and is attributed to the interfacial area exhibiting very high ESM amplitudes. This is verified by focusing on the interfacial area and highlighted in Figure 7b, where the amplitude distribution according to the highlighted area in Figure 5a is illustrated.

A significantly enlarged peak 5 is found. The obtained amplitude distributions indicate that, in Figure 4e, a fourth peak might be hidden in the amplitude values distributed around 40 to 50 pm V^{-1}. However, on larger scale images, the resolution is not sufficient to visualize the transition zone of 20 nm. The increased amplitude signal validates the higher lithium mobility inside the interfacial area, while a transport process between an amorphous polymer and ceramic is possible.

These results underline the importance of the interaction between ceramic particles and the polymer matrix to increase the overall conductivity. The outstanding influence of interfacial transition zones, reminiscent of the space charge areas between the polymer phase and embedded ceramic particles, as discussed in the literature [29–31], are confirmed.

4. Conclusions

A PEO_6–LiTFSI with 50 wt.% LLZO:Ta all-solid-state hybrid electrolyte film was examined by means of Electrochemical Strain Microscopy. We demonstrated that the individual materials exhibited significantly different signals in the friction force, Contact Resonance Frequency and amplitude signals. The polymer electrolyte exhibited a higher friction force with lower CRFs as expected. In agreement with the literature, higher ionic conductivity inside the ceramic led to significantly higher ESM amplitude values compared to the bulk polymer phase.

The amplitude distribution of a pure PEO_6–LiTFSI polymer electrolyte exhibited two peaks that can be attributed to crystalline (peak 1) and amorphous (peak 2) regions. The distribution on a LLZO:Ta pellet showed two peaks at higher amplitude values. Within the hybrid electrolyte film, those peaks were similar except for peak 1, as the introduction of particles suppresses the crystallization. High-resolution imaging revealed a region around the ceramic particles with increased amplitude values even higher than those of bulk LLZO:Ta, thereby, indicating the accumulation of lithium ions.

This interfacial area has a dimension of 20 nm inside the polymer phase adjacent to the ceramic particle as verified by friction force and CRF and schematically depicted in Figure 8. In the distribution of the amplitude values, another peak was observed that is attributed to this interfacial area.

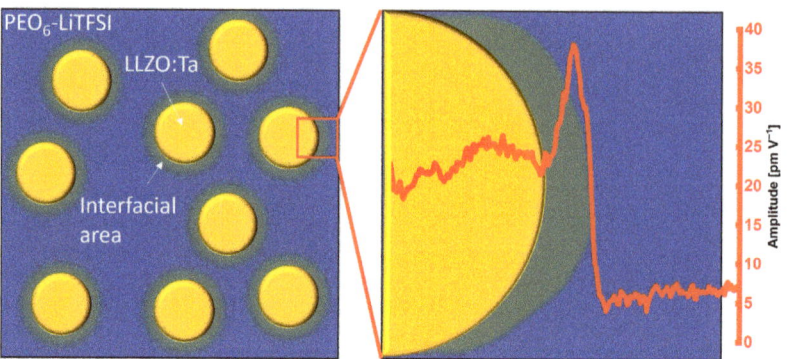

Figure 8. Schematic illustration of the bulk polymer phase with several ceramic particles and a segment of a single particle with the ESM amplitude line section of Figure 6. The interfacial area shown is not to scale.

The results presented in this manuscript are of significant importance to obtain a further understanding of the ionic transport mechanisms inside hybrid electrolytes. Based on these results, we propose two strategies for the further improvement of hybrid solid-state electrolytes.

First, coating the particles to decrease the interfacial resistance might open a path to take advantage of the intrinsically higher ionic conductivity of the ceramic. Second, to capitalize on the high lithium content of the interfacial area and increase the overall ionic conductivity of the hybrid electrolyte, a system with a continuous percolation path along the ceramic particles should be created. Both aspects are subject to further investigations.

Author Contributions: Conceptualization, P.M.V., M.W., R.S., H.T., H.K., R.-A.E. and F.H.; formal analysis, P.M.V.; investigation, P.M.V., M.W. and R.S.; resources, H.K., R.-A.E. and F.H.; data curation, P.M.V. and F.H.; writing - original draft preparation, P.M.V. and F.H.; writing - review and editing, P.M.V., M.W., R.S., H.T., H.K., R.-A.E. and F.H.; visualization, P.M.V.; supervision, H.T., H.K., R.-A.E. and F.H.; project administration, H.K.; funding acquisition, H.T., H.K., R.-A.E. and F.H. All authors have read and agreed to the published version of the manuscript.

Funding: This research was supported by the "US-German Cooperation on Energy Storage" under the project "LiSi-Lithium Solid Electrolyte Interfaces" (Project No.: 13XP0224B) by the US-Department of Energy (DOE) and the Federal Ministry of Education and Research of Germany (BMBF).

Data Availability Statement: The data presented in this study are available on request from the corresponding author.

Conflicts of Interest: The authors declare no conflict of interest.

Abbreviations

The following abbreviations are used in this manuscript:

ESM Electrochemical Strain Microscopy
AFM Atomic Force Microscopy
ASSB All-Solid-State Battery
FIB Focused Ion Beam
SEM Scanning Electron Microscopy
CRF Contact Resonance Frequency
NMR Nuclear Magnetic Resonance
BSE Back Scattered Electron

References

1. Doughty, D.; Roth, E.P. A General Discussion of Li Ion Battery Safety. *Electrochem. Soc. Interface* **2012**, *21*, 37–44. [CrossRef]
2. Choi, J.W.; Aurbach, D. Promise and reality of post-lithium-ion batteries with high energy densities. *Nat. Rev. Mater.* **2016**, *1*, 16013. [CrossRef]
3. Manthiram, A.; Yu, X.; Wang, S. Lithium battery chemistries enabled by solid-state electrolytes. *Nat. Rev. Mater.* **2017**, *2*, 294. [CrossRef]
4. Arya, A.; Sharma, A.L. Polymer electrolytes for lithium ion batteries: A critical study. *Ionics* **2017**, *23*, 497–540. [CrossRef]
5. Meyer, W.H. Polymer Electrolytes for LithiumIon Batteries. *Adv. Mater.* **1998**, *10*, 439–448. [CrossRef]
6. Aziz, S.B.; Woo, T.J.; Kadir, M.; Ahmed, H.M. A conceptual review on polymer electrolytes and ion transport models. *J. Sci. Adv. Mater. Devices* **2018**, *3*, 1–17. [CrossRef]
7. Kim, G.T.; Appetecchi, G.B.; Alessandrini, F.; Passerini, S. Solvent-free, PYR1ATFSI ionic liquid-based ternary polymer electrolyte systems. *J. Power Sources* **2007**, *171*, 861–869. [CrossRef]
8. Mertens, A.; Yu, S.; Schön, N.; Gunduz, D.C.; Tempel, H.; Schierholz, R.; Hausen, F.; Kungl, H.; Granwehr, J.; Eichel, R.A. Superionic bulk conductivity in $Li_{1.3}Al_{0.3}Ti_{1.7}(PO_4)_3$ solid electrolyte. *Solid State Ionics* **2017**, *309*, 180–186. [CrossRef]
9. Knauth, P. Inorganic solid Li ion conductors: An overview. *Solid State Ionics* **2009**, *180*, 911–916. [CrossRef]
10. Ma, F.; Zhao, E.; Zhu, S.; Yan, W.; Sun, D.; Jin, Y.; Nan, C. Preparation and evaluation of high lithium ion conductivity $Li_{1.3}Al_{0.3}Ti_{1.7}(PO_4)_3$ solid electrolyte obtained using a new solution method. *Solid State Ionics* **2016**, *295*, 7–12. [CrossRef]
11. Zhai, H.; Xu, P.; Ning, M.; Cheng, Q.; Mandal, J.; Yang, Y. A Flexible Solid Composite Electrolyte with Vertically Aligned and Connected Ion-Conducting Nanoparticles for Lithium Batteries. *Nano Lett.* **2017**, *17*, 3182–3187. [CrossRef]
12. Yang, L.; Wang, Z.; Feng, Y.; Tan, R.; Zuo, Y.; Gao, R.; Zhao, Y.; Han, L.; Wang, Z.; Pan, F. Flexible Composite Solid Electrolyte Facilitating Highly Stable "Soft Contacting" Li-Electrolyte Interface for Solid State Lithium-Ion Batteries. *Adv. Energy Mater.* **2017**, *7*, 1701437. [CrossRef]
13. Pervez, S.A.; Ganjeh-Anzabi, P.; Farooq, U.; Trifkovic, M.; Roberts, E.P.L.; Thangadurai, V. Fabrication of a Dendrite–Free all Solid–State Li Metal Battery via Polymer Composite/Garnet/Polymer Composite Layered Electrolyte. *Adv. Mater. Interfaces* **2019**, *334*, 1900186. [CrossRef]
14. Yu, S.; Schmohl, S.; Liu, Z.; Hoffmeyer, M.; Schön, N.; Hausen, F.; Tempel, H.; Kungl, H.; Wiemhöfer, H.D.; Eichel, R.A. Insights into a layered hybrid solid electrolyte and its application in long lifespan high-voltage all-solid-state lithium batteries. *J. Mater. Chem. A* **2019**, *7*, 3882–3894. [CrossRef]
15. Keller, M.; Varzi, A.; Passerini, S. Hybrid electrolytes for lithium metal batteries. *J. Power Sources* **2018**, *392*, 206–225. [CrossRef]
16. Chen, L.; Li, Y.; Li, S.P.; Fan, L.Z.; Nan, C.W.; Goodenough, J.B. PEO/garnet composite electrolytes for solid-state lithium batteries: From "ceramic-in-polymer" to "polymer-in-ceramic". *Nano Energy* **2018**, *46*, 176–184. [CrossRef]
17. Cheng, S.H.S.; Liu, C.; Zhu, F.; Zhao, L.; Fan, R.; Chung, C.Y.; Tang, J.; Zeng, X.; He, Y.B. (Oxalato)borate: The key ingredient for polyethylene oxide based composite electrolyte to achieve ultra-stable performance of high voltage solid-state $LiNi_{0.8}Co_{0.1}Mn_{0.1}O_2$/lithium metal battery. *Nano Energy* **2021**, *80*, 105562. [CrossRef]
18. Xu, Y.; Zhou, Y.; Li, T.; Jiang, S.; Qian, X.; Yue, Q.; Kang, Y. Multifunctional covalent organic frameworks for high capacity and dendrite-free lithium metal batteries. *Energy Storage Mater.* **2020**, *25*, 334–341. [CrossRef]
19. Langer, F.; Bardenhagen, I.; Glenneberg, J.; Kun, R. Microstructure and temperature dependent lithium ion transport of ceramic–polymer composite electrolyte for solid-state lithium ion batteries based on garnet-type $Li_7La_3Zr_2O_{12}$. *Solid State Ionics* **2016**, *291*, 8–13. [CrossRef]
20. Keller, M.; Appetecchi, G.B.; Kim, G.T.; Sharova, V.; Schneider, M.; Schuhmacher, J.; Roters, A.; Passerini, S. Electrochemical performance of a solvent-free hybrid ceramic-polymer electrolyte based on $Li_7La_3Zr_2O_{12}$ in $P(EO)_{15}$ LiTFSI. *J. Power Sources* **2017**, *353*, 287–297. [CrossRef]
21. Wirtz, M.; Linhorst, M.; Veelken, P.; Tempel, H.; Kungl, H.; Moerschbacher, B.M.; Eichel, R.A. Polyethylene oxide–$Li_{6.5}La_3Zr_{1.5}Ta_{0.5}O_{12}$ hybrid electrolytes: Lithium salt concentration and biopolymer blending. *Electrochem. Sci. Adv.* **2021**, *1*, e2000029. [CrossRef]
22. Croce, F.; Appetecchi, G.B.; Persi, L.; Scrosati, B. Nanocomposite polymer electrolytes for lithium batteries. *Nature* **1998**, *394*, 456–458. [CrossRef]

23. Appetecchi, G.B.; Croce, F.; Persi, L.; Ronci, F.; Scrosati, B. Transport and interfacial properties of composite polymer electrolytes. *Electrochim. Acta* **2000**, *45*, 1481–1490. [CrossRef]
24. Croce, F.; Persi, L.; Scrosati, B.; Serraino-Fiory, F.; Plichta, E.; Hendrickson, M. Role of the ceramic fillers in enhancing the transport properties of composite polymer electrolytes. *Electrochim. Acta* **2001**, *46*, 2457–2461. [CrossRef]
25. Dixit, M.B.; Zaman, W.; Bootwala, Y.; Zheng, Y.; Hatzell, M.C.; Hatzell, K.B. Scalable Manufacturing of Hybrid Solid Electrolytes with Interface Control. *ACS Appl. Mater. Interfaces* **2019**, *11*, 45087–45097. [CrossRef] [PubMed]
26. Dixit, M.B.; Zaman, W.; Hortance, N.; Vujic, S.; Harkey, B.; Shen, F.; Tsai, W.Y.; de Andrade, V.; Chen, X.C.; Balke, N.; et al. Nanoscale Mapping of Extrinsic Interfaces in Hybrid Solid Electrolytes. *Joule* **2020**, *4*, 207–221. [CrossRef]
27. Langer, F.; Palagonia, M.S.; Bardenhagen, I.; Glenneberg, J.; La Mantia, F.; Kun, R. Impedance Spectroscopy Analysis of the Lithium Ion Transport through the $Li_7La_3Zr_2O_{12}/P(EO)_{20}Li$ Interface. *J. Electrochem. Soc.* **2017**, *164*, A2298–A2303. [CrossRef]
28. Yu, S.; Xu, Q.; Tsai, C.L.; Hoffmeyer, M.; Lu, X.; Ma, Q.; Tempel, H.; Kungl, H.; Wiemhöfer, H.D.; Eichel, R.A. Flexible All-Solid-State Li-Ion Battery Manufacturable in Ambient Atmosphere. *ACS Appl. Mater. Interfaces* **2020**, *12*, 37067–37078. [CrossRef]
29. Zheng, J.; Tang, M.; Hu, Y.Y. Lithium Ion Pathway within $Li_7La_3Zr_2O_{12}$-Polyethylene Oxide Composite Electrolytes. *Angew. Chem.* **2016**, *55*, 12538–12542. [CrossRef]
30. Li, M.; Kolek, M.; Frerichs, J.E.; Sun, W.; Hou, X.; Hansen, M.R.; Winter, M.; Bieker, P. Investigation of Polymer/Ceramic Composite Solid Electrolyte System: The Case of PEO/LGPS Composite Electrolytes. *ACS Sustain. Chem. Eng.* **2021**, *9*, 11314–11322. [CrossRef]
31. Ranque, P.; Zagórski, J.; Devaraj, S.; Aguesse, F.; Del López Amo, J.M. Characterization of the interfacial Li-ion exchange process in a ceramic–polymer composite by solid state NMR. *J. Mater. Chem. A* **2021**, *9*, 17812–17820. [CrossRef]
32. Balke, N.; Jesse, S.; Kim, Y.; Adamczyk, L.; Tselev, A.; Ivanov, I.N.; Dudney, N.J.; Kalinin, S.V. Real space mapping of Li-ion transport in amorphous Si anodes with nanometer resolution. *Nano Lett.* **2010**, *10*, 3420–3425. [CrossRef] [PubMed]
33. Balke, N.; Jesse, S.; Morozovska, A.N.; Eliseev, E.; Chung, D.W.; Kim, Y.; Adamczyk, L.; García, R.E.; Dudney, N.; Kalinin, S.V. Nanoscale mapping of ion diffusion in a lithium-ion battery cathode. *Nat. Nanotechnol.* **2010**, *5*, 749–754. [CrossRef]
34. Giridharagopal, R.; Flagg, L.Q.; Harrison, J.S.; Ziffer, M.E.; Onorato, J.; Luscombe, C.K.; Ginger, D.S. Electrochemical strain microscopy probes morphology-induced variations in ion uptake and performance in organic electrochemical transistors. *Nat. Mater.* **2017**, *16*, 737–742. [CrossRef] [PubMed]
35. Simolka, M.; Kaess, H.; Friedrich, K.A. Comparison of fresh and aged lithium iron phosphate cathodes using a tailored electrochemical strain microscopy technique. *Beilstein J. Nanotechnol.* **2020**, *11*, 583–596. [CrossRef]
36. Schön, N.; Schierholz, R.; Jesse, S.; Yu, S.; Eichel, R.A.; Balke, N.; Hausen, F. Signal Origin of Electrochemical Strain Microscopy and Link to Local Chemical Distribution in Solid State Electrolytes. *Small Methods* **2021**, *5*, 2001279. [CrossRef]
37. Yablon, D.G.; Gannepalli, A.; Proksch, R.; Killgore, J.; Hurley, D.C.; Grabowski, J.; Tsou, A.H. Quantitative Viscoelastic Mapping of Polyolefin Blends with Contact Resonance Atomic Force Microscopy. *Macromolecules* **2012**, *45*, 4363–4370. [CrossRef]
38. Quartarone, E.; Mustarelli, P.; Magistris, A. PEO-based composite polymer electrolytes. *Solid State Ionics* **1998**, *110*, 1–14. [CrossRef]
39. Choi, J.H.; Lee, C.H.; Yu, J.H.; Doh, C.H.; Lee, S.M. Enhancement of ionic conductivity of composite membranes for all-solid-state lithium rechargeable batteries incorporating tetragonal $Li_7La_3Zr_2O_{12}$ into a polyethylene oxide matrix. *J. Power Sources* **2015**, *274*, 458–463. [CrossRef]

Article

Aqueous Organic Zinc-Ion Hybrid Supercapacitors Prepared by 3D Vertically Aligned Graphene-Polydopamine Composite Electrode

Ruowei Cui, Zhenwang Zhang, Huijuan Zhang, Zhihong Tang, Yuhua Xue and Guangzhi Yang *

School of Materials and Chemistry, University of Shanghai for Science and Technology, Shanghai 200093, China; crw2155@163.com (R.C.); 183762634@st.usst.edu.cn (Z.Z.); hjzhang@usst.edu.cn (H.Z.); zhtang@usst.edu.cn (Z.T.); xueyuhua@usst.edu.cn (Y.X.)
* Correspondence: yanggzh@usst.edu.cn; Tel.: +86-21-55270632

Abstract: A three-dimensional vertical-aligned graphene-polydopamine electrode (PDA@3DVAG) composite with vertical channels and conductive network is prepared by a method of unidirectional freezing and subsequent self-polymerization. When the prepared PDA@3DVAG is constructed as the positive electrode of zinc-ion hybrid supercapacitors (ZHSCs), excellent electrochemical performances are obtained. Compared with the conventional electrolyte, PDA@3DVAG composite electrode in highly concentrated salt electrolyte exhibits better multiplicity performance (48.92% at a current density of 3 A g^{-1}), wider voltage window (−0.8~0.8 V), better cycle performance with specific capacitance from 96.7 to 59.8 F g^{-1}, and higher energy density (46.14 Wh kg^{-1}).

Keywords: zinc-ion supercapacitor; three-dimensional vertically aligned graphene; polydopamine; highly concentrated salt electrolyte

Citation: Cui, R.; Zhang, Z.; Zhang, H.; Tang, Z.; Xue, Y.; Yang, G. Aqueous Organic Zinc-Ion Hybrid Supercapacitors Prepared by 3D Vertically Aligned Graphene-Polydopamine Composite Electrode. *Nanomaterials* 2022, 12, 386. https://doi.org/10.3390/nano12030386

Academic Editors: Federico Cesano, Simas Rackauskas and Mohammed Jasim Uddin

Received: 17 December 2021
Accepted: 24 January 2022
Published: 25 January 2022

Publisher's Note: MDPI stays neutral with regard to jurisdictional claims in published maps and institutional affiliations.

Copyright: © 2022 by the authors. Licensee MDPI, Basel, Switzerland. This article is an open access article distributed under the terms and conditions of the Creative Commons Attribution (CC BY) license (https://creativecommons.org/licenses/by/4.0/).

1. Introduction

The construction and manufacture of electrochemical energy storage systems with high power and energy density, fast charge and discharge rates, and excellent cycle performance are of great importance to the rational use of energy [1–3]. Traditional energy storage devices mainly include batteries and supercapacitors, which are widely used in high energy density and high power ranges, respectively [4]. As energy storage components, batteries have the advantages of high energy density and facilitate long-term storage of electrical energy [5]; however, the disadvantages of low power density, low charge and discharge efficiency, poor cycle performance limit their applications [6–8]. Supercapacitors have the advantages of high capacity, high specific power density, fas t charge and discharge rates and excellent cycle performance, but the low energy density limits their application [9,10]. The electrochemical hybrid supercapacitors combine the advantages of capacitive cathode and battery type anode material and have high energy and power density together with excellent cycle performance [11].

A few organic materials are being studied as electrode materials for supercapacitors, such as polyaniline (PANI) [12,13], polypyrrole (PPy) [14,15], poly(3,4-ethylenedioxythiophene) (PEDOT) [16], and quinone-based organics [3,17]. In recent years, the quinone-based organic of polydopamine (PDA), which contains a large amount of catechol and nitrogenous amino groups, has been widely studied as electrocatalysts and electrode materials for energy conversion and storage. Wang et al. [18] constructed a fibrous aqueous zinc-ion battery with PDA, which has a large specific capacity (372.3 mAh g^{-1} at 50 mA g^{-1}) and long-term cycle performance (80% capacity retention after 1700 cycles at 1 A g^{-1}). While for the organic cathode, two factors hinder its development and application. Firstly, organic molecules are easily soluble in the electrolyte, resulting in low cycle performance. Secondly, most organics have poor electrical conductivity, which limits their multiplicative performance [19]. To solve the solubilization of quinone-based organic materials, many methods have been developed, such as separation modification, polymerization, mesoporous matrix constraint, and so on [20,21]. As for the problem of poor conductivity, studies have found that the ion/electron conductivity is closely related to the pore curvature of the electrode and the composite electrode, based on a highly conductive three-dimensional graphene structure, which can effectively improve the multiplicative performance [22]. In particular, three-dimensional vertically oriented graphene (3DVAG) has a special structure with vertically open channels and low pore curvature, which contributes to fast ion/electron transfer and, therefore, excellent performance.

In this work, unidirectional freezing and thermal reduction are used to prepare 3DVAG with a 3D long-range ordered structure. Three-dimensional vertically aligned graphene-polydopamine (PDA@3DVAG) composite electrodes were prepared by loading PDA particles on the as-prepared 3DVAG substrates via oxidative self-polymerization. By constructing PDA@3DVAG as the positive electrode of zinc-ion hybrid supercapacitors (ZHSCs), excellent electrochemical performances are obtained in highly concentrated salt electrolytes for both three and two electrode systems.

2. Experimental

Figure 1 illustrates the preparation process of PDA@3DVAG electrode material. In brief, 3DVAG aerogel was prepared by directional freezing and subsequent thermal reduction. Firstly, LGO suspension (2 mL, 5 mg mL^{-1}) was mixed with ascorbic acid (VC, 20 mg), which was subsequently heated in an oil bath for the first hydrothermal reduction to obtain partially reduced graphene oxide (PrGO) hydrogels. Then, PrGO was further heated in a water bath, and impurities were removed by deionized water. Finally, after freeze-drying and reduction, 3DVAG was obtained. Subsequently, PDA@3DVAG was prepared by the liquid-phase graphene reduction and oxidative self-polymerization process. Finally, it was assembled into a buckled supercapacitor for electrode testing.

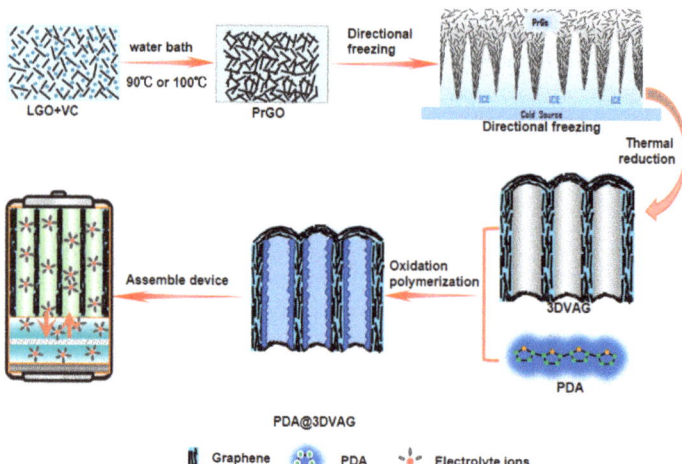

Figure 1. Illustration of preparation process of PDA@3DVAG composite electrode material.

2.1. Preparation of 3DVAG

Preparation of PDA@3DVAG 3D vertically oriented graphene was prepared by hydrothermal-assisted unidirectional freezing and subsequent thermal reduction. Firstly, 2 mL solution of graphene oxides suspension (5 mg mL^{-1}) and 20 mg of ascorbic acid were mixed and heated in an oil bath for the first hydrothermal reduction to obtain partially reduced graphene oxide (PrGO) hydrogel. Then, the PrGO was placed on the surface of a copper ingot impregnated with liquid nitrogen for 5 min of unidirectional freezing. After thawing at room temperature, PrGO was reduced in a water bath for 6 h. The obtained gel was washed with deionized water to remove soluble impurities. After being chemically reduced by 20 µL 80 wt.% of hydrazine hydrate and subsequently washing, 3DVAG was finally formed.

2.2. Preparation of PDA@3DVAG

PDA@3DVAG composite electrode material was prepared by liquid-phase graphene reduction and one-step oxidative self-polymerization. Firstly, 0.15 g dopamine hydrochloride was dissolved in 75 mL water, then the pH of the dopamine hydrochloride solution was adjusted to 8.5 by adding trimethyl methylamine. Subsequently, 3DVAG hydrogel was impregnated into the solution to self-polymerize for 24 h. Finally, PDA@3DVAG was obtained after washing and freeze-drying.

2.3. Materials and Electrochemical Characterization

The morphologies of the samples were characterized using field emission scanning electron microscopy (SEM, Dutch FEI) and energy dispersive spectrometer (EDS, Dutch FEI) mapping scanning. The structure of PDA@3DVAG was characterized by Fourier transform infrared (FTIR, Perkin Elmer Spectrum 100) spectroscopy spectra and X-ray diffraction (XRD, Bruker D8-Advanced diffractometer). UV–vis spectroscopy was performed on electrolyte solutions with a Perkin Elmer LAMBDA 750.

The cyclic voltammetry (CV) and galvanostatic charge–discharge (GCD) of the as-prepared PDA@3DVAG composite electrode materials were measured via a three-electrode system in this work. In 2M $ZnSO_4$ aqueous electrolyte, the electrochemical performances were tested with PDA@3DVAG composite electrode as working electrode, platinum sheet as counter electrode and Ag/AgCl electrode as reference electrode under the voltage window of 0~0.7 V. PDA@3DVAG composites were assembled into buckle supercapacitor for two-electrode testing. Porous activated carbon (AC) was used as the anode electrode, and CV tests were performed under the voltage window of −0.8~0 V.

The CV was performed by applying a linear voltage between the upper and lower limits of the two voltages for the working electrodes to perform a cyclic scan. The GCD examined the electrochemical response to controlled current, which is the response voltage versus time curve obtained by controlling a constant current.

The mass specific capacitance (C_m, F g^{-1}) was calculated as follows:

$$C_m = \frac{I \times \Delta t}{\Delta U \times m} \qquad (1)$$

where I was the constant discharge current (A), Δt was the discharge time (s), ΔU was the discharge voltage window (V), and m was the mass of active substance (g).

The mass energy density (E_m, Wh kg^{-1}) was calculated as follows:

$$E_m = \frac{1}{2} \times \frac{C_m \times (\Delta U)^2}{3.6} \qquad (2)$$

The mass power density (P_m, W kg^{-1}) was calculated as follows:

$$P_m = 3600 \times \frac{E_m}{\Delta t} \qquad (3)$$

3. Results and Discussion

3.1. Morphology and Structure of PDA@3DVAG

Figure 2 shows the SEM images and EDS analysis of PDA@3DVAG. The ordered porous structure can be observed in Figure 2a, where the channels show vertical orientation with a pore size of about 20–30 µm. Figure 2b,c shows the partial enlargement, where self-polymerized PDA particles are uniformly loaded on graphene substrates. As shown in Figure 2d, the EDS spectrum of PDA@3DVAG shows the main elements are C, N, and O.

The mass fraction of C element occupies 91.43%, while N and O elements occupy 3.2% and 5.37%, respectively. Figure 2c,d indicate the loading amount of PDA is small and a thin layer is coated on the surface of graphene, which ensures a good contact between PDA particles and graphene surfaces.

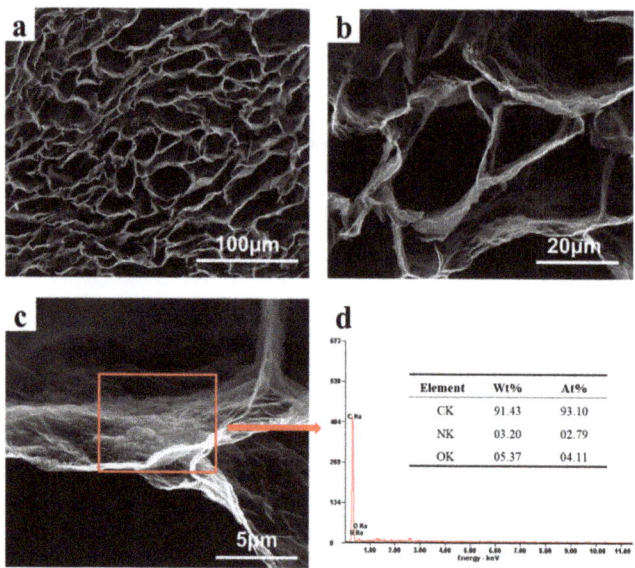

Figure 2. SEM images (**a**,**b**) and EDS analysis (**c**,**d**) of PDA@3DVAG.

Figure 3a illustrates the FTIR spectra of 3DVAG and PDA@3DVAG. Compared with 3DVAG, more functional group spectral peaks can be observed in the spectrum of PDA@3DVAG. The absorption peaks located at 3394, 3227, and 1721 cm^{-1} are found existing in the spectrum of PDA@3DVAG, which are induced by the stretching vibrations of the C−OH, N−H, and C=O group, respectively. The broad absorption peaks located at 1000~1800 cm^{-1} are induced by the vibrations of the indole group. The peaks located at 1609 and 1517 cm^{-1} are induced by the stretching vibrations of C=C on the benzene ring and C=N in the carbon−nitrogen five−membered ring, respectively [18,23]. The above results indicate that PDA is obtained by situ polymerization on the surface of graphene to form a PDA@3DVAG composite. As shown in Figure 3b, the XRD pattern of PDA@3DVAG reveals the characteristic peak of graphene structure (2θ = 26.1°), indicating a composite of graphene, which is consistent with FTIR.

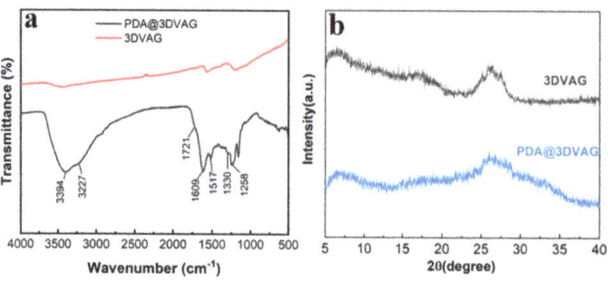

Figure 3. Chemical composition and structural characterization of PDA@3DVAG: (**a**) FTIR; (**b**) XRD.

3.2. Electrochemical Performance

Figure 4a shows the CV curve of the composite electrode at the scan rates as 2~100 mV s^{-1}, from which a pair of redox peaks can be clearly observed. The forward CV scan shows a broad oxidation peak with a peak potential of ~0.46 V vs. Ag/AgCl. In this experiment, Ag/AgCl is used as the reference electrode, and its potential is 0.2 V. The coordination potential of the quinone group with Zn ion is 0.46 V vs. Ag/AgCl by conversion. It is consistent with the potential of the oxidation peak in the CV curve, which is the current response of coordination reaction between the quinone group in PDA and Zn ion. A reduction peak appearing at ~0.26 V vs. Ag/AgCl has a potential difference of 0.2 V from the oxidation peak, which is consistent with the literature (the reduction potential is ~3.1 V vs. Li) [19]. The reduction peak reflects the reduction in the benzoquinone to a catechol group, resulting in the loss of electrochemical activity and the occurrence of Zn^{2+} ligand detachment. The above analysis shows PDA has a reversible Faradaic reaction process of Zn^{2+} embedding and de−embedding. Figure 4b shows the GCD curves of the composite electrode at current densities of 0.5~2 A g^{-1}. As the current density increases, the mass specific capacitance decreases. The highest specific capacitance is 142.4 F g^{-1} at the current density of 0.5 A g^{-1}.

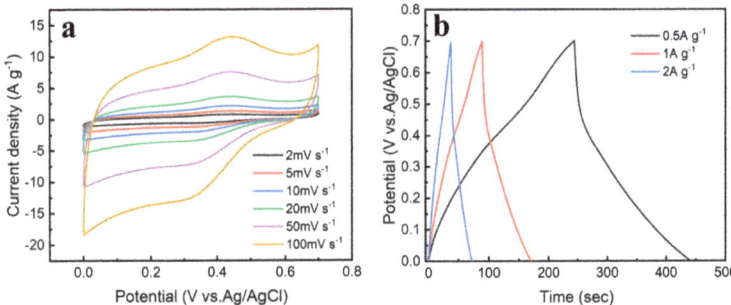

Figure 4. Electrochemical performances of PDA@3DVAG composites: (**a**) CV curves at different scan rates; (**b**) GCD curves at different current densities.

As shown in Figure 5a,b, the CV of the AC anode is performed under the voltage window of −0.8~0 V. The CV curve is rectangular−like, indicating that the anode stores zinc ions with electrostatic double−layer behavior. The PDA@3DVAG exhibits the apparent redox peaks in the CV curve and is a battery type as a cathode. The two electrode materials are assembled to construct an organic zinc−ion hybrid supercapacitor for further investigation of performances.

Figure 5c shows the CV curves of the ZHSCs at the scan rate of 20 mV s^{-1} for a voltage window of −0.8~0.7 V. Two pairs of redox peaks can be clearly observed in Figure 5c, one oxidation peak appears at 0.26 V, and another reduction peak appears at −0.2 V. The potential difference is 0.46 V, which is consistent with that (1.44 − 0.98 = 0.46 V) of the redox peak (Figure 5d) in the Zn//PDA@3DVAG zinc−ion battery (ZIBs). In the two sets of CV curves, the relative positions of the two pairs of redox peaks are close and their shapes and sizes are similar, indicating that the energy storage effect of the AC anode as the anode electrode is similar to that of metal zinc, while the mechanism is different. A reversible dissolution/deposition process occurs for metal zinc anode, while an electrostatic adsorption behavior for AC anode. Because of the low oxidation potential of AC anode and low redox potential of PDA, the assembled ZHSCs exhibit a low potential of −0.8~0.7 V, while the zinc cell voltage window can reach 0.5~1.7 V. However, the CV area of ZHSCs is larger than that of Zn−ion batteries, and the redox peaks at both locations are remarkable, indicating that the PDA is more electrochemically active in ZHSCs and has a greater ability to bind zinc ions.

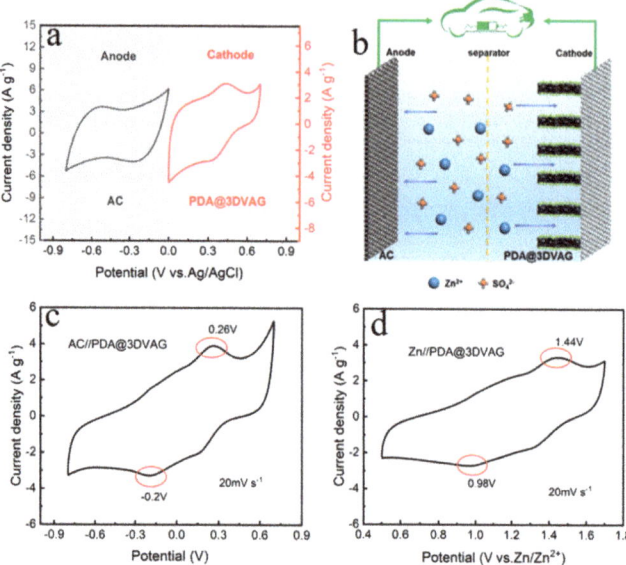

Figure 5. (**a**) CV curves of AC negative and PDA@3DVAG positive at a scan rate of 20 mV s^{-1}; (**b**) Schematic diagram of ZHSCs structure; (**c**) CV curves of AC//PDA@3DVAG ZHSCs; (**d**) CV curves of Zn//PDA@3DVAG ZIBs.

The PDA@3DVAG electrode is impregnated in a 2M ZnSO$_4$ electrolyte to observe its solubility. There is a color change from colorless and transparent to slightly brown after 24 h, as shown in Figure 6a. UV–vis spectroscopy of the electrolyte solution (Figure 6b) reveals an absorption peak at 281 nm obtained after 24 h of impregnation, which is consistent with the absorption peak of dopamine, indicating the presence of the soluble dopamine group in the conventional 2M ZnSO$_4$ solution [18]. In order to inhibit the dissolution of active substances, an ultra-high concentration zinc salt electrolyte (18 m ZnCl$_2$ + 6 m NH$_4$Cl, WIS) is used. As shown in Figure 6a, there is no obvious color change, and the UV–vis spectra can also reveal no absorption peak of the dopamine group at 281 nm, indicating that the highly concentrated salt electrolyte can significantly inhibit the dissolution of PDA.

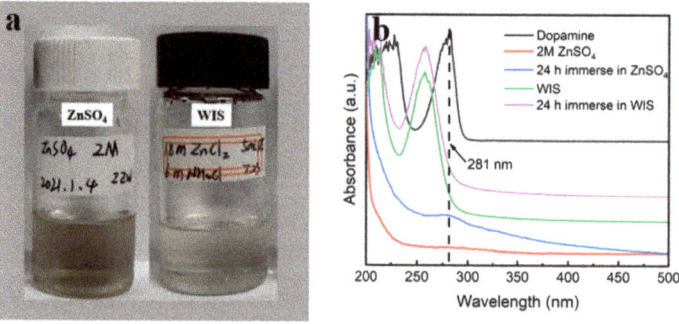

Figure 6. Solubility of PDA@3DVAG composite in two electrolytes: (**a**) camera picture; (**b**) UV–vis spectra.

The electrochemical performances of PDA@3DVAG are compared in the two aqueous electrolytes of 2M ZnSO$_4$ and highly concentrated salt. Figure 7a,c shows the CV curves of ZHSCs constructed with the two electrolytes at different scan rates from 2 to 100 mV s^{-1}.

As shown in Figure 7a, the PDA@3DVAG//2M ZnSO$_4$//AC ZHSCs shows two pairs of redox peaks under the −0.8~0.6 V voltage window, with oxidation peak potentials of ~−0.1 V and ~0.3 V, and reduction peak potentials of ~0.1 V and ~−0.3 V, corresponding to the coordination reaction of Zn^{2+} and H$^+$ with quinone group in the PDA, respectively. The highly concentrated salt electrolyte exhibits a wider voltage window (−0.8~0.8 V), as shown in Figure 7c. Due to the introduction of NH$_4^+$ in the electrolyte, the redox peaks in the CV curves of PDA@3DVAG//18 m ZnCl$_2$ + 6 m NH$_4$Cl//AC ZHSCs are −0.3 V and 0.35 V, and the reduction peak potentials are 0.15 V and −0.65 V, corresponding to the coordination reaction of Zn^{2+} and NH$_4^+$ with quinone group, respectively. According to the potential difference between AC anode and metal zinc, the redox peaks at 0.35 V and 0.15 V are coordination reactions occurring in Zn^{2+}, which is consistent with the three−electrode system.

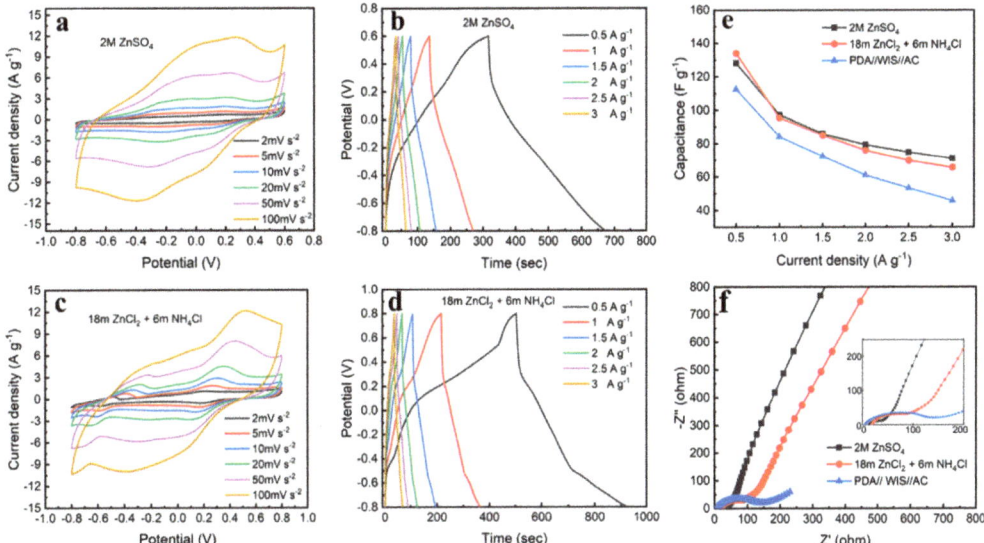

Figure 7. Electrochemical performances of evaluation: (**a**,**b**) CV and GCD curves of 2M ZnSO$_4$ ZHSCs; (**c**,**d**) CV and GCD curves of highly concentrated salt electrolytes ZHSCs; (**e**) multiplicative performance plots at different current densities; (**f**) electrochemical impedance spectra (enlarged figure of high−frequency range exhibited in inset).

Figure 7b,d shows the GCD curves of the two ZHSCs at different current densities. As shown in Figure 7b, the symmetrical GCD curves can be observed in the conventional electrolyte ZHSCs, which indicates the good electrochemical performance of ZHSCs in conventional electrolytes. As the current density increases from 0.5 to 3 A g$^{−1}$, the specific capacitance decreases from 128 F g$^{−1}$ to 71.1 F g$^{−1}$, with the capacitance retention of 55.54%. In the GCD curves of the highly concentrated salt electrolyte ZHSCs (Figure 7d), two pairs of charge and discharge plateaus can be clearly observed, corresponding to the redox peaks in CV. As the current density increases from 0.5 to 3 A g$^{−1}$, the specific capacitance decreases from 133.9 F g$^{−1}$ to 65.5 F g$^{−1}$, with the capacitance retention of 48.92%. The specific capacitances of the two electrode materials at different current densities are also shown in Figure 7e. It can also be seen that the PDA@3DVAG with vertical orientation exhibits better multiplier performance compared to the system of PDA//WIS//AC ZHSCs with the capacitance retention of 40.67%. This result can also be found in the electrochemical impedance spectrum (EIS), as shown in Figure 7f. The slope is higher in the AC−based electrode, which indicates double−layer capacitance and not diffusion, while the PDA/WIS/AC follows a diffusion pattern. Therefore, there is a fundamental difference between these systems. In the Nyquist plot, the semicircular arc appearing in the high−frequency range is the charge transfer resistance, which is mainly affected by the contact interface resistance between the electrolyte and the electrode. The sloping line in the low-frequency range reflects the ion transfer resistance of the pores inside the electrode. Both PDA@3DVAG//2M $ZnSO_4$//AC ZHSCs and PDA@3DVAG//18 m $ZnCl_2$ + 6 m NH_4Cl//AC ZHSCs have higher slopes in the low−frequency range due to the presence of carbon material in the positive electrode, but there is still ion diffusion resistance due to the presence of PDA pseudocapacitive material. The main energy storage contribution of PDA//WIS//AC is pseudocapacitance, so it exhibits large ion diffusion resistance. As shown in the enlarged figure of the high−frequency range, PDA@3DVAG exhibits a smaller ionic conductivity in WIS electrolytes compared to that in 2M $ZnSO_4$, mainly due to the low ionic conductivity of the highly concentrated electrolyte. However, it has a positive effect on pseudocapacitance, which has a priority of high energy density compared with traditional Electrochemical Double−Layer Capacitors (EDLCs). The EIS characterizations of the two electrolytes exhibit different ion transport resistances, with the conventional electrolyte having a smaller equivalent resistance than the highly concentrated salt electrolyte. In the low−frequency range, conventional electrolytes have a higher slope than highly concentrated electrolytes, indicating a lower resistance to ion transport, while PDA electrodes exhibit the poorest ion diffusion. From the three−dimensional vertical channels and the conductive network, PDA@3DVAG ensures good ion diffusion, achieves excellent multiplicity performance under highly concentrated electrolytes, and improves the voltage window and achieves better electrochemical performance [24,25].

Figure 8 shows the cycle performances of conventional and highly concentrated salt electrolytes for ZHSCs at the current density of 1 A g^{-1}. After 3000 cycles, the specific capacitance of the conventional electrolyte decreases from 95.8 to 42.3 F g^{-1}, with poor capacitance retention of only 44.2%, which may be caused by the dissolution of PDA during the long-term charge and discharge process as discussed before. The highly concentrated salt electrolyte shows a decrease in specific capacitance from 96.7 to 59.8 F g^{-1} after the long-term cycles, with a relatively high capacitance retention rate of 61.8%, which is improved by 17.6% compared with that of the conventional electrolyte.

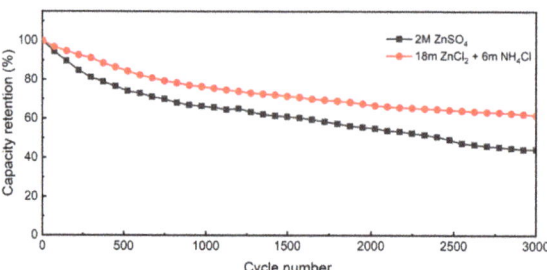

Figure 8. Cycle performances of different electrolytes for ZHSCs at the current density of 1 A g^{-1}.

Figure 9 shows the Ragone plot of ZHSCs constructed with highly concentrated salt electrolytes. It can be found that the PDA@3DVAG composite electrode shows both high energy and power densities. The energy density reaches 46.14 Wh kg^{-1} at the power density of 393.75 W kg^{-1} and 19.29 Wh kg^{-1} at 2183 W kg^{-1}, respectively. The energy density exceeds that of conventional electrochemical capacitors and some large−size batteries. As shown in Table 1, compared with the ZHSCs reported in the literature, the ZHSCs constructed with PDA@3DVAG composite electrodes exhibit an excellent energy density and good power density.

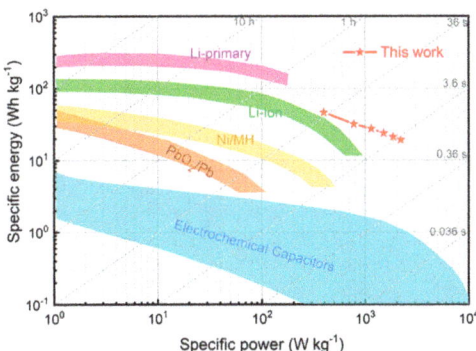

Figure 9. Ragone plot of ZHSCs constructed with highly concentrated salt electrolytes.

Table 1. Performance of ZHSCs constructed by PDA@3DVAG composites compared with other systems.

Electrode Material	Electrolyte	Voltage (V)	Energy Density (Wh kg^{-1})	Power Density (W kg^{-1})	Reference
PDA@3DVAG//AC	WIS	0.8	46.14	393.75	This work
NTC	1M H_2SO_4	0.8	4.5	40	[26]
V_2O_5//AC	2M $ZnSO_4$	2	34.6	1300	[27]
MnO_2–CNTs//MXene	2M $ZnSO_4$	1.9	29.7	2480	[28]
V_2O_5–ECF//ECF	6M LiCl	2	22.3	1500	[29]
$LiNi_{0.5}Mn_{1.5}O_4$//AC	1M $LiPF_6$	1.75	19	103	[30]
TiO_2@EEG//EEG	1M $LiPF_6$	1.5	10	2000	[31]
rGO/COF//rGO	1M H_2SO_4	1	10.3	50	[32]

4. Conclusions

The selection of material and the design of structure for electrode are of great importance for the performances of ZHSCs, which requires large surface area, good electric conductivity, certain channels with a proper diameter of pores, and so on for ion and mass transfer, abundant active chemical sites, etc. In this work, a method of unidirectional freezing and subsequent self-polymerization is used to obtain a PDA@3DVAG composite, which has a three-dimensional vertically aligned structure with long ordered channels and uniform pores. When the composite is assembled as electrode materials for an aqueous organic zinc-ion hybrid supercapacitor, the electrochemical performances are fairly well with a wide voltage window, good cycle performance, and high energy density, especially in highly concentrated salt electrolyte systems. The structure of three-dimensional vertically aligned graphene and the composite of PDA polymer may provide some references for the development of high-performance aqueous organic zinc-ion energy storage devices.

Author Contributions: Conceptualization, R.C.; data curation, R.C.; formal analysis, Z.Z.; funding acquisition, G.Y.; investigation, Z.T.; methodology, H.Z.; project administration, Y.X.; resources, Y.X.; software, Z.Z.; supervision, H.Z.; validation, Z.Z.; visualization, Z.T.; writing—original draft, R.C.; writing—review and editing, G.Y. All authors have read and agreed to the published version of the manuscript.

Funding: This work is financially supported by the National Natural Science Foundation of China (U1760119), the Science and Technology Commission of Shanghai Municipality (19DZ2271100), Shanghai Scientific and Technological Innovation Project (19JC1410400), and the Innovation Program of Shanghai Municipal Education Commission (2019-01-07-00-07-E00015).

Data Availability Statement: The data presented in this study are available from the corresponding author, upon reasonable request.

Conflicts of Interest: The authors declare no conflict of interest.

References

1. Dunn, B.; Kamath, H.; Tarascon, J.-M. Electrical Energy Storage for the Grid: A Battery of Choices. *Science* **2011**, *334*, 928–935. [CrossRef] [PubMed]
2. Mirzaeian, M.; Abbas, Q.; Ogwu, A.; Hall, P.; Goldin, M.; Mirzaeian, M.; Jirandehi, H.F. Electrode and electrolyte materials for electrochemical capacitors. *Int. J. Hydrogen Energy* **2017**, *42*, 25565–25587. [CrossRef]
3. Li, M.; Liu, J.; Zhang, T.; Song, X.; Chen, W.; Chen, L. 2D Redox-Active Covalent Organic Frameworks for Supercapacitors: Design, Synthesis, and Challenges. *Small* **2021**, *17*, e2005073. [CrossRef] [PubMed]
4. Xu, G.; Nie, P.; Dou, H.; Ding, B.; Li, L.; Zhang, X. Exploring metal organic frameworks for energy storage in batteries and supercapacitors. *Mater. Today* **2017**, *20*, 191–209. [CrossRef]
5. Zhang, W.; Zhao, Q.; Hou, Y.; Shen, Z.; Fan, L.; Zhou, S.; Lu, Y.; Archer, L.A. Dynamic interphase–mediated assembly for deep cycling metal batteries. *Sci. Adv.* **2021**, *7*, 3752. [CrossRef] [PubMed]
6. Obrovac, M.N.; Chevrier, V.L. Correction to Alloy Negative Electrodes for Li-Ion Batteries. *Chem. Rev.* **2015**, *115*, 2043. [CrossRef] [PubMed]
7. Tarascon, J.-M.; Armand, M. Issues and challenges facing rechargeable lithium batteries. *Nat. Cell Biol.* **2001**, *414*, 359–367. [CrossRef]
8. Pan, H.; Shao, Y.; Yan, P.; Cheng, Y.; Han, K.S.; Nie, Z.; Wang, C.; Yang, J.; Li, X.; Bhattacharya, P.; et al. Reversible aqueous zinc/manganese oxide energy storage from conversion reactions. *Nat. Energy* **2016**, *1*, 16039. [CrossRef]
9. Aravindan, V.; Gnanaraj, J.; Lee, Y.-S.; Madhavi, S. Insertion-Type Electrodes for Nonaqueous Li-Ion Capacitors. *Chem. Rev.* **2014**, *114*, 11619–11635. [CrossRef]
10. Jiao, Y.; Zhang, H.; Zhang, H.; Liu, A.; Liu, Y.; Zhang, S. Highly bonded T-Nb2O5/rGO nanohybrids for 4 V quasi-solid state asymmetric supercapacitors with improved electrochemical performance. *Nano Res.* **2018**, *11*, 4673–4685. [CrossRef]
11. Li, B.; Zheng, J.; Zhang, H.; Jin, L.; Yang, D.; Lv, H.; Shen, C.; Shellikeri, A.; Zheng, Y.; Gong, R.; et al. Electrode Materials, Electrolytes, and Challenges in Nonaqueous Lithium-Ion Capacitors. *Adv. Mater.* **2018**, *30*, e1705670. [CrossRef] [PubMed]
12. Zhong, F.; Ma, M.; Zhong, Z.; Lin, X.; Chen, M. Interfacial growth of free-standing PANI films: Toward high-performance all-polymer supercapacitors. *Chem. Sci.* **2021**, *12*, 1783–1790. [CrossRef] [PubMed]
13. Fan, Z.; Cheng, Z.; Feng, J.; Xie, Z.; Liu, Y.; Wang, Y. Ultrahigh volumetric performance of a free-standing compact N-doped holey graphene/PANI slice for supercapacitors. *J. Mater. Chem. A* **2017**, *5*, 16689–16701. [CrossRef]
14. Fan, Z.; Zhu, J.; Sun, X.; Cheng, Z.; Liu, Y.; Wang, Y. High Density of Free-Standing Holey Graphene/PPy Films for Superior Volumetric Capacitance of Supercapacitors. *ACS Appl. Mater. Interfaces* **2017**, *9*, 21763–21772. [CrossRef] [PubMed]
15. Zhuo, H.; Hu, Y.; Chen, Z.; Zhong, L. Cellulose carbon aerogel/PPy composites for high-performance supercapacitor. *Carbohydr. Polym.* **2019**, *215*, 322–329. [CrossRef] [PubMed]
16. Wang, H.; Diao, Y.; Lu, Y.; Yang, H.; Zhou, Q.; Chrulski, K.; D'Arcy, J.M. Energy storing bricks for stationary PEDOT supercapacitors. *Nat. Commun.* **2020**, *11*, 3882. [CrossRef]
17. Tisawat, N.; Samart, C.; Jaiyong, P.; Bryce, R.A.; Nueangnoraj, K.; Chanlek, N.; Kongparakul, S. Enhancement performance of carbon electrode for supercapacitors by quinone derivatives loading via solvent-free method. *Appl. Surf. Sci.* **2019**, *491*, 784–791. [CrossRef]
18. Wang, C.; He, T.; Cheng, J.; Guan, Q.; Wang, B. Bioinspired Interface Design of Sewable, Weavable, and Washable Fiber Zinc Batteries for Wearable Power Textiles. *Adv. Funct. Mater.* **2020**, *30*, 2004430. [CrossRef]
19. Liu, T.; Kim, K.C.; Lee, B.; Chen, Z.; Noda, S.; Jang, S.S.; Lee, S.W. Self-polymerized dopamine as an organic cathode for Li- and Na-ion batteries. *Energy Environ. Sci.* **2017**, *10*, 205–215. [CrossRef]
20. Zhao, Q.; Huang, W.; Luo, Z.; Liu, L.; Lu, Y.; Li, Y.; Li, L.; Hu, J.; Ma, H.; Chen, J. High-capacity aqueous zinc batteries using sustainable quinone electrodes. *Sci. Adv.* **2018**, *4*, eaao1761. [CrossRef]
21. Kundu, D.; Oberholzer, P.; Glaros, C.; Bouzid, A.; Tervoort, E.; Pasquarello, A.; Niederberger, M. Organic Cathode for Aqueous Zn-Ion Batteries: Taming a Unique Phase Evolution toward Stable Electrochemical Cycling. *Chem. Mater.* **2018**, *30*, 3874–3881. [CrossRef]
22. Ye, J.; Simon, P.; Zhu, Y. Designing ionic channels in novel carbons for electrochemical energy storage. *Natl. Sci. Rev.* **2020**, *7*, 191–201. [CrossRef]
23. Yue, X.; Liu, H.; Liu, P. Polymer grafted on carbon nanotubes as a flexible cathode for aqueous zinc ion batteries. *Chem. Commun.* **2019**, *55*, 1647–1650. [CrossRef] [PubMed]
24. Dong, S.; Xie, Z.; Fang, Y.; Zhu, K.; Gao, Y.; Wang, G.; Yan, J.; Cheng, K.; Ye, K.; Cao, D. Polydopamine-Modified Reduced Graphene Oxides as a Capable Electrode for High-Performance Supercapacitor. *Chemistryselect* **2019**, *4*, 2711–2715. [CrossRef]
25. Yang, Y.; Lin, Q.; Ding, B.; Wang, J.; Malgras, V.; Jiang, J.; Li, Z.; Chen, S.; Dou, H.; Alshehri, S.M.; et al. Lithium-ion capacitor based on nanoarchitectured polydopamine/graphene composite anode and porous graphene cathode. *Carbon* **2020**, *167*, 627–633. [CrossRef]
26. Yuksel, R.; Buyukcakir, O.; Panda, P.K.; Lee, S.H.; Jiang, Y.; Singh, D.; Hansen, S.; Adelung, R.; Mishra, Y.K.; Ahuja, R.; et al. Necklace-like Nitrogen-Doped Tubular Carbon 3D Frameworks for Electrochemical Energy Storage. *Adv. Funct. Mater.* **2020**, *30*, 1909725. [CrossRef]

27. Ma, X.; Wang, J.; Wang, X.; Zhao, L.; Xu, C. Aqueous V2O5/activated carbon zinc-ion hybrid capacitors with high energy density and excellent cycling stability. *J. Mater. Sci. Mater. Electron.* **2019**, *30*, 5478–5486. [CrossRef]
28. Wang, S.; Wang, Q.; Zeng, W.; Wang, M.; Ruan, L.; Ma, Y. A New Free-Standing Aqueous Zinc-Ion Capacitor Based on MnO2–CNTs Cathode and MXene Anode. *Nano-Micro Lett.* **2019**, *11*, 70. [CrossRef]
29. Li, L.; Peng, S.; Bin Wu, H.; Yu, L.; Madhavi, S.; Lou, X.W. A Flexible Quasi-Solid-State Asymmetric Electrochemical Capacitor Based on Hierarchical Porous V2O5Nanosheets on Carbon Nanofibers. *Adv. Energy Mater.* **2015**, *5*, 1500753. [CrossRef]
30. Arun, N.; Jain, A.; Aravindan, V.; Jayaraman, S.; Ling, W.C.; Srinivasan, M.; Madhavi, S. Nanostructured spinel LiNi 0.5 Mn 1.5 O 4 as new insertion anode for advanced Li-ion capacitors with high power capability. *Nano Energy* **2015**, *12*, 69–75. [CrossRef]
31. Wang, F.; Wang, C.; Zhao, Y.; Liu, Z.; Chang, Z.; Fu, L.; Zhu, Y.; Wu, Y.; Zhao, D. A Quasi-Solid-State Li-Ion Capacitor Based on Porous TiO2Hollow Microspheres Wrapped with Graphene Nanosheets. *Small* **2016**, *12*, 6207–6213. [CrossRef] [PubMed]
32. Wang, C.; Liu, F.; Chen, J.; Yuan, Z.; Liu, C.; Zhang, X.; Xu, M.; Wei, L.; Chen, Y. A graphene-covalent organic framework hybrid for high-performance supercapacitors. *Energy Storage Mater.* **2020**, *32*, 448–457. [CrossRef]

Article

Sustainable and Printable Nanocellulose-Based Ionogels as Gel Polymer Electrolytes for Supercapacitors

Rosa M. González-Gil [1,2,†], Mateu Borràs [3,4,†], Aiman Chbani [3], Tiffany Abitbol [5], Andreas Fall [5], Christian Aulin [5,6], Christophe Aucher [3,*] and Sandra Martínez-Crespiera [1,*]

1. Applied Chemistry and Materials, ARTS Department, Leitat Technological Center, C/Pallars, 186-179, 08005 Barcelona, Spain; rosamaria.gonzalez@icn2.cat
2. Novel Energy-Oriented Materials Group, Catalan Institute of Nanoscience and Nanotechnology, ICN2 (CSIC-BIST), Edifici ICN2, Campus UAB, 08193 Barcelona, Spain
3. Energy and Engineering, ARTS Department, Leitat Technological Center, C/de la Innovació, 2, 08225 Barcelona, Spain; mateu@bioo.tech (M.B.); achbani@leitat.org (A.C.)
4. Arkyne Technologies SL (Bioo), C/de la Tecnología, 17, 08840 Barcelona, Spain
5. Bioeconomy and Health, RISE Research Institutes of Sweden, Drottning Kristinas väg 61, 114 28 Stock-holm, Sweden; tiffany.abitbol@ri.se (T.A.); andreas.fall@ri.se (A.F.); Christian.aulin@Holmen.com (C.A.)
6. Holmen Iggesund, 825 80 Iggesund, Sweden
* Correspondence: caucher@leitat.org (C.A.); sandramartinez@leitat.org (S.M.-C.); Tel.: +34-93-788-23-00 (S.M.-C.)
† These authors contributed equally to this work.

Citation: González-Gil, R.M.; Borràs, M.; Chbani, A.; Abitbol, T.; Fall, A.; Aulin, C.; Aucher, C.; Martínez-Crespiera, S. Sustainable and Printable Nanocellulose-Based Ionogels as Gel Polymer Electrolytes for Supercapacitors. *Nanomaterials* **2022**, *12*, 273. https://doi.org/10.3390/nano12020273

Academic Editors: Federico Cesano, Simas Rackauskas and Mohammed Jsaim Uddin

Received: 22 December 2021
Accepted: 12 January 2022
Published: 15 January 2022

Publisher's Note: MDPI stays neutral with regard to jurisdictional claims in published maps and institutional affiliations.

Copyright: © 2022 by the authors. Licensee MDPI, Basel, Switzerland. This article is an open access article distributed under the terms and conditions of the Creative Commons Attribution (CC BY) license (https://creativecommons.org/licenses/by/4.0/).

Abstract: A new gel polymer electrolyte (GPE) based supercapacitor with an ionic conductivity up to 0.32–0.94 mS cm^{-2} has been synthesized from a mixture of an ionic liquid (IL) with nanocellulose (NC). The new NC-ionogel was prepared by combining the IL 1-ethyl-3-methylimidazolium dimethyl phosphate (EMIMP) with carboxymethylated cellulose nanofibers (CNFc) at different ratios (CNFc ratio from 1 to 4). The addition of CNFc improved the ionogel properties to become easily printable onto the electrode surface. The new GPE based supercapacitor cell showed good electrochemical performance with specific capacitance of 160 F g^{-1} and an equivalent series resistance (ESR) of 10.2 Ω cm^{-2} at a current density of 1 mA cm^{-2}. The accessibility to the full capacitance of the device is demonstrated after the addition of CNFc in EMIMP compared to the pristine EMIMP (99 F g^{-1} and 14.7 Ω cm^{-2}).

Keywords: nanocellulose; ionic liquid; ionogel; gel polymer electrolyte; renewable energy storage; supercapacitors

1. Introduction

From all the different types of energy storage systems (ESS), the interest in supercapacitors (SCs) has grown thanks to their energy storage mechanism by electrical double layer, which increases cycle life and allows the use of safer ionic conductors and electrode materials than in traditional batteries [1]. Their promising properties such as high-power density, fast charge–discharge rate and long-cycle stability, have turned SCs into a potentially useful tool for application in flexible and wearable electronics [2–4]. In this sense, the use of these quasi-solid electrolytes could play an important role in the production of printed supercapacitors providing flexibility, transparency, stability, and safety to the devices [5].

Different types of electrolytes can be used for ESS, including liquid, solid state electrolytes (SSEs) [6] and gel polymer electrolytes (GPEs). GPEs consist basically of a liquid electrolyte, that can be an ionic liquid (IL), or an ionic salt trapped inside a three-dimensional (3D) polymer network [7,8]. ILs, generally a room temperature molten salt, are usually used in GPEs to improve the properties of the ESSs, because of their great electrochemical stability over a wide electrochemical window with minimum decomposition, high ionic conductivity, non-volatility, non-flammability, and good thermal and chemical stabilities [9,10]. Moreover, ionogel electrolytes (IL based gels) present interesting advantages against liquid leakage and corrosion problems, increasing electrochemical stability, and facilitating packaging and manufacturing processes, due to both their liquid-like ionic conductivity and solid-like structural and mechanical features with great flexibility [10].

Recently, in response to an increased demand concerning the sustainable and cost-effective development of electrochemical energy storage devices, great efforts are being devoted to replacing the existing fossil energy resources by new ESSs based on green sustainable materials. As a class of green materials, nanocellulose (NC) has received extensive attention due to its natural abundance, recyclability, facile and diverse modification, large specific area, and dimensional stability [11,12]. NC can be classified into the following categories depending on the source, preparation methods and fiber morphology: (1) cellulose nanofibers (CNF), which refers to the nanoscale cellulose fibrils obtained by fibrillating cellulose under the action of high pressure or mechanical forces, (2) cellulose nanocrystals (CNC) obtained by acid hydrolysis conditions to form nano-sized, rod shaped particles with high crystallinity, and (3) bacterial nanocellulose (BC) with a high aspect ratio obtained by a microbial fermentation process [12,13]. Compared with other biomass-derived green materials, such as lignin, chitin, etc., NC shows great advantages in the field of ESSs [14–20]: versatile surface modification via functionalization of its hydroxyl groups that allow the integration of other active materials; great thermal stability and extraordinary mechanical properties that can improve the safety properties of the ESSs, through formation of the previous mentioned gel polymer electrolytes (GPEs) [20].

Despite the NC advantages described above, the processing of cellulose-based materials for practical application in energy storage fields, like GPEs, is limited due to their highly ordered structure and crystallinity. However, nanocellulose can be solubilized in some ionic liquids by the disruption of the hydrogen bonding network and the formation of new strong hydrogen bonds with the IL anions [21]. Hence, ILs anions play a predominant role during this process, different than cations with weaker interactions (van der Waals and/or electrostatic interactions) [21,22]. In this way, the use of NC or other sources of cellulose as polymeric matrix for ionogel preparation has been studied.to develop green and biofriendly electrolyte alternatives to fluorinated polymers. Thiemann et al., synthesized an ionic liquid GPE by gelation of microcellulose thin films with 1-ethyl-3-methylimidazolium methylphosphonate [23]. These cellulose ionogels showed high specific capacitances (5–15 $\mu F \cdot cm^{-2}$), high transparency and flexibility. In another example, a cellulose-based dual network ionogel electrolyte was synthesized by Rana et al. by dissolving phosphorylated microcrystalline cellulose in 1,3-dimethylimidazolium methyl phosphate [DMIM][MeO(H)PO$_3$] followed by post-polymerization. The as-synthesized ionogel electrolyte exhibited a high ionic conductivity (2.6–22.4 $mS \cdot cm^{-1}$) over a wide temperature range (up to 120 °C) [22].

In this work, we have worked with 1-ethyl-3-methylimidazolium dimethyl phosphate (EMIMP) as the ionic liquid because of the well-known ability of phosphates, among other anions, to dissolve carbohydrates [21,22]. While ionogels from natural cellulose or microcellulose have already been obtained easily [21–24], we selected carboxymethylated cellulose nanofibers (CNFc) for our studies due to its superior properties: better mechanical performance, and processability; together with the presence of carboxylic groups on the surface, which can boost the strong inter- and intra- molecular interactions inside the resulting ionogel [25]. We present the combination of the EMIMP with CNFc as hosting renewable polymer to provide an easy-to-prepare transparent ionogel electrolyte with exceptional thermal and electrochemical capabilities. For that purpose, four different ratios of CNFc to IL were studied and tested in a coin-cell supercapacitor device as electrolyte (CNFc ratio from 1 to 4) to observe the change in electrochemical performance. Physicochemical characterizations (composition, morphology, rheology, and thermal stability) and electrochemical characterizations (capacity, ionic conductivity and cyclability) were performed to investigate its properties and understand the functionality of the novel nanocellulose-based ionogel electrolytes.

2. Materials and Methods

2.1. Materials

Carboxymethylated cellulose nanofibers were produced by RISE Research Institutes of Sweden from a commercial sulfite softwood pulp (Domsjö Dissolving plus; Domsjö Fabriker AB, Domsjö, Sweden). Ethanol (Rectapur®) was purchased from VWR (VWR AB International, Spånga, Sweden). Monochloroacetic acid (99%, ACS reagent, $ClCH_2COOH$), acetic acid (ACS reagent), iso-propanol (ACS reagent), sodium hydroxide (ACS reagent), sodium hydrogen carbonate (ACS reagent) and methanol (ACS reagent) were purchased from Sigma Aldrich (Stockholm, Sweden). 1-ethyl-3-methylimidazolium dimethyl phosphate (>98% grade) (EMIMP) was purchased at Iolitec (Heilbronn, Germany) and used as received. Active Carbon YP50F (Kuraray, Kurashiki, Japan), conductive carbon TIMCAL C 65 (Cabot), 1-Methyl-2-pyrrolidone (NMP, Scharlau, Sant Feliu del Llobregat, Spain) and 5130 PVDF Solef (Solvay, Barcelona, Spain) were used in the formulation of the active layers of the supercapacitors. The fabricated electrodes were assembled in CR2032 coin cells from MTI by using cellulose separator Kodoshi (Nippon Kodoshi Corporation, Kochi, Japan) and 1-ethyl-3-methylimidazolium dimethyl phosphate (EMIMP, >98%, Iolitec) or EMIMP:CNFc mixtures (96:4 to 99:1 in ratio) as electrolyte. 5 samples electrolytes are reported including EMIMP pure electrolyte and 4 mixtures adding 1%, 2%, 3% and 4% of CNF in EMIMP ionic liquid. It has not been possible to produced jelly electrolyte with higher rate of CNF due to the loss of the mechanical properties not allowing to print and process the component. The preparation of the electrode is identic for each symmetric coin cell and described in Section 2.3. Each electrode is prepared by coating of the jelly electrolyte directly on the top of the dried carbon electrode and then the CR2032 coin cell is assembled with staking 2 symmetric electrode separated by Kodoshi nanocellulose separator (thickness 50 µm) without further addition of electrolyte or solvent.

2.2. Carboxymethylated Cellulose Nanofibers (CNFc) Fabrication

CNFc fabrication was done according to the general methodology described in the literature [26]. Most commonly, CNFc is prepared with a total charge of ~600 µeq/g, corresponding to a carboxmethyl degree of substitution (DS) of 0.1 [27]. In this work, a higher charge DS 0.3 CNFc was produced as has been recently described [28]. Generally, resulted pulp is solvent exchanged to ethanol followed by carboxymethylation reaction, which begins by impregnating the pulp with a solution of monochloric acetic acid in isopropanol, followed by transfer of the pulp to a heated methanolic solution of NaOH in isopropanol and reaction under reflux for 1 h. Compared to the reaction conditions for DS 0.1, to achieve DS 0.3, the dosage of monochloroacetic acid is increased 4.4-fold and NaOH 2.7-fold. After reaction, the carboxymethylated pulp is washed with $NaHCO_3$ to neutralize, and finally washed with water to remove excess reagents, until the wash has a conductivity below 10 µS/cm. To generate CNFc, the carboxymethylated pulp is microfluidized (Microfluidizer M-110EH, Microfluidics Corp., Newton, MA, USA) at 1700 bar (1 pass) to give a suspension of DS 0.3 CNFc at 1 wt%. The CNFc was then subjected to lyophilization and morturation before use in Leitat (Barcelona, Spain).

2.3. CNFc Based Ionogel Preparation

In Figure 1 it is shown a scheme about the NC-ionogel preparation process. For this, freeze-dried CNFc from an aqueous dispersion (at 3 wt%) was mechanically mixed with different ratios of EMIMP (p/p, EMIMP/CNFc: 99:1, 98:2, 97:3 and 96:4, referred as EMIMP/CNFc 99:1, EMIMP/CNFc 98:2, EMIMP/CNFc 97:3 and EMIMP/CNFc 96:4) with heating (100 °C) for 16 h to obtain a yellowish viscous gel (final appearance pictures are depicted in Figure S1 from Supplementary Materials). The as prepared ionogel could be used without any additional step in ambient air, showing long-term stability since no destabilization phenomena, like phase-separation, have been observed after approx. 2 years.

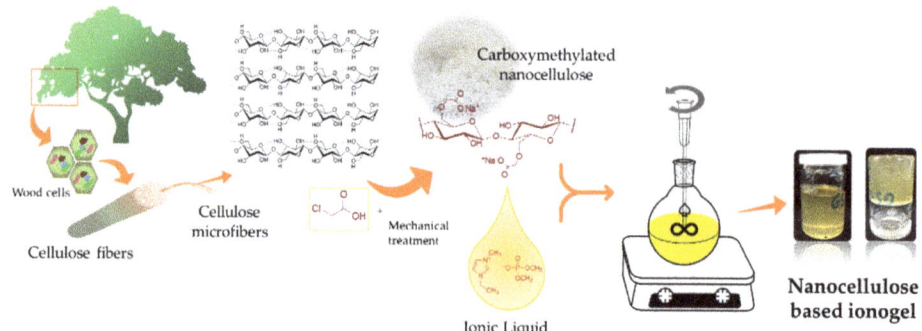

Figure 1. Schematic preparation of CNFc-based ionogels.

2.4. Supercapacitor Preparation

The active ink used in the supercapacitor application was fabricated by mixing NMP, active carbon YP50F, conductive carbon TiMCALC65 and binder PVDF at dry weight ratio of 80:10:10 by using a Dipermat (VMA-GETZMANN GmbH, Reischsof, Germany) with a 3D printed head reproducing the shape of an industrial planetary mixer but adapted for lower volume of ink. The prepared ink was printed onto a 20 µm aluminum foil by bare coating (gap of 100 µm) and dried for 2 h at 120 °C in vacuum obtaining a dry coating of 50 µm thickness and active material loading of 1.8 mg cm^{-2}. For the CNFc-ionogel printing, heating to 80–100 °C allowed direct printing using a Doctor Blade technique (gap of 150 µm bar) onto a carbon electrode surface as shown in Figure 2.

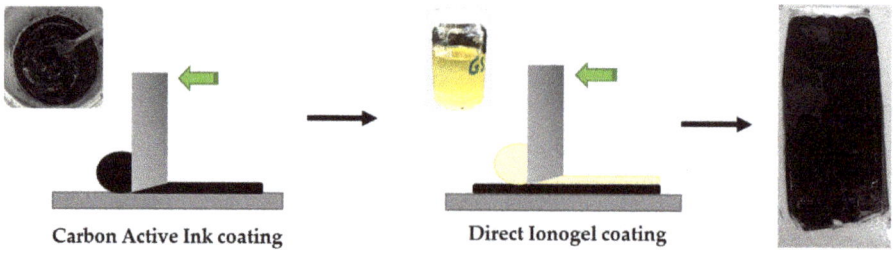

Figure 2. Direct printing process by Doctor Blade technique of the CNFc-ionogel onto the electrodes.

2.5. Characterization

X-ray diffraction (XRD) patterns were recorded for each sample on a diffractometer (Malvern PANalytical X'pert Pro MPD, X-ray Diffraction Facility, *Institut Català de Nanociència i Nanotecnologia*) at room temperature from 10 to 50 2θ. Fourier transform infrared spectroscopy were recorded with a Shimadzu Iraffinity-1S spectrometer in ATR mode. The specimens were measured directly with a scan range from 400 to 4000 cm^{-1}. The thermogravimetric analysis (TGA) of the samples (5–10 mg) were carried out by using a TGA/DSC (Mettler-Toledo, University of Alicante, Alicante, Spain), equipped with autosampler and gas-controlled unit, from room temperature to 900 °C with heating rate of 20 °C/min under N$_2$ conditions. Scanning Electron Microscopy (SEM) were performed in a Zeiss Merlin FE-SEM (Carl Zeiss Microscopy, Autonomous University of Barcelona, Bellaterra, Spain) with samples mounted directly on SEM stubs with adhesive carbon tape and metalized with Au thin film (E5000 Sputter Coater. Polaroid Equipment Limited. Watford). Images were recorded at different steps of magnification. The rheologic properties of EMIMP and all fabricated electrolytes were analyzed by using the rheometer Bohlin CVO 100–901 in the share rate range from 0.0517 to 888 s^{-1}.

The electrochemical testing was performed with a multichannel potentiostat VMP3 (Bio-Logic) run with the EC-Lab software. Cyclic voltammetry was performed between 0 < V < 2.0, and with a scan rate between 10 < mV s^{-1} < 50, to investigate side reactions, evaluate the potential window of stability and the device behavior. Galvanostatic measurements at a constant areal capacity of mA·cm^{-2} and between 0 < V < 2.0 V) were used to calculate the specific capacity (mAh·g^{-1} of AC, energy (unit), power (unit), and equivalent series resistance (ESR) is calculated from the ohmic drop between of the developed supercapacitors. Electrochemical impedance spectrum measurements were carried out with frequency range of 50 mHz to 1 MHz. All these analyses were carried out by VMP3 from Biologic. Finally, a long cycling at 1 mA·s^{-1} for 10,000 cycles were carried out by Neware cycler. The electrochemical CR2032 coin-cell type used for the tests is built from the stack of two symmetric electrodes separated by cellulose membrane separators and with 40 µL of EMIMP or 150 µm (wet) bar-coated EMIMP:CNFc gel as electrolyte ensuring it excess.

The cell capacitance (C_{cell}) of a symmetrical capacitor can be calculated according to Equation (1) [29]:

$$C_{cell} = \frac{C_e}{2}; C_e = 2C_{cell} \qquad (1)$$

where $C_e = C_+ = C_-$ are the specific capacitance of the two electrodes in mAh·g^{-1} of activated carbon, and the cell capacitance of a supercapacitor can be calculated from the charge-discharge experiment Equation (2):

$$C_{cell} = \frac{I}{mdV/dt} \qquad (2)$$

This can be then used, when combined with C_{cell} equation to obtain the electrode capacity, and subsequently obtain information on the capacity of an investigated material Equation (3):

$$C_e = \frac{4I}{mdV/dt} \qquad (3)$$

where I (A) is the discharge current, m (g) the total mass of active materials of two electrodes and dV/dt the slope of the discharge curve between 80% and 40% of the cut-off voltage (V). From the total specific cell capacitance of two-electrode system (C_{cell}), the maximum energy ($E_{max.}$) and maximum power ($P_{max.}$) of a supercapacitor cell can be calculated according to Equations (4) and (5):

$$E_{max.} = \frac{1}{2}C_{cell}V^2 \qquad (4)$$

$$P_{max.} = \frac{V^2}{4ESR} \qquad (5)$$

where, V (V) is the cell voltage minus the ohmic drop, and ESR (Ω) is the total equivalent series resistance of the supercapacitor. From the EIS, the ionic conductivity of the different gels has been calculated using the following formula Equation (6):

$$\sigma = \frac{l}{R_1 A} \qquad (6)$$

where σ is the ionic conductivity (S m^{-1}), l (m) is the thickness of the membrane, A (m^2) is the contact area between the electrode and the electrolyte and R_1 (Ω) is the ionic resistance at high frequency corresponding to the first interception of the circle in the Nyquist plots.

3. Results and Discussion

An extensive characterization of the prepared CNFc-ionogels have been realized to study their composition, homogeneity, thermal and rheological properties. XRD patterns for each IL/CNFc ratio, pristine CNFc (lyophilized) and EMIMP are shown in Figure 3a. Broad signals are obtained in all cases, obtaining major intensities at 12° and 25° for all the EMIMP/CNFc mixtures, which are related to the ionic liquid EMIMP (triangle signals), with a minor signal present as a broad band at 15° corresponding to (002) nanocellulose crystalline plane (star signals). The major component of ionic liquid ensures a good ionic conductivity and a good gel flexibility due to their mainly amorphous nature. Fourier transform infrared spectroscopy (FTIR) of each CNFc-ionogel were also recorded and are depicted in Figure 3b, where the IR profiles of pure CNFc and EMIMP are also added. As in case of previous XRD studies, the EMIMP has the major intensity, showing its typical bands of -CH from the imidazolium ring (3060 and 1570 cm^{-1}), the -CH$_2$ st and -CH$_3$ st (2920–2970 cm^{-1}), -CH$_3$ bend (1450 cm^{-1}) and -CH$_2$ bend (780 cm^{-1}) from the ethyl chains, the -C-N st at 1240 cm^{-1} and finally, the -P-O-C (930–1050 cm^{-1}) and -P=O (1090–1200 cm^{-1}) vibrations of the diethylphosphate group. CNFc (highlighted in blue) shows an -OH band at 3400 cm^{-1}, which increases with the higher ratio of CNFc in the mixtures, a -C=O peak related with the carboxylated groups at 1600 cm^{-1}, which appears slightly shifted to 1650 cm^{-1} for all ionogel mixtures, and a -C-O-C streaching at 1000 cm^{-1} from CNFc glucose units, which can not be appreciated in the mixtures due to overlapping with EMIMP P=O and P-O-C signals.

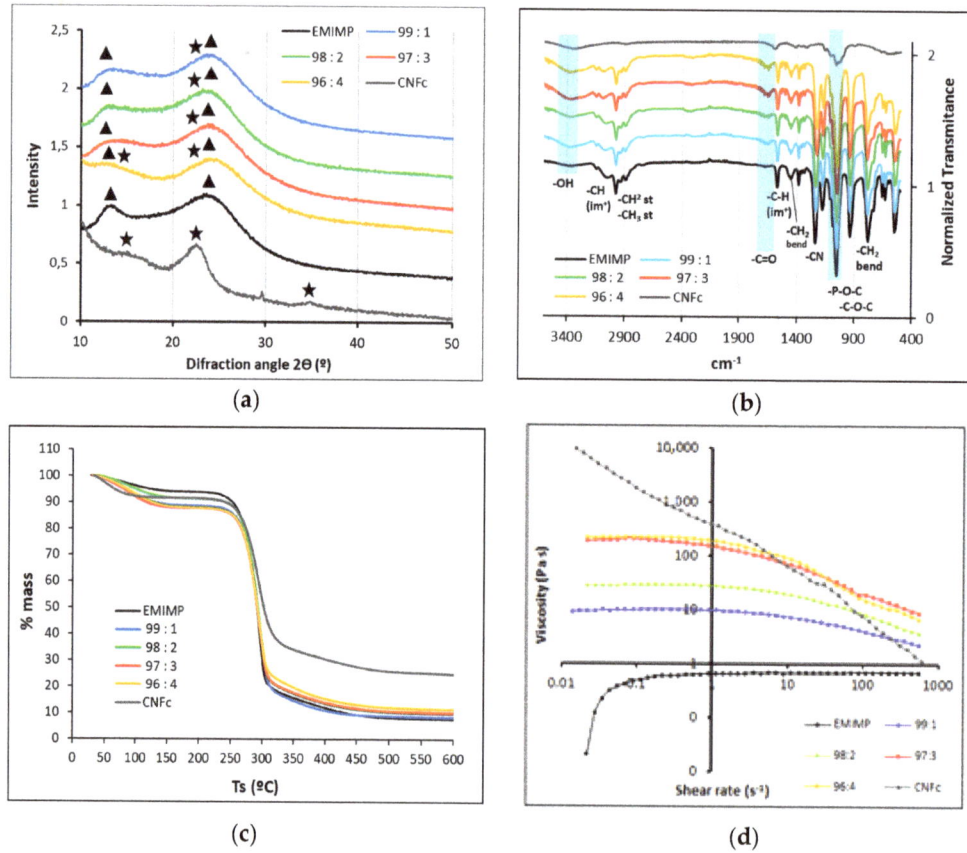

Figure 3. (a) XRD; (b) FTIR; (c) TGA and (d) Rheological data from CNFc-EMIMP GPE, including pristine CNFc (gel form at 1%wt) and EMIMP.

Thermal properties (TGA) were studied for each CNFc-ionogel and are depicted in Figure 3c. The TGA profiles shown in all cases a two-step weight loss at temperatures below 100 °C and above 250 °C, which can be reasonably attributed to evaporation of adsorbed water (less than a 15% wt) and the degradation of CNFc and the EMIMP, respectively. This data shows the high thermal stability of the new gel electrolyte. DSC experiments have been also studied to determine the degradation temperatures for each ionogel in comparison to pure EMIMP (Figure S2 and Table S1, Supplementary Materials), obtaining similar results between them (290 °C for EMIMP, to 293 °C for EMIMP/CNFc 96:4). The addition of smalls amounts of nanocellulose has a minor affection on the decomposition temperatures, improving them slightly. ILs are mostly Newtonian fluids and more viscous than most common molecular solvents [30]. Their ionogels show similar viscoelastic properties under stress, providing a higher-quality electrode/electrolyte contact, in comparison with other rigid solid-state electrolytes [10,31]. In the case of this work, viscosity measurements (Figure S3, Supplementary Materials) reveal that almost all the prepared NC-ionogels have similar viscosity characteristics, except for the mixture EMIMP/CNFc 96:4, that reveals a thixotropic behavior, typical of NC. Regarding the apparent viscosity dependence of the applied shear stress (Figure 3d), all NC-ionogels, and CNFc (1% wt, gel), show a reduction in the viscosity, which is more pronounced in the ionogels with a higher CNFc content (EMIMP/CNFc 96:4 and EMIMP/CNFc 97:3), with no changes in case of pure EMIMP. This characteristic, named shear-thinning [32], is common of pseudoplastic fluids and consents on a more effective particles flow through the liquid phase, reducing the viscosity of the solution. It is intrinsic of the nanocellulose and its derivatives [33,34], and it is responsible for the good printability of the presented ionogels, while the desired mechanical properties were maintained.

High resolution scanning electronic microscopy (FE-SEM) was done for the ionogel EMIMP/CNFc 96:4 to study its morphology, showing a smooth and homogeneous surface. Moreover, an EDS mapping of nitrogen, phosphorous and sodium elements that are present in the ionic liquid (N and P) and the nanocellulose (Na) was also performed to confirm the homogeneity of the GPE. (See Figure 4).

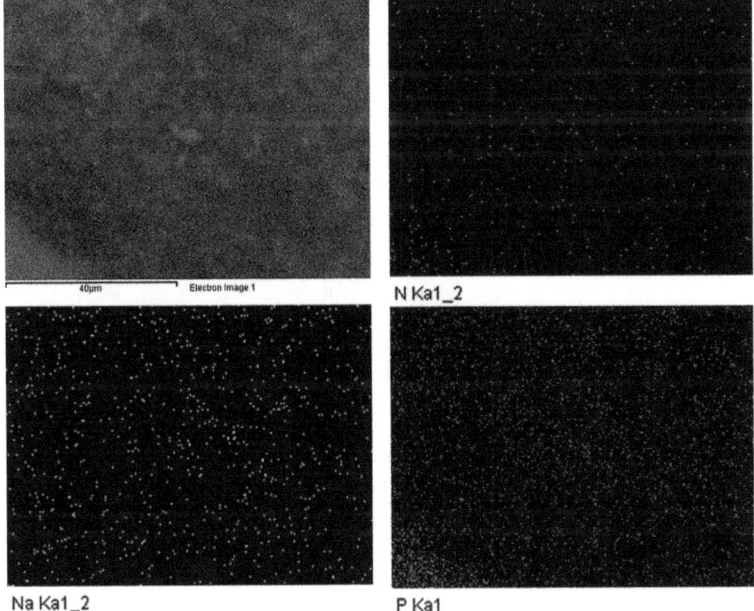

Figure 4. FE-SEM image of EMIMP/CNFc 96:4 gel and its EDS mapping on N, Na and P elements.

In the same way, to use these NC-based ionogels as electrolytes for supercapacitors, their electrochemical performance and properties were first evaluated. Symmetric cells were assembled in CR2032 coin cell to test the electrochemical performance of the EMIMP electrolyte, alone and with the different CNFc ratios, using a cellulosic Kodoshi separator. The cyclic voltammogram at 10 mV·s^{-1} (See Figure 5a) of the supercapacitor with EMIMP electrolyte shows a near rectangle shape, approximated to the ideal situation of double layer capacitor (or supercapacitor). There is no visible redox peak from Faradaic current over the potential region. However, a peak onset appears as 2 V is approached. At high scan rates (See Figure 5b), the shape of the CV curve was distorted due to an increase of the internal resistance and the peak onset near 2 V was shifted to higher voltage, due to the increase in cell resistance at higher scan rate.

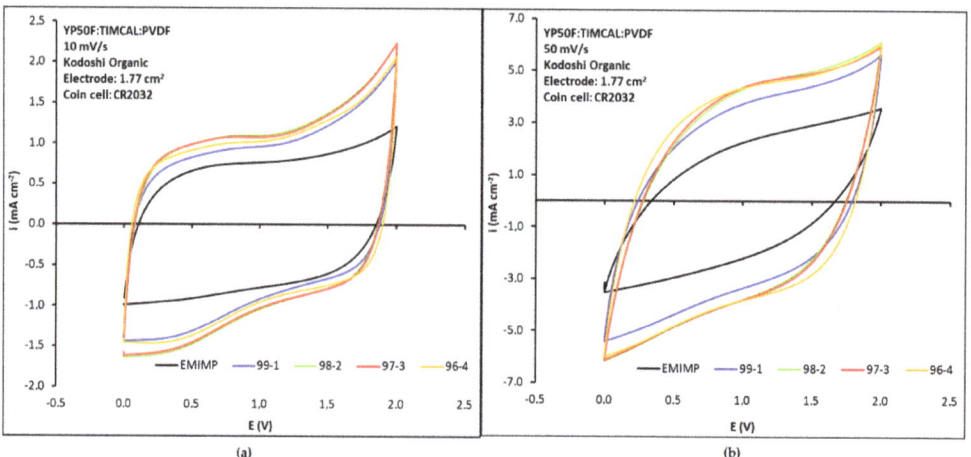

Figure 5. CV (cyclic voltammetry) curves at (**a**) 10 mV s^{-1}; (**b**) 50 mV s^{-1} scan rates for each EMIMP/CNFc ionogel (99:1, 98:2, 97:3 and 96:4). Pristine EMIMP is also added for comparison.

When CNFc is added to the EMIMP ionic liquid, the capacity of the device increases with the addition of CNFc, the peak onset near 2 V in the oxidation zone becomes more evident and a new one between 0 and 0.5 V in the reduction zone appears (Figure 5a). At high scan rates (Figure 5b), the shape of the CV curve was distorted due to an increase of the internal resistance but to lesser extent than with EMIMP alone and the peaks onset near 2 V and between 0 and 0.5 V disappear. However, no clear differences in capacity and resistance between the electrolytes with different ratios of EMIMP/CNFc can be appreciated from the cyclic voltammetry studies.

In order to clarify if the peak onset observed in the cyclic voltammetry was caused by the electrolyte degradation or because of its own behavior (as has been seen before in different IL electrolytes) [35], a floating test was carried out [36]. The supercapacitors with the different studied electrolytes were charged at 2V in a stepwise ramp and the voltage was maintained for a period of 24 h, as shown Figure S4 from Supplementary Materials. The leakage current obtained at each time was recorded and could be used to directly assess the importance of the electrochemical degradation of the electrolytes in the fully charged state. The results showed that the leakage current of the EMIMP alone was 6 times higher than the one obtained by adding CNFc (i.e., 120 µA vs 20 µA). The data obtained for the different EMIMP/CNFc electrolytes were low enough to determine that the peak onset showed in the cyclic voltammogram was not caused by electrolyte degradation [37]. Moreover, these results demonstrated that the addition of CNFc increase the EMIMP voltage resistance. The capacity and resistance of the supercapacitors with EMIMP and with the gel electrolytes (by the introduction of different rates of CNFc in the EMIMP) were evaluated by galvanostatic charge-discharge tests. All these parameters were calculated using the equations described in Section 2 and the results are showed in Table S1 in Supplementary Materials and Figure 6. In the case of capacities calculated from cyclic voltammetry, as shown Table S4 from Supplementary Materials, the capacitance tendency in all cases is decreasing at higher scan rate. Higher specific capacity is measured after addition of CNFc, which is online with galvanostatic charge-discharge results. The supercapacitors with CNFc in the electrolyte exhibit a smaller ESR in all intensity ranges than the supercapacitors with only EMIMP (4.0–10.2 Ω for different rates of EMIMP:CNFc and 14.7 Ω for EMIMP alone). These results also indicate that EMIMP/CNFc gel electrolytes (in all the studied ranges) have higher ionic conductivity than EMIMP alone, and this has been confirmed by the electrochemical impedance spectroscopy showed in Figure 7. In addition, according to Equations (2) and (3), the capacity for the electrolyte EMIMP and EMIMP/CNFc with different ratios were calculated to be 99 and 132–160 F g^{-1} of YP50F active material at current density of 1 mA cm^{-2}. The introduction of CNF in the electrolyte significantly increases the capacity of the electrode, and this increase on capacity and decrease on ESR is also significant between the four gel electrolytes with different CNFc ratios. Maximum of capacity is reached for EMIMP/CNFc 97:3 ionogel but presenting similar capacity than the EMIMP/CNFc 96:4 ratio. However, the ESR of EMIMP/CNFc 96:4 is clearly lower than the other gel electrolytes that present similar values. In addition, we previously demonstrated that this increase on capacity does not come from the electrolyte (or CNFc) degradation by the floating test, rather this increment on capacity suggests that part of the CNFc present in the electrolyte is displaced to the electrode surface (due to his high polarity), creating an extra porous structure where more ions are allocated during the charge process, and therefore, increasing the capacity of the device. It has not been tested above 96:4 due to the poor mechanical of gel above this range.

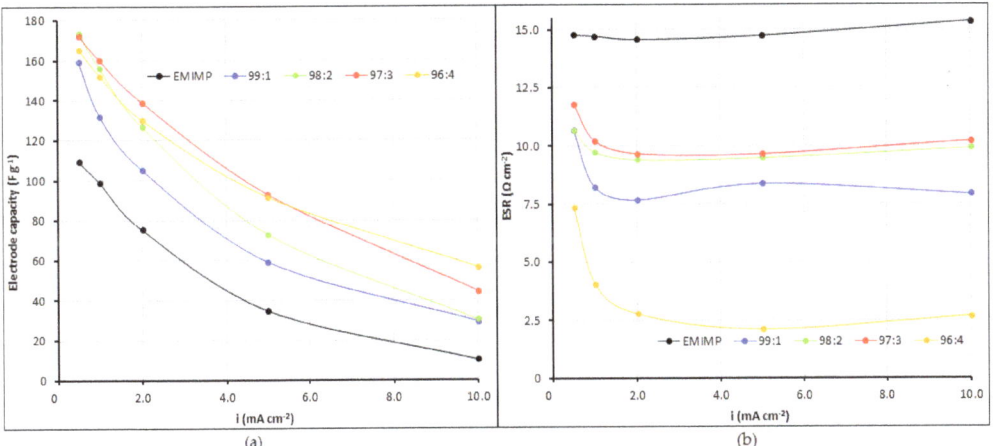

Figure 6. (**a**) Electrode capacity and (**b**) internal resistance (ESR) of the supercapacitors with EMIMP and EMIMP/CNFc mixtures (99:1, 98:2, 97:3 and 96:4) as electrolyte at different charge-discharge current rates.

The capacity variation with the cycles is shown in Figure S5. The supercapacitors with EMIMP show a good capacity retention of 80% after 10,000 cycles at 1 mA cm^{-2} between 0 < V <2. The supercapacitors with CNFc had a quick capacity decay in the first 2000 cycles (~0.020 mAh g^{-1} cycle^{-1} of capacity decay) and then present a moderate degradation (~0.001 mAh g^{-1} cycle^{-1} of capacity decay) reaching the 10,000 cycles with a capacity retention below 50%. It is possible that during the cycles, the CNFc cannot be retained in the matrix of the electrolyte and that creates a larger SEI in the first cycle, starting to cover part of the carbon porous and therefore limiting the capacity of the electrodes.

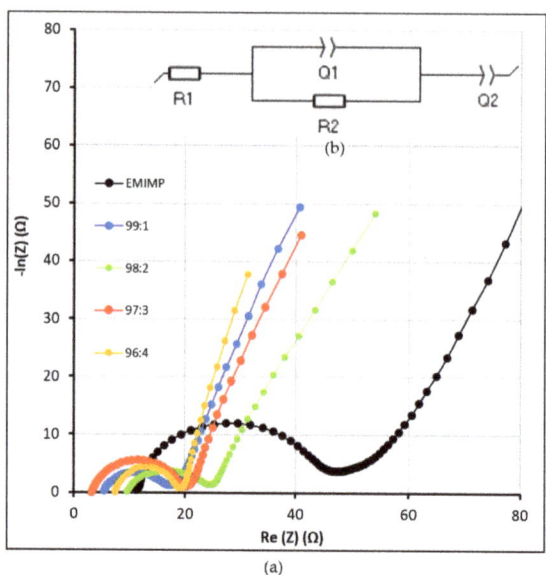

Figure 7. (**a**) Nyquist plots of the different EMIMP/CNFc gel electrolytes and EMIMP ionic liquid; (**b**) Equivalent circuit of the system.

To rule out a possible dissolution of CNFc to carboxymethyl cellulose (CMC) units, an ionogel with a 96:4 ratio EMIMP/CMC was prepared and assessed. Hence, in the case of use CMC instead CNFc, the electrochemical performance of these supercapacitors, shown in Table S3, provides specific capacitance values of 123 F g^{-1}, being like EMIMP/CNFc 99:1 ionogel values. Moreover, it has been observed the formation of a solid precipitates by mixing CMC to EMIMP, presumably due to a lack of interaction between the CMC and IL because of the diminution of available -OH groups. As it has been mentioned, the presence of these hydroxyl groups plays an important role in the dissolution of cellulose in ILs, due to they are the responsible of the formation of new hydrogen bonds between the NC and the ILs anions [38,39].

Finally, electrochemical impedance spectroscopy (EIS) for the supercapacitors with EMIMP/CNFc gel electrolytes and EMIMP are shown in Figure 7a. The equivalent circuit was made with EC-Lab (Figure 7b). The Nyquist plots show a semi-circle at high frequencies region and a line perpendicular to the X-axis at low frequencies that correspond the ionic diffusion and the double layer formation at the surface of the conductive solid phase. R1 is mostly corresponding to the ionic conductivity in the dielectric, both gel and membrane separator. R2 is referring to the charge transfer both ionic and electronic in the bulk of the electrode. Q1 and Q2 are constant phase element simulating the real behaviour of a pure capacitor. The first thing to be considered is a charge transfer resistance (R_{ct}, that includes the ionic resistance), which could be intensively investigated in the semi-circle loop appearing in the middle frequency region. A decrease in resistances is shown in supercapacitors with CNFc compared to supercapacitors with only EMIMP. This decrease in resistances has been observed above all in the value of the charge transfer resistance (R_{ct}). These results are in line with the ones obtained in the CV and galvanostatic tests. The real and imaginary capacitance versus the frequency is given in supporting in Figure S7 and using the definition from P.L. Taberna and et al. [40]. The real part of the capacitance shows the variation of the available stored energy with the frequency. The imaginary part of the capacitance represents the losses that occur during charge storage. In these graphs, can observe that the maximum capacitance is reached faster at higher frequencies in the case of the blended electrolyte containing CNFc versus EMIMP electrolyte. No real difference is observed within the different percentage of CNF used in this studied. No degradation of the electrolyte is demonstrated by the floating test, also the same active material and electrode systems have been used for all electrolyte testing, then we assume no intrinsic supplementary capacity is reached by adding CNFc to EMIMP but a faster access to this maximal capacity is demonstrated.

All the values of ionic conductivity (σ, mS cm^{-1}) and resistances (R1 and R2, Ω) are included in Table 1. The increase in conductivity with the addition of CNFc was not expected because of the important increment on the electrolyte viscosity (almost 0 to 60 Pa ·s), Figure 3d and Figure S3 from Supplementary Materials). However, as can be seen in Table 1, adding nanocellulose has helped to increase the ionic conductivity of the GPEs, causing a decrease in the internal resistance of the system (0.32–0.94 mS cm^{-1} for EMIMP/CNFc ionogels against 0.26 mS cm^{-1} for EMIMP). Plenty of polar functional groups in the system helped to increase the dielectric constant of the gel electrolytes, which is proportional to the capacity assuming all electrode surface area and distances between the two electrodes are the same for each cell assembled. This also increases the interaction between ions and polymer chains and helps to dissociate these ion pairs into separate ions. As a result, more free ions are generated and consequently, the ionic conductivity is enhanced. Moreover, the abundant polar functional groups supplied by cellulose offer an ionic conduction channel by aiding facile ion hopping onto nearby polar groups However no tendency could be measured with the increased ratio of CNFc.

Table 1. Ionic conductivity and resistances of the different EMIMP/CNFc gel electrolytes and pristine EMIMP ionic liquid.

Electrolyte	σ (mS cm^{-1})	R1 (Ω)	R2 (Ω)
EMIMP	0.26 ± 0.03	11.14	36.02
EMIMP/CNFc 99:1	0.57 ± 0.03	5.07	12.76
EMIMP/CNFc 98:2	0.32 ± 0.03	8.92	15.77
EMIMP/CNFc 97:3	0.94 ± 0.03	3.08	17.03
EMIMP/CNFc 96:4	0.44 ± 0.03	6.5	13

4. Conclusions

Novel GPEs have been prepared by the addition of different CNFc ratios (from 1 to 4) to EMIMP, showing excellent thermal and mechanical properties. Contrary to the usual liquid electrolytes, these new NC based GPEs not only improve safety issues, but also show a good printability, very interesting for printed electronics applications. Moreover, enhanced electrochemical properties have also been measured. In case of specific capacitance, EMIMP/CNFc ionogels have shown a significant increase (160 F g^{-1}) compared to the pristine IL (99 F g^{-1}), and its internal resistances have considerable decreased from 14.7 Ω cm^{-2} to 10.2 Ω cm^{-2} at a current density of 1 mA cm^{-2}. Although the viscosity (Figure 3d) of the electrolyte increases after CNF addition, which should promote lower ionic conductivity, all electrochemical measurements show both higher ionic conductivity and capacity with CNF addition. These data confirm that the addition of polar groups coming from nanocellulose play a beneficial role. However, one of the points to be improved in future experiments is to maintain the capacitance over the cycles (50 % decrease in capacitance in the first 2000 cycles). A drastic loss of the capacity is observed for all samples containing CNF after long cycling (Supplementary Materials Figure S5), although the leakage current test showed no electrolyte degradation (Figure S4). With these results we assume the addition of CNF is beneficial at device level, reducing the initial internal resistance and increasing the quality of the interface between the electrolyte and the electrode. But this effect disappears within the cycling, where samples with and without CNFc show similar capacity after 2.000 cycles A possible passivation of the electrode that it is reducing both capacity and ionic conductivity is not discarded and needs to be further studied. Moreover, investigations about the use of CNFc vs CMC have been also performed, showing that the use of the nanofibrils of cellulose form (CNFc) provides better solubility in EMIMP and higher electrochemical performances. With all this results we can conclude that these new GPEs have shown great potential for applications in energy storage devices.

Supplementary Materials: The following are available online at https://www.mdpi.com/article/10.3390/nano12020273/s1. Figure S1: Final aspect of the as-prepared ionogels pictures; Figure S2: Differencial scanning calorimetry thermogram (DSC) of the EMIMP/CNFc mixtures (99:1, 98:2, 97:3 and 96:4). Pristine CNFc and EMIMP were added for comparison purposes; Figure S3: Rheological results at room temperature vs. time of the EMIMP/CNFc mixtures (99:1, 98:2, 97:3 and 96:4). Pristine CNFc and EMIMP are added for comparison purposes; Figure S4: Current (μA) vs. time for supercapacitors with EMIMP and EMIMP/CNFc (99:1, 98:2, 97:3 and 96:4) as electrolytes for a period of 24h; Figure S5. The capacity variation with the cycles and for supercapacitors with EMIMP and EM-IMP/CNFc (99:1, 98:2, 97:3 and 96:4) as electrolytes; Figure S6: Galvanostatic charge-discharge curve of EMIMP and ionogels with different ratio of CNFc at i = 1 mA cm^{-2}; Figure S7. (a) Real part of the capacitance vs Frequency and (b) Imaginary part of the capacitance versus Frequencies of EMIMP and ionogels with different ratio of CNFc. Table S1: Temperature decomposition for each EMIMP/CNFc mixtures (99:1, 98:2, 97:3 and 96:4) mixtures, obtained from differential scanning calorimetry thermograms (DSC). Pristine CNFc and EMIMP temperatures are also added; Table S2: Summary of the key performance parameters on different current density such as specific capacitance, ESR, Emax and Pmax; Table S3: Summary of the key performance parameters on different current density such as specific capacitance and ESR of EMIMP/CMC vs EMIMP/CNFc 99:1; Table S4: Summary of specific capacitance of EMIMP and ionogels with different ratio of CNFc at 10 mV s^{-1} and 50 mV s^{-1}. scan rate for CV.

Author Contributions: Conceptualization, S.M.-C., R.M.G.-G. and M.B.; investigation, R.M.G.-G., M.B., A.C. and T.A.; data curation, R.M.G.-G., M.B. and A.C.; writing—original draft preparation, R.M.G.-G., M.B. and A.C.; writing—review and editing R.M.G.-G., M.B., A.C., T.A., A.F., C.A. (Christian Aulin), C.A. (Christophe Aucher) and S.M.-C.; visualization, R.M.G.-G., M.B., A.C., T.A., A.F., C.A. (Christian Aulin), C.A. (Christophe Aucher) and S.M.-C.; supervision, S.M.-C. and C.A. (Christophe Aucher); project administration, S.M.-C.; funding acquisition, S.M.-C. All authors have read and agreed to the published version of the manuscript.

Funding: This research was funded by European Union's Horizon 2020 research and innovation program under grant agreement No 761000 (GREENSENSE project).

Institutional Review Board Statement: Not applicable.

Informed Consent Statement: Not applicable.

Data Availability Statement: The data presented in this study are available on request from the corresponding author.

Acknowledgments: Authors thank to Thermal Analysis Lab at the University of Alicante's Technical Research Services (SSTTI) for your support in TGA Analysis. They also thank to Electron Microscopy Services from Autonomous University of Barcelona (UAB) for the FE-SEM and EDS mapping experiments; and to X-ray Diffraction Facility from *Institut Català de Nanociència i Nanotecnologia* (ICN2) for their support on XRD analysis.

Conflicts of Interest: The authors declare no conflict of interest.

References

1. Fu, W.; Turcheniuk, K.; Naumov, O.; Mysyk, R.; Wang, F.; Liu, M.; Kim, D.; Ren, X.; Magasinski, A.; Yu, M.; et al. Materials and technologies for multifunctional, flexible or integrated supercapacitors and batteries. *Mater. Today* **2021**, *48*, 176–197. [CrossRef]
2. Jang, S.; Kang, J.; Kwak, S.; Seol, M.-L.; Meyyappan, M.; Nam, I. Methodologies for Fabricating Flexible Supercapacitors. *Micromachines* **2021**, *12*, 163. [CrossRef] [PubMed]
3. Dong, L.; Xu, C.; Li, Y.; Huang, Z.-H.; Kang, F.; Yang, Q.-H.; Zhao, X. Flexible electrodes and supercapacitors for wearable energy storage: A review by category. *J. Mater. Chem. A* **2016**, *4*, 4659–4685. [CrossRef]
4. Wang, F.; Wu, X.; Yuan, X.; Liu, Z.; Zhang, Y.; Fu, L.; Zhu, Y.; Zhou, Q.; Wu, Y.; Huang, W. Latest advances in supercapacitors: From new electrode materials to novel device designs. *Chem. Soc. Rev.* **2017**, *46*, 6816–6854. [CrossRef] [PubMed]

5. Wang, F.; Wang, X.; Chang, Z.; Wu, X.; Liu, X.; Fu, L.; Zhu, Y.; Wu, Y.; Huang, W. A Quasi-Solid-State Sodium-Ion Capacitor with High Energy Density. *Adv. Mater.* **2015**, *27*, 6962–6968. [CrossRef]
6. Kou, Z.Y.; Lu, Y.; Miao, C.; Li, J.Q.; Liu, C.J.; Xiao, W. High-performance sandwiched hybrid solid electrolytes by coating polymer layers for all-solid-state lithium-ion batteries. *Rare Met.* **2021**, *40*, 3175–3184. [CrossRef]
7. He, C.; Cheng, J.; Liu, Y.; Zhang, X.; Wang, B. Thin-walled hollow fibers for flexible high energy density fiber-shaped supercapacitors. *Energy Mater.* **2021**, *1*, 100010. [CrossRef]
8. Yong, H.; Park, H.; Jung, C. Quasi-solid-state gel polymer electrolyte for a wide temperature range application of acetonitrile-based supercapacitors. *J. Power Sources* **2020**, *447*, 227390. [CrossRef]
9. Shahzad, S.; Shah, A.; Kowsari, E.; Iftikhar, F.J.; Nawab, A.; Piro, B.; Akhter, M.S.; Rana, U.A.; Zou, Y. Ionic Liquids as Environmentally Benign Electrolytes for High-Performance Supercapacitors. *Glob. Chall.* **2019**, *3*, 1800023. [CrossRef]
10. Wu, J.; Xia, G.; Li, S.; Wang, L.; Ma, J. A Flexible and Self-Healable Gelled Polymer Electrolyte Based on a Dynamically Cross-Linked PVA Ionogel for High-Performance Supercapacitors. *Ind. Eng. Chem. Res.* **2020**, *59*, 22509–22519. [CrossRef]
11. Guo, R.; Zhang, L.; Lu, Y.; Zhang, X.; Yang, D. Research progress of nanocellulose for electrochemical energy storage: A review. *J. Energy Chem.* **2020**, *51*, 342–361. [CrossRef]
12. Klemm, D.; Cranston, E.D.; Fischer, D.; Gama, M.; Kedzior, S.A.; Kralisch, D.; Kramer, F.; Kondo, T.; Lindström, T.; Nietzsche, S.; et al. Nanocellulose as a natural source for groundbreaking applications in materials science: Today's state. *Mater. Today* **2018**, *21*, 720–748. [CrossRef]
13. Trache, D.; Tarchoun, A.F.; Derradji, M.; Hamidon, T.S.; Masruchin, N.; Brosse, N.; Hussin, M.H. Nanocellulose: From Fundamentals to Advanced Applications. *Front. Chem.* **2020**, *8*, 392. [CrossRef]
14. Chen, G.; Fang, Z. Application of Nanocellulose in Energy Materials and Devices. In *Nanocellulose: From Fundamentals to Advanced Materials*; Wiley-VCH: Weinheim, Germany, 2019; pp. 397–421. [CrossRef]
15. Chen, W.; Yu, H.; Lee, S.Y.; Wei, T.; Li, J.; Fan, Z. Nanocellulose: A promising nanomaterial for advanced electrochemical energy storage. *Chem. Soc. Rev.* **2018**, *47*, 2837–2872. [CrossRef] [PubMed]
16. Jose, J.; Thomas, V.; Vinod, V.; Abraham, R.; Abraham, S. Nanocellulose based functional materials for supercapacitor applications. *J. Sci. Adv. Mater. Devices* **2019**, *4*, 333–340. [CrossRef]
17. Sezali, N.A.A.; Ong, H.L.; Jullok, N.; Villagracia, A.R.; Doong, R.A. A Review on Nanocellulose and Its Application in Supercapacitors. *Macromol. Mater. Eng.* **2021**, *306*, 2100556. [CrossRef]
18. Xu, T.; Du, H.; Liu, H.; Liu, W.; Zhang, X.; Si, C.; Liu, P.; Zhang, K.; Xu, T.; Liu, H.; et al. Advanced Nanocellulose-Based Composites for Flexible Functional Energy Storage Devices. *Adv. Mater.* **2021**, *33*, 2101368. [CrossRef] [PubMed]
19. Kim, J.H.; Lee, D.; Lee, Y.H.; Chen, W.; Lee, S.Y. Nanocellulose for Energy Storage Systems: Beyond the Limits of Synthetic Materials. *Adv. Mater.* **2019**, *31*, 1804826. [CrossRef]
20. Chen, C.; Hu, L. Nanocellulose toward Advanced Energy Storage Devices: Structure and Electrochemistry. *Acc. Chem. Res.* **2018**, *51*, 3154–3165. [CrossRef] [PubMed]
21. Zhang, J.; Wu, J.; Yu, J.; Zhang, X.; He, J.; Zhang, J. Application of ionic liquids for dissolving cellulose and fabricating cellulose-based materials: State of the art and future trends. *Mater. Chem. Front.* **2017**, *1*, 1273–1290. [CrossRef]
22. Rana, H.H.; Park, J.H.; Gund, G.S.; Park, H.S. Highly conducting, extremely durable, phosphorylated cellulose-based ionogels for renewable flexible supercapacitors. *Energy Storage Mater.* **2020**, *25*, 70–75. [CrossRef]
23. Thiemann, S.; Sachnov, S.J.; Pettersson, F.; Bollström, R.; Österbacka, R.; Wasserscheid, P.; Zaumseil, J. Cellulose-Based Ionogels for Paper Electronics. *Adv. Funct. Mater.* **2014**, *24*, 625–634. [CrossRef]
24. da S. Oliveira, R.; Bizeto, M.A.; Camilo, F.F. Production of self-supported conductive films based on cellulose, polyaniline and silver nanoparticles. *Carbohydr. Polym.* **2018**, *199*, 84–91. [CrossRef]
25. Zander, N.E.; Dong, H.; Steele, J.; Grant, J.T. Metal Cation Cross-Linked Nanocellulose Hydrogels as Tissue Engineering Substrates. *ACS Appl. Mater. Interfaces* **2014**, *6*, 18502–18510. [CrossRef] [PubMed]
26. Wågberg, L.; Decher, G.; Norgren, M.; Lindström, T.; Ankerfors, M.; Axnäs, K. The build-up of polyelectrolyte multilayers of microfibrillated cellulose and cationic polyelectrolytes. *Langmuir* **2008**, *24*, 784–795. [CrossRef] [PubMed]
27. Attias, N.; Reid, M.; Mijowska, S.C.; Dobryden, I.; Isaksson, M.; Pokroy, B.; Grobman, Y.J.; Abitbol, T. Nanocellulose–Mycelium Hybrid Materials: Biofabrication of Nanocellulose–Mycelium Hybrid Materials (Adv. Sustainable Syst. 2/2021). *Adv. Sustain. Syst.* **2021**, *5*, 2170003. [CrossRef]
28. Rosén, T.; He, H.; Wang, R.; Zhan, C.; Chodankar, S.; Fall, A.; Aulin, C.; Larsson, P.T.; Lindström, T.; Hsiao, B.S. Cross-Sections of Nanocellulose from Wood Analyzed by Quantized Polydispersity of Elementary Microfibrils. *ACS Nano* **2020**, *14*, 16743–16754. [CrossRef]
29. Chen, T.; Dai, L. Flexible supercapacitors based on carbon nanomaterials. *J. Mater. Chem. A* **2014**, *2*, 10756–10775. [CrossRef]
30. Makino, W.; Kishikawa, R.; Mizoshiri, M.; Takeda, S.; Yao, M. Viscoelastic properties of room temperature ionic liquids. *J. Chem. Phys.* **2008**, *129*, 104510. [CrossRef]
31. Pal, P.; Ghosh, A. Solid-state gel polymer electrolytes based on ionic liquids containing imidazolium cations and tetrafluoroborate anions for electrochemical double layer capacitors: Influence of cations size and viscosity of ionic liquids. *J. Power Sources* **2018**, *406*, 128–140. [CrossRef]
32. Pinto, F.; Meo, M. Design and Manufacturing of a Novel Shear Thickening Fluid Composite (STFC) with Enhanced out-of-Plane Properties and Damage Suppression. *Appl. Compos. Mater.* **2016**, *24*, 643–660. [CrossRef]

33. Nechyporchuk, O.; Belgacem, M.N.; Pignon, F. Current Progress in Rheology of Cellulose Nanofibril Suspensions. *Biomacromolecules* **2016**, *17*, 2311–2320. [CrossRef] [PubMed]
34. Moberg, T.; Sahlin, K.; Yao, K.; Geng, S.; Westman, G.; Zhou, Q.; Oksman, K.; Rigdahl, M. Rheological properties of nanocellulose suspensions: Effects of fibril/particle dimensions and surface characteristics. *Cellulose* **2017**, *24*, 2499–2510. [CrossRef]
35. Lei, Z.; Liu, Z.; Wang, H.; Sun, X.; Lu, L.; Zhao, X.S. A high-energy-density supercapacitor with graphene–CMK-5 as the electrode and ionic liquid as the electrolyte. *J. Mater. Chem. A* **2013**, *1*, 2313–2321. [CrossRef]
36. Andreas, H.A. Self-Discharge in Electrochemical Capacitors: A Perspective Article. *J. Electrochem. Soc.* **2015**, *162*, A5047–A5053. [CrossRef]
37. Zhao, Q.; Liu, X.; Stalin, S.; Khan, K.; Archer, L.A. Solid-state polymer electrolytes with in-built fast interfacial transport for secondary lithium batteries. *Nat. Energy* **2019**, *4*, 365–373. [CrossRef]
38. Fukaya, Y.; Sugimoto, A.; Ohno, H. Superior solubility of polysaccharides in low viscosity, polar and halogen-free 1,3-dialkylimidazolium formates. *Biomacromolecules* **2006**, *7*, 3295–3297. [CrossRef]
39. Verma, C.; Mishra, A.; Chauhan, S.; Verma, P.; Srivastava, V.; Quraishi, M.A.; Ebenso, E.E. Dissolution of cellulose in ionic liquids and their mixed cosolvents: A review. *Sustain. Chem. Pharm.* **2019**, *13*, 100162. [CrossRef]
40. Taberna, P.L.; Portet, C.; Simon, P. Electrode Surface Treatment and Electrochemical Impedance Spectroscopy Study on Carbon/Carbon Supercapacitors. *Appl. Phys. A* **2006**, *82*, 639–646. [CrossRef]

Fabrication of Mn_3O_4-CeO_2-rGO as Nanocatalyst for Electro-Oxidation of Methanol

Mohammad Bagher Askari [1], Seyed Mohammad Rozati [1,*] and Antonio Di Bartolomeo [2,*]

[1] Department of Physics, Faculty of Science, University of Guilan, Rasht P.O. Box 41335-1914, Iran; mbaskari@phd.guilan.ac.ir
[2] Department of Physics "E. R. Caianiello" and Interdepartmental Center NANOMATES, University of Salerno, 84084 Fisciano, SA, Italy
* Correspondence: smrozati@guilan.ac.ir (S.M.R.); adibartolomeo@unisa.it (A.D.B.)

Abstract: Recently, the use of metal oxides as inexpensive and efficient catalysts has been considered by researchers. In this work, we introduce a new nanocatalyst including a mixed metal oxide, consisting of manganese oxide, cerium oxide, and reduced graphene oxide (Mn_3O_4-CeO_2-rGO) by the hydrothermal method. The synthesized nanocatalyst was evaluated for the methanol oxidation reaction. The synergetic effect of metal oxides on the surface of rGO was investigated. Mn_3O_4-CeO_2-rGO showed an oxidation current density of 17.7 mA/cm^2 in overpotential of 0.51 V and 91% stability after 500 consecutive rounds of cyclic voltammetry. According to these results, the synthesized nanocatalyst can be an attractive and efficient option in the methanol oxidation reaction process.

Keywords: Mn_3O_4-CeO_2-rGO; nanocatalyst; methanol oxidation; cyclic voltammetry

1. Introduction

The increasing energy need and the scarcity of fossil fuels have pushed countries towards the use of clean and natural fuel sources and alternative energy resources [1,2]. The excessive consumption of fossil fuels has caused global warming and an increase in greenhouse gases, thus posing very serious environmental risks to human health and the planet [3–5].

The use of clean, available, natural, and cost-effective fuels is one of the best suggestions to overcome these environmental crises [6]. In recent years, the use of renewable energy sources such as sunlight and wind energy may have become more widely available and accessible than other renewable sources [7]; furthermore, various sciences have introduced attractive and portable devices, which open new avenues in the field of energy storage and conversion. The use of various types of electrochemical batteries, supercapacitors, and fuel cells are among these devices [8,9].

Fuel cells convert directly chemical energy to electrical energy and have high efficiency [10], while supercapacitors and electrochemical batteries are used to store energy [11]. One of the most important challenges for researchers is the introduction of inexpensive materials for use in electrodes of supercapacitors and electrochemical batteries as well as fuel cells.

The introduction of high-efficiency and cost-effective materials in the field of energy storage has almost been achieved, and various studies have confirmed this claim [12–14]. However, in the field of catalysts, especially catalysts in alcohol fuel cells, the introduction of high-efficiency and low-cost catalysts has not been fully possible so far. The materials based on platinum, palladium, ruthenium, and other rare and precious metals are the best catalysts in methanol oxidation [15], and finding a catalyst that can compete with these materials requires more effort and research. It should be noted that many cost-effective catalysts have been introduced in the field of methanol oxidation, although small amounts of precious metals have been often used in the catalyst structure. Moreover, composing and hybridizing the Pt, Pd, Ru, etc., with materials such as carbon, metal oxides, metal sulfides, and conductive polymers have been reported in many works [16–20].

For the introduction of cheap platinum-free catalysts, metal oxides have recently received much attention [21]. Although these materials suffer from poor electrical conductivity and never have the electrocatalytic activity that can compete with commercial catalysts for the oxidation of methanol, they could have relatively good catalytic activity in the oxidation of methanol by some modifications [22–24]. Among the transition metal oxides, manganese oxides such as MnO_2, Mn_2O_3, and Mn_3O_4 have received significant attention in different applications due to their cost-effective, eco-friendly, and electrochemical properties [25]. For example, the development of metal oxides including Mn_3O_4 nanoparticles [25], hollow-like Mn_3O_4 nanostructures on graphene matrix [26], Fe-doped Mn_3O_4 nanoboxes [27], and Ce-doped Mn_3O_4 [28] for methanol oxidation, oxygen reduction, and oxygen evolution reactions has been considered by researchers.

Cerium oxide has also recently received much attention for catalyst and sensing applications. The placement of CeO_2 in the structure of catalysts has always improved their electrochemical properties. Applications of this material in the form of multi-component composites that have studied the synergetic effect of CeO_2 with other materials in the catalyst include the use of $NiCoO_2$@CeO_2 nanoboxes in ultrasensitive electrochemical immunosensing based on oxygen evolution reaction [29] or the application of CeO_2@CoP as a supercapacitor electrode [30]. Other studies have been reported including the use of CeO_2-ZnO in low-temperature solid oxide fuel cells [31] and extensive use of CeO_2-based material in electrochemical water splitting [32]. In addition, CeO_2 has been reported as a filler in proton-exchange membranes, for which one can refer to the research of Vinothkannan et al. [33]. CeO_2 is one of the metal oxides used in the catalyst materials for the methanol oxidation process, for example, the application of Pt/CeO_2 [34], Pd-CeO_2 [35], and Au@CeO_2 [36] in the MOR process. In these cases, CeO_2 has been used as a catalyst along with precious and rare metals. However, the use of this material along with inexpensive materials has also been studied as a catalyst for methanol oxidation, including the use of CeO_2-decorated rGO [37] and CeO_2/NiO hollow spheres [38]. According to the literature, CeO_2 has always been considered in most electrochemical fields, including various types of electrochemical sensors or energy production and storage devices.

Moreover, we investigated $NiCo_2O_4$ nanorods/reduced graphene oxide (rGO) [39], MoS_2/Ni_3S_2/rGO [40], NiO-Co_3O_4-rGO [41], and $MgCo_2O_4$/rGO [42] as catalysts for methanol oxidation in previous works, and our results show that the synergetic effect between metal oxides/sulfides and reduced graphene oxide improves the catalytic activity of the catalyst. These catalysts were synthesized by the hydrothermal method. In this method, time and temperature are two very important factors that by controlling these parameters, a synthesis with very good efficiency and desired morphology can be achieved. In many studies, by changing these two parameters, nanomaterials with different sizes, morphology, and porosity have been obtained [43,44].

In this work, we study a hybrid consisting of a mixed metal oxide (Mn_3O_4 and CeO_2) and reduced graphene oxide (rGO) as a catalyst for use in the methanol oxidation process. We investigate the synergetic effect of rGO with Mn_3O_4-CeO_2 on the electrochemical properties of catalyst in the methanol oxidation process. RGO was added to increase the active surface area and the conductivity of the catalyst. Electrochemical tests indicate the synergetic effect of composite components on the oxidation of methanol. It seems that this catalyst, due to its good capability in the methanol oxidation process, can be evaluated as a stable and relatively efficient catalyst in the anode structure of methanol fuel cells.

2. Materials and Methods

2.1. Materials and Instruments

Cerium nitrate ($Ce(NO_3)_3 \cdot 6H_2O$), potassium permanganate ($KMnO_4$), and urea were used as precursors for the synthesis of catalysts and used with high purity. Precursors were purchased from Merck company (Merck, Darmstadt, Germany). X-ray diffraction analysis was performed by a Thermo Scientific device (Thermo Fisher Scientific, ARL Equinox 3000, Waltham, MA, USA), and scanning electron microscopy (SEM) was performed by FEI Quanta 200 FEG-Netherlands. Electrochemical tests were conducted by potentiostat–galvanostat AUTOLAB PGSTAT 302 N ("Metrohm Autolab bv", Utrecht, The Netherlands).

2.2. Synthesis of Mn_3O_4-CeO_2 and Mn_3O_4-CeO_2-rGO Nanocatalysts

For the synthesis of Mn_3O_4-CeO_2, 1.2 g of cerium nitrate, 2.5 g of potassium permanganate, and 4 g of urea were completely dissolved in 30 mL of deionized water for 30 min. After that, the solution entered into the autoclave and was placed in the furnace for 14 h at a temperature of 120 °C. The resulting materials were washed several times with deionized water and dried at a temperature of 40 °C and then calcinated at 350 °C.

The hybrid of Mn_3O_4-CeO_2 with reduced graphene oxide was prepared with the same synthesis used for Mn_3O_4-CeO_2, except that 50 mg GO, synthesized by the Hummers method [45], was added to the solution.

Graphene oxide (GO) was synthesized by the Hummers method. In this regard, 1 g graphite powder was dispersed in 25 mL H_2SO_4, and then 1 g sodium nitrate was dissolved in the solution and stirred for 1 h. After that, the beaker was placed in an ice-water bath, and 3.6 g of potassium permanganate was added to the solution and stirred for 2 h. The temperature was increased to 35 °C and stirred for another 2 h, and 23 mL of deionized water was added to the above mixture and stirred for 30 min at 90 °C. Finally, the reaction was stopped by adding 70 mL of deionized water and 8 mL of H_2O_2 solution (30%). The product was washed with HCl (3%) and deionized water three times and dried in an oven at 50 °C.

2.3. Electrochemical Studies

Electrochemical studies were performed with a three-electrode system including Ag/AgCl, platinum wire, and working electrodes. The catalyst-modified glass carbon electrode (GCE) was used as a working electrode. To modify the GCE, a slurry containing 5 mg of each of the catalysts was prepared in 1 mL of a deionized water/isopropyl alcohol solution. A certain amount of slurry was placed on the GCE surface. The electrode was dried at 20 °C for 20 min and prepared for electrochemical tests.

3. Results and Discussion

3.1. Characterization

The crystal structure of Mn_3O_4-CeO_2-rGO nanocatalyst was examined by XRD analysis. As shown in Figure 1, the characteristic peaks of CeO_2 were seen at the diffraction angles of about 28.9°, 47.9°, 56.8°, and 69.7°, which belonged to (111), (220), (311), and (400), crystal planes, respectively, which is in full compliance with JCPDS card no 34-0394 [46]. In addition, at diffraction angles of 17.9°, 38.1°, 36.1°, 44.4°, 50.1°, 59.8°, and 64.6°, characteristic peaks of Mn_3O_4 could be seen, which were related to the (101), (103), (211), (004), (105), (224), and (400) crystal planes, respectively. This diffraction pattern is in full compliance with JCPDS card no 24-0734 [47]. From XRD results, the synthesis of Mn_3O_4-CeO_2-rGO was confirmed.

The surface morphology of the catalysts was examined by SEM images. Figure 2a–c refer to the Mn_3O_4-CeO_2 and show the porous morphology and shortcuts that are useful for methanol to reach the depth of the catalyst. In the SEM images of Mn_3O_4-CeO_2 (a–c), which were prepared at the scale of 100 nm, uniform dispersion and interconnectedness of Mn_3O_4 and CeO_2 can be seen. In these images, holes are seen that are shortcuts for electrolyte and methanol to penetrate the core of the catalyst. The presence of porosity and the same shortcuts facilitate the process of methanol oxidation. The SEM images of Mn_3O_4-CeO_2-rGO (Figure 2d–f) show the uniform dispersion of Mn_3O_4-CeO_2 on the surface of rGO nanosheets. Mn_3O_4-CeO_2 covered almost the entire surface of rGO, and rGO plates were visible in some places, as shown in Figure 2d–f. In these pictures, these nanosheets are shown with an arrow or a red line around the rGO sheets.

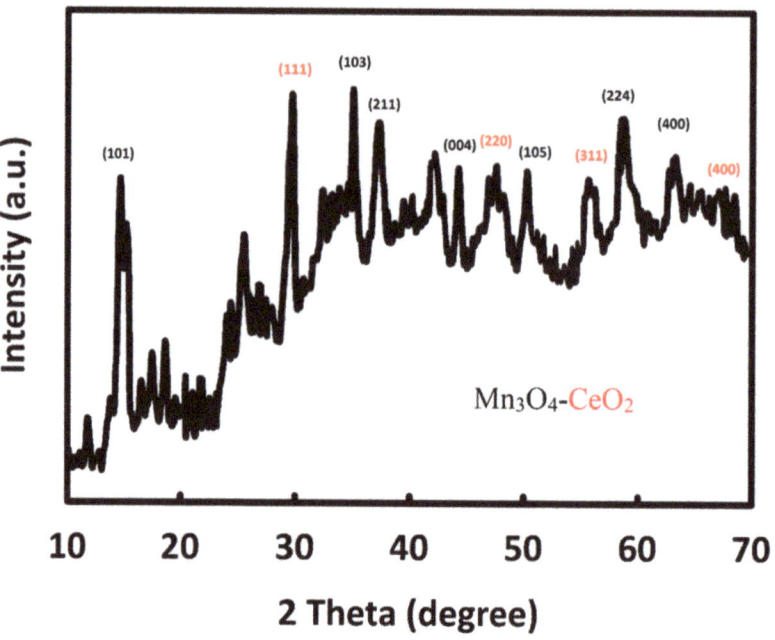

Figure 1. The XRD pattern of Mn_3O_4-CeO_2-rGO.

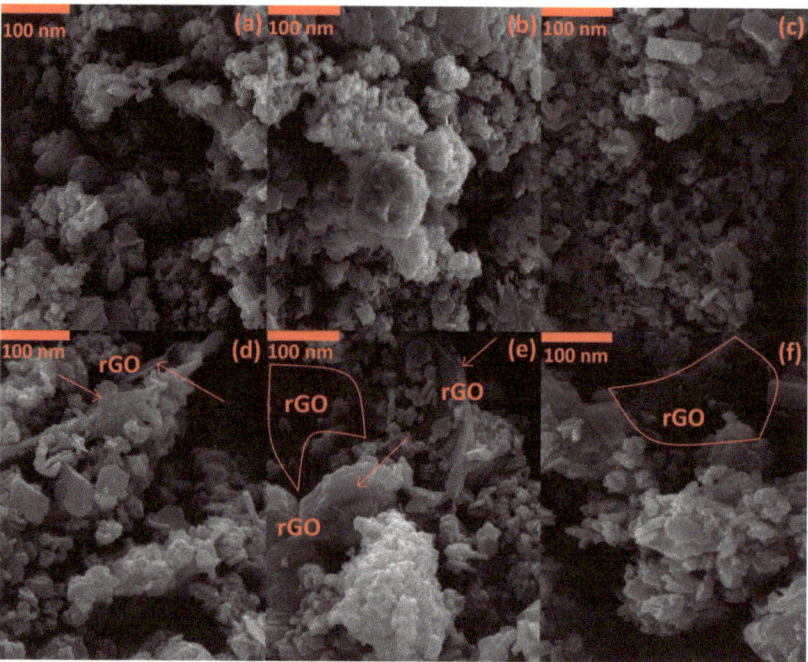

Figure 2. The SEM images of Mn_3O_4-CeO_2 (**a**–**c**) and Mn_3O_4-CeO_2-rGO (**d**–**f**).

3.2. Electro-Catalytic Investigations

The capability of Mn_3O_4-CeO_2 and Mn_3O_4-CeO_2-rGO catalysts was evaluated in the methanol oxidation process by performing cyclic voltammetry (CV) in 1 M KOH in the presence and absence of methanol. For this purpose, CV analysis of Mn_3O_4-CeO_2 and Mn_3O_4-CeO_2-rGO was performed in 1 M KOH solution in the potential range of 0 to 0.8 V at a scan rate of 10 mV/s. As shown in Figure 3a, both catalysts had capacitive behavior, and the current density for the Mn_3O_4-CeO_2-rGO catalyst was significantly higher than that of Mn_3O_4-CeO_2 due to the presence of rGO in its structure. By adding methanol (0.2 M) to the 1 M KOH solution and performing a CV test, we could see methanol oxidation peaks in both catalysts (Figure 3b), which is evidence of the relatively good electrocatalytic activity of both catalysts in the MOR process. The comparison of the behavior of two catalysts in MOR indicated the effective role of rGO in the structure of Mn_3O_4-CeO_2-rGO. In addition to increasing the electrical conductivity of the catalyst, rGO increased the active surface area of the catalyst [48], and, as a result, the current density increased, and the overvoltage reduced in the methanol oxidation reaction (MOR) process. Methanol oxidation peaks for Mn_3O_4-CeO_2-rGO and Mn_3O_4-CeO_2 were seen at 0.51 and 0.53 V, respectively.

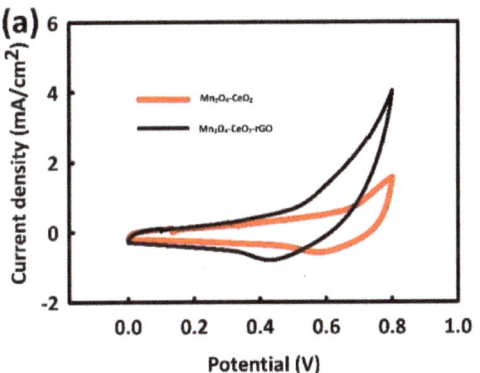
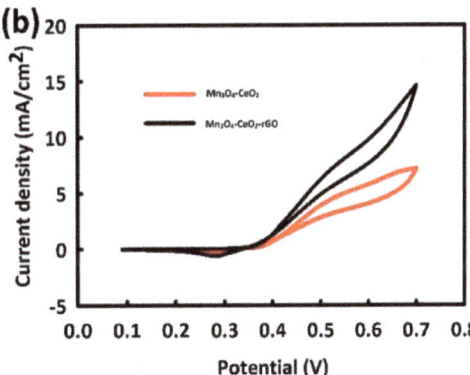

Figure 3. CV curves of Mn_3O_4-CeO_2 and Mn_3O_4-CeO_2-rGO in the absence (**a**) and in the presence of 0.2 M methanol and (**b**) in 1 M KOH.

To obtain the optimal concentration of methanol in the MOR process, CV analysis of Mn_3O_4-CeO_2 and Mn_3O_4-CeO_2-rGO was performed at different concentrations of methanol (0.2, 0.4, 0.6, 0.8, and 1 M) and 1 M KOH. CV analysis of the GCE modified with Mn_3O_4-CeO_2 (Figure 4a) showed an upward trend of methanol oxidation peak to a concentration of 0.6 M, and from this concentration onwards, we could see a decrease in oxidation peak. For Mn_3O_4-CeO_2-rGO, the oxidation peak trend was ascending to a concentration of 0.8 M (Figure 4b), and at a concentration of 1 M methanol, the current density decreased. The reason for the decrease in current density from an optimal concentration onwards is probably due to the saturation of the catalyst surface by the by-products of methanol oxidation. The likely reason is that the catalysts containing rGO are saturated later; the more effective surface of this catalyst is due to the presence of rGO in its structure. Figure 4c shows a plot of methanol concentration in terms of maximum current density. The behavior of nanocatalysts at different concentrations can be compared according to this plot.

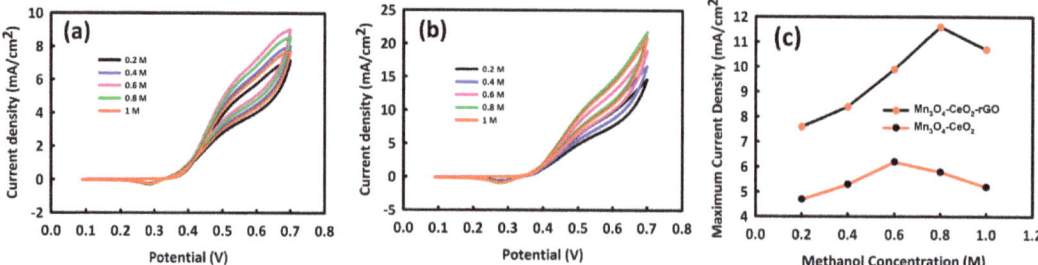

Figure 4. CV curves of Mn$_3$O$_4$-CeO$_2$ (**a**) and Mn$_3$O$_4$-CeO$_2$-rGO (**b**) at different concentrations of methanol, and (**c**) methanol concentration in terms of maximum current density for Mn$_3$O$_4$-CeO$_2$-rGO and Mn$_3$O$_4$-CeO$_2$.

By selecting the optimal concentrations of methanol for Mn$_3$O$_4$-CeO$_2$ and Mn$_3$O$_4$-CeO$_2$-rGO, which are 0.6 and 0.8 M, respectively, we then investigated the behavior of these catalysts at different scan rates in the presence of a 1 M KOH solution as an electrolyte (Figure 5a,b). It was observed that with increasing scan rate, the oxidation peak current density for Mn$_3$O$_4$-CeO$_2$ had an increasing trend until the scan rate of 70 mV/s, and from this scan rate onwards, the current density decreased. For Mn$_3$O$_4$-CeO$_2$-rGO, it was also observed that the anodic peak current density increased up to a scan rate of 90 mV/s, and after that, the current density decreased. It is likely that at higher scan rates, the electrolyte and methanol do not have enough time to fully engage with the catalyst and penetrate to its core, reducing current density [49]. The current density for Mn$_3$O$_4$-CeO$_2$ at scan rates of 10, 70, and 110 mV/s were 5.9, 8.8, and 7.9 mA/cm^2, respectively, and for Mn$_3$O$_4$-CeO$_2$-rGO at scan rates of 10, 90, and 110 mV/s were 11.6, 17.7, and 16.05 mA/cm^2 (the selected scan rates are the lowest, optimal, and the maximum scan rates, respectively). The graph of catalyst behavior in different scan rates in terms of maximum current density is plotted in Figure 5c.

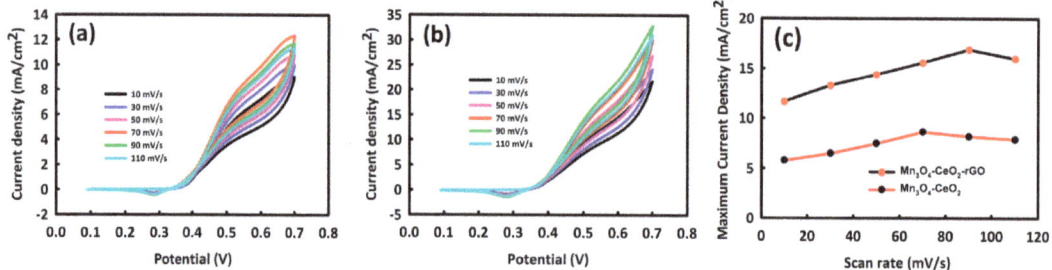

Figure 5. CV curves of Mn$_3$O$_4$-CeO$_2$ (**a**) and Mn$_3$O$_4$-CeO$_2$-rGO (**b**) at different scan rates and (**c**) different scan rates in terms of maximum current density for the Mn$_3$O$_4$-CeO$_2$-rGO and Mn$_3$O$_4$-CeO$_2$.

The proposed mechanism of methanol oxidation by the catalyst is as follows (Equations (1)–(4)):

$$Mn_3O_4 - CeO_2 - rGO + CH_3OH \rightarrow Mn_3O_4 - CeO_2 - rGO - CH_3OH_{ads} \qquad (1)$$

$$Mn_3O_4 - CeO_2 - rGO - CH_3OH_{ads} + 4OH^- \rightarrow \\ Mn_3O_4 - CeO_2 - rGO - (CO)_{ads} + 4H_2O + 4e^- \qquad (2)$$

$$Mn_3O_4 - CeO_2 - rGO + OH^- \rightarrow Mn_3O_4 - CeO_2 - rGO - OH_{ads} + e^- \qquad (3)$$

$$Mn_3O_4 - CeO_2 - rGO - CO_{ads} + Mn_3O_4 - CeO_2 - rGO - OH_{ads} + OH^- \rightarrow$$
$$Mn_3O_4 - CeO_2 - rGO + CO_2 + H_2O + e^- \quad (4)$$

The mechanism of methanol oxidation in alkaline media at the surface of these catalysts can be divided into three stages, including adsorption of methanol, adsorption of hydroxyl ions, and breaking the C–H and O–H bonds of methanol to produce products such as CH_2OH, CHOH, COH, and CO [50]. In the adsorption process, likely the synergetic effect of Mn_3O_4, CeO_2, and rGO improves this process by creating more active sites. Hydroxyl ions adsorbed on the surface of the catalyst also help to oxidize the adsorbed CO, thereby regenerating the active sites of the catalyst. Since the active sites of catalysts are important in the MOR process, with the addition of rGO, the overall performance of the catalyst increases.

To evaluate the stability of Mn_3O_4-CeO_2 and Mn_3O_4-CeO_2-rGO catalysts, 500 CV cycles were performed at optimal methanol concentrations and scan rates (0.6 M in scan rate of 70 mV/s for Mn_3O_4-CeO_2 and 0.8 M in scan rate of 90 mV/s for Mn_3O_4-CeO_2-rGO). As seen in Figure 6a, Mn_3O_4-CeO_2 showed stability of about 87%, while the stability achieved 91% for Mn_3O_4-CeO_2-rGO (Figure 6b). Examination of the current density against the number of cycles is shown in Figure 6c, indicating Mn_3O_4-CeO_2 reached relatively good stability from about 200 cycles onwards, and Mn_3O_4-CeO_2-rGO achieved stability in current density approximately in the 250th cycle. The greatest reduction in current density in the initial cycles for both Mn_3O_4-CeO_2 and Mn_3O_4-CeO_2-rGO catalysts was seen. It seems that in the initial cycles, methanol and electrolyte did not find enough time to penetrate the catalyst nucleus, and over time, and in subsequent cycles, methanol and catalyst came into full contact with each other, and after a slight decrease, stability in the current density was observed [51]. Moreover, the CV shape of catalysts remained stable after different cycles, and no change was seen in isopotential points, which is the reason for the very good structural stability of the catalyst [52].

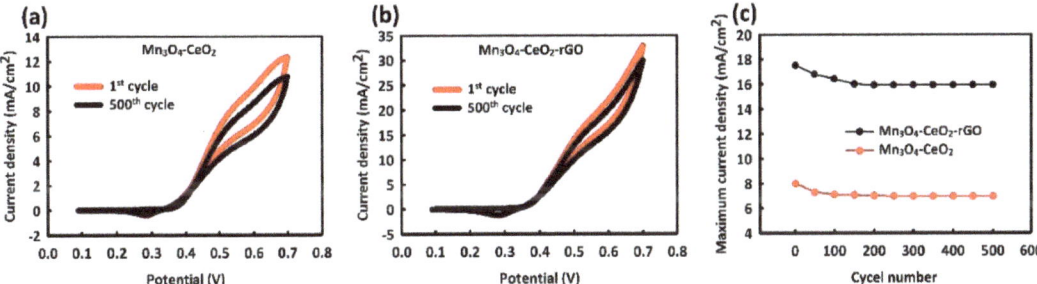

Figure 6. The first and 500th CV cycles of Mn_3O_4-CeO_2 (**a**) and Mn_3O_4-CeO_2-rGO (**b**) and plot of maximum current density versus cycle number (**c**).

The effect of temperature in the MOR process on Mn_3O_4-CeO_2 and Mn_3O_4-CeO_2-rGO catalysts was evaluated by performing a linear sweep voltammetry LSV test at an optimal concentration of methanol and scan rate. As seen in Figure 7a,c, the current density in both catalysts increased with increasing temperature, indicating that the increasing temperature facilitates the process of methanol adsorption on the surface of the catalyst. The rate of increase of current density with growing temperature for Mn_3O_4-CeO_2-rGO was slightly higher than Mn_3O_4-CeO_2; here too, it is likely that rGO played an effective role in increasing the active surface area of the catalyst. The two parameters of maximum current density and temperature were linearly related to each other, which can be seen for Mn_3O_4-CeO_2 and Mn_3O_4-CeO_2-rGO in Figure 7b,d, respectively.

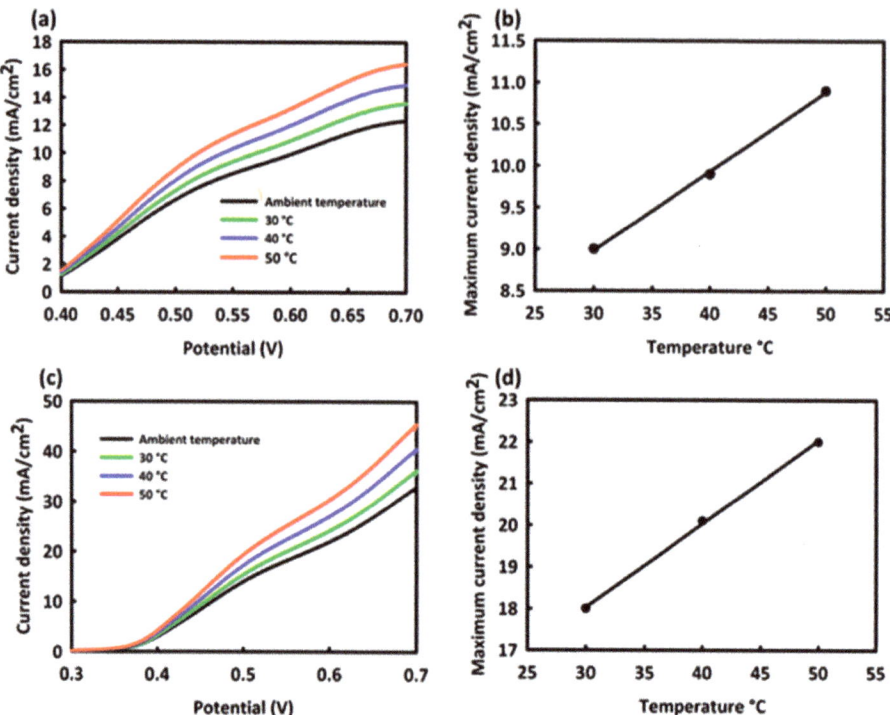

Figure 7. CV curves at different temperatures and plots of maximum current density against temperature (at 30–50 °C) for Mn_3O_4-CeO_2 (**a**,**c**) and Mn_3O_4-CeO_2-rGO (**b**,**d**).

4. Conclusions

In the energy-dependent modern world, we see every day the introduction of new and high-efficiency materials for use in the structure of energy storage and production devices. In this regard, we synthesized a stable and inexpensive nanocatalyst based on metal oxides (Mn_3O_4-CeO_2-rGO) with a one-step and easy synthesis hydrothermal method for use in the methanol oxidation process. The structure and morphology of nanocatalyst were evaluated by XRD and SEM. In this study, the synergetic effect and effective role of rGO in the catalyst structure were investigated. Mn_3O_4-CeO_2-rGO with 91% cyclic stability after 500 consecutive CV cycles and a maximum current density of 17.7 mA/cm² at an overvoltage of 0.51 V (at scan rate and optimum concentration) can be an attractive and new option for use as a catalyst for methanol oxidation.

Author Contributions: Conceptualization, M.B.A. and A.D.B.; methodology, M.B.A. and S.M.R.; software, S.M.R.; validation, M.B.A. and A.D.B.; formal analysis, M.B.A., S.M.R. and A.D.B.; investigation, M.B.A. and S.M.R.; resources, M.B.A. and S.M.R.; data curation, M.B.A. and S.M.R.; writing—original draft preparation, M.B.A. and S.M.R.; writing—review and editing, A.D.B.; visualization, M.B.A. and S.M.R.; supervision, A.D.B.; project administration, M.B.A. and S.M.R.; funding acquisition, M.B.A. and A.D.B. All authors have read and agreed to the published version of the manuscript.

Funding: M. B. Askari was supported by a grant from Basic Science Research Fund (No. BSRF-Phys-399-14). The APC was funded by ADB.

Data Availability Statement: The data presented in this study are available upon request from the corresponding authors.

Conflicts of Interest: The authors declare no conflict of interest.

References

1. Peter, S.C. Reduction of CO_2 to chemicals and fuels: A solution to global warming and energy crisis. *ACS Energy Lett.* **2018**, *3*, 1557–1561. [CrossRef]
2. Salarizadeh, P.; Askari, M.B. MoS_2–ReS_2/rGO: A novel ternary hybrid nanostructure as a pseudocapacitive energy storage material. *J. Alloys Compd.* **2021**, *874*, 159886. [CrossRef]
3. Martins, F.; Felgueiras, C.; Smitkova, M.; Caetano, N. Analysis of fossil fuel energy consumption and environmental impacts in European countries. *Energies* **2019**, *12*, 964. [CrossRef]
4. Rasoulinezhad, E.; Taghizadeh-Hesary, F.; Taghizadeh-Hesary, F. How is mortality affected by fossil fuel consumption, CO_2 emissions and economic factors in CIS region? *Energies* **2020**, *13*, 2255. [CrossRef]
5. Withey, P.; Johnston, C.; Guo, J. Quantifying the global warming potential of carbon dioxide emissions from bioenergy with carbon capture and storage. *Renew. Sustain. Energy Rev.* **2019**, *115*, 109408. [CrossRef]
6. Usman, M.; Jahanger, A.; Makhdum, M.S.A.; Balsalobre-Lorente, D.; Bashir, A. How do financial development, energy consumption, natural resources, and globalization affect Arctic countries' economic growth and environmental quality? An advanced panel data simulation. *Energy* **2022**, *241*, 122515. [CrossRef]
7. Burke, P.J.; Widnyana, J.; Anjum, Z.; Aisbett, E.; Resosudarmo, B.; Baldwin, K.G. Overcoming barriers to solar and wind energy adoption in two Asian giants: India and Indonesia. *Energy Policy* **2019**, *132*, 1216–1228. [CrossRef]
8. Sharma, P.; Minakshi Sundaram, M.; Watcharatharapong, T.; Jungthawan, S.; Ahuja, R. Tuning the Nanoparticle Interfacial Properties and Stability of the Core–Shell Structure in Zn-Doped $NiMoO_4$@ AWO_4. *ACS Appl. Mater. Interfaces* **2021**, *13*, 56116–56130. [CrossRef]
9. Askari, M.B.; Rozati, S.M.; Salarizadeh, P.; Saeidfirozeh, H.; Di Bartolomeo, A. A remarkable three-component RuO_2-$MnCo_2O_4$/rGO nanocatalyst towards methanol electrooxidation. *Int. J. Hydrogen Energy* **2021**, *46*, 36792–36800. [CrossRef]
10. Minakshi, M.; Singh, P.; Sharma, N.; Blackford, M.; Ionescu, M. Lithium extraction—Insertion from/into $LiCoPO_4$ in aqueous batteries. *Ind. Eng. Chem. Res.* **2011**, *50*, 1899–1905. [CrossRef]
11. Liu, H.; Liu, X.; Wang, S.; Liu, H.K.; Li, L. Transition metal based battery-type electrodes in hybrid supercapacitors: A review. *Energy Storage Mater.* **2020**, *28*, 122–145. [CrossRef]
12. Akdemir, M.; Imanova, G.; Karakaş, D.E.; Kıvrak, H.D.; Kaya, M. High Efficiency Biomass-Based Metal-Free Catalyst as a Promising Supercapacitor Electrode for Energy Storage. 2021. Available online: https://papers.ssrn.com/sol3/papers.cfm?abstract_id=3908407 (accessed on 20 August 2021).
13. Iqbal, S.; Khatoon, H.; Pandit, A.H.; Ahmad, S. Recent development of carbon based materials for energy storage devices. *Mater. Sci. Energy Technol.* **2019**, *2*, 417–428. [CrossRef]
14. Nguyen, L.H.; Gomes, V.G. High efficiency supercapacitor derived from biomass based carbon dots and reduced graphene oxide composite. *J. Electroanal. Chem.* **2019**, *832*, 87–96.
15. Askari, M.B.; Salarizadeh, P.; Beheshti-Marnani, A. A hierarchical hybrid of $ZnCo_2O_4$ and rGO as a significant electrocatalyst for methanol oxidation reaction: Synthesis, characterization, and electrocatalytic performance. *Int. J. Energy Res.* **2020**, *44*, 8892–8903. [CrossRef]
16. Chen, S.; Huang, D.; Liu, D.; Sun, H.; Yan, W.; Wang, J.; Fan, W. Hollow and porous $NiCo_2O_4$ nanospheres for enhanced methanol oxidation reaction and oxygen reduction reaction by oxygen vacancies engineering. *Appl. Catal. B Environ.* **2021**, *291*, 120065. [CrossRef]
17. Huang, H.; Wei, Y.; Yang, Y.; Yan, M.; He, H.; Jiang, Q.; Zhu, J. Controllable synthesis of grain boundary-enriched Pt nanoworms decorated on graphitic carbon nanosheets for ultrahigh methanol oxidation catalytic activity. *J. Energy Chem.* **2021**, *57*, 601–609. [CrossRef]
18. Jin, D.; Li, Z.; Wang, Z. Hierarchical $NiCo_2O_4$ and $NiCo_2S_4$ nanomaterials as electrocatalysts for methanol oxidation reaction. *Int. J. Hydrogen Energy* **2021**, *46*, 32069–32080. [CrossRef]
19. Narayanan, N.; Bernaurdshaw, N. Reduced graphene oxide supported $NiCo_2O_4$ nano-rods: An efficient, stable and cost-effective electrocatalyst for methanol oxidation reaction. *ChemCatChem* **2020**, *12*, 771–780. [CrossRef]
20. Pattanayak, P.; Pramanik, N.; Kumar, P.; Kundu, P.P. Fabrication of cost-effective non-noble metal supported on conducting polymer composite such as copper/polypyrrole graphene oxide (Cu_2O/PPy–GO) as an anode catalyst for methanol oxidation in DMFC. *Int. J. Hydrogen Energy* **2018**, *43*, 11505–11519. [CrossRef]
21. Sharma, P.; Minakshi Sundaram, M.; Watcharatharapong, T.; Laird, D.; Euchner, H.; Ahuja, R. Zn metal atom doping on the surface plane of one-dimesional $NiMoO_4$ nanorods with improved redox chemistry. *ACS Appl. Mater. Interfaces* **2020**, *12*, 44815–44829. [CrossRef]
22. Askari, M.B.; Beheshti-Marnani, A.; Seifi, M.; Rozati, S.M.; Salarizadeh, P. Fe_3O_4@ MoS_2/RGO as an effective nano-electrocatalyst toward electrochemical hydrogen evolution reaction and methanol oxidation in two settings for fuel cell application. *J. Colloid Interface Sci.* **2019**, *537*, 186–196. [CrossRef] [PubMed]
23. Hameed, R.M.A.; Amin, R.S.; El-Khatib, K.M.; Fetohi, A.E. Preparation and characterization of Pt–CeO_2/C and Pt–TiO_2/C electrocatalysts with improved electrocatalytic activity for methanol oxidation. *Appl. Surf. Sci.* **2016**, *367*, 382–390. [CrossRef]
24. Rezaee, S.; Shahrokhian, S. Facile synthesis of petal-like NiCo/NiO-CoO/nanoporous carbon composite based on mixed-metallic MOFs and their application for electrocatalytic oxidation of methanol. *Appl. Catal. B Environ.* **2019**, *244*, 802–813. [CrossRef]

25. Nagajyothi, P.; Ramaraghavulu, R.; Munirathnam, K.; Yoo, K.; Shim, J. One-pot hydrothermal synthesis: Enhanced MOR and OER performance using low-cost Mn_3O_4 electrocatalyst. *Int. J. Hydrogen Energy* **2021**, *46*, 13946–13951. [CrossRef]
26. Das, S.K.; Kamila, S.; Satpati, B.; Kandasamy, M.; Chakraborty, B.; Basu, S.; Jena, B.K. Hollow Mn_3O_4 nanospheres on graphene matrix for oxygen reduction reaction and supercapacitance applications: Experimental and theoretical insight. *J. Power Sources* **2020**, *471*, 228465. [CrossRef]
27. Li, T.; Hu, Y.; Liu, K.; Yin, J.; Li, Y.; Fu, G.; Tang, Y. Hollow yolk-shell nanoboxes assembled by Fe-doped Mn_3O_4 nanosheets for high-efficiency electrocatalytic oxygen reduction in Zn-Air battery. *Chem. Eng. J.* **2022**, *427*, 131992. [CrossRef]
28. Wang, W.; Chen, J.Q.; Tao, Y.R.; Zhu, S.N.; Zhang, Y.X.; Wu, X.C. Flowerlike Ag-supported Ce-doped Mn_3O_4 nanosheet heterostructure for a highly efficient oxygen reduction reaction: Roles of metal oxides in Ag surface states. *ACS Catal.* **2019**, *9*, 3498–3510. [CrossRef]
29. Cao, L.; Cai, J.; Deng, W.; Tan, Y.; Xie, Q. $NiCoO_2@CeO_2$ nanoboxes for ultrasensitive electrochemical immunosensing based on the oxygen evolution reaction in a neutral medium: Application for interleukin-6 detection. *Anal. Chem.* **2020**, *92*, 16267–16273. [CrossRef]
30. Xing, H.; Long, G.; Zheng, J.; Zhao, H.; Zong, Y.; Li, X.; Zheng, X. Interface engineering boosts electrochemical performance by fabricating $CeO_2@CoP$ Schottky conjunction for hybrid supercapacitors. *Electrochim. Acta* **2020**, *337*, 135817. [CrossRef]
31. Qiao, Z.; Xia, C.; Cai, Y.; Afzal, M.; Wang, H.; Qiao, J.; Zhu, B. Electrochemical and electrical properties of doped CeO_2-ZnO composite for low-temperature solid oxide fuel cell applications. *J. Power Sources* **2018**, *392*, 33–40. [CrossRef]
32. Li, T.; Yin, J.; Sun, D.; Zhang, M.; Pang, H.; Xu, L.; Xue, J. Manipulation of Mott–Schottky Ni/CeO_2 Heterojunctions into N-Doped Carbon Nanofibers for High-Efficiency Electrochemical Water Splitting. *Small* **2022**, 2106592. [CrossRef] [PubMed]
33. Vinothkannan, M.; Hariprasad, R.; Ramakrishnan, S.; Kim, A.R.; Yoo, D.J. Potential bifunctional filler (CeO_2–ACNTs) for nafion matrix toward extended electrochemical power density and durability in proton-exchange membrane fuel cells operating at reduced relative humidity. *ACS Sustain. Chem. Eng.* **2019**, *7*, 12847–12857. [CrossRef]
34. Tao, L.; Shi, Y.; Huang, Y.C.; Chen, R.; Zhang, Y.; Huo, J.; Wang, S. Interface engineering of Pt and CeO_2 nanorods with unique interaction for methanol oxidation. *Nano Energy* **2018**, *53*, 604–612. [CrossRef]
35. Tan, Q.; Shu, C.; Abbott, J.; Zhao, Q.; Liu, L.; Qu, T.; Wu, G. Highly dispersed Pd-CeO_2 nanoparticles supported on N-doped core–shell structured mesoporous carbon for methanol oxidation in alkaline media. *ACS Catal.* **2019**, *9*, 6362–6371. [CrossRef]
36. Van Dao, D.; Adilbish, G.; Le, T.D.; Nguyen, T.T.; Lee, I.H.; Yu, Y.T. Au@CeO_2 nanoparticles supported Pt/C electrocatalyst to improve the removal of CO in methanol oxidation reaction. *J. Catal.* **2019**, *377*, 589–599.
37. Salarizadeh, P.; Askari, M.B.; Mohammadi, M.; Hooshyari, K. Electrocatalytic performance of CeO_2-decorated rGO as an anode electrocatalyst for the methanol oxidation reaction. *J. Phys. Chem. Solids* **2020**, *142*, 109442. [CrossRef]
38. Li, W.; Song, Z.; Deng, X.; Fu, X.Z.; Luo, J.L. Decoration of NiO hollow spheres composed of stacked nanosheets with CeO_2 nanoparticles: Enhancement effect of CeO_2 for electrocatalytic methanol oxidation. *Electrochim. Acta* **2020**, *337*, 135684. [CrossRef]
39. Askari, M.B.; Salarizadeh, P. Superior catalytic performance of $NiCo_2O_4$ nanorods loaded rGO towards methanol electro-oxidation and hydrogen evolution reaction. *J. Mol. Liq.* **2019**, *291*, 111306. [CrossRef]
40. Salarizadeh, P.; Askari, M.B.; Di Bartolomeo, A. MoS_2/Ni_3S_2/Reduced Graphene Oxide Nanostructure as an Electrocatalyst for Alcohol Fuel Cells. *ACS Appl. Nano Mater.* **2022**, *5*, 3361–3373. [CrossRef]
41. Askari, M.B.; Salarizadeh, P.; Beheshti-Marnani, A.; Di Bartolomeo, A. NiO-Co_3O_4-rGO as an Efficient Electrode Material for Supercapacitors and Direct Alcoholic Fuel Cells. *Adv. Mater. Interfaces* **2021**, *8*, 2100149. [CrossRef]
42. Askari, M.B.; Salarizadeh, P.; Di Bartolomeo, A.; Beitollahi, H.; Tajik, S. Hierarchical nanostructures of $MgCo_2O_4$ on reduced graphene oxide as a high-performance catalyst for methanol electro-oxidation. *Ceram. Int.* **2021**, *47*, 16079–16085. [CrossRef]
43. Emadi, H.; Salavati-Niasari, M.; Sobhani, A. Synthesis of some transition metal (M: $_{25}$Mn, $_{27}$Co, $_{28}$Ni, $_{29}$Cu, $_{30}$Zn, $_{47}$Ag, $_{48}$Cd) sulfide nanostructures by hydrothermal method. *Adv. Colloid Interface Sci.* **2017**, *246*, 52–74. [CrossRef] [PubMed]
44. Majid, F.; Shahin, A.; Ata, S.; Bibi, I.; Malik, A.; Ali, A.; Nazir, A. The effect of temperature on the structural, dielectric and magnetic properties of cobalt ferrites synthesized via hydrothermal method. *Z. Für Phys. Chem.* **2021**, *235*, 1279–1296. [CrossRef]
45. Hummers, W.S.; Offeman, R.E. Preparation of graphitic oxide. *J. Am. Chem. Soc.* **1958**, *80*, 1339. [CrossRef]
46. Zhao, Y.; Chen, T.; Ma, R.; Du, J.; Xie, C. Synthesis of flower-like CeO_2/BiOCl heterostructures with enhanced ultraviolet light photocatalytic activity. *Micro Nano Lett.* **2018**, *13*, 1394–1398. [CrossRef]
47. Lan, D.; Qin, M.; Yang, R.; Wu, H.; Jia, Z.; Kou, K.; Zhang, F. Synthesis, characterization and microwave transparent properties of Mn_3O_4 microspheres. *J. Mater. Sci. Mater. Electron.* **2019**, *30*, 8771–8776. [CrossRef]
48. Chu, D.; Li, F.; Song, X.; Ma, H.; Tan, L.; Pang, H.; Xiao, B. A novel dual-tasking hollow cube $NiFe_2O_4$-NiCo-LDH@rGO hierarchical material for high preformance supercapacitor and glucose sensor. *J. Colloid Interface Sci.* **2020**, *568*, 130–138. [CrossRef] [PubMed]
49. Askari, M.B.; Salarizadeh, P.; Di Bartolomeo, A. $NiCo_2O_4$-rGO/Pt as a robust nanocatalyst for sorbitol electrooxidation. *Int. J. Energy Res.* **2022**, *46*, 6745–6754. [CrossRef]
50. Fajín, J.L.; Cordeiro, M.N.D. Insights into the Mechanism of Methanol Steam Reforming for Hydrogen Production over Ni–Cu-Based Catalysts. *ACS Catal.* **2021**, *12*, 512–526. [CrossRef]

51. Askari, M.B.; Salarizadeh, P.; Seifi, M.; Di Bartolomeo, A. $ZnFe_2O_4$ nanorods on reduced graphene oxide as advanced supercapacitor electrodes. *J. Alloys Compd.* **2021**, *860*, 158497. [CrossRef]
52. Askari, M.B.; Salarizadeh, P.; Di Bartolomeo, A.; Şen, F. Enhanced electrochemical performance of $MnNi_2O_4$/rGO nanocomposite as pseudocapacitor electrode material and methanol electro-oxidation catalyst. *Nanotechnology* **2021**, *32*, 325707. [CrossRef] [PubMed]

Article

Tailoring Amine-Functionalized Ti-MOFs via a Mixed Ligands Strategy for High-Efficiency CO_2 Capture

Yinji Wan [1,†], Yefan Miao [1,†], Tianjie Qiu [2], Dekai Kong [1], Yingxiao Wu [2], Qiuning Zhang [1], Jinming Shi [2], Ruiqin Zhong [1,*] and Ruqiang Zou [2,*]

[1] State Key Laboratory of Heavy Oil Processing, China University of Petroleum-Beijing, No. 18 Fuxue Road, Changping District, Beijing 102249, China; wanyinji0613@163.com (Y.W.); 18332751996@163.com (Y.M.); kongdekai6@163.com (D.K.); 13180275342@163.com (Q.Z.)
[2] Beijing Key Laboratory for Theory and Technology of Advanced Battery Materials, School of Materials Science and Engineering, Peking University, No. 5 Yiheyuan Road, Haidian District, Beijing 100871, China; qtjie@pku.edu.cn (T.Q.); yingxiaowucup@163.com (Y.W.); jinmings@pku.edu.cn (J.S.)
* Correspondence: zhong2004@foxmail.com (R.Z.); rzou@pku.edu.cn (R.Z.)
† These authors contributed equally to this work.

Abstract: Amine-functionalized metal-organic frameworks (MOFs) are a promising strategy for the high-efficiency capture and separation of CO_2. In this work, by tuning the ratio of 1,3,5-benzenetricarboxylic acid (H_3BTC) to 5-aminoisophthalic acid (5-NH_2-H_2IPA), we designed and synthesized a series of amine-functionalized highly stable Ti-based MOFs (named MIP-207-NH_2-n, in which n represents 15%, 25%, 50%, 60%, and 100%). The structural analysis shows that the original framework of MIP-207 in the MIP-207-NH_2-n (n = 15%, 25%, and 50%) MOFs remains intact when the mole ratio of ligand H_3BTC to 5-NH_2-H_2IPA is less than 1 to 1 in the resulting MOFs. By the introduction of amino groups, MIP-207-NH_2-25% demonstrates outstanding CO_2 capture performance up to 3.96 and 2.91 mmol g^{-1}, 20.7% and 43.3% higher than those of unmodified MIP-207 at 0 and 25 °C, respectively. Furthermore, the breakthrough experiment indicates that the dynamic CO_2 adsorption capacity and CO_2/N_2 separation factors of MIP-207-NH_2-25% are increased by about 25% and 15%, respectively. This work provides an additional strategy to construct amine-functionalized MOFs with the maintenance of the original MOF structure and high performance of CO_2 capture and separation.

Keywords: Ti-MOFs; amine functionalization; CO_2 capture; separation; breakthrough experiment

1. Introduction

More than 85% of the worldwide energy demand is provided by the combustion of fossil fuels [1,2], but at the cost of considerable CO_2 (3 × 10^{13} kg CO_2 per year) being emitted into the atmosphere, thus leading to the daunting greenhouse effect [3–5]. The carbon capture and storage/sequestration (CCS) technology therefore has been proposed to mitigate emissions of atmospheric CO_2. For CCS technology, the breakthrough of novel adsorbents with a large CO_2 working capacity as well as a high CO_2 selectivity and easy regeneration is the core [6–8].

Metal-organic frameworks (MOFs) have been widely used for various applications owing to their ordered crystallinity, high specific surface area, and versatile tunability of chemical environments [9–17]. In particular, they can serve as attractive platforms for CO_2 adsorption and separation to mitigate the greenhouse effect [18–25]. It is widely acknowledged that amine-functionalized MOFs are one of the most effective ways to capture CO_2, because this method has the advantages of a large working CO_2 capacity as well as a high CO_2 selectivity and a low energy penalty for regeneration [7,26]. Currently, many MOF materials have been functionalized by the direct synthesis or post-synthesis modification method to graft amine [27–30]. For example, Kim et al. [28] prepared a robust tetraamine-functionalized Mg-MOF by the post-synthesis strategy. The tetraamine-functionalized framework showed an excellent CO_2 trapping efficiency under low CO_2

partial pressure in the flue stream. Han et al. [31] used a one-step hydrothermal method to synthesize MIL-101(Cr)-NH_2 nanoparticles. Compared with MIL-101(Cr), the CO_2 adsorption capacity of MIL-101(Cr)-NH_2 increased from 1.85 to 2.25 mmol g^{-1}. Moreover, the separation factor of CO_2/N_2 was enhanced from 7.5 to 11.6 at 1 atm and 35 °C. Recently, Zhong et al. [32] introduced three kinds of organic amine molecules into the channels of MIL-101(Cr) by the post-modification method. The results showed that the CO_2/CO selectivity of the Tris(2-aminoethyl) amine-modified MIL-101 was 103 times higher than that of its pure MIL-101 counterpart.

In this work, the MIP-207, fabricated by Ti–O clusters and the H_3BTC ligand [21,33], was selected as the porous material, mainly due to its large specific surface area and high chemical stability even in highly acidic media (pH $\leq$ 0). More importantly, there are uncoordinated and isolated -COOH groups toward the channels of MIP-207 because of the meta-connection mode of the H_3BTC ligand, so the chemical environment of MIP-207 cavities can be easily modulated by the mixed linkers strategy [33]. There are quite a few reports on amine-grafted highly stable MIP-207 for highly efficient CO_2 capture [33]. Additionally, the accurate grafting of amine molecules without any framework destruction of the host frameworks is less of a concern. Herein, we for the first time prepared a series of amine-functionalized highly stable MIP-207 materials and further tailored the content of -NH_2 by the mixed linkers strategy for capturing CO_2 from N_2. It should be pointed out that amine-functionalized MIP-207 materials cannot be obtained when the mole ratio of H_3BTC to 5-NH_2-H_2IPA exceeds 1 (See Figure 1). The physiochemical properties of the as-prepared materials were systematically characterized and analyzed, and the CO_2 adsorption and separation were also investigated. Based on the excellent CO_2 capture performance of MIP-207-NH_2-25%, the breakthrough experiments further evaluated the dynamic adsorption capacity and separation factors under different gas flow rates.

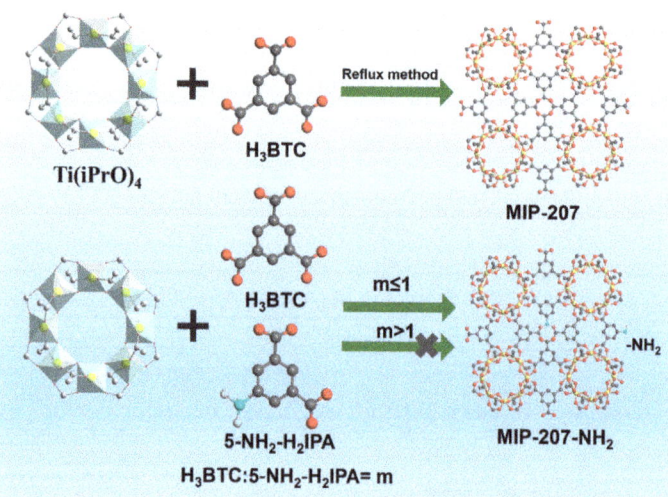

Figure 1. Schematic Diagram of MIP-207 and amine-functionalized MIP-207; Ti is shown in yellow, C in gray, O in red, N in light blue, and H in white.

2. Experimental Section

2.1. Synthesis of MIP-207

All the reagents used were commercially purchased without further purification. MIP-207 was synthesized in a similar method to the one reported [33]. 1,3,5-benzenetricarboxylic acid (H_3BTC), acetic acid, and acetic anhydride were purchased from Maclin, Shanghai, China. Tetraisopropyl titanate was purchased from Sinopharm Chemical Reagent Co., Ltd.,

Beijing, China. H_3BTC (840 mg, 4 mmol), acetic acid (10 mL) and acetic anhydride (10 mL) were added and mixed into a 50 mL round-bottom flask at ambient temperature. Then, tetraisopropyl titanate (800 μL, 2.7 mmol) was added under stirring. The mixture was refluxed at 120 °C for 12 h. After cooling to room temperature, the crude product was separated and washed with boiling anhydrous acetone. Finally, the product was collected by centrifugation and placed in an 80 °C oven for 12 h.

2.2. Synthesis of Amine-Functionalized MIP-207

The amine-functionalized MIP-207 was prepared with a pre-synthesis modification method. Part of the H_3BTC ligand was replaced with a certain amount of 5-NH_2-H_2IPA, which accounted for 15%, 25%, 50%, 60%, and 100% of the total H_3BTC, respectively. A series of amine-functionalized MIP-207 materials were synthesized according to the above synthetic steps of MIP-207, and the as-prepared materials were denoted as MIP-207-NH_2-n (n = 15%, 25%, 50%, 60%, and 100%), respectively. It must be pointed out that the structure of MIP-207-NH_2-100% is completely different from that of MIP-207; it is still named MIP-207-NH_2-100% simply for the purpose of comparison.

2.3. Sample Characterization

Powder X-ray diffraction (PXRD) patterns of the samples were recorded on a Rigaku D/max 2400 X-ray diffractometer equipped with Cu K_α radiation operating at 45 kV and 200 mA. Scanning electron microscopy (SEM) tests were conducted on a Hitachi S4800 electron microscope to observe the morphologies of the samples. A N_2 physisorption test was carried out on a Quantachrome Autosorb-iQ (Quantachrome Instruments, Boynton Beach, FL, USA) at −196 °C. The elemental analysis (EA) of the samples was performed on an Elementar Analysensysteme GmbH Vario EL (Analytical Instrumentation Department of the Heraeus technology group, Frankfurt, Germany) analyzer to accurately analyze the percentage content of the C, H, and N elements of samples. The specific surface area of the samples was analyzed according to the Brunauer-Emmett-Teller (BET) method and pore size distribution was calculated using the non-local density functional theory (NLDFT) model.

2.4. Gas Adsorption Measurements

All the gases (N_2 and CO_2) used were of ultrahigh purity (99.999%) in this study. N_2 and CO_2 adsorption isotherms were measured by a Quantachrome Autosorb-iQ gas adsorption analyzer up to 1 bar, and the temperatures of 0 and 25 °C were both maintained with an ethylene glycol/H_2O bath by a cooling and heating system. Before the measurement, about 100 mg of the adsorbent was degassed at 150 °C for 8 h in vacuum condition. The adsorption and desorption of CO_2 cyclic stability was carried out on an SDT Q600 analyzer (TA Instruments, New Castle, DE, USA). Firstly, the sample fully absorbed CO_2 at 35 °C for 1 h, and then it was injected with N_2 gas at 150 °C for 2 h. The breakthrough experiments were performed on a homemade setup to simulate the actual mixture gas (20 vol% CO_2, 20 vol% N_2, and balanced gas He) separation to evaluate the dynamic CO_2/N_2 adsorption performance; the setup diagram of the breakthrough experiment can be found in our previous work [34].

3. Results and Discussion

3.1. Structural Analysis of Samples

As shown in Figure S1, compared with simulated MIP-207, the characteristic diffraction peak positions and relative intensities of MIP-207 after reflux treatment at 120 °C for 12 h fit very well, illustrating that MIP-207 with a high purity was synthesized. Also, Figure S1 exhibits that the structure of MIP-207 was maintained well after activation at 150 °C. To evaluate the influence of the content of -NH_2 on the crystal structure of MIP-207, the PXRD patterns of MIP-207-NH_2 were obtained, and the results are shown in Figure 2. The comparison of PXRD patterns among 207-NH_2-n (n = 0, 15%, 25%, 50%) supported that they were of the same pure phase. However, the internal crystal structure of the amine-modified MIP-207 composites started to change when the mole ratio of H_3BTC to 5-NH_2-H_2IPA was more than 1. As shown in Figure 2, there were still two characteristic diffraction peaks of 5° and 11.5° in the MIP-207-NH_2-60%, but peak relative intensities were significantly reduced, showing that most of the crystal structure of MIP-207 in the composites was changed. When the ligand H_3BTC was totally replaced by 5-NH_2-H_2IPA, the characteristic diffraction peaks (Figure S2) of the sample were totally different from those of the original MIP-207, indicating that another crystalline phase was formed due to the transformation of the coordination mode.

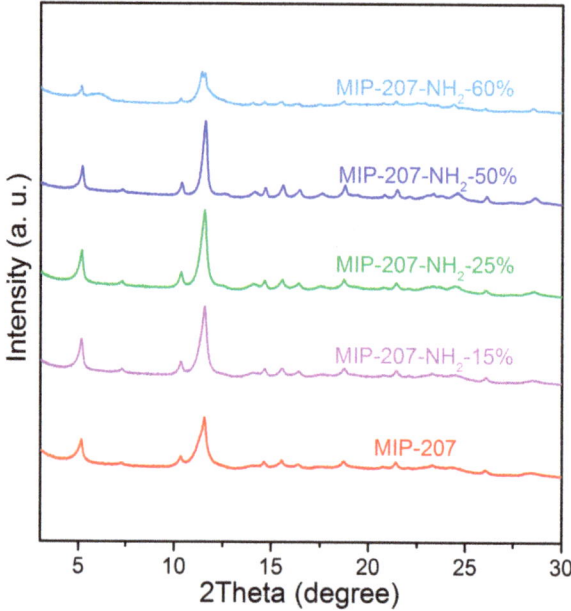

Figure 2. PXRD pattern of samples.

The SEM images of pristine MIP-207 are presented in Figure S3a,b, and the stacked nanoparticles with a size range of 20–25 nm can be observed. As shown in Figure S3c–e, as the amount of exchange ligand 5-NH_2-H_2IPA increases, the stacking of nanoparticles becomes loose, and the particle size was also in the range of 20–25 nm in the MIP-207-NH_2-n (n = 15%, 25%, 50%) composites, which is consistent with XRD results obtained by the Scherrer equation (Table 1). However, the original morphology of MIP-207 is basically not observed in the MIP-207-NH_2-60% (Figure S3f), and particle size sharply reduced to about 15 nm. Overall, based on the above PXRD and SEM analysis, the crystal structure and texture of MIP-207 in the amine-modified MIP-207 composites can be maintained with the amount of added 5-NH_2-H_2IPA being less than or equal to 50%.

As shown in Table 1, the N element was not found in the parent MIP-207. The N element was detected and the N content of the amine-modified MIP-207 composites increased with the increase of the added 5-NH_2-H_2IPA ligand, demonstrating that -NH_2 was introduced into the framework of MIP-207 through a mixed linkers strategy. Notably, Table 1 indicates that the N element content in the MIP-207-NH_2-n (n = 15%, 25%, 50%) composites was lower than the theoretical value, which is attributed to the electronic effect of the functional group of the ligand. Generally, the electron-donating groups such as -NH_2, -OH, and -CH_3 are difficult to connect with the second building units of MIP-207 [33]. The theoretical N content of MIP-207-NH_2-100% is 3.54%. This value is close to the actual value, while the structure completely changed according to the PXRD results.

Table 1. The content of C, H, and N elements of samples and particle size obtained by Scherrer equation.

Samples	C (%)	H (%)	The Actual N (%)	The Theoretical N (%)	Particle Size (nm)
MIP-207	35.37	2.91	0	0	21.9
MIP-207-NH$_2$-15%	35.46	2.76	0.19	0.47	22.3
MIP-207-NH$_2$-25%	35.03	2.72	0.26	0.83	23.6
MIP-207-NH$_2$-50%	35.07	2.89	0.45	1.70	28.4
MIP-207-NH$_2$-60%	35.11	3.04	0.93	2.06	15.5
MIP-207-NH$_2$-100%	41.48	3.99	3.24	3.54	-

Note: The theoretical value of the N element is calculated assuming that 5-NH$_2$-H$_2$IPA completely reacts.

The results of the measurement of N$_2$ adsorption and desorption isotherms are shown in Figure 3 and Table 2, and the N$_2$ adsorption–desorption curves of MIP-207 (Figure 3) conform to the typical I-type isotherm characteristics in the low-pressure zone (0–0.6 atm) relating to microporous characteristics [35]. With the increase of pressure, there was a hysteresis loop in the adsorption–desorption curves, indicating the existence of mesopores, which may be caused by the accumulation of materials. The specific surface area of MIP-207 was 563 m^2 g^{-1}, where the specific surface area was mainly micropores (534 m^2 g^{-1}), which confirms that the mesopores are caused by stacked pores. Figure S4 shows that the average pore size of MIP-207 was mainly distributed at 0.57 and 0.82 nm. Obviously, the BET area and pore volume of the MIP-207-NH$_2$-n (n = 15%, 25%, 50%) materials were higher than that of the unmodified MIP-207 (Table 2), mainly because the mass and volume of the -NH$_2$ group is smaller than the -COOH group, so the BET area of MIP-207-NH$_2$-n (n = 15%, 25%, 50%) increased. Among them, the BET area of MIP-207-NH$_2$-25% was the highest, reaching 735 m^2 g^{-1}. On the contrary, the BET area of MIP-207-NH$_2$-60% reduced in comparison with MIP-207. In addition, compared with the MIP-207-NH$_2$-n (n = 15%, 25%, 50%) (Table 2), the pore volume of MIP-207-NH$_2$-60% (0.44 cm^3 g^{-1}) decreased. The probable reason is that the original structure of MIP-207 cannot be maintained with the -NH$_2$ increasing to over 60%. The specific surface area of MIP-207-NH$_2$-100% sharply decreased and the micropores almost disappeared (Figure S5 and Table S1), further demonstrating the changes from the MIP-207 framework in MIP-207-NH$_2$-100%.

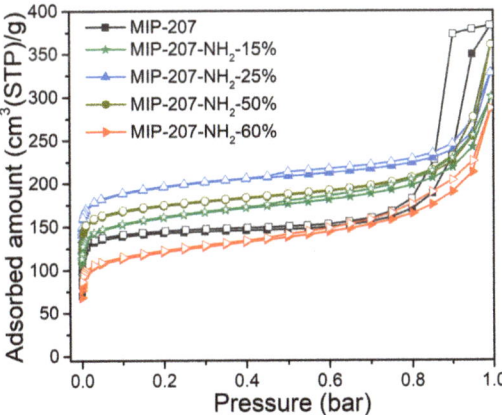

Figure 3. N$_2$ adsorption and desorption isotherms at −196 °C.

Table 2. The summary of specific surface area, pore volume, and particle size of samples.

Samples	BET Area (m^2 g^{-1})	Micropore Area (m^2 g^{-1})	Total Pore Volume (cm^3 g^{-1})	Micropore Volume (cm^3 g^{-1})
MIP-207	563	534	0.36	0.21
MIP-207-NH$_2$-15%	576	468	0.46	0.20
MIP-207-NH$_2$-25%	735	659	0.51	0.27
MIP-207-NH$_2$-50%	654	569	0.56	0.23
MIP-207-NH$_2$-60%	435	321	0.44	0.14

3.2. Gas Adsorption Performance of Materials

The CO_2 adsorption data of as-prepared materials over N_2 are presented in Figure 4 and Table 3. As can be seen from Table 3, the CO_2 adsorption capacity of MIP-207-NH$_2$-25% was up to 3.96 and 2.91 mmol g^{-1} at 0 and 25 °C, which means an improvement of 20.7% and 43.3% compared with the pure MIP-207, respectively. Moreover, the CO_2 capture performance of MIP-207-NH$_2$-25% outperforms most reported amine-modified MOF CO_2 adsorbents (Table 3). Similarly, the CO_2 adsorption capacity of MIP-207-NH$_2$-50% was higher than that of the unmodified MIP-207. The increase of CO_2 adsorption capacity is mainly due to the amine-grafted MIP-207 materials with a high specific area (Figure S6) and many Lewis basic sites (LBS), which greatly enhance their affinity for CO_2 [36,37]. Unfortunately, as the added exchange ligand 5-NH$_2$-H$_2$IPA went above 50%, the CO_2 working capacity in the MIP-207-NH$_2$-60% adsorbent sharply decreased. One reasonable explanation is that excess 5-NH$_2$-H$_2$IPA slows down the rate of the crystal nucleation formation of MIP-207 and disturbs the self-assembly process. When the ligand reactant is completely 5-NH$_2$-H$_2$IPA, the resulting product cannot even form the original crystal nucleus structure of MIP-207. It can be seen that the adsorption performance is a result of both the adsorption sites and the spatial framework of materials.

Table 3. The summary of BET area and CO_2 adsorption results in this work and reported amine-functionalized MOFs.

Materials	Surface Area (m^2 g^{-1})	CO_2 Uptake at Testing Condition	CO_2/N_2 (CO) Selectivity	Q_{st} (kJ mol^{-1})	Ref.
MIP-207	563	3.28/2.03 mmol g^{-1} @ 0/25 °C and 1 bar	59	-	This work
MIP-207-NH$_2$-15%	576	3.12/2.21 mmol g^{-1} @ 0/25 °C and 1 bar	-	30–35	This work
MIP-207-NH$_2$-25%	735	3.96/2.91 mmol g^{-1} @ 0/25 °C and 1 bar	77	30–35	This work
MIP-207-NH$_2$-50%	654	3.49/2.36 mmol g^{-1} @ 0/25 °C and 1 bar	-	30–35	This work
MIP-207-NH$_2$-60%	435	2.02/1.04 mmol g^{-1} @ 0/25 °C and 1 bar	-	30–35	This work
ZIF-8 (40)	844	0.11 mmol g^{-1} @ 45 °C and 0.15 bar	-	55	[19]
ED@Cu$_3$(BTC)$_2$-1	444	4.28/2.15 mmol g^{-1} @ 0/25 °C and 1 bar	21.5	39	[29]
ED@Cu$_3$(BTC)$_2$-2	163	1.03/0.54 mmol g^{-1} @ 0/ and 1 bar	2.68	-	[29]
MAF-23	-	2.5 mmol g^{-1} @ 25 °C and 1 bar	87	34.9 ± 0.9	[38]
ED@MIL-101	1584.6	3.93/1.93 mmol g^{-1} @ 0/25 °C and 1 bar	17.3	-	[32]
TEDA@MIL-101	1806.9	3.81/1.65 mmol g^{-1} @ 0/25 °C and 1 bar	15.5	-	[32]
MIL-101(Cr)-NH$_2$	2800 ± 200	3.4 mmol g^{-1} @ 15 °C and 1 bar	26.5	54.6	[31]
PM24@MOF	2550	2.9 mmol g^{-1} @ 0/25 °C and 1 bar	84	84	[39]
R-PM24@MOF	2410	3.6 mmol g^{-1} @ 0/25 °C and 1 bar	143	50	[39]

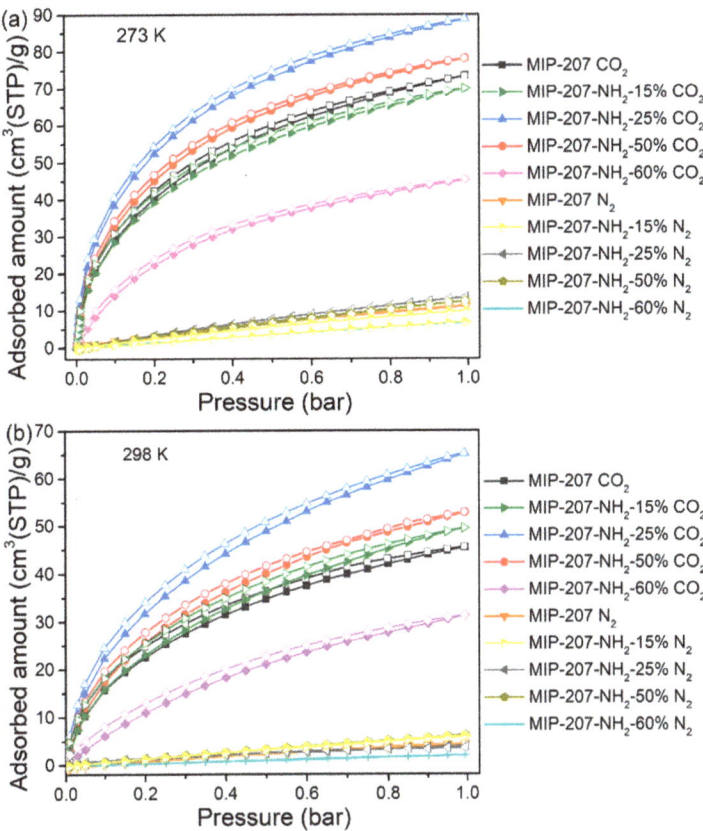

Figure 4. CO_2 and N_2 adsorption and desorption isotherms at (**a**) 0 °C and (**b**) 25 °C.

To figure out the CO_2 adsorption separation performance, the selectivity of CO_2/N_2 was calculated by the IAST model (Supporting Information). As shown in Figure 5a, compared with the separation factor of MIP-207 (59), the CO_2/N_2 separation factor of MIP-207-NH_2-15% was 69, which was 17% higher than that of MIP-207. Additionally, MIP-207-NH_2-25% exhibited the highest CO_2/N_2 separation factor (77), which was 33% higher than MIP-207, mainly because the introduction of the -NH_2 group into the MIP-207 channels could produce more adsorption sites, leading to an enhanced affinity toward CO_2. However, the structure of MIP-207-NH_2-60% possessed more -NH_2 and the separation factor of MIP-207-NH_2-60% was only 22, illustrating that the spatial framework of MIP-207 in MIP-207-NH_2-60% was destroyed, thus increasing the non-selective uptake. In addition to focusing on the adsorption and separation performance, it is also necessary to take energy consumption into account during the regeneration process in industrial applications [32]. The isosteric heat of CO_2 adsorption (Q_{st}) of the materials was obtained from the CO_2 adsorption isotherms at 0 °C. As shown in Figure 5b, the MIP-207-NH_2-n (n = 15%, 25%, 50%) adsorbents had an isosteric adsorption heat of about 30–35 kJ mol^{-1}, which exhibits a medium-strength interaction with CO_2. It can also be found that the Q_{st} of MIP-207-NH_2-60% was significantly lower than that of the above four materials, which indicates that the framework structure of MIP-207-NH_2-60% has a negative effect on the adsorption capacity of CO_2. Considering that the MIP-207-NH_2-25% exhibited superior CO_2 adsorption performance with a remarkable adsorption heat, a test of the cyclic stability of CO_2 was performed and the results are displayed in Figure 6. The CO_2 adsorption capacity of MIP-207-NH$_2$-25% after six cycles did not significantly decrease, indicating that MIP-207-NH$_2$-25% has an outstanding CO$_2$ adsorption–desorption stability.

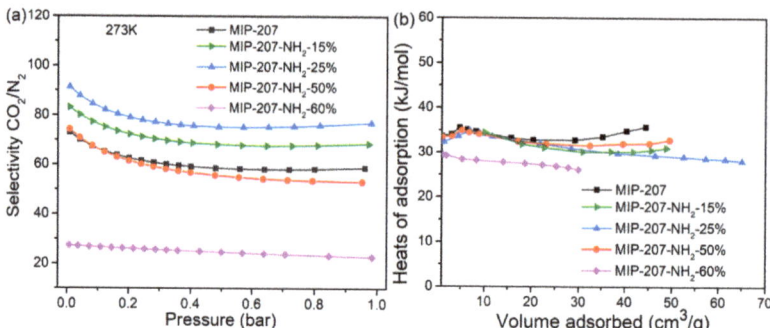

Figure 5. (**a**) CO$_2$/N$_2$ selectivity at 0 °C and (**b**) CO$_2$ adsorption enthalpy curves.

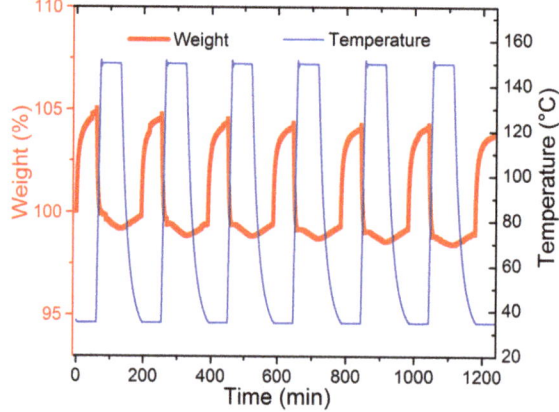

Figure 6. CO$_2$ adsorption and desorption cycle of MIP-207-NH$_2$-25%.

The results of the dynamic CO$_2$/N$_2$ adsorption of both MIP-207-NH$_2$-25% and MIP-207 (as a comparison) are shown in Figure 7. There was an obvious difference in breakthrough time between CO$_2$ and N$_2$ under different gas flow rates. After an initial period where the N$_2$ and CO$_2$ were fully absorbed, the N$_2$ preferentially penetrated the adsorption bed, followed by the CO$_2$. The outlet concentration of N$_2$ exceeded the inlet concentration because CO$_2$ adsorption equilibrium was not reached. Finally, CO$_2$ began to be eluted and the concentration of N$_2$ and CO$_2$ gradually reached the feed concentration value (c/c_0 = 1), indicating that the adsorption bed was saturated. The interval of breakthrough time between CO$_2$ and N$_2$ in MIP-207-NH$_2$-25% was longer than that of MIP-207 at any gas flow rate, especially at 10 sccm, which fully demonstrates that MIP-207-NH$_2$-25% has a better dynamic separation performance. The reason is that the electric field of the MIP-207-NH$_2$-25% framework has the stronger interaction with CO$_2$ due to the presence of LBSs and the hydrogen bond [37,39–41]. Moreover, from Figure 7, MIP-207-NH$_2$-25% has a larger slope than MIP-207 under different gas flow rates, indicating that the mass transfer resistance of gas in MIP-207-NH$_2$-25% is smaller, which is more conducive to gas diffusion and spread. The possible reason for this is that MIP-207-NH$_2$-25% has a larger pore volume.

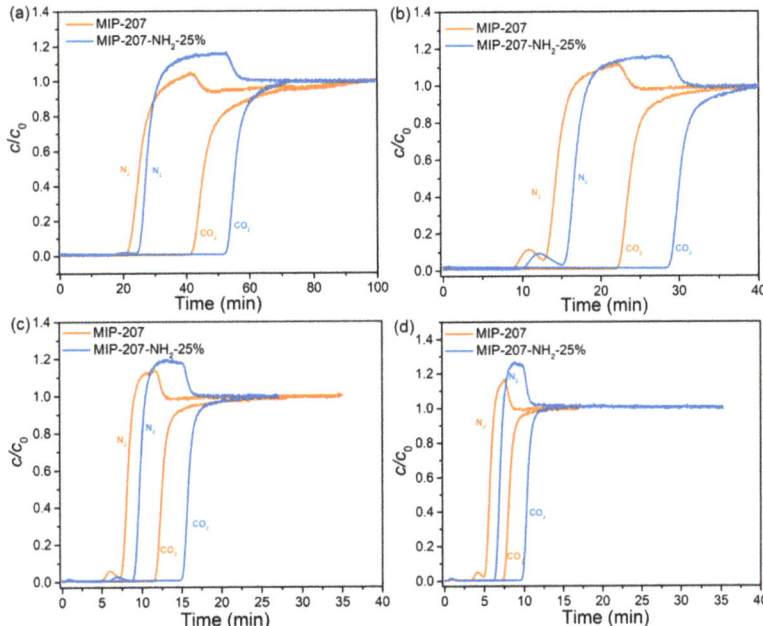

Figure 7. CO_2 and N_2 breakthrough curves of MIP-207 and MIP-207-NH_2-25% at different gas flow rates: (**a**) 10 sccm, (**b**) 20 sccm, (**c**) 50 sccm, and (**d**) 100 sccm, respectively.

The dynamic equilibrium adsorption capacity and separation factor of both MIP-207 and MIP-207-NH_2-25% were calculated based on previous reports [34,42]. As shown in Figure 8, it can be found that the CO_2 equilibrium adsorption capacity of MIP-207-NH_2-25% was higher than that of MIP-207 under the four mixed gas flow rates (Figure 8a,b). At 10 and 20 sccm, the CO_2/N_2 separation factors of MIP-207-NH_2-25% were 2.36 and 2.03, which were higher than the 2.01 and 1.93 of MIP-207, respectively. The difference of separation factors between MIP-207 and MIP-207-NH_2-25% cannot be clearly observed at the mixture gas flow rates of 50 and 100 sccm (Figure 8a,b). This is because the residence time of gas in the adsorption bed decreases with the increase of flow rate.

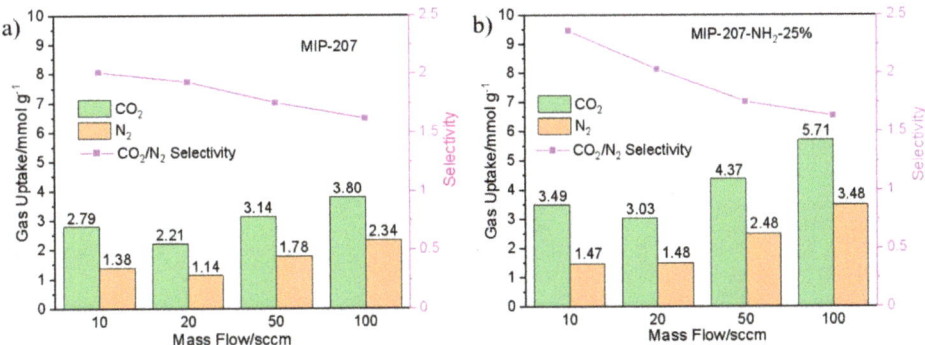

Figure 8. Dynamic CO_2 and N_2 adsorption and separation performance of (**a**) MIP-207 and (**b**) MIP-207-NH_2-25%.

4. Conclusions

In summary, the amine-modified highly stable MIP-207 with different -NH_2 content was successfully prepared by the mixed linkers method. The texture and structure framework of the original MIP-207 were maintained in the MIP-207-NH_2-n (n = 15%, 25%, 50%) composites. The CO_2 adsorption and breakthrough experiments show that MIP-207-NH_2-25% demonstrates the superior CO_2 capture and separation performance. The highly efficient CO_2 uptake is attributed to the introduction of -NH_2 into the framework of MIP-207, leading to the increase of specific surface area and more Lewis basic adsorption sites, thereby enhancing the CO_2 working capacity and CO_2/N_2 selectivity. This work provides an additional avenue to prepare highly stable amine-functionalized MOFs for high efficiency CO_2 capture.

Supplementary Materials: The following are available online at https://www.mdpi.com/article/10.3390/nano11123348/s1, Figure S1: PXRD patterns of MIP-207 and MIP-207 after activation at 150 °C. Figure S2: PXRD pattern of MIP-207-NH_2-100%. Figure S3: SEM images of (**a**,**b**) MIP-207, (**c**) MIP-207-NH_2-15%, (**d**) MIP-207-NH_2-25%, (**e**) MIP-207-NH_2-50%, and (**f**) MIP-207-NH_2-60%. Figure S4: The pore size distribution curves of MIP-207. Figure S5: N_2 adsorption and desorption isotherms of MIP-207-NH_2-100%. Figure S6: The relationship of between CO_2/N_2 adsorption and specific surface area of samples. Table S1: The BET data of MIP-207-NH_2-100%.

Author Contributions: Data curation, Y.W. (Yinji Wan), Y.M., D.K., Y.W. (Yingxiao Wu) and J.S.; formal analysis, Y.W. (Yinji Wan); funding acquisition, R.Z. (Ruiqin Zhong) and R.Z. (Ruqiang Zou); investigation, T.Q. and Q.Z.; methodology, T.Q.; resources, R.Z. (Ruqiang Zou); supervision, R.Z. (Ruiqin Zhong); writing—original draft, Y.W. (Yinji Wan); writing—review and editing, R.Z. (Ruiqin Zhong) and R.Z. (Ruqiang Zou). All authors have read and agreed to the published version of the manuscript.

Funding: This research was funded by the National Natural Science Foundation of China, grant numbers 51772329 and 51972340.

Data Availability Statement: All data are available upon reasonable request.

Acknowledgments: The authors extend much thanks to the reviewers for their valuable suggestions that have helped improve our paper substantially.

Conflicts of Interest: The authors declare no conflict of interest.

References

1. Haszeldine, R.S. Carbon Capture and Storage: How Green Can Black Be? *Science* **2009**, *325*, 1647–1652. [CrossRef] [PubMed]
2. Boyd, P.G.; Chidambaram, A.; Garcia-Diez, E.; Ireland, C.P.; Daff, T.D.; Bounds, R.; Gladysiak, A.; Schouwink, P.; Moosavi, S.M.; Maroto-Valer, M.M.; et al. Data-driven design of metal-organic frameworks for wet flue gas CO_2 capture. *Nature* **2019**, *576*, 253–256. [CrossRef] [PubMed]
3. Sovacool, B.K.; Griffiths, S.; Kim, J.; Bazilian, M. Climate change and industrial F-gases: A critical and systematic review of developments, sociotechnical systems and policy options for reducing synthetic greenhouse gas emissions. *Renew. Sustain. Energy Rev.* **2021**, *141*, 110759. [CrossRef]
4. Chao, C.; Deng, Y.M.; Dewil, R.; Baeyens, J.; Fan, X.F. Post-combustion carbon capture. *Renew. Sustain. Energy Rev.* **2021**, *138*, 19. [CrossRef]
5. Zeng, Y.; Zou, R.; Zhao, Y. Covalent Organic Frameworks for CO_2 Capture. *Adv. Mater.* **2016**, *28*, 2855–2873. [CrossRef] [PubMed]
6. Younas, M.; Rezakazemi, M.; Daud, M.; Wazir, M.B.; Ahmad, S.; Ullah, N.; Inamuddin; Ramakrishna, S. Recent progress and remaining challenges in post-combustion CO_2 capture using metal-organic frameworks (MOFs). *Prog. Energy Combust. Sci.* **2020**, *80*, 100849. [CrossRef]
7. Samanta, A.; Zhao, A.; Shimizu, G.K.H.; Sarkar, P.; Gupta, R. Post-Combustion CO_2 Capture Using Solid Sorbents: A Review. *Ind. Eng. Chem. Res.* **2012**, *51*, 1438–1463. [CrossRef]
8. D'Alessandro, D.M.; Smit, B.; Long, J.R. Carbon dioxide capture: Prospects for new materials. *Angew. Chem. Int. Ed.* **2010**, *49*, 6058–6082. [CrossRef]
9. Wei, Y.S.; Zhang, M.; Zou, R.; Xu, Q. Metal-Organic Framework-Based Catalysts with Single Metal Sites. *Chem. Rev.* **2020**, *120*, 12089–12174. [CrossRef]
10. Zou, R.; Li, P.-Z.; Zeng, Y.-F.; Liu, J.; Zhao, R.; Duan, H.; Luo, Z.; Wang, J.-G.; Zou, R.; Zhao, Y. Bimetallic Metal-Organic Frameworks: Probing the Lewis Acid Site for CO_2 Conversion. *Small* **2016**, *12*, 2334–2343. [CrossRef]
11. Li, H.; Wang, K.; Sun, Y.; Lollar, C.T.; Li, J.; Zhou, H.-C. Recent advances in gas storage and separation using metal-organic frameworks. *Mater. Today* **2018**, *21*, 108–121. [CrossRef]

12. Furukawa, H.; Cordova, K.E.; O'Keeffe, M.; Yaghi, O.M. The Chemistry and Applications of Metal-Organic Frameworks. *Science* **2013**, *341*, 1230444. [CrossRef]
13. Zhou, H.-C.; Long, J.R.; Yaghi, O.M. Introduction to Metal-Organic Frameworks. *Chem. Rev.* **2012**, *112*, 673–674. [CrossRef] [PubMed]
14. Bose, P.; Bai, L.; Ganguly, R.; Zou, R.; Zhao, Y. Rational Design and Synthesis of a Highly Porous Copper-Based Interpenetrated Metal-Organic Framework for High CO_2 and H_2 Adsorption. *ChemPlusChem* **2015**, *80*, 1259–1266. [CrossRef]
15. Zheng, B.; Bai, J.; Duan, J.; Wojtas, L.; Zaworotko, M.J. Enhanced CO_2 Binding Affinity of a High-Uptake rht-Type Metal-Organic Framework Decorated with Acylamide Groups. *J. Am. Chem. Soc.* **2011**, *133*, 748–751. [CrossRef]
16. Hong, D.-Y.; Hwang, Y.K.; Serre, C.; Ferey, G.; Chang, J.-S. Porous Chromium Terephthalate MIL-101 with Coordinatively Unsaturated Sites: Surface Functionalization, Encapsulation, Sorption and Catalysis. *Adv. Funct. Mater.* **2009**, *19*, 1537–1552. [CrossRef]
17. Trickett, C.A.; Helal, A.; Al-Maythalony, B.A.; Yamani, Z.H.; Cordova, K.E.; Yaghi, O.M. The chemistry of metal-organic frameworks for CO_2 capture, regeneration and conversion. *Nat. Rev. Mater.* **2017**, *2*, 17045. [CrossRef]
18. Ding, M.; Flaig, R.W.; Jiang, H.-L.; Yaghi, O.M. Carbon capture and conversion using metal-organic frameworks and MOF-based materials. *Chem. Soc. Rev.* **2019**, *48*, 2783–2828. [CrossRef]
19. Martinez, F.; Sanz, R.; Orcajo, G.; Briones, D.; Yangueez, V. Amino-impregnated MOF materials for CO_2 capture at post-combustion conditions. *Chem. Eng. Sci.* **2016**, *142*, 55–61. [CrossRef]
20. Kong, L.; Zou, R.; Bi, W.; Zhong, R.; Mu, W.; Liu, J.; Han, R.P.S.; Zou, R. Selective adsorption of CO_2/CH_4 and CO_2/N_2 within a charged metal–organic framework. *J. Mater. Chem. A* **2014**, *2*, 17771–17778. [CrossRef]
21. Zhu, J.; Li, P.-Z.; Guo, W.; Zhao, Y.; Zou, R. Titanium-based metal-organic frameworks for photocatalytic applications. *Coord. Chem. Rev.* **2018**, *359*, 80–101. [CrossRef]
22. Wang, L.; Zou, R.; Guo, W.; Gao, S.; Meng, W.; Yang, J.; Chen, X.; Zou, R. A new microporous metal-organic framework with a novel trinuclear nickel cluster for selective CO_2 adsorption. *Inorg. Chem. Commun.* **2019**, *104*, 78–82. [CrossRef]
23. Wang, Q.; Xia, W.; Guo, W.; An, L.; Xia, D.; Zou, R. Functional Zeolitic-Imidazolate-Framework-Templated Porous Carbon Materials for CO_2 Capture and Enhanced Capacitors. *Chem. Eur. J.* **2013**, *8*, 1879–1885. [CrossRef]
24. Xiang, Z.; Hu, Z.; Cao, D.; Yang, W.; Lu, J.; Han, B.; Wang, W. Metal-Organic Frameworks with Incorporated Carbon Nanotubes: Improving Carbon Dioxide and Methane Storage Capacities by Lithium Doping. *Angew. Chem. Int. Ed.* **2011**, *50*, 491–494. [CrossRef]
25. Ma, S.; Sun, D.; Simmons, J.M.; Collier, C.D.; Yuan, D.; Zhou, H.-C. Metal-organic framework from an anthracene derivative containing nanoscopic cages exhibiting high methane uptake. *J. Am. Chem. Soc.* **2008**, *130*, 1012–1016. [CrossRef]
26. Sumida, K.; Rogow, D.L.; Mason, J.A.; McDonald, T.M.; Bloch, E.D.; Herm, Z.R.; Bae, T.-H.; Long, J.R. Carbon Dioxide Capture in Metal-Organic Frameworks. *Chem. Rev.* **2012**, *112*, 724–781. [CrossRef]
27. Cao, L.Y.; Lin, Z.K.; Peng, F.; Wang, W.W.; Huang, R.Y.; Wang, C.; Yan, J.W.; Liang, J.; Zhang, Z.M.; Zhang, T.; et al. Self-Supporting Metal-Organic Layers as Single-Site Solid Catalysts. *Angew. Chem. Int. Ed.* **2016**, *55*, 4962–4966. [CrossRef]
28. Kim, E.J.; Siegelman, R.L.; Jiang, H.Z.H.; Forse, A.C.; Lee, J.-H.; Martell, J.D.; Milner, P.J.; Falkowski, J.M.; Neaton, J.B.; Reimer, J.A.; et al. Cooperative carbon capture and steam regeneration with tetraamine-appended metal-organic frameworks. *Science* **2020**, *369*, 392–396. [CrossRef] [PubMed]
29. Zhong, R.; Yu, X.; Meng, W.; Han, S.; Liu, J.; Ye, Y.; Sun, C.; Chen, G.; Zou, R. A solvent 'squeezing' strategy to graft ethylenediamine on $Cu_3(BTC)_2$ for highly efficient CO_2/CO separation. *Chem. Eng. Sci.* **2018**, *184*, 85–92. [CrossRef]
30. Bae, Y.S.; Farha, O.K.; Hupp, J.T.; Snurr, R.Q. Enhancement of CO_2/N_2 selectivity in a metal-organic framework by cavity modification. *J. Mater. Chem.* **2009**, *19*, 2131–2134. [CrossRef]
31. Han, G.; Qian, Q.H.; Rodriguez, K.M.; Smith, Z.P. Hydrothermal Synthesis of Sub-20 nm Amine-Functionalized MIL-101(Cr) Nanoparticles with High Surface Area and Enhanced CO_2 Uptake. *Ind. Eng. Chem. Res.* **2020**, *59*, 7888–7900. [CrossRef]
32. Zhong, R.; Yu, X.; Meng, W.; Liu, J.; Zhi, C.; Zou, R. Amine-Grafted MIL-101(Cr) via Double-Solvent Incorporation for Synergistic Enhancement of CO_2 Uptake and Selectivity. *ACS Sustain. Chem. Eng.* **2018**, *6*, 16493–16502. [CrossRef]
33. Wang, S.; Reinsch, H.; Heymans, N.; Wahiduzzaman, M.; Martineau-Corcos, C.; De Weireld, G.; Maurin, G.; Serre, C. Toward a Rational Design of Titanium Metal-Organic Frameworks. *Matter* **2020**, *2*, 440–450. [CrossRef]
34. Zhong, R.; Liu, J.; Huang, X.; Yu, X.; Sun, C.; Chen, G.; Zou, R. Experimental and theoretical investigation of a stable zinc-based metal-organic framework for CO_2 removal from syngas. *CrystEngComm* **2015**, *17*, 8221–8225. [CrossRef]
35. Gandara, F.; Furukawa, H.; Lee, S.; Yaghi, O.M. High methane storage capacity in aluminum metal-organic frameworks. *J. Am. Chem. Soc.* **2014**, *136*, 5271–5274. [CrossRef] [PubMed]
36. Darunte, L.A.; Oetomo, A.D.; Walton, K.S.; Sholl, D.S.; Jones, C.W. Direct Air Capture of CO_2 Using Amine Functionalized MIL-101(Cr). *ACS Sustain. Chem. Eng.* **2016**, *4*, 5761–5768. [CrossRef]
37. Lin, Y.; Lin, H.; Wang, H.; Suo, Y.; Li, B.; Kong, C.; Chen, L. Enhanced selective CO_2 adsorption on polyamine/MIL-101(Cr) composites. *J. Mater. Chem. A* **2014**, *2*, 14658–14665. [CrossRef]
38. Liao, P.-Q.; Zhou, D.-D.; Zhu, A.-X.; Jiang, L.; Lin, R.-B.; Zhang, J.-P.; Chen, X.-M. Strong and Dynamic CO_2 Sorption in a Flexible Porous Framework Possessing Guest Chelating Claws. *J. Am. Chem. Soc.* **2012**, *134*, 17380–17383. [CrossRef]

39. Yoo, D.K.; Yoon, T.U.; Bae, Y.S.; Jhung, S.H. Metal-organic framework MIL-101 loaded with polymethacrylamide with or without further reduction: Effective and selective CO_2 adsorption with amino or amide functionality. *Chem. Eng. J.* **2020**, *380*, 122496. [CrossRef]
40. Ding, R.; Zheng, W.; Yang, K.; Dai, Y.; Ruan, X.; Yan, X.; He, G. Amino-functional ZIF-8 nanocrystals by microemulsion based mixed linker strategy and the enhanced CO_2/N_2 separation. *Sep. Purif. Technol.* **2020**, *236*, 116209. [CrossRef]
41. Zhang, M.; Guo, Y. Rate based modeling of absorption and regeneration for CO_2 capture by aqueous ammonia solution. *Appl. Energy* **2013**, *111*, 142–152. [CrossRef]
42. Xiang, S.; He, Y.; Zhang, Z.; Wu, H.; Zhou, W.; Krishna, R.; Chen, B. Microporous metal-organic framework with potential for carbon dioxide capture at ambient conditions. *Nat. Commun.* **2012**, *3*, 954. [CrossRef] [PubMed]

 nanomaterials

Article

Catalytic Hydrogen Evolution of NaBH₄ Hydrolysis by Cobalt Nanoparticles Supported on Bagasse-Derived Porous Carbon

Yiting Bu [1,2,†], Jiaxi Liu [1,†], Hailiang Chu [1], Sheng Wei [1,2], Qingqing Yin [1], Li Kang [1], Xiaoshuang Luo [1], Lixian Sun [1,2,*], Fen Xu [1,*], Pengru Huang [1,3], Federico Rosei [4], Aleksey A. Pimerzin [5], Hans Juergen Seifert [6], Yong Du [7] and Jianchuan Wang [7]

1. Guangxi Key Laboratory of Information Materials and Guangxi Collaborative Innovation Center of Structure and Property for New Energy and Materials, School of Material Science & Engineering, Guilin University of Electronic Technology, Guilin 541004, China; ytb1172701255@163.com (Y.B.); jxliu2019@126.com (J.L.); chuhailiang@guet.edu.cn (H.C.); ws1801101003@163.com (S.W.); yqq15870030656@163.com (Q.Y.); kangli000hello@163.com (L.K.); shirley_lxs@126.com (X.L.); pengruhuang@guet.edu.cn (P.H.)
2. School of Mechanical & Electrical Engineering, Guilin University of Electronic Technology, Guilin 541004, China
3. Department of Materials Science and Engineering, National University of Singapore, Singapore 117575, Singapore
4. Centre for Energy, Materials and Telecommunications, Institut National de la Recherche Scientifique, 1650 Boulevard Lionel-Boulet Varennes, Québec, QC J3X 1S2, Canada; rosei@emt.inrs.ca
5. Chemical Department, Samara State Technical University, 443100 Samara, Russia; al.pimerzin@gmail.com
6. Karlsruhe Institute of Technology, Institute for Applied Materials-Applied Materials Physics, Hermann-von-Helmholtz-Platz 1, 76344 Eggenstein-Leopoldshafen, Germany; hans.Seifert@kit.edu
7. State Key Laboratory of Powder Metallurgy, Central South University, Changsha 410083, China; yong-du@csu.edu.cn (Y.D.); jcw728@126.com (J.W.)
* Correspondence: sunlx@guet.edu.cn (L.S.); xufen@guet.edu.cn (F.X.)
† Yiting Bu and Jiaxi Liu contributed equally.

Abstract: As a promising hydrogen storage material, sodium borohydride (NaBH₄) exhibits superior stability in alkaline solutions and delivers 10.8 wt.% theoretical hydrogen storage capacity. Nevertheless, its hydrolysis reaction at room temperature must be activated and accelerated by adding an effective catalyst. In this study, we synthesize Co nanoparticles supported on bagasse-derived porous carbon (Co@xPC) for catalytic hydrolytic dehydrogenation of NaBH₄. According to the experimental results, Co nanoparticles with uniform particle size and high dispersion are successfully supported on porous carbon to achieve a Co@150PC catalyst. It exhibits particularly high activity of hydrogen generation with the optimal hydrogen production rate of 11086.4 $mL_{H2} \cdot min^{-1} \cdot g_{Co}^{-1}$ and low activation energy (E_a) of 31.25 kJ mol^{-1}. The calculation results based on density functional theory (DFT) indicate that the Co@xPC structure is conducive to the dissociation of [BH₄]⁻, which effectively enhances the hydrolysis efficiency of NaBH₄. Moreover, Co@150PC presents an excellent durability, retaining 72.0% of the initial catalyst activity after 15 cycling tests. Moreover, we also explored the degradation mechanism of catalyst performance.

Keywords: sodium borohydride; hydrolysis; porous carbon; Co nanoparticles; durability

1. Introduction

Two of the seventeen Sustainable Development Goals (SDGs), namely 7 (affordable and clean energy) and 13 (climate action), dictate the urgency of transitioning from highly polluting and non-sustainable fossil fuels to renewable energy sources [1–3]. Hydrogen is considered a very promising green new energy carrier because of its high-energy and zero-emission applications [4–7]. However, the physical properties of hydrogen make it have high energy consumption and high risk for application on-board. Therefore, we need to find alternative ways to store and transport H₂. Compared with high-pressure gas storage and cryogenic liquid storage, solid state hydrogen storage materials [8] such as

MgH$_2$, NaAlH$_4$, NaBH$_4$, etc., have attracted more and more attention and research due to high hydrogen capacity, safe operation and relative abundance.

Among many hydrogen storage materials, NaBH$_4$ is considered to be a promising one because of its theoretical hydrogen storage capacity of 10.8 wt.% and good stability in alkaline solutions [9]. Nevertheless, the hydrogen production rate is very low and unsatisfactory without a proper catalyst at room temperature [10]. Therefore, it is necessary to activate and accelerate the NaBH$_4$ hydrolysis reaction at room temperature through the use of high-efficiency catalysts. Guella et al. [11] followed the reaction of NaBH$_4$ in the presence of catalysts and its perdeuterated analogue NaBD$_4$ in H$_2$O, D$_2$O and H$_2$O/D$_2$O mixtures. The results revealed that NaBH$_4$ can react with water to generate hydrogen (Equation (1)), of which four equivalents of H come from BH$_4^-$, and four equivalents of D from the decomposition of D$_2$O. The reaction formula is summarized as follows:

$$BH_4^- + 4D_2O \rightarrow B(OD)_4^- + 4HD \qquad (1)$$

In recent years, relatively inexpensive transition metals [12] (such as Co [13], Ni [14], Cu [15] and Fe [16]) have been observed to exhibit excellent performances in catalytic sodium borohydride hydrolysis, especially metal nanoparticles prepared with transition metals as catalysts. In particular, Co-based catalysts are considered to be particularly attractive in the hydrolysis of NaBH$_4$ for hydrogen production due to their high activity and relatively low cost [17–19]. Metal catalysts are usually prepared by chemical reduction using NaBH$_4$ and ammonia borane. However, metal nanoparticles are prone to agglomeration during the reduction and catalytic process, resulting in reduced catalyst activity and poor cycle stability [20]. In order to prevent agglomeration, the most used method is to choose suitable support materials, including SiO$_2$ [21], γ-Al$_2$O$_3$ [22], hydroxyapatite [23], carbon materials [24,25], etc. In particular, the addition of appropriate promoters could be used to enhance the dispersibility of Co nanoparticles and increase the surficial active sites [18]. Therefore, using the reduction method, the surface of supporting materials could be loaded with reduced Co nanoparticles, so that the agglomeration of the composite catalyst is inhibited and its specific surface area increased. Compared with unsupported metal catalysts, catalysts composed of support materials and metal particles have a larger specific surface area [26], which increases the contact area between metal particles and NaBH$_4$ aqueous solution. In addition, supporting materials increase the stability of metal particles, greatly increasing the durability [27–29]. Among the supporting materials mentioned above, carbon materials such as porous carbon (PC) are the most attractive support due to their chemical inertness and large surface area [24]. At present, the reported porous carbon materials [30] loaded with Co nanoparticles mainly include carbon nanotubes, graphene, activated carbon, organic drugs [31] and polymer materials [13]. Among them, carbon nanotubes, graphene, activated carbon and polymer materials are relatively expensive. Meanwhile, most organic drugs are also not conducive to practical promotion because they contain a certain level of toxicity. Therefore, PC prepared from biomass is an excellent candidate as a support because it not only has a wide range of raw materials (especially biomass such as bagasse) but is also simple to prepare.

In our previous work [31], we synthesized nitrogen-doped mesoporous graphitic carbon-coated cobalt nanoparticles (Co@NMGC) with a core–shell structure by carbonizing carbon derived from ethylenediaminetetraacetic acid (EDTA). However, the specific surface area of Co@NMGC is only 124.55 m$^2 \cdot$g^{-1}, and the cobalt nanoparticles coated with graphite carbon cannot effectively contact and react with NaBH$_4$, resulting in an optimal hydrogen production rate of Co@NMGC of only 3575 mL min^{-1} g^{-1}. Based on these considerations, using bagasse as raw material for PC, we designed and prepared a kind of PC to load Co nanoparticles (Co@xPC) for NaBH$_4$ hydrolysis. The structural characteristics and catalytic performance of several PC-supported Co nanoparticle catalysts were studied in detail. After optimization, a Co-based catalyst with high efficiency, excellent activity and high durability was achieved. The Co@xPC structure was beneficial to improve the hydrolysis efficiency of NaBH$_4$, confirmed by the theoretical calculation based on density functional theory (DFT).

2. Materials and Methods

2.1. Materials

Bagasse was collected from fruit shops on East West Street of Guangxi, China, which have abundant sugarcane. We purchased sodium borohydride (NaBH$_4$, AR), ZnCl$_2$ (AR), cobalt(II) chloride hexahydrate (CoCl$_2$·6H$_2$O, AR) and Mg(NO$_3$)$_2$·6H$_2$O (AR) from Alfa Aesar Co., Ltd. (Tianjin, China). Hydrochloric acid (HCl) was from Xilong Chemical Co., Ltd. (Shantou, China). Ultrapure water, obtained from a Millipore System (Millipore Q, Burlington, MA, USA), was used throughout the experiments.

2.2. Synthesis of Co@xPC Catalyst

Bagasse was freeze-dried with a freezer, then crushed with a pulverizer. The crushed bagasse powder and activator were completely ground in a mortar according to the mass ratio (1:2). The activator was a mixture of Mg(NO$_3$)$_2$·6H$_2$O and ZnCl$_2$ with a mass ratio of 1:1. The well-ground mixture was moved to a crucible and kept at 800 °C for 2 h at a heating rate of 3 °C min^{-1} under N$_2$ flow. The resulting black sample was soaked in 3 M HCl at 80 °C for 24 h to remove inorganic salts. Then, the black sample was washed with ultrapure water to neutrality and dried at 100 °C for 24 h, and the resulting sample was named PC. In the preparation process, appropriate CoCl$_2$·6H$_2$O was dissolved in 20 mL ultrapure water, and different amounts of PC (50, 100, 150 and 200 mg) were added at the same time. The above solution was treated under ultrasonic conditions for 1 h, and then 20 mL 3 wt.% NaBH$_4$ solution was slowly added under constant stirring. The black solid was separated with vacuum suction filtration and washed with ethanol and ultrapure water several times, respectively. The sample was dried in a vacuum oven at 60 °C. According to the different amounts of added PC, the corresponding catalysts were marked as Co@50PC, Co@100PC, Co@150PC and Co@200PC, respectively. At the same time, pure Co particle catalysts without PC were also prepared in the same conditions for comparison.

2.3. Catalyst Characterization

The morphology, surface structure and element distribution of the catalysts were analyzed through scanning electron microscopy (SEM, JSM-6360LV, JEOL Ltd., Tokyo, Japan) combined with energy dispersive spectroscopy (EDS) at 20kV. The surface interactions and electronic states between the elements of the catalyst were obtained by using X-ray photoelectron spectroscopy (XPS, Thermo Electron ESCALAB 250, Shanghai, China). The nitrogen adsorption–desorption isotherms of the Co@xPC catalysts were tested at 77 K using a gas adsorption analyzer (Autosorb iQ2, Quantachrome sorptometer, Osaka, Japan). The specific surface areas and pore size distribution of the Co@xPC catalysts were obtained by the Brunauer–Emmett–Teller (BET) method and the Barrett–Joyner–Halenda (BJH) method, respectively. The crystallographic structure and chemical composition of the catalysts were obtained by X-ray diffraction (XRD, 1820, Philips, Amsterdam, the Netherlands), inductively coupled plasma atomic emission spectroscopy (ICP-AES, Optima 8000, PerkinElmer, Chiba, Japan), laser confocal Raman spectroscopy (LabRAM HR Evolution, Horiba JY, Edison, NJ, USA) and Fourier transform infrared spectroscopy (FT-IR, Nicolet 6700, Shanghai, China).

2.4. Measurement Method of Hydrogen Production

The catalytic activity of Co@xPC was evaluated by measuring the hydrogen production rate of Co@xPC in alkaline NaBH$_4$ hydrolysis. In the self-made reactor, the classic water displacement method was used to measure the hydrogen production rate of NaBH$_4$, as in our previous work [31]. The volume at constant time intervals was determined by using a balance to record the weight of the replaced water [31,32], and the volume of hydrogen produced was measured in an equivalent way. Usually, a magnetic rotor and

0.1 g Co@xPC catalyst were placed at the bottom of a wide-mouthed bottle. At 25 °C, 10 mL mixed solution containing 1.5 wt.% $NaBH_4$ and 5.0 wt.% NaOH was quickly injected into the wide-mouthed bottle, and a constant temperature water bath was used to maintain the test system at a constant temperature. The tests were carried out at 15, 25, 35, 45 and 55 °C, and other experimental conditions remain unchanged to obtain E_a. In the stability tests, the reaction was repeated 15 times, and the reacted catalyst was named Co@xPC-15th. Similarly, the reaction of pure Co particle catalyst was repeated 5 times, and the reacted catalyst was named Co-5th. In the stability test, after the hydrolysis was over, the supernatant was poured from the wide-mouthed bottle and then 10 mL $NaBH_4$ alkali solution was added to start the next test. Since the catalyst was magnetic, a magnet was used to recover the catalyst after the hydrolysis reaction was completed. During the experiment, the hydrogen generation rate for all catalysts was based on the amount of Co.

2.5. DFT Calculations

In this work, all calculations are performed with DFT [33] using the projected augmented wave method, which was implemented in the Vienna Ab-initio Simulation Package [34–36]. The generalized gradient approximation of Perdew–Burke–Ernzerhof was used for the exchange–correlation interaction [37,38]. The wave functions are expressed in the plane wave basis set with an energy cutoff of 450 eV. A vacuum region of 15 Å was set to eliminate undesirable interactions between the periodic sheets of the graphene patches. The Brillouin zone was sampled by a Monkhorst–Pack special k-point mesh of $5 \times 5 \times 1$ for geometry optimizations and energy calculations. All atoms were allowed to relax until the final energy and the forces on each atom converged to 10^{-4} eV and 0.02 eV/Å, respectively. The quantitative charge changes of the Co_4 and graphene patch were described using a grid-based Bader charge transfer analysis method [39]. The adsorption energy was calculated as follows:

$$E_{ads} = E_{total} - E_{graphene} - E_{Co_4} \quad (2)$$

where E_{Co_4} is the calculated energy of the Co_4 cluster, $E_{graphene}$ is the energy of the graphene surface and E_{total} is the total energy of the absorption system that contains both the cluster and the graphene patch.

During the hydrolysis process, the remaining one or more hydrogen atoms from the previous step are accepted when far away from the central unit cell of the surface. As the total number of atoms in the system must be conserved in order to generate a relative energy diagram, thus the infinite distance approach (IDA) was considered in the calculation. IDA energy is equal to the number of the adsorbed hydrogen atom(s), and it can be defined as follows [40]:

$$E_{IDA} = E_{H/slab} - E_{slab} \quad (3)$$

where $E_{H/slab}$ is the total energy of the hydrogen adsorbed slab and E_{slab} is the total energy of the pure slab.

3. Results and Discussion

3.1. Characterization of the as-Prepared Catalysts

First, the PC was prepared by carbonizing the powder mixture of bagasse, zinc chloride and magnesium sulfate at 800 °C for 2 h under N_2 flow with a heating rate of 3 °C min^{-1}. Then, the PC was dispersed in the aqueous solution by ultrasound. Under stirring conditions, cobalt chloride hexahydrate was added to the suspension so that Co^{2+} was well dispersed around the PC. After that, the $NaBH_4$ aqueous solution was slowly dropped into the system to reduce Co^{2+} to Co on the PC surface (Figure 1). Finically, the catalysts of Co@50PC, Co@100PC, Co@150PC and Co@200PC were obtained according to the different amounts of PC (50, 100, 150 and 200 mg).

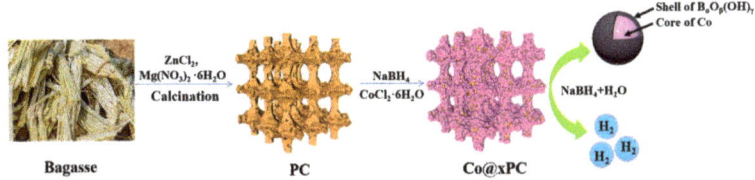

Figure 1. Schematic diagram of the preparation and catalytic process of Co@xPC.

In Figure 2a, the PC shows a clear porous structure on its surface and on the scaffold. Co nanoparticles are flaky and aggregate into branches (Figure 2b). After supporting Co nanoparticles (Figure 2c–f), with the increase in the addition of PC, the dispersibility of Co nanoparticles in Co@xPC was significantly enhanced and gradually changed from the mixed state of flakes and particles to Co nanoparticles with uniform size. Compared with the agglomerated Co in Figure 2b, the dispersion of Co in Co@xPC has been significantly improved. Particularly in Co@150PC and Co@200PC, Co nanoparticles were evenly dispersed on the surface of PC (Figure S1) without obvious agglomeration (Figure 2e–f). The average particle size of Co nanoparticles in Co@150PC and Co@200PC was determined to be 50 and 45 nm, respectively, much smaller than that in Co, Co@50PC and Co@100PC samples. Furthermore, compared with Co nanoparticles in Figure 2b, the presence of PC support could effectively prevent the agglomeration of catalysts in the preparation and catalytic reaction process, resulting in a smaller size and more even dispersion of Co particles. Through careful examination, the Co nanoparticles of Co@150PC (51.82 wt.%) were found to be more uniform than Co@200PC (44.65 wt.%) and the relative content of Co was higher in Co@150PC, confirmed by ICP analysis. The XRD patterns of PC, Co, Co@50PC, Co@100PC, Co@150PC, Co@200PC and Co@150PC-15th are shown in Figure 2g and Figure S2. For the PC, two broad and weak peaks observed at approximately 24° and 42° are attributed to the diffraction of the (002) and (100) planes of graphite [41], respectively. For the neat Co nanoparticles and Co@xPC samples, no observable peaks for the metallic cobalt are found, ascribed to the amorphous state of metallic cobalt [42]. A similar phenomenon was also observed in the XRD patterns of phosphorus-modified spirulina microalgae strains supporting a Co-B catalyst [19]. As described before, Co@150PC exhibits more uniform Co nanoparticles (Figure S3) and relatively high Co content among the as-prepared catalysts. Therefore, only Co@150PC is characterized in the following parts. As shown in Figure 2h, a typical type IV isotherm is observed for the PC, Co and Co@150PC catalysts. The specific surface area (Table 1) of PC was determined to be 1527.5 $m^2 \cdot g^{-1}$ with a pore-size distribution in a narrow 3–4 nm range (Figure 2i). The specific surface areas of the Co and Co@150PC catalysts were 87.1 and 274.1 $m^2 \cdot g^{-1}$, respectively. Therefore, the addition of PC effectively increases the specific surface area of Co@150PC.

From Figure 3a, the C 1s and Co 2p photoionization signals are observed in the XPS survey spectrum, which confirmed the presence of boron, carbon and cobalt species in the Co@150PC. The binding energy of the C 1s spectrum in Figure S4 can be divided into three fitting peaks at 284.8 eV, 286.2 eV and 288.9 eV, which are, respectively, related to C-C, C-O and C=O [43–45]. In Figure 3b, the Co 2p spectrum shows the six fitting peaks. The peaks of 778.8 eV (Co $2p_{3/2}$) and 793.6 eV (Co $2p_{1/2}$) correspond to the reduced metallic Co in the Co@150PC. Two peaks were observed at 780.8 eV (Co $2p_{3/2}$) and 796.4 eV (Co $2p_{1/2}$), which corresponded to the spin-orbit peaks of Co $2p_{3/2}$ and Co $2p_{1/2}$ of CoO [31]. This indicated the presence of CoO in the Co@150PC catalyst. Moreover, satellite peaks corresponding to CoO were also observed at 786.2 eV and 802.8 eV. The quantitative analysis of the Co 2p XPS spectrum indicated that the Co^0:Co^{2+} atomic ratio was 6.21:1.00, showing that most of the cobalt in Co@150PC exists in a reduced state. In summary, Co nanoparticles in Co@150PC mainly existed as metallic cobalt and also contained a small amount of CoO, which possibly resulted from cobalt nanoparticles easily combining with atmospheric oxygen in the preparation and storage of the catalyst [17].

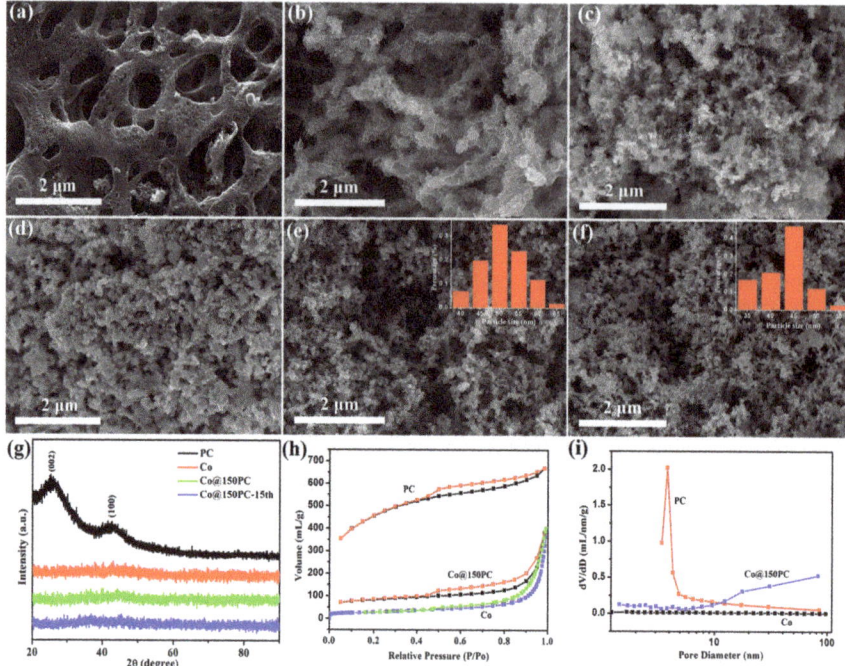

Figure 2. SEM images of PC (**a**), Co (**b**), Co@50PC (**c**), Co@100PC (**d**), Co@150PC (**e**) and Co@200PC (**f**). XRD patterns of the as-prepared catalysts (**g**). N$_2$ adsorption–desorption isotherms (**h**) and corresponding BJH pore-size distribution plots (**i**) of PC, Co and Co@150PC. The insets in (**e**,**f**) are the size distribution of metal Co nanoparticles.

Table 1. The nitrogen adsorption–desorption measurement parameters of pure PC, Co and Co@150PC.

Catalyst	Specific Surface Area (m²·g⁻¹)	Pore Volume (cm³·g⁻¹)	Average Pore Diameter (nm)
PC	1527.499	0.530	3.837
Co	87.098	0.528	0.478
Co@150PC	274.101	0.348	1.429

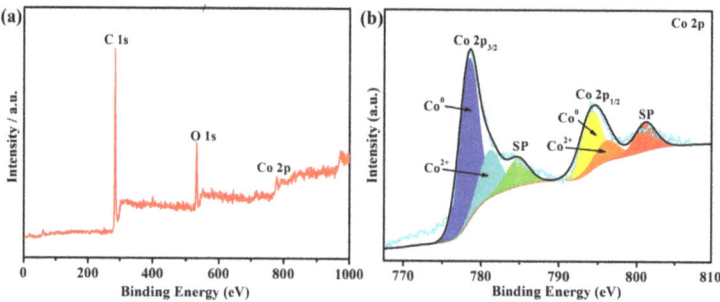

Figure 3. XPS spectra of Co@150PC: (**a**) survey spectrum, (**b**) Co 2p spectrum.

3.2. Catalytic Activity Tests of Co@xPC for Hydrolysis of NaBH$_4$

The NaBH$_4$ hydrolysis performance catalyzed by Co and Co@xPC was tested under the same conditions to explore the influence of the addition of PC on the catalytic performance of

Co (Figure 4a–b). The hydrogen production rate was only 3693.94 mL$_{H2}$·min^{-1}·g$_{Co}$$^{-1}$ when no PC was added. This low value was probably obtained due to the severe agglomeration of Co particles without support materials (Figure 2b). This caused a decrease in the number of active sites on the Co surface, which in turn affected the catalytic performance. Correspondingly, when increasing the amount of PC, the hydrogen production rate of Co@xPC first increased rapidly because the presence of PC can effectively inhibit the agglomeration of Co (Figure 2c–f). Then, the hydrogen production rate suddenly decreased after reaching a maximum value with Co@150PC. This may be because the amount of PC added was too high, and the Co content in Co@200PC was reduced too much, slowing down its catalytic rate of NaBH$_4$. Compared with previous work (Table 2), the addition of PC with a large specific surface area obtained from bagasse is more effective than other carbon materials in improving the performance of Co nanoparticles.

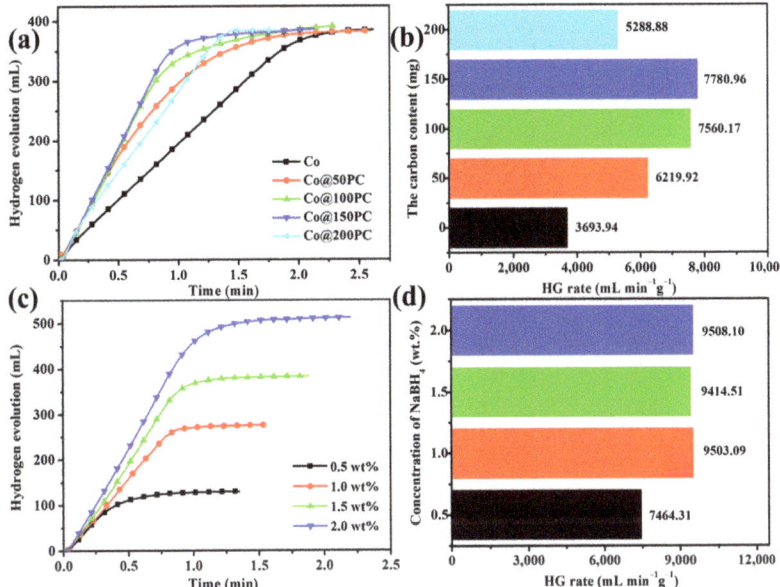

Figure 4. Hydrogen production rate of the as-prepared samples with (**a**,**b**) amount of PC added in Co, (**c**,**d**) NaBH$_4$ concentration.

Table 2. Comparison of the performance of different catalysts in catalyzing the NaBH$_4$ hydrolysis reaction.

Catalyst Sample	Maximum Hydrogen Production Rate (mL$_{H2}$·min^{-1}·g$_M$$^{-1}$)	E_a (kJ mol^{-1})	Durability	References
Co-Fe3O4@C	1403	49.2	59.3% after 5 cycles	[42]
CNSs@Pt0.1Co0.9	8943	38.0	85.12% after 5 cycles	[46]
Co-B/C	8033.89	56.72	-	[47]
Co-B/C	3887.1	56.37	25% after 6 cycles	[48]
Co-Mo-B/CC	1280.8	51.0	75.1% after 3 cycles	[49]
Ru–Co/C	9360	36.83	70% after 8 cycles	[50]
Co-B/MWCNT	5100	40.40	-	[51]
Co-B/N-C-700	2649	37.57	-	[52]
Co/PGO	5955	55.2	73% after 5 cycles	[53]
Co@NMGC	3575	35.2	82.5% after 20 cycles	[31]
Modified CCS/Co	11,600	33.4	-	[54]
CAs/Co	11,220	38.4	96.4% after 5 cycles	[55]
Co/C	530	44.1	-	[56]
Co@150PC	11,086.4	31.25	72% after 15 cycles	This work

The effect of different NaBH$_4$ concentrations on the hydrogen production rate catalyzed by Co@150PC catalyst with a Co loading of 51.82 wt.% (characterized by ICP) was studied. In Figure 4c,d, it can be seen that when the concentration of NaBH$_4$ was 0.5 wt.%, the reaction rate was significantly slower due to the low concentration of NaBH$_4$. However, as the concentration of NaBH$_4$ increased, the hydrogen production rate did not increase significantly, which indicated that the concentration of NaBH$_4$ had no obvious effect on the hydrolysis catalyzed by Co@150PC. Therefore, a zero-order reaction is ascribed to the hydrolysis of NaBH$_4$ to produce hydrogen using a Co@150PC catalyst [57,58] and the rate-determining step should have been the hydrolysis reaction [59].

The hydrogen production rate of NaBH$_4$ catalyzed by different amounts of Co@150PC (0.05, 0.10, 0.15 and 0.20 g) was also tested. Figure 5a shows the relationship between the hydrogen production rate and the amount of catalyst added, indicating that the time required to complete the reaction decreases rapidly as the amount of catalyst added increases. Moreover, the rate of the NaBH$_4$ hydrolysis reaction shows a good linear fit with respect to the amount of catalyst added (Figure 5b).

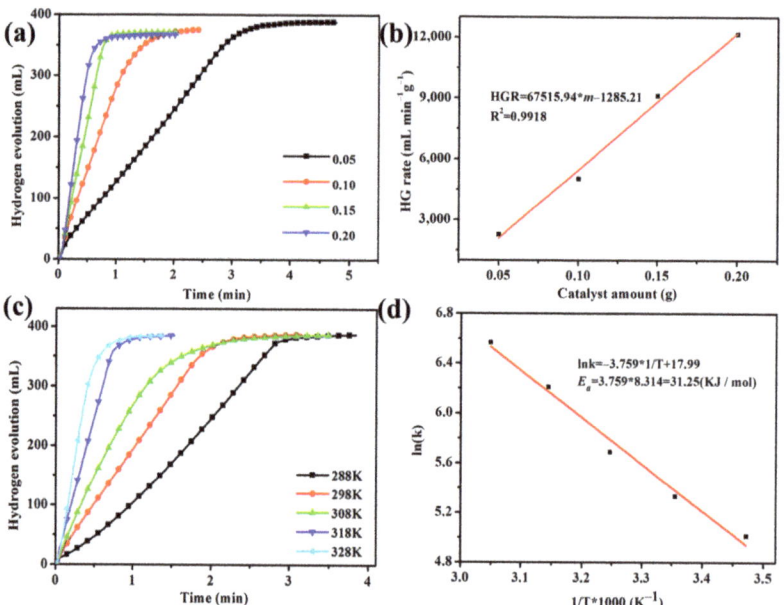

Figure 5. Hydrogen production rate of the as-prepared samples with (**a,b**) catalyst amount, (**c**) hydrogen generation kinetics curves employing Co@150PC at different solution temperatures and (**d**) Arrhenius plot.

To further explore the catalytic activity of the Co@150PC catalyst, Co@150PC catalyst (100 mg) was used to hydrolyze NaBH$_4$ at different temperatures from 15 °C to 55 °C. Figure 5c shows that the rate of hydrogen generation increases significantly with the increase in temperature. The activation energy (E_a) of the NaBH$_4$ hydrolysis reaction catalyzed by Co@150PC can be obtained from the Arrhenius equation:

$$\ln k = \ln A - \left(\frac{E_a}{RT}\right) \qquad (4)$$

According to the linear slope in Figure 5d, E_a for the NaBH$_4$ hydrolysis reaction catalyzed by the Co@150PC catalyst was calculated to be 31.25 kJ mol^{-1}, lower than those of most reported Co-based catalysts (Table 2).

3.3. Catalytic Stability Tests of Co@150PC for Hydrolysis of NaBH₄

In addition to having excellent catalytic activity, low cost and environmentally friendly properties, the hydrolysis catalyst is also required to have good stability. Therefore, the stability studies of Co@150PC catalyst were conducted. The histogram of the hydrogen generation rate and the different cycle times was drawn to study the changes in the hydrogen generation rate during the stability test (Figure 6). Obviously, the hydrogen generation rate decreased slowly as the number of tests increased. The hydrogen production rate of the Co@150PC catalyst is 7982.54 $mL_{H2} \cdot min^{-1} \cdot g_{Co}^{-1}$ at the 15th cycle with a 72.0% hydrogen generation rate of the initial cycle retained. Compared with previous reported values (Table 2), the cycle stability of Co@150PC is significantly improved and the decline in performance is relatively small compared with the Co-based catalysts supported by other carbon materials. To prove that the presence of PC could effectively improve the cycling performance of the catalyst, we performed a repeatability study on the pure Co catalyst under the same conditions (Figure 6). The hydrogen production rate catalyzed by neat Co was 1684.30 $mL_{H2} \cdot min^{-1} \cdot g_{Co}^{-1}$ after five hydrolysis cycles, which was only 45.6% of the hydrogen generation rate of the first cycle.

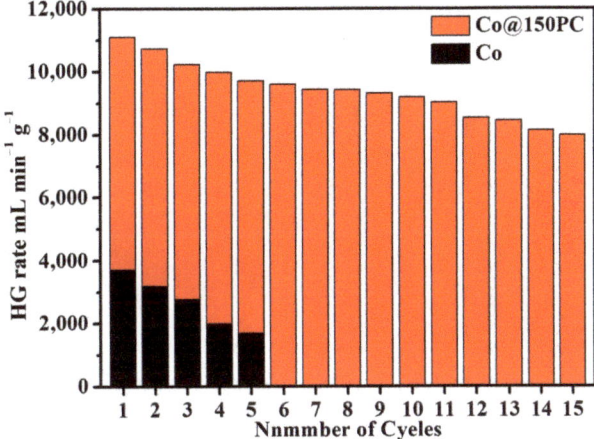

Figure 6. Histogram of hydrogen generation rate and the different cycle times of Co@150PC and Co in the cycle test.

SEM images showed that the structure of Co was agglomerated from flakes into a large number of irregular blocks (Figure 7a). Moreover, compared with Co@150PC-15th (Figure 7b), Co-5th has poor dispersion and uneven size. Therefore, PC effectively inhibits the agglomeration of Co during the hydrolysis reaction process, which reduces the decrease in the number of active sites on the catalyst surface, thus ensuring the high stability of the catalytic performance of Co@150PC.

3.4. DFT Calculations of Co@150PC

By constructing a model of Co₄ clusters on graphene (named Co₄@graphene) to perform DFT calculations, we can further understand the catalysis of the Co@xPC structure in the NaBH₄ hydrolysis process. Although the models of Co₄ clusters and graphene in this experiment are much smaller than the observed Co@xPC nanoparticles, the relative energy of the NaBH₄ hydrolysis process can be discussed. Therefore, the sequential dissociation of BH_x (x = 0→4) molecules on the Co(111) and Co₄@graphene surface was calculated. First, the structure optimization showed that Co₄ clusters could be anchored on the graphene surface to maintain a stable structure with an adsorption energy of −3.34 eV. As shown in different charge density maps (Figure 8a), there is a charge accumulation between graphene and the Co_4 cluster. The graphene gains 0.794 electrons from the Co_4 cluster, indicating that the redistribution of the electron potential of the Co_4@graphene structure is weakening the inert B-H bond and activating $[BH_4]^-$. From the density of states (DOS) plots, an increased electron state of $*BH_4$ DOS at the Fermi level on Co_4@graphene compared with the Co(111) surface can be observed (Figure 8b), indicating that the electrons in Co_4@graphene can efficiently back donate to the unoccupied orbital of $[BH_4]^-$, thus activating the $[BH_4]^-$ molecule. According to Figure 8c, the rate-determining step (RDS) is the BH dissociation step common to both systems in the pathways. Obviously, H abstracting from $*$ BH on the Co(111) and Co_4@graphene surface is endothermic, which requires 0.911 and 0.527 eV of energies, respectively. Hence, the above results fully suggested that the Co_4@graphene structure is favorable to the dissociation of $[BH_4]^-$ molecules.

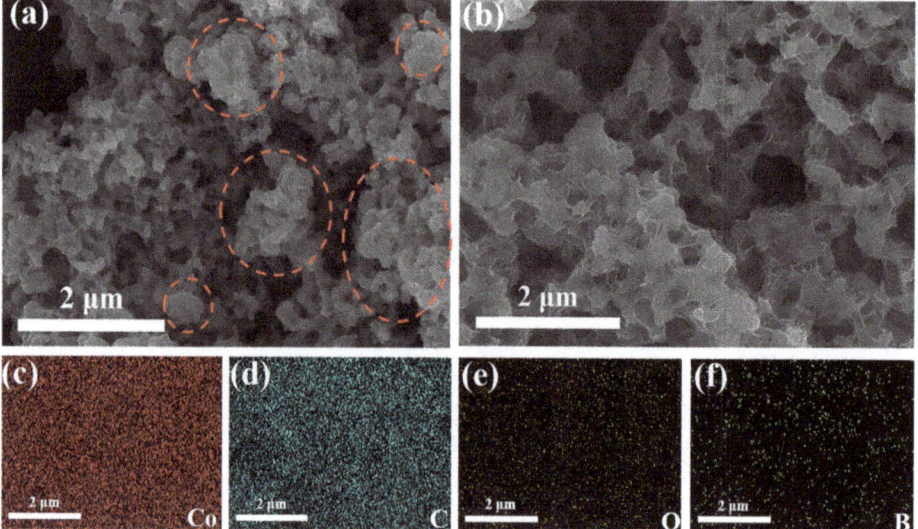

Figure 7. SEM images of Co-5th (**a**), Co@150PC-15th (**b**). Corresponding EDS mapping of Co@150PC-15th (**c–f**).

3.5. Mechanism Analysis on Performance Decrease of Co@150PC during Cycles

To explore the reasons for the decrease in catalyst stability, the used catalysts were characterized after 15 cycles. The dispersed Co particles in Figure 2e have obvious agglomeration after 15 tests. As shown in Figure 7b, the Co nanoparticles on the porous carbon surface became significantly larger, sticking to each other and agglomerating together. By comparing the EDS mappings of Co@150PC (Figure S3) and Co@150PC-15th (Figure 7f), it was found that the B content in Co@150PC-15th increased significantly. ICP-AES analysis further found that the content of B in Co@150PC increased from 0.41 wt.% to 10.9 wt.% after 15 cycles. This indicates that as the number of cycles increased, the content of B continued to increase. At the same time, a $Co^0:Co^{2+}$ atomic ratio is determined to be 1.90:1.00 according to XPS of Co 2p in Co@150PC-15th (Figure 9a). That is to say, most of the cobalt in Co@150PC-15th exists in a reduced state. In the XRD pattern (Figure S5), compared with Co@150PC, Co@150PC-15th also did not show the characteristic peak of Co-B around 46°. This is probably because of the presence of heat in the reaction during the test and because the surface of Co nanoparticles is covered by a film consisting of strongly adsorbed (poly)borates [46].

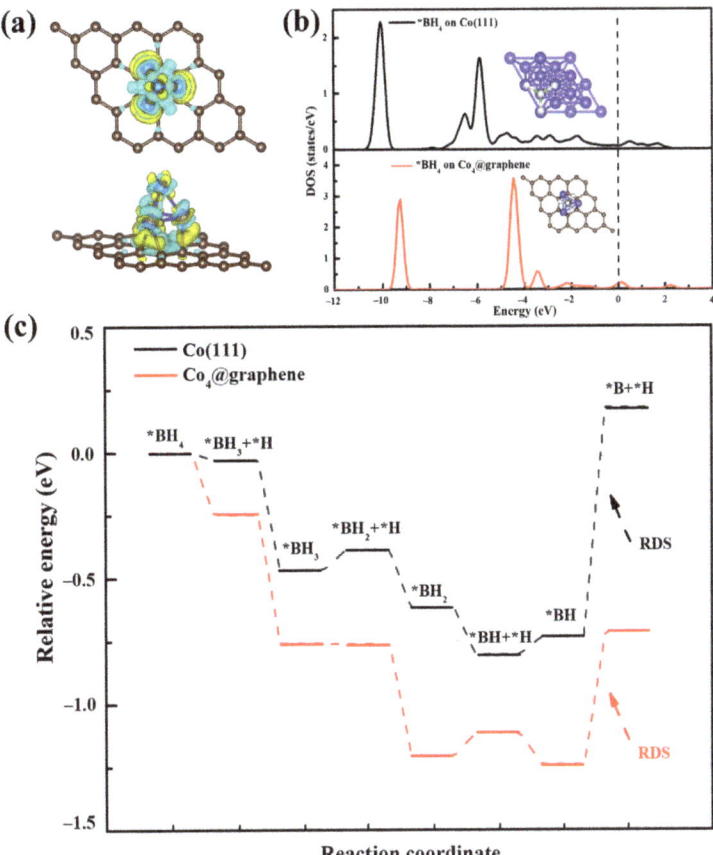

Figure 8. (a) Different charge densities of Co$_4$@graphene. The isovalue value is 0.005 e/Å^3, where the cyan and yellow regions indicate a charge depletion and accumulation, respectively. (b) Density of states of the absorbed *BH$_4$ on Co(111) and Co$_4$@graphene. The dashed line indicates the Fermi level. (c) Potential energy diagram of Co(111) and Co$_4$@graphene.

To verify the above results, we conducted FT-IR and Raman tests. In the FT-IR spectra (Figure 9b), we can see that at 525–630 cm^{-1}, 1349 cm^{-1} and 875 cm^{-1}, Co, Co@150PC and Co@150PC-15th all have pulse vibrations, indicating that they all contain borate [60,61]. Among them, the weaker peak intensity of Co and Co@150PC indicates that these untested catalysts have less borate content, and the peak intensity of Co@150PC-15th also proved that the borate content is higher with the increase in cycles of the experiments. Similarly, in the Raman spectra (Figure 9c), Co@150PC-15th had an obvious Raman characteristic peak [60] of B(OH)$_4^-$ at 754 cm^{-1}, while Co and Co@150PC were present in negligible quantities.

On this basis we infer that Co nanoparticles in the prepared catalyst first existed in the form of metallic cobalt. As the number of experiments increased, the borate content on the surface of the Co nanoparticles gradually increased, forming a layer of a strongly adsorbed (poly)borate shell (Figure 9d), similarly to previous observations [13,47,62]. The existence of this shell reduces the contact area between Co nanoparticles and NaBH$_4$. At the same time, the mutual fusion between the strongly adsorbed (poly)borate shells on the surface of adjacent Co nanoparticles makes the Co nanoparticles stick together, which reduces the dispersion of Co nanoparticles and further affects the catalytic performance of Co@150PC.

Based on the above research, it is found that Co@150PC catalyst can significantly improve the hydrolysis performance of NaBH$_4$. It is attributable to the supporting and inhibiting agglomeration effect of PC with a large specific surface area on Co nanoparticles during the synthesis of Co@150PC catalyst, which effectively reduces the particle size of Co particles and distributes them more uniformly to provide more active sites. Similarly, due to the role of PC with a large specific surface area on the loading of Co nanoparticles and the inhibition of agglomeration, the catalyst Co@150PC can effectively retain the good catalytic performance. The calculation results of DFT also proved that Co nanoparticles can be anchored on the surface of PC to maintain a stable structure, and the Co@xPC structure is conducive to the dissociation of [BH$_4$]$^-$ molecules to realize the rapid water release of hydrogen from NaBH$_4$.

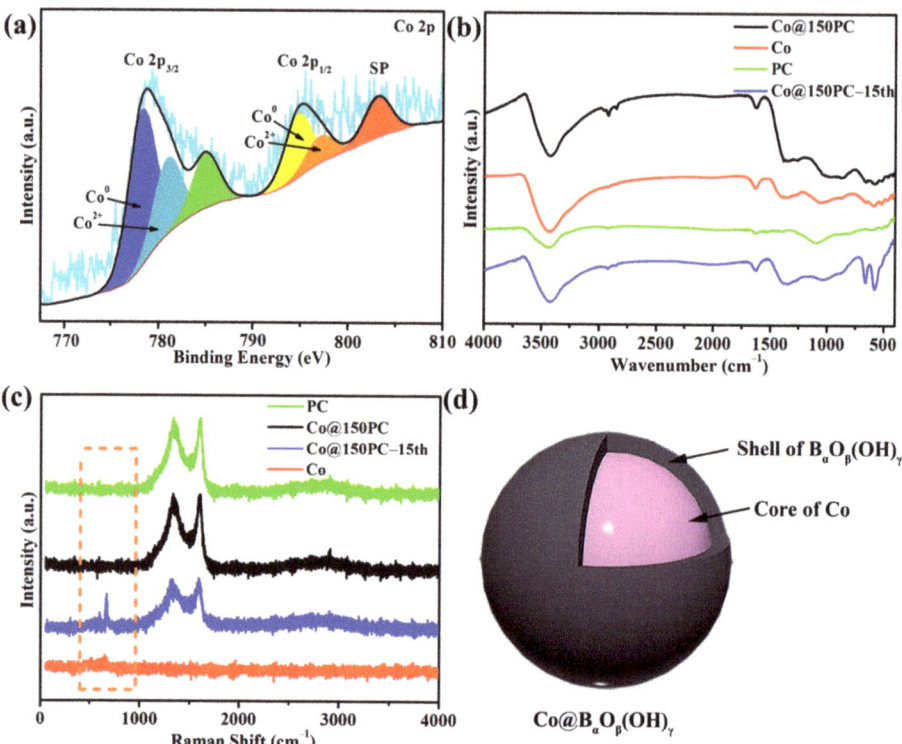

Figure 9. Co 2p XPS spectrum of Co@150PC-15th (**a**). FT-IR spectra (**b**), Raman spectra of the as-prepared catalysts (**c**). Schematic diagram of the structure of Co@B$_\alpha$O$_\beta$(OH)$_\gamma$, a core-shell structure with metallic cobalt as the core and (poly)borate as the outer shell (**d**).

4. Conclusions

The Co@xPC catalyst was synthesized to catalyze NaBH$_4$ hydrolysis for hydrogen evolution. During the synthesis of the Co@xPC catalyst, the PC support can inhibit the agglomeration of Co nanoparticles, which effectively reduced the particle size of the Co nanoparticles and distributed them more uniformly to provide more active sites. Therefore, a superior hydrogen production rate of 11086.4 mL$_{H2}$·min^{-1}·g$_{Co}$$^{-1}$ and a low activation energy of 31.52 kJ mol^{-1} can be obtained for Co@150PC. Furthermore, it still maintained about 72.0% of the initial hydrogen production rate after 15 cycles. The DFT calculation results also indicated that Co@xPC can activate [BH$_4$]$^-$ molecules to promote the rapid dissociation of NaBH$_4$ to release hydrogen. We also found the main reasons for the decrease in the catalytic performance of Co@150PC was the accumulation of borate as a by-product of the reaction during the test, which led to the formation of a strong adsorption (poly)borate shell on the surface of the Co nanoparticles. At the same time, the mutual fusion between the strongly adsorbed (poly)borate shells also weakened the dispersibility of Co nanoparticles, which further decreased the catalytic performance of Co@150PC.

Supplementary Materials: The following are available online at https://www.mdpi.com/article/10.3390/nano11123259/s1, Figure S1. TEM image of Co@150PC; Figure S2. XRD patterns of Co@50PC, Co@100PC and Co@200PC; Figure S3. Corresponding EDS mapping Co@150PC (a–d); Figure S4. C 1s XPS spectrum of Co@150PC; Figure S5. XRD patterns of Co@150PC and Co@150PC-15th.

Author Contributions: Author contributions: Conceptualization, Y.B.; methodology, Y.B.; software, Y.B., J.L., P.H. and X.L.; validation, Y.B. and J.L.; formal analysis, Y.B., L.S. and F.X.; investigation, Y.B., L.S. and F.X.; resources, L.S. and F.X.; data curation, Y.B., Q.Y. and L.K.; writing—original draft preparation, Y.B. and J.L.; writing—review and editing, Y.B., H.C., S.W., F.R., A.A.P., H.J.S., Y.D. and J.W.; visualization, Y.B.; supervision, L.S. and H.C.; project administration, L.S. and F.X.; funding acquisition, L.S. and F.X. All authors have read and agreed to the published version of the manuscript.

Funding: This work has been supported by the National Key Research and Development Program of China (2018YFB1502105 and 2018YFB1502103), the National Natural Science Foundation of China (Grant No. U20A20237, 51971068, 22179026, and 51871065), the Guangxi Natural Science Foundation (No. 2018GXNSFFA281005 and 2018GXNSFDA281051), the Scientific Research and Technology Development Program of Guangxi (AD17195073, AA19182014 and AA17202030-1), the Guangxi Collaborative Innovation Centre of Structure and Property for New Energy and Materials, Guangxi Bagui Scholar Foundation, Chinesisch-Deutsche Kooperationsgruppe (GZ1528), Guangxi Advanced Functional Materials Foundation and Application Talents Small Highlands, the Innovation Project of GUET Graduate Education (2018YJCX88, 2019YCXS114 and 2020YCXS119) and the Study Abroad Program for Graduate Student of Guilin University of Electronic Technology (GDYX2019020).

Data Availability Statement: No data.

Acknowledgments: We acknowledge cooperation with BIOCOKE Lab Co., Ltd., Zaoxiao Zhang and Zhen Wu from Xi'an Jiaotong University. F.R. is grateful to the Canada Research Chairs program for partial salary support.

Conflicts of Interest: The authors declare no conflict of interest.

References

1. Lu, Y.; Chen, J. Prospects of organic electrode materials for practical lithium batteries. *Nat. Rev. Chem.* **2020**, *4*, 127–142. [CrossRef]
2. Das, P.; Wu, Z.S.; Li, F.; Liang, J.; Cheng, H.M. Two-dimensional energy materials: Opportunities and perspectives. *Energy Storage Mater.* **2019**, *22*, 15–17. [CrossRef]
3. Huang, X.; Xiao, X.Z.; Wang, X.Z.; Yao, Z.D.; Wang, C.T.; Fan, X.L.; Chen, L.X. Highly synergetic catalytic mechanism of Ni@g-C$_3$N$_4$ on the superior hydrogen storage performance of Li-Mg-B-H system. *Energy Storage Mater.* **2018**, *13*, 199–206. [CrossRef]
4. Cui, B.Y.; Wu, G.M.; Qiu, S.J.; Zou, Y.J.; Yan, E.H.; Xu, F.; Sun, L.X.; Chu, H.L. Ruthenium Supported on Cobalt-Embedded Porous Carbon with Hollow Structure as Efficient Catalysts toward Ammonia-Borane Hydrolysis for Hydrogen Production. *Adv. Sustain. Syst.* **2021**, *5*, 2100209. [CrossRef]
5. Wen, Z.Y.; Fu, Q.; Wu, J.; Fan, G.Y. Ultrafine Pd nanoparticles supported on soft nitriding porous carbon for hydrogen production from hydrolytic dehydrogenation of dimethyl amine-borane. *Nanomaterials* **2020**, *10*, 1612. [CrossRef]
6. Qiu, S.J.; Chu, H.L.; Zou, Y.L.; Xiang, C.L.; Xu, F.; Sun, L.X. Light metal borohydrides/amides combined hydrogen storage systems: Composition, structure and properties. *J. Mater. Chem. A* **2017**, *5*, 25112–25130. [CrossRef]
7. Armaroli, N.; Balzani, V. The future of energy supply: Challenges and opportunities. *Angew. Chem. Int. Ed.* **2007**, *46*, 52–66. [CrossRef]
8. Abe, J.O.; Popoola, A.P.I.; Ajenifuja, E.; Popoola, O.M. Hydrogen energy, economy and storage: Review and recommendation. *Int. J. Hydrog. Energy* **2019**, *44*, 15072–15086. [CrossRef]
9. Zhong, H.; Ouyang, L.Z.; Ye, J.S.; Liu, J.W.; Wang, H.; Yao, X.D.; Zhu, M. An one-step approach towards hydrogen production and storage through regeneration of NaBH$_4$. *Energy Storage Mater.* **2017**, *7*, 222–228. [CrossRef]
10. Fangaj, E.; Ceyhan, A.A. Apricot Kernel shell waste treated with phosphoric acid used as a green, metal-free catalyst for hydrogen generation from hydrolysis of sodium borohydride. *Int. J. Hydrog. Energy* **2020**, *45*, 17104–17117. [CrossRef]

11. Guella, G.; Zanchetta, C.; Patton, B.; Miotello, A. New insights on the mechanism of palladium-catalyzed hydrolysis of sodium borohydride from B-11 NMR measurements. *J. Phys. Chem. B* **2006**, *110*, 17024–17033. [CrossRef]
12. Muir, S.S.; Yao, X.D. Progress in sodium borohydride as a hydrogen storage material: Development of hydrolysis catalysts and reaction systems. *Int. J. Hydrog. Energy* **2011**, *36*, 5983–5997. [CrossRef]
13. Zhang, H.M.; Xu, G.C.; Zhang, L.; Wang, W.F.; Miao, W.K.; Chen, K.L.; Cheng, L.N.; Li, Y.; Han, S.M. Ultrafine cobalt nanoparticles supported on carbon nanospheres for hydrolysis of sodium borohydride. *Renew. Energy* **2020**, *162*, 345–354. [CrossRef]
14. Kılınç, D.; Şahin, O. Synthesis of polymer supported Ni(II)-Schiff Base complex and its usage as a catalyst in sodium borohydride hydrolysis. *Int. J. Hydrog. Energy* **2018**, *43*, 10717–10727. [CrossRef]
15. Kılınç, D.; Saka, C.; Şahin, O. Hydrogen generation from catalytic hydrolysis of sodium borohydride by a novel Co(II)–Cu(II) based complex catalyst. *J. Power Sources* **2012**, *217*, 256–261. [CrossRef]
16. Ocon, J.D.; Tuan, T.N.; Yi, Y.M.; de Leon, R.L.; Lee, J.K.; Lee, J. Ultrafast and stable hydrogen generation from sodium borohydride in methanol and water over Fe–B nanoparticles. *J. Power Sources* **2013**, *243*, 444–450. [CrossRef]
17. Shi, L.M.; Xie, W.; Jian, Z.Y.; Liao, X.M.; Wang, Y.J. Graphene modified Co–B catalysts for rapid hydrogen production from NaBH$_4$ hydrolysis. *Int. J. Hydrog. Energy* **2019**, *44*, 17954–17962. [CrossRef]
18. Guo, J.; Hou, Y.J.; Li, B.; Liu, Y.L. Novel Ni–Co–B hollow nanospheres promote hydrogen generation from the hydrolysis of sodium borohydride. *Int. J. Hydrog. Energy* **2018**, *43*, 15245–15254. [CrossRef]
19. Marchionni, A.; Bevilacqua, M.; Filippi, J.; Folliero, M.G.; Innocenti, M.; Lavacchi, A.; Miller, H.A.; Pagliaro, M.V.; Vizza, F. High volume hydrogen production from the hydrolysis of sodium borohydride using a cobalt catalyst supported on a honeycomb matrix. *J. Power Sources* **2015**, *299*, 391–397. [CrossRef]
20. Ai, L.H.; Gao, X.Y.; Jiang, J. In situ synthesis of cobalt stabilized on macroscopic biopolymer hydrogel as economical and recyclable catalyst for hydrogen generation from sodium borohydride hydrolysis. *J. Power Sources* **2014**, *257*, 213–220. [CrossRef]
21. Yang, C.C.; Chen, M.S.; Chen, Y.W. Hydrogen generation by hydrolysis of sodium borohydride on CoB/SiO$_2$ catalyst. *Int. J. Hydrog. Energy* **2011**, *36*, 1418–1423. [CrossRef]
22. Ye, W.; Zhang, H.M.; Xu, D.Y.; Ma, L.; Yi, B.L. Hydrogen generation utilizing alkaline sodium borohydride solution and supported cobalt catalyst. *J. Power Sources* **2007**, *164*, 544–548. [CrossRef]
23. Rakap, M.; Ozkar, S. Hydroxyapatite-supported cobalt(0) nanoclusters as efficient and cost-effective catalyst for hydrogen generation from the hydrolysis of both sodium borohydride and ammonia-borane. *Cataly. Today* **2012**, *183*, 17–25. [CrossRef]
24. Chu, H.L.; Li, N.P.; Qiu, X.Y.; Wang, Y.; Qiu, S.J.; Zeng, J.L.; Zou, Y.J.; Xu, F.; Sun, L.X. Poly(N-vinyl-2-pyrrolidone)-stabilized ruthenium supported on bamboo leaf-derived porous carbon for NH$_3$BH$_3$ hydrolysis. *Int. J. Hydrog. Energy* **2019**, *44*, 29255–29262. [CrossRef]
25. Lu, Y.H.; Zhang, S.L.; Yin, J.M.; Bai, C.C.; Zhang, J.H.; Li, Y.X.; Yang, Y.; Ge, Z.; Zhang, M.; Wei, L.; et al. Mesoporous activated carbon materials with ultrahigh mesopore volume and effective specific surface area for high performance supercapacitors. *Carbon* **2017**, *124*, 64–71. [CrossRef]
26. Wei, S.; Xue, S.S.; Huang, S.S.; Huang, C.S.; Chen, B.Y.; Zhang, H.Z.; Sun, L.X.; Xu, F.; Xia, Y.P.; Cheng, R.G.; et al. Multielement synergetic effect of NiFe$_2$O$_4$ and h-BN for improving the dehydrogenation properties of LiAlH$_4$. *Inorg. Chem. Front.* **2021**, *8*, 3111–3126. [CrossRef]
27. Bandal, H.A.; Jadhav, A.R.; Chaugule, A.A.; Chung, W.J.; Kim, H. Fe$_2$O$_3$ hollow nanorods/CNT composites as an efficient electrocatalyst for oxygen evolution reaction. *Electrochim. Acta* **2016**, *222*, 1316–1325. [CrossRef]
28. White, R.J.; Luque, R.; Budarin, V.L.; Clark, J.H.; Macquarrie, D.J. Supported metal nanoparticles on porous materials. Methods and applications. *J. Cheminform.* **2009**, *38*, 481–494. [CrossRef]
29. Alaf, M.; Gultekin, D.; Akbulut, H. Electrochemical properties of free-standing Sn/SnO$_2$/multi-walled carbon nano tube anode papers for Li-ion batteries. *Appl. Surf. Sci.* **2013**, *275*, 244–251. [CrossRef]
30. Heena, H.; Lombardo, L.; Luo, W.; Kin, W.; Zuttel, A. Hydrogen storage properties of various carbon supported NaBH$_4$ prepared via metathesis. *Int. J. Hydrog. Energy* **2018**, *43*, 7108–7116.
31. Li, J.H.; Hong, X.Y.; Wang, Y.L.; Luo, Y.M.; Huang, P.R.; Li, B.; Zhang, K.X.; Zou, Y.J.; Sun, L.X.; Xu, F.; et al. Encapsulated cobalt nanoparticles as a recoverable catalyst for the hydrolysis of sodium borohydride. *Energy Storage Mater.* **2020**, *27*, 187–197. [CrossRef]
32. ElSheikh, A.M.A.; Backovic, G.; Oliveira, R.C.P.; Sequeira, C.A.C.; McGregor, J.; Sljukic, B.; Santos, D.M.F. Carbon-Supported Trimetallic Catalysts (PdAuNi/C) for Borohydride Oxidation Reaction. *Nanomaterials* **2021**, *14*, 1441. [CrossRef] [PubMed]
33. Fang, F.; Chen, Z.L.; Wu, D.Y.; Liu, H.; Dong, C.K.; Song, Y.; Sun, D.L. Subunit volume control mechanism for dehydrogenation performance of AB$_3$-type superlattice intermetallics. *J. Power Sources* **2019**, *427*, 145–153. [CrossRef]
34. Blochl, P.E. Projector augmented-wave method. *Phys. Rev. B* **1994**, *50*, 17953–17979. [CrossRef] [PubMed]
35. Kresse, G.; Hafner, J. Ab initio molecular-dynamics simulation of the liquid-metal–amorphous-semiconductor transition in germanium. *Phys. Rev. B* **1994**, *49*, 14251–14269. [CrossRef] [PubMed]
36. Kresse, G.; Furthmüller, J. Efficient iterative schemes for ab initio total-energy calculations using a plane-wave basis set. *Phys. Rev. B* **1996**, *54*, 11169–11186. [CrossRef] [PubMed]
37. Perdew, J.P.; Burke, K.; Ernzerhof, M. Generalized gradient approximation made simple. *Phys. Rev. Lett.* **1997**, *78*, 3865–3868. [CrossRef]

38. Zhong, S.L.; Ju, S.L.; Shao, Y.F.; Chen, W.; Zhang, T.F.; Huang, Y.Q.; Zhang, H.Y.; Xia, G.L.; Yu, X.B. Magnesium hydride nanoparticles anchored on MXene sheets as high capacity anode for lithium-ion batteries. *J. Energy Chem.* **2021**, *62*, 431–439. [CrossRef]
39. Tang, W.; Sanville, E.; Henkelman, G. A grid-based Bader analysis algorithm without lattice bias. *J. Phys. Condens. Mat.* **2009**, *21*, 084204. [CrossRef]
40. Akça, A.; Genç, A.E.; Kutlu, B. BH_4 dissociation on various metal (111) surfaces: A DFT study. *Appl. Surf. Sci.* **2019**, *473*, 681–692. [CrossRef]
41. Zou, Y.J.; Yin, Y.; Gao, Y.B.; Xiang, C.L.; Chu, H.L.; Qiu, S.J.; Yan, E.H.; Xu, F.; Sun, L.X. Chitosan-mediated Co–Ce–B nanoparticles for catalyzing the hydrolysis of sodium borohydride. *Int. J. Hydrog. Energy* **2018**, *43*, 4912–4921. [CrossRef]
42. Chen, B.; Chen, S.J.; Bandal, H.A.; Appiah-Ntiamoah, R.; Jadhav, A.R.; Kim, H.; Appiah-Ntiamoah, A.R.; Jadhav, H.K. Cobalt nanoparticles supported on magnetic core-shell structured carbon as a highly efficient catalyst for hydrogen generation from $NaBH_4$ hydrolysis. *Int. J. Hydrog. Energy* **2018**, *43*, 9296–9306. [CrossRef]
43. Liu, W.L.; Ju, S.L.; Yu, X.B. Phosphorus-Amine-Based Synthesis of Nanoscale Red Phosphorus for Application to Sodium-Ion Batteries. *ACS Nano* **2020**, *14*, 974–984. [CrossRef]
44. Li, L.; Wang, H.L.; Xie, Z.J.; An, C.H.; Jiang, G.X.; Wang, Y.J. 3D graphene-encapsulated nearly monodisperse Fe_3O_4 nanoparticles as high-performance lithium-ion battery anodes. *J. Alloy. Compd.* **2020**, *815*, 152337. [CrossRef]
45. Liang, C.; Chen, Y.; Wu, M.; Wang, K.; Zhang, W.K.; Gan, Y.P.; Huang, H.; Chen, J.; Xia, Y.; Zhang, J.; et al. Green synthesis of graphite from CO_2 without graphitization process of amorphous carbon. *Nat. Commun.* **2021**, *12*, 119. [CrossRef]
46. Zhang, H.M.; Zhang, L.; Rodriguez-Perez, I.A.; Miao, W.K.; Chen, K.L.; Wang, W.F.; Li, Y.; Han, S.M. Carbon nanospheres supported bimetallic Pt-Co as an efficient catalyst for $NaBH_4$ hydrolysis. *Appl. Surf. Sci.* **2021**, *540*, 148296. [CrossRef]
47. Baydaroglu, F.; Ozdemir, E.; Hasimoglu, A. An effective synthesis route for improving the catalytic activity of carbon-supported Co-B catalyst for hydrogen generation through hydrolysis of $NaBH_4$. *Int. J. Hydrog. Energy* **2014**, *39*, 1516–1522. [CrossRef]
48. Peng, C.L.; Li, T.S.; Zou, Y.J.; Xiang, C.L.; Xu, F.; Zhang, J.; Sun, L.X. Bacterial cellulose derived carbon as a support for catalytically active Co-B alloy for hydrolysis of sodium borohydride. *Int. J. Hydrog. Energy* **2021**, *46*, 666–675. [CrossRef]
49. Wei, Y.S.; Wang, R.; Meng, L.Y.; Wang, Y.; Li, G.D.; Xin, S.G.; Zhao, X.S.; Zhang, K. Hydrogen generation from alkaline $NaBH_4$ solution using a dandelion-like Co-Mo-B catalyst supported on carbon cloth. *Int. J. Hydrog. Energy* **2016**, *42*, 9945–9951. [CrossRef]
50. Wang, F.H.; Wang, Y.A.; Zhang, Y.J.; Luo, Y.M.; Zhu, H. Highly dispersed RuCo bimetallic nanoparticles supported on carbon black: Enhanced catalytic activity for hydrogen generation from $NaBH_4$ methanolysis. *J. Mater. Sci.* **2018**, *53*, 6831–6841. [CrossRef]
51. Huang, Y.Q.; Wang, Y.; Zhao, R.X.; Shen, P.K.; Wei, Z.D. Accurately measuring the hydrogen generation rate for hydrolysis of sodium borohydride on multiwalled carbon nanotubes/Co-B catalysts. *Int. J. Hydrog. Energy* **2008**, *33*, 7110–7115. [CrossRef]
52. Xu, J.N.; Du, X.X.; Wei, Q.L.; Huang, Y.M. Efficient Hydrolysis of Sodium Borohydride by Co-B Supported on Nitrogen-doped Carbon. *Chem. Sel.* **2020**, *5*, 6683–6690. [CrossRef]
53. Zhang, H.M.; Feng, X.L.; Cheng, L.N.; Hou, X.W.; Li, Y.; Han, S.M. Non-noble Co anchored on nanoporous graphene oxide, as an efficient and long-life catalyst for hydrogen generation from sodium borohydride. *Colloids Surf. A Physicochem. Eng. Asp.* **2019**, *563*, 112–119. [CrossRef]
54. Makiabadi, M.; Shamspur, T.; Mostafavi, A. Performance improvement of oxygen on the carbon substrate surface for dispersion of cobalt nanoparticles and its effect on hydrogen generation rate via $NaBH_4$ hydrolysis. *Int. J. Hydrog. Energy* **2020**, *45*, 1706–1718. [CrossRef]
55. Zhu, J.; Li, R.; Niu, W.L.; Wu, Y.J.; Gou, X.L. Fast hydrogen generation from $NaBH_4$ hydrolysis catalyzed by carbon aerogels supported cobalt nanoparticles. *Int. J. Hydrog. Energy* **2013**, *38*, 10864–10870. [CrossRef]
56. Xu, D.Y.; Dai, P.; Liu, X.M.; Cao, C.Q.; Guo, Q.J. Carbon-supported cobalt catalyst for hydrogen generation from alkaline sodium borohydride solution. *J. Power Sources* **2008**, *182*, 616–620. [CrossRef]
57. Netskina, O.V.; Kochubey, D.I.; Prosvirin, I.P.; Malykhin, S.E.; Komova, O.V.; Kanazhevskiy, V.V.; Chukalkin, Y.G.; Bobrovskii, V.I.; Kellermann, D.G.; Ishchenko, A.V.; et al. Cobalt-boron catalyst for $NaBH_4$ hydrolysis: The state of the active component forming from cobalt chloride in a reaction medium. *Mol. Catal.* **2017**, *441*, 100–108. [CrossRef]
58. Retnamma, R.; Novais, A.Q.; Rangel, C.M. Kinetics of hydrolysis of sodium borohydride for hydrogen production in fuel cell applications: A review. *Int. J. Hydrog. Energy* **2011**, *36*, 9772–9790. [CrossRef]
59. Demirci, U.B.; Miele, P. Reaction mechanisms of the hydrolysis of sodium borohydride: A discussion focusing on cobalt-based catalysts. *CR Chim.* **2014**, *17*, 707–716. [CrossRef]
60. Jia, Y.Z.; Gao, S.Y.; Xia, S.P.; Li, J. FT-IR spectroscopy of supersaturated aqueous solutions of magnesium borate. *Spectrochim. Acta Part A* **2000**, *56*, 1291–1297.
61. Sasaki, K.; Toshiyuki, K.; Guo, B.L.; Ideta, K.; Hayashi, Y.; Hirajima, T.; Miyawaki, J. Calcination effect of borate-bearing hydroxyapatite on the mobility of borate. *J. Hazard. Mater.* **2018**, *344*, 90–97. [CrossRef]
62. Akdim, O.; Demirci, U.B.; Miele, P. Deactivation and reactivation of cobalt in hydrolysis of sodium borohydride. *Int. J. Hydrog. Energy* **2012**, *36*, 13669–13675. [CrossRef]

Article

$Li_4(OH)_3Br$-Based Shape Stabilized Composites for High-Temperature TES Applications: Selection of the Most Convenient Supporting Material

Imane Mahroug [1,2,3,4,*], Stefania Doppiu [1], Jean-Luc Dauvergne [1], Angel Serrano [1] and Elena Palomo del Barrio [1,5]

1. Centre for Cooperative Research on Alternative Energies (CIC energiGUNE), Basque Research and Technology Alliance (BRTA), Alava Technology Park, 01510 Vitoria-Gasteiz, Spain; sdoppiu@cicenergigune.com (S.D.); jldauvergne@cicenergigune.com (J.-L.D.); aserrano@cicenergigune.com (A.S.); epalomo@cicenergigune.com (E.P.d.B.)
2. INP Bordeaux, UMR 5295, I2M, CNRS, Avenue Pey-Berland 16, 33607 Pessac, France
3. UMR 5295, I2M, CNRS, University of Bordeaux, Esplanade des Arts et Métiers, 33405 Talence, France
4. Department of Applied physics, University of the Basque Country UPV-EHU, 48940 Leioa, Spain
5. Ikerbasque, Basque Foundation for Science, 48013 Bilbao, Spain
* Correspondence: imahroug@cicenergigune.com

Citation: Mahroug, I.; Doppiu, S.; Dauvergne, J.-L.; Serrano, A.; Palomo del Barrio, E. $Li_4(OH)_3Br$-Based Shape Stabilized Composites for High-Temperature TES Applications: Selection of the Most Convenient Supporting Material. *Nanomaterials* **2021**, *11*, 1279. https://doi.org/10.3390/nano11051279

Academic Editors: Federico Cesano, Simas Rackauskas and Mohammed Jasim Uddin

Received: 16 April 2021
Accepted: 11 May 2021
Published: 13 May 2021

Publisher's Note: MDPI stays neutral with regard to jurisdictional claims in published maps and institutional affiliations.

Copyright: © 2021 by the authors. Licensee MDPI, Basel, Switzerland. This article is an open access article distributed under the terms and conditions of the Creative Commons Attribution (CC BY) license (https://creativecommons.org/licenses/by/4.0/).

Abstract: Peritectic compound $Li_4(OH)_3Br$ has been recently proposed as phase change material (PCM) for thermal energy storage (TES) applications at approx. 300 °C Compared to competitor PCM materials (e.g., sodium nitrate), the main assets of this compound are high volumetric latent heat storage capacity (>140 kWh/m^3) and very low volume changes (<3%) during peritectic reaction and melting. The objective of the present work was to find proper supporting materials able to shape stabilize $Li_4(OH)_3Br$ during the formation of the melt and after its complete melting, avoiding any leakage and thus obtaining a composite apparently always in the solid state during the charge and discharge of the TES material. Micro-nanoparticles of MgO, Fe_2O_3, CuO, SiO_2 and Al_2O_3 have been considered as candidate supporting materials combined with the cold-compression route for shape-stabilized composites preparation. The work carried out allowed for the identification of the most promising composite based on MgO nanoparticles through a deep experimental analysis and characterization, including chemical compatibility tests, anti-leakage performance evaluation, structural and thermodynamic properties analysis and preliminary cycling stability study.

Keywords: peritectic compound $Li_4(OH)_3Br$; phase change materials; thermal energy storage; shape stabilized composites; supporting materials; oxides

1. Introduction

High-temperature thermal energy storage (HT-TES) is part of the storage solution that is expected to be deployed in future energy systems. HT-TES is currently used in concentrating solar thermal power plants to warrant dispatchability. Moreover, using HT-TES is also envisaged in conventional thermal power plants to provide them with greater operational flexibility. It is also expected for HT-TES to provide a second life to coal-fired plants, which are being closed for environmental reasons, and that participate as well in the emergence of stand-alone energy storage plants in the grid, where it has a cost advantage over other technologies. In the industrial sector, in addition to the already known uses of recovery and valorization of waste heat and improvement of the overall efficiency of cogeneration systems or steam boilers, the use of HT-TES associated with increasing solarization and/or electrification of heat and cold production will be added.

The development of HT-TES technologies has been closely linked to the development and deployment of concentrating solar thermal power plants [1,2]. The technology that currently dominates the market is molten salt due to the ability of nitrate mixtures to operate at temperatures up to 450–560 °C at reasonable cost, despite the drawbacks of the risk of salt solidification and inherent corrosion problems. At lower temperatures (up to 400 °C), thermal storage in concrete is also a commercially available option. Fixed packed-bed systems, which use rocks or solid industrial by-products as granular filler material, are in the process of being commercialized. Compared to molten salts and concrete, their main advantages are much lower investment costs and the possibility to operate in a very wide temperature range (up to 1000 °C).

All of the above technologies store/deliver energy by increasing/decreasing the temperature of the storage material (sensible heat storage) and therefore suffer from a lack of compactness (low storage capacity per unit volume). In this regard, latent heat storage technologies based on high-temperature phase change materials (HT-PCMs) could meet the desired compactness objectives while keeping costs affordable. Inorganic anhydrous salts and their mixtures are the most commonly investigated HT-PCMs [3–6]. They store thermal energy in an almost isothermal manner during melting, and they return it back during the reverse process of solidification. Moreover, they are usually characterized by high enthalpies of fusion, high density, excellent thermochemical stability and low price. However, they often suffer from some shortcomings. The most important ones are corrosivity and low thermal conductivity (<1 W/m/K). Corrosion involves using expensive corrosion-resistant materials for the storage tank and heat exchanger, while low thermal conductivity has to be compensated by oversizing the latter, thus increasing investment cost.

To overcome such problems, different techniques for encapsulating high-temperature salts are being investigated. They can be classified into two main categories, namely core–shell microencapsulation [7–11], where the shell acts as a container to prevent liquid leakage, and so-called shape-stabilized composite materials (ss-composites) [8,9,11–13], where a porous supporting material encapsulates the salt and retain the liquid phase by capillary forces and surface tension. Compared with core–shell microencapsulation, ss-composites have clear advantages regarding production cost and performance. Indeed, they are generally produced by simple melting infiltration in a porous support or by cold compression of a mixture of the supporting material micro and/or nanoparticles and salt powders. Moreover, they usually have the ability of self-management of the volume changes of the salt during phase transitions, which is one of the major concerns of core–shell microencapsulation, and they lead to higher apparent thermal conductivity enhancement.

The selection of the supporting material is critical for successful ss-composites [12]. The thermal stability in the planned working temperature range and the chemical compatibility with encapsulated salt are the most basic requirements, which directly determine the usability of the supporting material. Good wettability with loaded salt and high specific surface area is also of great importance because it determines maximum salt loading and, therefore, the latent heat storage capacity of the final material. To a lesser extent, high thermal conductivity is also advisable to reduce the size of heat exchangers or, alternatively, to maximize the size of pellets/grains in fixed packed-bed storage systems. Obviously, economic and safety aspects are also relevant, therefore, safe (non-corrosive, non-flammable, non-explosive) and inexpensive supporting materials that are easy to obtain and process are required as well.

Supporting material used in ss-composites studied so far can be classified into three main groups:

- Carbon-based supporting materials, such as expanded graphite and graphite foams [14–28]. They have proven to be compatible with nitrites and chlorides and have high salt absorption capacity (>85 wt.%). Furthermore, they are excellent in heat transfer enhancement because of their high thermal conductivity (up to 100 W/m/K reported). However, they are characterized by poor wettability with salts, and they also tend to oxidize at temperatures above approx. 600 °C. The ss-composites using expanded graphite are usually prepared by the uniaxial or isostatic cold-compression

route, whereas the vacuum-assisted melting infiltration method is used in the preparation of graphite foam-based composites.
- Clay mineral supporting materials such as expanded perlite, expanded vermiculite and diatomite [9,29–36]. They demonstrate chemical compatibility with nitrites, chlorides and sulfates as well as high salt absorption capacity (>85 wt.% for expanded perlite and vermiculite; 55–70 wt.% for diatomite). Moreover, the wettability with molten salts is good, and they can support temperatures above 1000 °C. However, they have low values of thermal conductivity (<0.15 W/m/K), and the melting infiltration route is needed for ss-composite preparation, which is more expensive than the cold compression method. Another type of clay mineral used as an additive for PCM composites is natural halloysite nanoclay. These materials are characterized by good thermal stability, a high adsorption capacity and low cost. Halloysite nanoclay is used generally as a nucleating agent to mitigate the supercooling phenomena of the hydrate PCMs and is applied for cold storage [37,38]
- Other supporting materials including refractory oxides (MgO, Al_2O_3, SiO_2, mullite), SiC and $Ca(OH)_2$ [11,13,39–57]. The compatibility and good wettability with nitrates, carbonates, chlorides and sulfates have been proven for most of them. Maximum salt loading is lower than in previous cases, but still significant (up to 70 wt.%). On the contrary, they show excellent thermal stability up to 1400–1600 °C (only 570 °C for calcium hydroxide). In addition, they have relatively high thermal conductivity (3–65 W/m/K), and corresponding ss-composites are prepared by the cold-compression route.

The present work deals with the peritectic compound $Li_4(OH)_3Br$ recently proposed as an HT-PCM for TES applications at approx. 300 °C [58–60]. Compared to sodium nitrate, which is the reference HT-PCM for this temperature level [61], $Li_4(OH)_3Br$ has two main advantages that make it particularly attractive [60]. On the one hand, the volumetric latent heat storage capacity of $Li_4(OH)_3Br$ (141.3 kWh/m^3) is 54% higher than that of $NaNO_3$, meaning that $Li_4(OH)_3Br$ offers the opportunity for reducing significantly the volume of the storage tank. On the other hand, whereas the volume expansion on melting of $NaNO_3$ is quite high (almost 11%), that of $Li_4(OH)_3Br$ is only 3%. This work aims at selecting suitable supporting materials for $Li_4(OH)_3Br$-based ss-composites obtained by the cold-compression route, which is the most advantageous from an economic point of view. Candidate supporting materials considered are micro and/or nanoparticles of MgO, Fe_2O_3, CuO, SiO_2 and Al_2O_3. An experimental screening, including chemical compatibility with $Li_4(OH)_3Br$ analysis, anti-leakage performance and maximum salt loading evaluation as well as thermal cycling stability of corresponding ss-composites, was performed to select the best candidate.

2. Materials and Methods

2.1. Materials

High purity anhydrous lithium hydroxide (CAS: 1310-65-2, purity 98%) and lithium bromide (CAS: 7550-35-8, purity 99+%), both provided by Acros Organics (Geel, Belgium), were used in the preparation of the peritectic compound $Li_4(OH)_3Br$. The synthesis was performed following the method proposed by Mahroug et al. [60]. A powder mixture of LiOH and LiBr (5 g approx.) was prepared under a protective argon atmosphere by weighing the right weight fraction of each component (mole ratio 75LiOH:25LiBr) using a Sartorius balance (±0.1 mg). The mixture was then homogenized by ball milling for 15 min using a Spex mixer mill (875 rpm, Spexsampleprep, Metuchen, NJ, USA) using stainless-steel vials and stainless-steel balls (3 balls of 3 mm BPR = 0.5) under mild conditions. The mixture was then introduced inside corundum crucibles put inside stainless steel reactors and sealed under argon. The synthesis was performed inside the furnace applying the following temperature program: (i) a heating ramp at 10 °C/min from ambient temperature up to 30 °C above the melting temperature of $Li_4(OH)_3Br$; (ii) an isothermal step of 1 h; (iii) a cooling step up to room temperature with a cooling rate of 1.8 °C/min. Key thermophysical properties of $Li_4(OH)_3Br$ are gathered in Table 1.

Table 1. Main storage-related properties of $Li_4(OH)_3Br$ [60].

Peritectic Temperature (°C)	289
Melting point (°C)	340
Thermal conductivity at room temperature (W/m/K)	0.47
Specific heat in solid close to the peritectic temperature (J/g/K)	1.68
Density in solid close to the peritectic temperature (g/cc)	1.85

As supporting material for shape stabilization, several commercial oxides were tested with different particle sizes (nano/microparticles). General information about the tested materials is presented in Table 2.

Table 2. General information about the tested oxides.

Material	MgO	Fe_2O_3	CuO	SiO_2	Al_2O_3
Supplier	Alfa Aesar Kandel, Germany	Sigma Aldrich St. Louis, MO, USA	Alfa Aesar Kandel, Germany	Sigma Aldrich St. Louis, MO, USA	Sigma Aldrich St. Louis, MO, USA
CAS number	1309-48-4	1309-37-1	1317-38-0	7631-86-9	1344-28-1
Purity (%)	99+%	≥99%	99.7%	>95%	
Particle size	100 nm	<5 μm	<74 μm	12 nm	13 nm
ρ (g/cm^3)	3.58	5.12	6.315	2.2–2.6	3.95

The morphology of the supporting materials was studied by Scanning Electron Microscopy (SEM) using a Quanta 200 FEG (FEI, Hillsboro, OR, USA) scanning electron microscope operated in high vacuum mode at 20 kV, with a back scattered electron detector (BSED). Particles size and morphology of the oxide supporting materials (MgO, Fe_2O_3, CuO, Al_2O_3 and SiO_2) were studied by analyzing SEM images of the samples using ImageJ 2.0 software [62]. As can be seen in Figure 1, Al_2O_3 and SiO_2 nanopowders formed spherical clusters with well-distributed cluster sizes (Figures 1d and 1e, respectively); the particle size provided by the supplier was considered in the study. In the case of MgO nanopowder, Figure 1a shows crystal agglomerates of MgO with crystal size less than 1 μm. Fe_2O_3 (see Figure 1b) presented spherical powder with uniform distribution of the particle size (<4 μm), whereas CuO showed a large distribution of particle size (<13 μm).

$Li_4(OH)_3Br$-based composites were prepared by the uniaxial cold compression route. The storage material $Li_4(OH)_3Br$ was initially grounded and sifted using a 200 μm sieve, and then it was mixed with the oxide according to specific mass ratios to obtain a total mass of 1 g of composite. The powder mixture was then physically mixed for 20 min using a ball mill (Spex mixer mill 875 RPM) without balls. The purpose of the physical mixing was to ensure the homogeneity of the oxide/salt mixture. Finally, a pellet of 13 mm diameter was made by cold compression of the powder mixture under a pressure of 5 tons for 1 min. The pellet was then placed in a corundum crucible inside a closed stainless steel reactor under Ar atmosphere and sintered in a muffle furnace over the melting temperature of the salt according to the following temperature program: a first heating step at 10 °C/min up to 350 °C, followed by an isothermal step at 350 °C for 1 h, and finally the sample was cooled down to room temperature at around 2 K/min cooling rate.

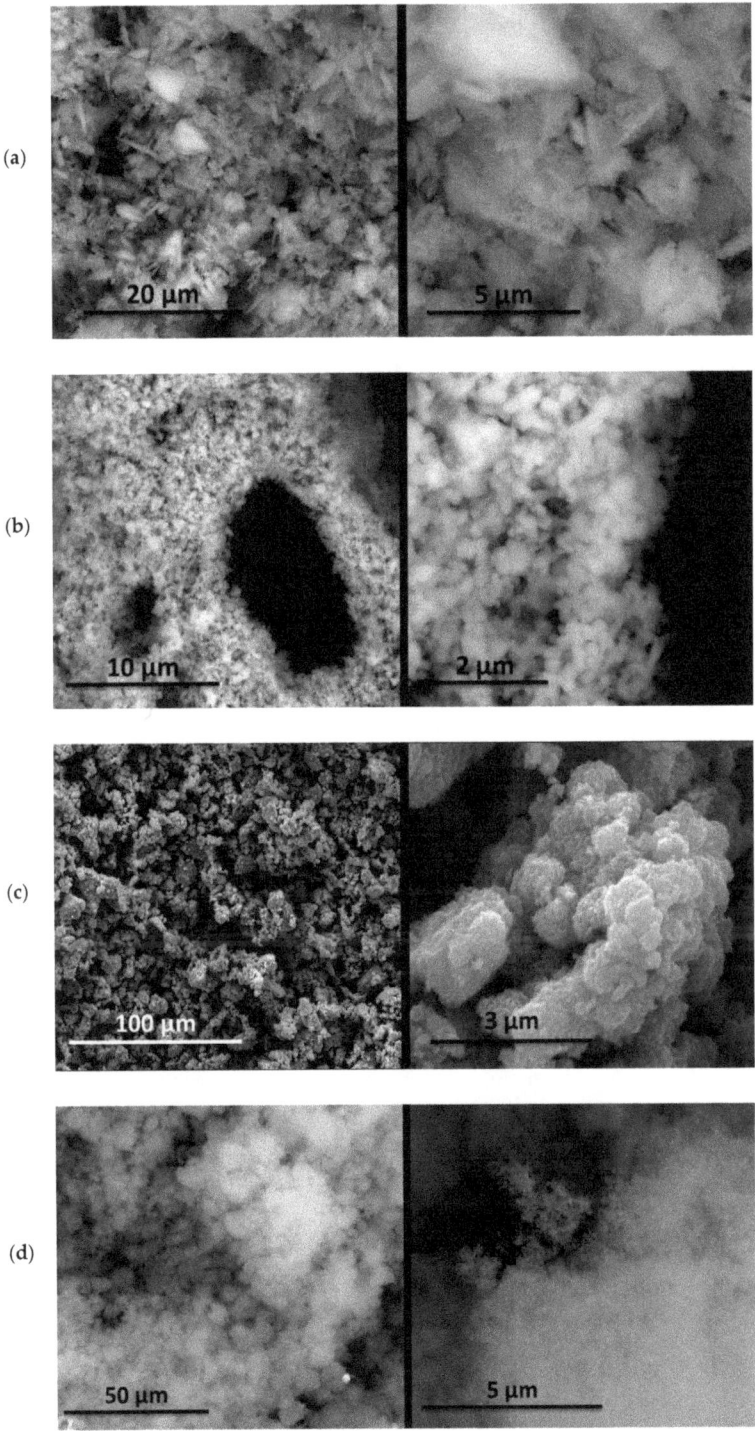

Figure 1. *Cont.*

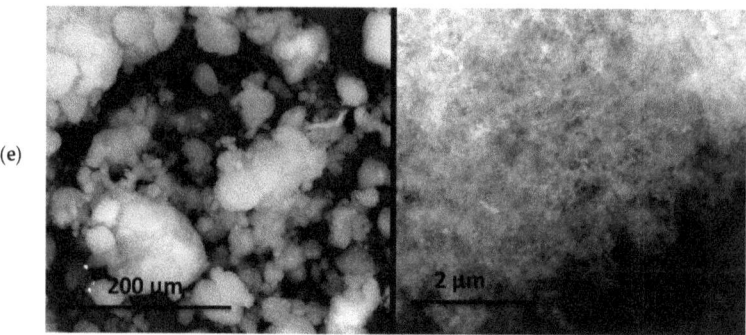

Figure 1. SEM images of the tested oxides: (**a**) MgO; (**b**) Fe$_2$O$_3$; (**c**) CuO; (**d**) Al$_2$O$_3$; (**e**) SiO$_2$.

2.2. Screening Methodology

The experimental screening carried out to select the best supporting material focused on fundamental aspects such as chemical compatibility of the supporting material with Li$_4$(OH)$_3$Br, anti-leakage performance of the corresponding ss-composite and maximum salt loading allowed as well thermal properties and thermal cycling stability of the final composites. A three steps methodology was established to progressively discard either useless or poorly performing supporting material:

1. Chemical compatibility test. It consists of preparing a mixture of 90 wt.% Li$_4$(OH)$_3$Br and 10 wt.% oxide. The powder mixture is then subjected to a heating process up to 400 °C (Tm$_{salt}$ + 60 °C) for 24 h inside a closed stainless-steel reactor under Ar atmosphere. After these extreme heating conditions, chemical compatibility is investigated by means of differential scanning calorimetry (DSC) analysis and X-ray diffraction analysis to detect eventual side-reactions or changes/degradations in the storage properties of Li$_4$(OH)$_3$Br.
2. Anti-leakage performance analysis and maximum salt loading allowed. Pellets of Li$_4$(OH)$_3$Br/oxide composite materials with different oxide loadings are prepared following the cold-compression method described in Section 2.2. They are then submitted to the following thermal treatment: a heating step at 10 °C/min up to 350 °C, followed by an isothermal step at 350 °C for 1 h and finally a cooling step at around 2 K/min up to room temperature. The effectiveness of the composite in retaining the liquid phase of Li$_4$(OH)$_3$Br is qualified by visual inspection of the pellets during the test. Those composites allowing higher salt content while displaying good anti-leakage performance are moved to the last step.
3. Stability of the composites under thermal cycling conditions. In this step, the phase transition properties of composites that passed previous tests are determined before and after 50 heating and cooling cycles. Thermal cycling tests are carried out in a muffle furnace under argon atmosphere, between 250 °C and 350 °C, applying heating/cooling rates of 10 K/min and 2 K/min, respectively. Determination of both cycling and storage properties are carried out using differential scanning calorimetry (TA DSC 2500 model (New Castle, DE, USA)). The composite showing better stability and heat storage capacity is finally selected as supporting material.

2.3. Thermal and Structural Characterizations

Thermal properties including reaction temperature and enthalpy were measured using differential scanning calorimetry (TA DSC 2500 model). The DSC measurements were performed for samples in the form of cohesive solids to preserve the shape stabilization effect. Samples of about 20 mg were analyzed at a heating rate of 1 °C/min and a cooling rate of 10 K/min in the temperature range of 40–300 °C. Hermetically sealed aluminum DSC crucibles were used for the measurement. The temperature and enthalpy were calibrated using sapphire and indium standards; argon was used as a purge gas (50 mL/min). The accuracy was estimated to be ±1 K for the temperature and ±3 J/g for the enthalpy. The phase transition temperature was considered as the onset temperature. The phase change enthalpy was calculated by peak integration in the heating run. Structural analysis of the materials was performed by X-ray diffraction analysis using a Bruker D8 Discover (Billerica, MA, USA) equipped with a LYNXEYE detector with monochromatic Cu Kα_1 radiation of λ = 1.54056 Å. Patterns were recorded in a 2θ angular range 10–80° with a step size of 0.02° and a step time of 1 s. The measurements were performed at room temperature.

3. Results

The peritectic salt $Li_4(OH)_3Br$ was deeply studied in a previous paper as TES material [60]. The study showed that upon heating, $Li_4(OH)_3Br$ undergoes different reversible phase transitions. A first solid state transformation occurs at around 230 °C. A second solid state transition occurs at 279 °C, and finally the peritectic reaction occurs at 289 °C. All these transformations are reflected in the DSC curve of this stoichiometric compound, as shown in Figure 2. This is the reference DSC in terms of transition temperature and reaction enthalpy, which were taken into consideration when analyzing the compatibility and performance of the different supporting materials. The values assigned to the peritectic reaction in this work represent the sum of both the second solid state transition and the peritectic transition.

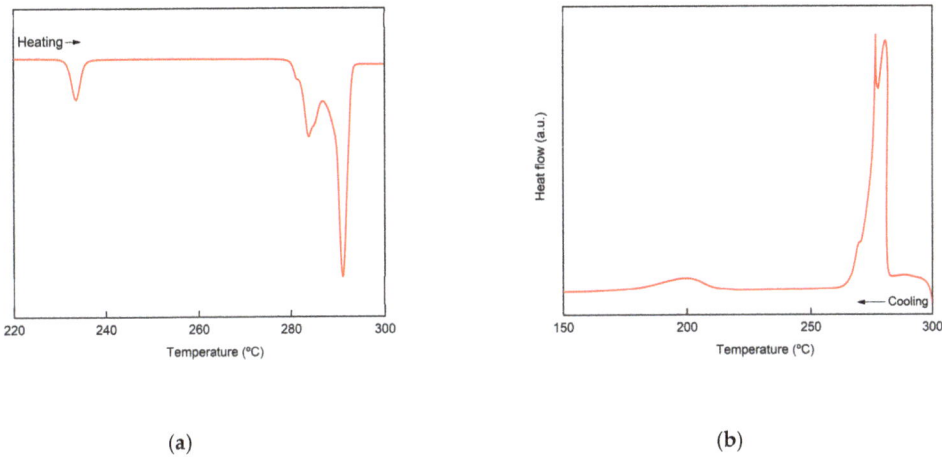

Figure 2. DSC curves of $Li_4(OH)_3Br$ (**a**) upon heating and (**b**) upon cooling.

3.1. Chemical Compatibility of the Supporting Material with Molten $Li_4(OH)_3Br$

As already mentioned in Section 2.2 (step 1 of the screening), different mixtures of $Li_4(OH)_3Br$/oxide were prepared using 10 wt.% of the support materials reported in Table 2 and heated at 400 °C for 24 h inside closed stainless-steel reactors under Ar atmosphere. Then, all the samples were subjected to structural analysis using X-ray diffraction in order to test the compatibility with the salt and detect possible side products formed due to the reaction between the salt and the oxide. The DSC curves of different mixtures ($Li_4(OH)_3Br$ + 10 wt.% oxide) after compatibility tests obtained both upon heating and cooling are reported in Figure 3.

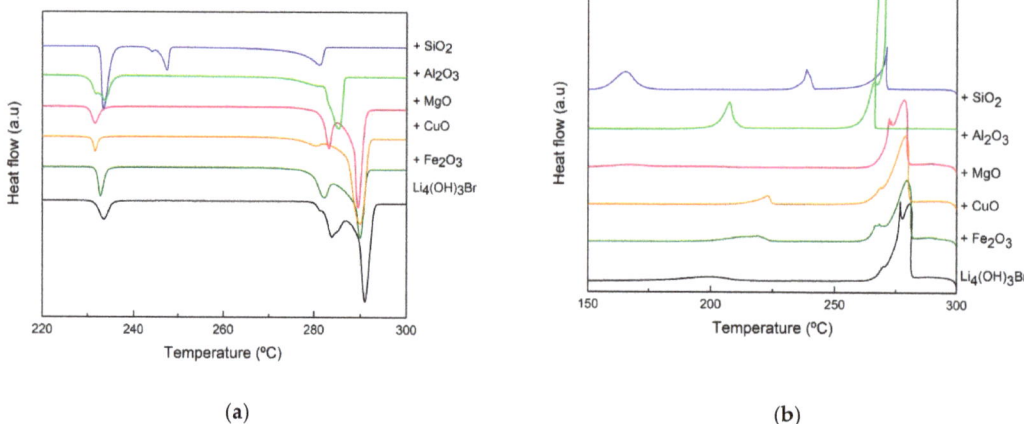

Figure 3. DSC curves of Li$_4$(OH)$_3$Br/oxide mixtures after compatibility tests recorded at 1 °C/min (**a**) upon heating and (**b**) upon cooling.

The temperatures and enthalpies of the reaction corresponding to the different composites are presented in Table 3. The enthalpy of the peritectic reaction relative to the mass fraction of the salt in the composite $\Delta H_{\text{calculated}}$ is calculated as follows:

$$\Delta H_{\text{Calculated}} = \Delta H_{\text{Li}_4(\text{OH})_3\text{Br}} * X_{\text{Li}_4(\text{OH})_3\text{Br}} \quad (1)$$

where $X_{\text{Li}_4(\text{OH})_3\text{Br}}$ represents the mass fraction of the salt, and $\Delta H_{\text{Li}_4(\text{OH})_3\text{Br}}$ is the enthalpy of the pure peritectic salt.

Table 3. Temperatures and enthalpies corresponding to the peritectic reaction of different Li$_4$(OH)$_3$Br/oxide mixtures after compatibility tests.

Composition	T_{onset} (°C)	$\Delta H_{\text{Experimental}}$ (J/g)	$\Delta H_{\text{Calculated}}$ (J/g)	Enthalpy Loss (%)
Pure Li$_4$(OH)$_3$Br	289	247	247	
90Li$_4$(OH)$_3$Br-10Fe$_2$O$_3$	288	197		10
90Li$_4$(OH)$_3$Br-10CuO	287	215	222	3
90Li$_4$(OH)$_3$Br-10MgO	288	209		6
90Li$_4$(OH)$_3$Br-10Al$_2$O$_3$	282	137		34

Analyzing Figure 3a, the DSC curves of sample 90Li$_4$(OH)$_3$Br-10SiO$_2$ showed a narrowing of the peak at the peritectic transition temperature together with the broadening of the first peak, in addition to the appearance of a new DSC peak at 246 °C. The three thermal events were reversible upon cooling (see Figure 3b). This behavior is an indication of a chemical reaction between the salt and the silica nanopowder. For this reason, the silicon dioxide was discarded at this level. Alumina nanopowder was also discarded due to the huge loss of the enthalpy of the peritectic reaction (−34%), as shown in Table 3. In addition, the DSC heating curve showed the displacement of both the second solid state reaction and the peritectic reaction to lower temperatures (−7 °C). The reason for this phenomenon may be due to the very small particle size of Al$_2$O$_3$ powder. In fact, the use of nanometric power (13 nm) with a very high specific surface area will create a large number of cavities inside the composite pellet, which leads to the possible confinement of a quantity of the peritectic salt inside these cavities, inducing the Gibbs–Thomson effect. Additionally, the salt trapped inside the cavities and the vacancies can cause a total or partial suppression of the peritectic reaction, which may be the reason for the significant decrease in the peritectic reaction enthalpy. In the cases of CuO, MgO and Fe$_2$O$_3$ oxides, good thermal stability of the composites was noticed. The temperature of the peritectic phase transition remained unchanged. Considering the systematic error ($\pm$ 3 J/g), the enthalpy of the peritectic reaction was stable in the case of the Li$_4$(OH)$_3$Br/CuO composite; however, it showed a slight decrease (6% loss) in the cases of MgO and (10% loss) Fe$_2$O$_3$ composites. CuO, MgO and Fe$_2$O$_3$ metal oxides were selected for further investigation (shape stability performance and thermal cycling stability).

Analyzing the XRD results of Li$_4$(OH)$_3$Br and the composite materials reported in Figure 4, it can be clearly seen that the patterns corresponding to the samples 90Li$_4$(OH)$_3$Br-10Fe$_2$O$_3$, 90Li$_4$(OH)$_3$Br-10MgO, 90Li$_4$(OH)$_3$Br-10CuO showed only the peaks related to the peritectic salt plus the Fe$_2$O$_3$, MgO and CuO, respectively. No new peaks related to the formation of new phases were detected despite the harsh conditions applied for the compatibility tests, which indicates good chemical compatibility between the salt and the tested oxides.

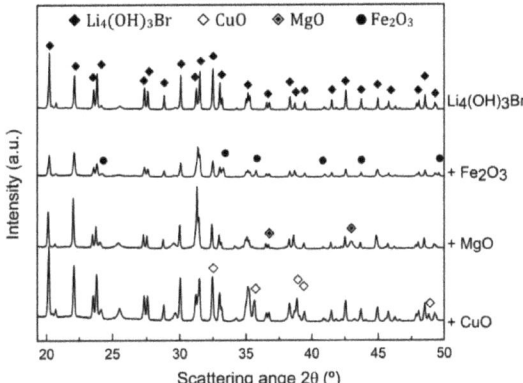

Figure 4. XRD results of Li$_4$(OH)$_3$Br/oxide mixtures after compatibility tests.

3.2. Anti-Leakage Effectiveness and Maximum Salt Loading

Capillary forces are mainly responsible for retaining the liquid salt during the phase change process. The parameters that determines the capillary effect are the size and topology of the pores, as well as the adhesion forces between the liquid salt and the pore wall, the latter of which is estimated by the effective contact angle between the liquid salt and the supporting material (the higher the liquid adhesion, the lower the effective contact angle) [63,64]. The work required to displace a liquid tube outside a cylindrical pore is given by

$$dW \equiv \Delta P_c \times dV = \frac{2\gamma_{lv} \cos\theta}{r} dV = S_p \gamma_{lv} \cos\theta \, dV \qquad (2)$$

where ΔP_c is the capillary pressure, dV is the infinitesimal volume of the liquid tube displaced along the axis of a cylindrical pore of radius r, γ_{lv} is the liquid–vapor surface tension, $S_p = 2/r$ represents the surface area per unit of pore volume and θ is the effective contact angle. It can be concluded that the greater the wettability and the higher the surface area (smaller pore size for same total porosity), the better the anti-leakage efficiency.

From this, several conclusions of practical interest follow for the interpretation of the results of the leakage tests performed: (1) decreasing the size of the oxide particles at constant oxide loading improves anti-leakage efficiency of the ss-composite; (2) the same happens when increasing the oxide loading at constant particle size, although at the expense of losing storage capacity; and (3) for equal oxide loading and particles size, oxides with higher wettability with the salt lead to ss-composites with better anti-leakage efficiency.

The anti-leakage performance of composites $Li_4(OH)_3Br/MgO$, $Li_4(OH)_3Br/CuO$ and $Li_4(OH)_3Br/Fe_2O_3$ was analyzed, as described in Section 2.2 (step 2 of the screening). For each composite, different pellets (13 mm in diameter, 5 mm in thickness) with oxide content ranging from 20 wt.% to 60 wt.% were prepared in order to determine the maximum salt loading allowed.

The following are shown in Table 4:

- $Li_4(OH)_3Br/MgO$ composite shows a minor salt leakage at 30–40 wt.% content of MgO. The sample with 50 wt.% MgO presents no sign of salt leakage, and the pellet shape is perfectly preserved showing a smooth surface without cracks. The sample with 60 wt.% MgO shows good structural stability without salt leakage; however, the pellet has cracked after sintering, which could be due to the high amount of MgO nanoparticles and the lesser amount of the salt, which ensure structural bonding after solidification. The advantages expected of using nanostructure MgO powder with 100 nm particle size were to have shape stabilization at a small loading of MgO thanks to the high specific surface area of MgO nanopowder, which generates a high surface tension between the salt and MgO; however, despite the nanometric particle size used, the form stability was ensured at a minimum content of 50 wt.% MgO, unlike $Li_4(OH)_3Br/Fe_2O_3$, and this could be due to the fact that the wettability of MgO by the molten salt is not as high as in the case of Fe_2O_3 micropowder.
- $Li_4(OH)_3Br/Fe_2O_3$ composite presents a significant salt leakage at 20 wt.% of Fe_2O_3. The samples with 30/40/50 wt.% Fe_2O_3 present excellent structural stability without any signs of salt leakage. Even though the nanostructure supporting materials prove to afford good anti-leakage efficiency of ss-composite at lesser content compared to materials with micrometric particle size, Fe_2O_3 with particle sizes <5 µm shows an excellent structural stability at only 30 wt.% loading compared to 50 wt.% MgO with a particle size of 100 nm. This can indicate the excellent wettability of Fe_2O_3 microparticles by the molten salt. In order to afford the maximum enthalpy of phase transition, the minimum content of 30 wt.% Fe_2O_3 was chosen for further investigations.
- $Li_4(OH)_3Br/CuO$ composite shows a constant improvement of the structural stability and no sign of leakage while increasing the content of the CuO from 30 to 60 wt.%. Samples with 30–40 wt.% of CuO show a significant amount of salt leakage with segregation of salt after sintering, which could be due to the difference in density of the two components. At 50 wt.% CuO, a small leakage of the salt can be observed. While increasing the CuO loading up to 60 wt.%, the shape stabilization is perfectly ensured and no salt leakage was observed. The high CuO loading (60 wt.%) required for the shape stability of the composite could be explained by (i) the large particle size of this material (<74 µm) giving a smaller surface area and thus less surface tension between the molten peritectic salt and CuO required for liquid salt retention inside the structure of the composite; (ii) and/or the modest wettability of CuO by the molten salt. The minimum loading required to guarantee the shape stabilization of the composite is 60 wt.% CuO, although this is at the expense of salt loading. This quite large amount of the supporting material will decrease considerably the storage capacity of the composite. The result was not satisfactory from a thermal storage application point of view, and for this reason, CuO was discarded at this level.

Table 4. Li$_4$(OH)$_3$Br-based ss-composites with different oxide loading after sintering showing salt leakage assessment.

wt.% Oxide	20	30	40	50	60
Li$_4$(OH)$_3$Br/MgO					
leakage assessment		Serious	Minor	No	No
Li$_4$(OH)$_3$Br/CuO					
leakage assessment		Serious	Serious	Minor	No
Li$_4$(OH)$_3$Br/Fe$_2$O$_3$					
leakage assessment	Serious	No	No	No	

3.3. Thermal and Microstructural Characterization and Stability of Li$_4$(OH)$_3$Br-Based Shape Stabilized Composites

The composites (70Li$_4$(OH)$_3$Br-30Fe$_2$O$_3$; 50Li$_4$(OH)$_3$Br-50MgO), which satisfied the criterion of form stabilization, were characterized by DSC. The objective was to investigate the influence of the shape stabilization on the thermal properties of the salt.

The morphology of the ss-composites was investigated by SEM. Figure 5a,b present the microstructures of 50Li$_4$(OH)$_3$Br-50MgO and 70Li$_4$(OH)$_3$Br-30Fe$_2$O$_3$ composites respectively. In both cases, it can be seen that Li$_4$(OH)$_3$Br salt is embedded in the oxide particles. This morphology prevents the leakage of the molten salt outside the structure of the composite by capillary force and surface tension. Analyzing the SEM images, two regions can be distinguished, one region with a smooth surface presenting smooth lamellar undulations which correspond to the peritectic salt, and another region with granular morphology corresponding to the oxide particles. Both structures of the composites present an open porosity.

The results of the DSC in terms of transition temperatures and reaction enthalpies are reported in Table 5.

Both DSC curves of 70Li$_4$(OH)$_3$Br-30Fe$_2$O$_3$ presented in Figure 6 and of 90Li$_4$(OH)$_3$Br-10Fe$_2$O$_3$ presented in Figure 3 show a shift in the peritectic transition to lower temperatures; the shift increases with increasing Fe$_2$O$_3$ content (peritectic transition temperature is 288 °C and 281 °C for 90Li$_4$(OH)$_3$Br-10Fe$_2$O$_3$ and 70Li$_4$(OH)$_3$Br-30Fe$_2$O$_3$, respectively). This can be explained by the possible salt confinement in the interparticle voids of the composites (thanks to the great wettability of Fe$_2$O$_3$ by the molten salt), which increases with the augmentation of the salt loading. A drop in the enthalpy of the peritectic reaction is also noticed, showing a higher loss for the sample with higher Fe_2O_3 loading (10% and 17% enthalpy loss for $90Li_4(OH)_3Br-10Fe_2O_3$ and $70Li_4(OH)_3Br-30Fe_2O_3$, respectively). This can be explained by the fact that the salt trapped inside the interparticle voids of the composite does not contribute to the total reaction enthalpy. Considering the high surface area of MgO compared to Fe_2O_3 (particle size is 100 nm and <5 μm for MgO and Fe_2O_3, respectively), the aforementioned phenomena is expected to be more pronounced; however, due to the moderate wettability of MgO by the molten salt, the Gibbs–Thomson effect is less likely to occur.

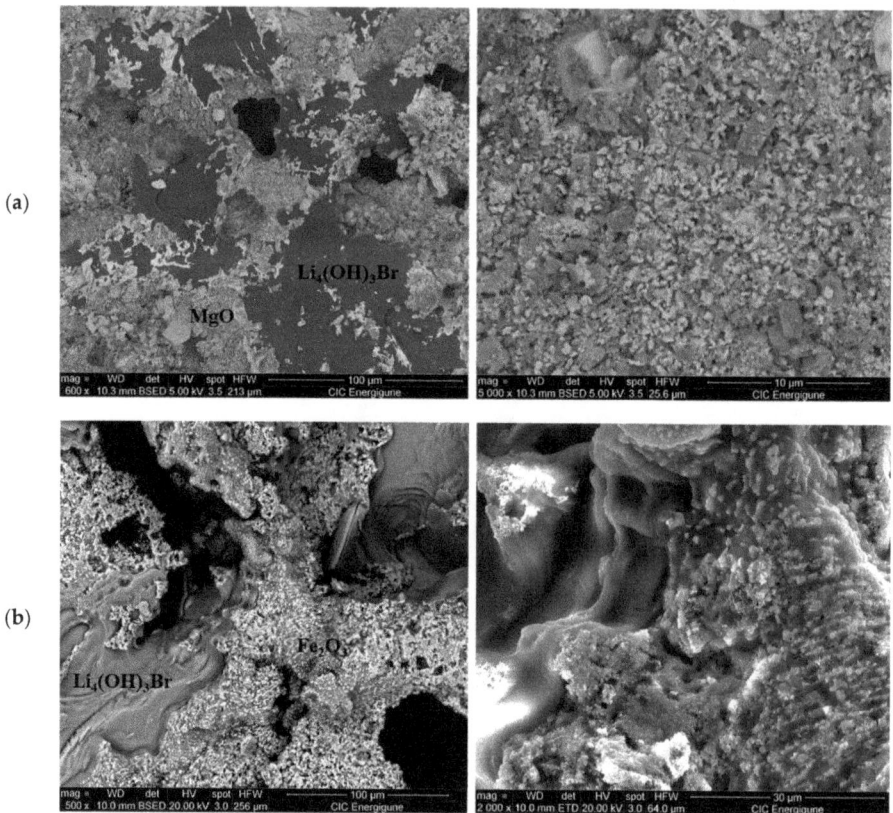

Figure 5. SEM images of the ss-composite: (**a**) $50Li_4(OH)_3Br-50MgO$; (**b**) $70Li_4(OH)_3Br-30Fe_2O_3$.

Table 5. Temperatures and enthalpies corresponding to the peritectic reaction of different $Li_4(OH)_3Br$-based ss-composites before and after thermal cycling.

Composition	T_{onset} (°C)	$\Delta H_{Experimental}$ (J/g)	$\Delta H_{Calculated}$ (J/g)	Enthalpy Loss (%)
Pure $Li_4(OH)_3Br$	289	247	247	
$70Li_4(OH)_3Br-30Fe_2O_3$-0Cycle	281	132	173	17
$70Li_4(OH)_3Br-30Fe_2O_3$-50Cycles	282	93	173	33
$50Li_4(OH)_3Br-50MgO$-0Cycle	285	114	124	4
$50Li_4(OH)_3Br-50MgO$-50Cycles	287	123	124	0.5

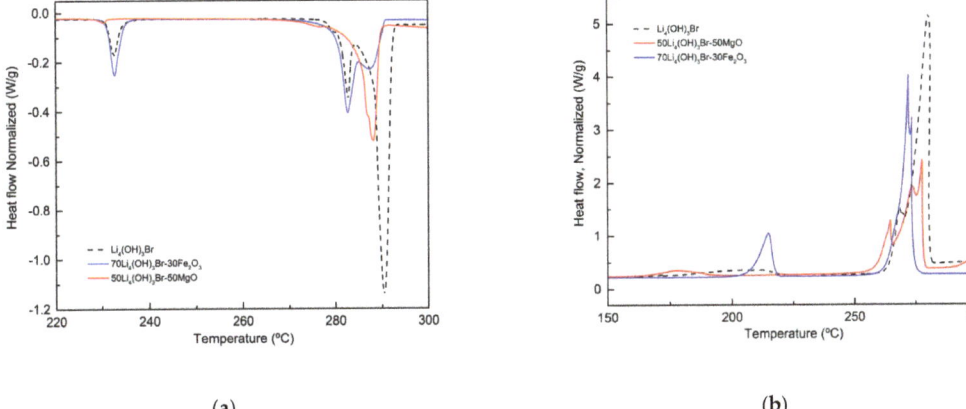

Figure 6. DSC curves of $Li_4(OH)_3Br$-based shape stabilized composites (**a**) upon heating and (**b**) upon cooling.

Analyzing Table 5, the most promising material is the composite with MgO as support. A slight enthalpy loss is observed, namely 4% compared to the 17% for the composite with Fe_2O_3. This enthalpy loss observed could be a sign of inhomogeneity of the composite structure, which means that the sample analyzed by DSC contained more supporting material than the active material (salt).

The composites with the best performances (with 50 wt.% of MgO and 30 wt.% of Fe_2O_3) were subjected to a lifecycle analysis test (up to 50 cycles) to determine the reactivity and shape stabilization performance after prolonged charge–discharge cycles. The results of these experiments are reported in Figure 7.

Figure 7. Images of the composite of $Li_4(OH)_3Br$-based shape stabilized composite before and after thermal cycling tests: (**a**) $50Li_4(OH)_3Br$-50MgO; (**b**) $70Li_4(OH)_3Br$-30Fe_2O_3.

No leaks of the salt outside the composite were observed after 50 charge/discharge cycles for both samples. These results are very promising; the number of cycles applied could indeed be considered representative regarding the stability of the shape-stabilized composite. When considering the chemical stability, both composites confirmed the trend already observed after one cycle.

The MgO-based composite was stable and inert, as confirmed by the results relevant to the reaction energy after 50 cycles (see Table 5) with enthalpy loss of only 0.5%. The heating DSC curves of $50Li_4(OH)_3Br$-50MgO composites showed good reproducibility, as can be seen in Figure 8a. On the other hand, the Fe_2O_3-based composite showed a significant decrease in the reaction energy of 33%, which excludes its possible further utilization. This result was probably due to a slow and progressive reaction of the salt with the oxide, because of the strong interaction between the two materials, ending with the gradual degradation of the salt. This was manifested by a change in the heating DSC curves, as presented in Figure 8b after 50 heating/cooling cycles.

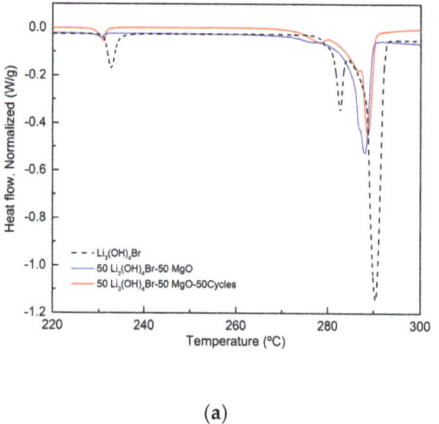
(a)

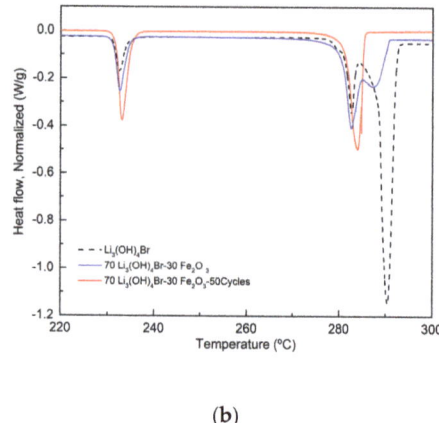
(b)

Figure 8. DSC curves of Li$_4$(OH)$_3$Br-based shape stabilized composites after cycling stability tests: (**a**) 50Li$_4$(OH)$_3$Br-50MgO; (**b**) 70Li$_4$(OH)$_3$Br-30Fe$_2$O$_3$.

4. Conclusions and Perspectives

The work carried out in this study allowed us to perform the selection of the most promising oxide-based supporting materials, with the best shape-stabilized performances, for the peritectic compound Li$_4$(OH)$_3$Br for application in thermal energy storage (TES). The combination of thermal treatment, structural analysis, thermodynamic characterization and cycling tests allowed us to have a clear overview of (i) the compatibility of the different materials tested and the possible by-products formed in terms of reacting behavior upon heating; (ii) the influence of the ceramic materials added on the energy of the peritectic reaction; (iii) the shape stabilization capability as a function of the concentration of inert oxide; and (iv) the effect of the cycling tests (up to 50 cycles) on both the reaction energy and the leakage occurrence. The results allowed us to select MgO as the most promising oxide showing good behavior for the two parameters considered. All the other oxides studied showed some type of reaction (more or less pronounced upon heating and upon cycling), leading to their rejection as possible candidate supporting materials. Even though the performance of MgO is good, the high amount used (50 wt.%) in the composite causes a considerable loss of energy density (only half of the reaction energy is available). For this reason, further work is now proceeding to apply different strategies to find new routes to decrease the amount used for maintaining the same performance.

Author Contributions: Conceptualization, I.M., S.D. and E.P.d.B.; methodology, I.M., S.D. and E.P.d.B.; software, I.M. and J.-L.D.; validation, I.M. and S.D.; formal analysis, I.M., S.D. and E.P.d.B.; investigation, I.M. and A.S.; resources, I.M. and S.D.; data curation, I.M.; writing—original draft preparation, I.M.; writing—review and editing, I.M., J.-L.D., S.D. and E.P.d.B.; visualization, I.M.; supervision, S.D. and E.P.d.B.; project administration, S.D.; funding acquisition, S.D. and E.P.d.B. All authors have read and agreed to the published version of the manuscript.

Funding: This research was funded by the Basque Government through the project Elkartek CICe2020 KK-2020/00078 and supported by the Polytechnique National Institute of Bordeaux (Bordeaux INP).

Institutional Review Board Statement: Not applicable.

Informed Consent Statement: Not applicable.

Acknowledgments: The authors acknowledge Cristina Luengo Vilumbrales and Maria Jáuregui for the help and commitment with the experimental measurements.

Conflicts of Interest: The authors declare no conflict of interest.

References

1. Gil, A.; Medrano, M.; Martorell, I.; Lázaro, A.; Dolado, P.; Zalba, B.; Cabeza, L.F. State of the art on high temperature thermal energy storage for power generation. Part 1—Concepts, materials and modellization. *Renew. Sustain. Energy Rev.* **2010**, *14*, 31–55. [CrossRef]
2. Medrano, M.; Gil, A.; Martorell, I.; Potau, X.; Cabeza, L.F. State of the art on high-temperature thermal energy storage for power generation. Part 2—Case studies. *Renew. Sustain. Energy Rev.* **2010**, *14*, 56–72. [CrossRef]
3. Wei, G.; Wang, G.; Xu, C.; Ju, X.; Xing, L.; Du, X.; Yang, Y. Selection principles and thermophysical properties of high temperature phase change materials for thermal energy storage: A review. *Renew. Sustain. Energy Rev.* **2018**, *81*, 1771–1786. [CrossRef]
4. Stutz, B.; Le Pierres, N.; Kuznik, F.; Johannes, K.; Del Barrio, E.P.; Bédécarrats, J.-P.; Gibout, S.; Marty, P.; Zalewski, L.; Soto, J.; et al. Storage of thermal solar energy. *Comptes Rendus Phys.* **2017**, *18*, 401–414. [CrossRef]
5. Cárdenas, B.; León, N. High temperature latent heat thermal energy storage: Phase change materials, design considerations and performance enhancement techniques. *Renew. Sustain. Energy Rev.* **2013**, *27*, 724–737. [CrossRef]
6. Kenisarin, M.M. High-temperature phase change materials for thermal energy storage. *Renew. Sustain. Energy Rev.* **2010**, *14*, 955–970. [CrossRef]
7. Milián, Y.E.; Gutiérrez, A.; Grágeda, M.; Ushak, S. A review on encapsulation techniques for inorganic phase change materials and the influence on their thermophysical properties. *Renew. Sustain. Energy Rev.* **2017**, *73*, 983–999. [CrossRef]
8. Cárdenas-Ramírez, C.; Jaramillo, F.; Gómez, M. Systematic review of encapsulation and shape-stabilization of phase change materials. *J. Energy Storage* **2020**, *30*, 101495. [CrossRef]
9. Leng, G.; Qiao, G.; Jiang, Z.; Xu, G.; Qin, Y.; Chang, C.; Ding, Y. Micro encapsulated & form-stable phase change materials for high temperature thermal energy storage. *Appl. Energy* **2018**, *217*, 212–220. [CrossRef]
10. Jacob, R.; Bruno, F. Review on shell materials used in the encapsulation of phase change materials for high temperature thermal energy storage. *Renew. Sustain. Energy Rev.* **2015**, *48*, 79–87. [CrossRef]
11. Ge, Z.; Ye, F.; Ding, Y. Composite materials for thermal energy storage: Enhancing performance through microstructures. *ChemSusChem* **2014**, *7*, 1318–1325. [CrossRef]
12. Jiang, F.; Zhang, L.; She, X.; Li, C.; Cang, D.; Liu, X.; Xuan, Y.; Ding, Y. Skeleton materials for shape-stabilization of high temperature salts based phase change materials: A critical review. *Renew. Sustain. Energy Rev.* **2020**, *119*, 109539. [CrossRef]
13. Liu, S.; Yang, H. Porous ceramic stabilized phase change materials for thermal energy storage. *RSC Adv.* **2016**, *6*, 48033–48042. [CrossRef]
14. Acem, Z.; Lopez, J.; Del Barrio, E.P. $KNO_3/NaNO_3$-graphite materials for thermal energy storage at high temperature: Part I.—Elaboration methods and thermal properties. *Appl. Therm. Eng.* **2010**, *30*, 1580–1585. [CrossRef]
15. Lopez, J.; Acem, Z.; Del Barrio, E.P. $KNO_3/NaNO_3$-graphite materials for thermal energy storage at high temperature: Part II.—Phase transition properties. *Appl. Therm. Eng.* **2010**, *30*, 1586–1593. [CrossRef]
16. Liu, J.; Wang, Q.; Ling, Z.; Fang, X.; Zhang, Z. A novel process for preparing molten salt/expanded graphite composite phase change blocks with good uniformity and small volume expansion. *Sol. Energy Mater. Sol. Cells* **2017**, *169*, 280–286. [CrossRef]
17. Kim, T.; Singh, D.; Zhao, W.; Yua, W.; France, D.M. An Investigation on the effects of phase change material on material components used for high temperature thermal energy storage System. In *AIP Conference Proceedings*; AIP publishing: Melville, NY, USA, 2016; Volume 1734. [CrossRef]
18. Singh, D.; Kim, T.; Zhao, W.; Yu, W.; France, D.M. Development of graphite foam infiltrated with $MgCl_2$ for a latent heat based thermal energy storage (LHTES) system. *Renewable Energy* **2016**, *94*, 660–667. [CrossRef]
19. Zhao, W.; France, D.M.; Yu, W.; Kim, T.; Singh, D. Phase change material with graphite foam for applications in high-temperature latent heat storage systems of concentrated solar power plants. *Renew. Energy* **2014**, *69*, 134–146. [CrossRef]
20. Singh, D.; Zhao, W.; Yu, W.; France, D.M.; Kim, T. Analysis of a graphite foam-NaCl latent heat storage system for supercritical CO_2 power cycles for concentrated solar power. *Solar Energy* **2015**, *118*, 232–242. [CrossRef]
21. Zhao, Y.; Wang, R.; Wang, L.; Yu, N. Development of highly conductive $KNO_3/NaNO_3$ composite for TES (thermal energy storage). *Energy* **2014**, *70*, 272–277. [CrossRef]
22. Yao, C.; Kong, X.; Li, Y.; Du, Y.; Qi, C. Numerical and experimental research of cold storage for a novel expanded perlite-based shape-stabilized phase change material wallboard used in building. *Energy Convers. Manag.* **2018**, *155*, 20–31. [CrossRef]
23. Xiao, X.; Zhang, P.; Li, M. Thermal characterization of nitrates and nitrates/expanded graphite mixture phase change materials for solar energy storage. *Energy Convers. Manag.* **2013**, *73*, 86–94. [CrossRef]
24. Lachheb, M.; Adili, A.; Albouchi, F.; Mzali, F.; Ben Nasrallah, S. Thermal properties improvement of lithium nitrate/graphite composite phase change materials. *Appl. Therm. Eng.* **2016**, *102*, 922–931. [CrossRef]
25. Ren, Y.; Xu, C.; Yuan, M.; Ye, F.; Ju, X.; Du, X. $Ca(NO_3)_2$-$NaNO_3$/expanded graphite composite as a novel shape-stable phase change material for mid- to high-temperature thermal energy storage. *Energy Convers. Manag.* **2018**, *163*, 50–58. [CrossRef]
26. Tian, H.; Wang, W.; Ding, J.; Wei, X.; Huang, C. Preparation of binary eutectic chloride/expanded graphite as high-temperature thermal energy storage materials. *Sol. Energy Mater. Sol. Cells* **2016**, *149*, 187–194. [CrossRef]
27. Tian, H.; Wang, W.; Ding, J.; Wei, X.; Song, M.; Yang, J. Thermal conductivities and characteristics of ternary eutectic chloride/expanded graphite thermal energy storage composites. *Appl. Energy* **2015**, *148*, 87–92. [CrossRef]

28. Liu, J.; Xie, M.; Ling, Z.; Fang, X.; Zhang, Z. Novel MgCl$_2$-KCl/expanded graphite/graphite paper composite phase change blocks with high thermal conductivity and large latent heat. *Sol. Energy* **2018**, *159*, 226–233. [CrossRef]
29. Xu, G.; Leng, G.; Yang, C.; Qin, Y.; Wu, Y.; Chen, H.; Cong, L.; Ding, Y. Sodium nitrate-diatomite composite materials for thermal energy storage. *Sol. Energy* **2017**, *146*, 494–502. [CrossRef]
30. Deng, Y.; Li, J.; Qian, T.; Guan, W.; Wang, X. Preparation and characterization of KNO$_3$/diatomite shape-stabilized composite phase change material for high temperature thermal energy storage. *J. Mater. Sci. Technol.* **2017**, *33*, 198–203. [CrossRef]
31. Qian, T.; Li, J.; Min, X.; Deng, Y.; Guan, W.; Ning, L. Diatomite: A promising natural candidate as carrier material for low, middle and high temperature phase change material. *Energy Convers. Manag.* **2015**, *98*, 34–45. [CrossRef]
32. Qin, Y.; Leng, G.; Yu, X.; Cao, H.; Qiao, G.; Dai, Y.; Zhang, Y.; Ding, Y. Sodium sulfate–diatomite composite materials for high temperature thermal energy storage. *Powder Technol.* **2015**, *282*, 37–42. [CrossRef]
33. Qin, Y.; Yu, X.; Leng, G.H.; Zhang, L.; Ding, Y.L. Effect of diatomite content on diatomite matrix based composite phase change thermal storage material. *Mater. Res. Innov.* **2014**, *18*, S2-453–S2-456. [CrossRef]
34. Deng, Y.; Li, J.; Nian, H. Expanded vermiculite: A promising natural encapsulation material of LiNO$_3$, NaNO$_3$, and KNO$_3$ phase change materials for medium-temperature thermal energy storage. *Adv. Eng. Mater.* **2018**, *20*. [CrossRef]
35. Li, R.; Zhu, J.; Zhou, W.; Cheng, X.; Li, Y. Thermal compatibility of sodium nitrate/expanded perlite composite phase change materials. *Appl. Therm. Eng.* **2016**, *103*, 452–458. [CrossRef]
36. Li, R.; Zhu, J.; Zhou, W.; Cheng, X.; Li, Y. Thermal properties of sodium nitrate-expanded vermiculite form-stable composite phase change materials. *Mater. Des.* **2016**, *104*, 190–196. [CrossRef]
37. Stoporev, A.; Mendgaziev, R.; Artemova, M.; Semenov, A.; Novikov, A.; Kiiamov, A.; Emelianov, D.; Rodionova, T.; Fakhrullin, R.; Shchukin, D. Ionic clathrate hydrates loaded into a cryogel—Halloysite clay composite for cold storage. *Appl. Clay Sci.* **2020**, *191*, 105618. [CrossRef]
38. Peng, L.; Sun, Y.; Gu, X.; Liu, P.; Bian, L.; Wei, B. Thermal conductivity enhancement utilizing the synergistic effect of carbon nanocoating and graphene addition in palmitic acid/halloysite FSPCM. *Appl. Clay Sci.* **2021**, *206*, 106068. [CrossRef]
39. Guo, Q.; Wang, T. Preparation and characterization of sodium sulfate/silica composite as a shape-stabilized phase change material by sol-gel method. *Chin. J. Chem. Eng.* **2014**, *22*, 360–364. [CrossRef]
40. Li, Y.; Guo, B.; Huang, G.; Kubo, S.; Shu, P. Characterization and thermal performance of nitrate mixture/SiC ceramic honeycomb composite phase change materials for thermal energy storage. *Appl. Therm. Eng.* **2015**, *81*, 193–197. [CrossRef]
41. Jiang, Y.; Sun, Y.; Jacob, R.D.; Bruno, F.; Li, S. Novel Na$_2$SO$_4$-NaCl-ceramic composites as high temperature phase change materials for solar thermal power plants (Part I). *Sol. Energy Mater. Sol. Cells* **2018**, *178*, 74–83. [CrossRef]
42. Guo, Q.; Wang, T. Study on preparation and thermal properties of sodium nitrate/silica composite as shape-stabilized phase change material. *Thermochim. Acta* **2015**, *613*, 66–70. [CrossRef]
43. Sang, L.; Li, F.; Xu, Y. Form-stable ternary carbonates/MgO composite material for high temperature thermal energy storage. *Sol. Energy* **2019**, *180*, 1–7. [CrossRef]
44. Li, C.; Li, Q.; Cong, L.; Li, Y.; Liu, X.; Xuan, Y.; Ding, Y. Carbonate salt based composite phase change materials for medium and high temperature thermal energy storage: A microstructural study. *Sol. Energy Mater. Sol. Cells* **2019**, *196*, 25–35. [CrossRef]
45. Li, C.; Li, Q.; Ding, Y. Investigation on the thermal performance of a high temperature packed bed thermal energy storage system containing carbonate salt based composite phase change materials. *Appl. Energy* **2019**, *247*, 374–388. [CrossRef]
46. Li, C.; Li, Q.; Cong, L.; Jiang, F.; Zhao, Y.; Liu, C.; Xiong, Y.; Chang, C.; Ding, Y. MgO based composite phase change materials for thermal energy storage: The effects of MgO particle density and size on microstructural characteristics as well as thermophysical and mechanical properties. *Appl. Energy* **2019**, *250*, 81–91. [CrossRef]
47. Guo, Y.R.; Liu, Y.; Zhang, G.Q.; Deng, Z.F.; Xu, G.Z.; Li, B.R. Effect of SiO$_2$ on the thermal stability of carbonate/MgO composite for thermal energy storage. *IOP Conf. Series Mater. Sci. Eng.* **2019**, *504*, 012011. [CrossRef]
48. Li, J.; Li, C.; Qiao, G.; Zhao, Y.; Huang, Y.; Peng, X.; Lei, X.; Ding, Y. Effects of MgO particle size and density on microstructure development of MgO based composite phase change materials. *Energy Procedia* **2019**, *158*, 4517–4522. [CrossRef]
49. Ge, Z.; Ye, F.; Cao, H.; Leng, G.; Qin, Y.; Ding, Y. Carbonate-salt-based composite materials for medium- and high-temperature thermal energy storage. *Particuology* **2014**, *15*, 77–81. [CrossRef]
50. Lu, Y.; Zhang, G.; Hao, J.; Ren, Z.; Deng, Z.; Xu, G.; Yang, C.; Chang, L. Fabrication and characterization of the novel shape-stabilized composite PCMs of Na$_2$CO$_3$-K$_2$CO$_3$/MgO/glass. *Sol. Energy* **2019**, *189*, 228–234. [CrossRef]
51. Li, C.; Li, Q.; Li, Y.; She, X.; Cao, H.; Zhang, P.; Wang, L.; Ding, Y. Heat transfer of composite phase change material modules containing a eutectic carbonate salt for medium and high temperature thermal energy storage applications. *Appl. Energy* **2019**, *238*, 1074–1083. [CrossRef]
52. Ye, F.; Ge, Z.; Ding, Y.; Yang, J. Multi-walled carbon nanotubes added to Na$_2$CO$_3$/MgO composites for thermal energy storage. *Particuology* **2014**, *15*, 56–60. [CrossRef]
53. Jiang, Z.; Jiang, F.; Li, C.; Leng, G.; Zhao, X.; Li, Y.; Zhang, T.; Xu, G.; Jin, Y.; Yang, C.; et al. A form stable composite phase change material for thermal energy storage applications over 700 °C. *Appl. Sci.* **2019**, *9*, 814. [CrossRef]
54. Zhu, J.; Li, R.; Zhou, W.; Zhang, H.; Cheng, X. Fabrication of Al$_2$O$_3$-NaCl composite heat storage materials by one-step synthesis method. *J. Wuhan Univ. Technol. Sci. Ed.* **2016**, *31*, 950–954. [CrossRef]
55. Suleiman, B.; Yu, Q.; Ding, Y.; Li, Y. Fabrication of form stable NaCl-Al$_2$O$_3$ composite for thermal energy storage by cold sintering process. *Front. Chem. Sci. Eng.* **2019**, *13*, 727–735. [CrossRef]

56. Li, Y.; Guo, B.; Huang, G.; Shu, P.; Kiriki, H.; Kubo, S.; Ohno, K.; Kawai, T. Eutectic compound (KNO_3/$NaNO_3$: PCM) quasi-encapsulated into SiC-honeycomb for suppressing natural convection of melted PCM. *Int. J. Energy Res.* **2015**, *39*, 789–804. [CrossRef]
57. Yu, Q.; Jiang, Z.; Cong, L.; Lu, T.; Suleiman, B.; Leng, G.; Wu, Z.; Ding, Y.; Li, Y. A novel low-temperature fabrication approach of composite phase change materials for high temperature thermal energy storage. *Appl. Energy* **2019**, *237*, 367–377. [CrossRef]
58. Achchaq, F.; del Barrio, E.P.; Lebraud, E.; Péchev, S.; Toutain, J. Development of a new LiBr/LiOH-based alloy for thermal energy storage. *J. Phys. Chem. Solids* **2019**, *131*, 173–179. [CrossRef]
59. Achchaq, F.; Del Barrio, E.P. A proposition of peritectic structures as candidates for thermal energy storage. *Energy Procedia* **2017**, *139*, 346–351. [CrossRef]
60. Mahroug, E.P.I.; Doppiu, S.; Dauvergne, J.L.; Echeverria, M.; Toutain, J. Study of peritectic compound $Li_4(OH)_3Br$ for high temperature thermal energy storage in solar power applications. *Sol. Energy Mater. Sol. Cells* **2021**, in press.
61. Bauer, T.; Laing, D.; Tamme, R. Characterization of Sodium Nitrate as Phase Change Material. *Int. J. Thermophys.* **2012**, *33*, 91–104. [CrossRef]
62. Rueden, C.T.; Schindelin, J.; Hiner, M.C.; Dezonia, B.E.; Walter, A.E.; Arena, E.T.; Eliceiri, K.W. ImageJ2: ImageJ for the next generation of scientific image data. *BMC Bioinform.* **2017**, *18*, 529. [CrossRef] [PubMed]
63. Jannot, Y.; Acem, Z. A quadrupolar complete model of the hot disc. *Meas. Sci. Technol.* **2007**, *18*, 1229–1234. [CrossRef]
64. Ukrainczyk, N.; Kurajica, S.; Šipušiae, J. Thermophysical comparison of five commercial paraffin waxes as latent heat storage materials. *Chem. Biochem. Eng. Q.* **2010**, *24*, 129–137.

Article

On the Use of Carbon Cables from Plastic Solvent Combinations of Polystyrene and Toluene in Carbon Nanotube Synthesis

Alvin Orbaek White [1,2,*], Ali Hedayati [1,3], Tim Yick [1], Varun Shenoy Gangoli [1], Yubiao Niu [4], Sean Lethbridge [4], Ioannis Tsampanakis [1], Gemma Swan [1], Léo Pointeaux [1,5], Abigail Crane [1], Rhys Charles [5], Jainaba Sallah-Conteh [1], Andrew O. Anderson [1], Matthew Lloyd Davies [5,6], Stuart. J. Corr [7,8,9,10] and Richard E. Palmer [4]

1. Energy Safety Research Institute, Swansea University, Bay Campus, Swansea SA1 8EN, UK; alihe@chalmers.se (A.H.); 748963@Swansea.ac.uk (T.Y.); V.S.Gangoli@Swansea.ac.uk (V.S.G.); 958003@Swansea.ac.uk (I.T.); 9901886@Swansea.ac.uk (G.S.); 2124215@Swansea.ac.uk (L.P.); 910979@Swansea.ac.uk (A.C.); 825585@Swansea.ac.uk (J.S.-C.); 934622@Swansea.ac.uk (A.O.A.)
2. Chemical Engineering, Faculty of Science and Engineering, Swansea University, Bay Campus, Swansea SA1 8EN, UK
3. TECNALIA, Basque Research and Technology Alliance (BRTA), Alava Science and Technology Park, Leonardo da Vinci 11, 01510 Vitoria-Gasteiz, Spain
4. Nanomaterials Lab, Mechanical Engineering, Faculty of Science and Engineering, Swansea University, Bay Campus, Swansea SA1 8EN, UK; Yubiao.Niu@Swansea.ac.uk (Y.N.); 906934@Swansea.ac.uk (S.L.); R.E.Palmer@Swansea.ac.uk (R.E.P.)
5. SPECIFIC, Materials Science and Engineering, Faculty of Science and Engineering, Swansea University, Bay Campus, Swansea SA1 8EN, UK; R.Charles@Swansea.ac.uk (R.C.); M.L.Davies@Swansea.ac.uk (M.L.D.)
6. School of Chemistry and Physics, University of KwaZulu-Natal, Durban 4041, South Africa
7. Department of Cardiovascular Surgery, Houston Methodist Hospital, Houston, TX 77030, USA; sjcorr@houstonmethodist.org
8. Department of Bioengineering, Rice University, Houston, TX 77005, USA
9. Department of Biomedical Engineering, University of Houston, Houston, TX 77204, USA
10. Swansea University Medical School, Institute of Life Science 2, Swansea University, Singleton Park, Swansea SA2 8PP, UK
* Correspondence: Alvin.OrbaekWhite@Swansea.ac.uk

Abstract: For every three people on the planet, there are approximately two Tonnes (Te) of plastic waste. We show that carbon recovery from polystyrene (PS) plastic is enhanced by the coaddition of solvents to grow carbon nanotubes (CNTs) by liquid injection chemical vapour deposition. Polystyrene was loaded up to 4 wt% in toluene and heated to 780 °C in the presence of a ferrocene catalyst and a hydrogen/argon carrier gas at a 1:19 ratio. High resolution transmission electron microscopy (HRTEM), scanning electron microscopy (SEM), thermogravimetric analysis (TGA) and Raman spectroscopy were used to identify multiwalled carbon nanotubes (MWCNTs). The PS addition in the range from 0 to 4 wt% showed improved quality and CNT homogeneity; Raman "Graphitic/Defective" (G/D) values increased from 1.9 to 2.3; mean CNT diameters increased from 43.0 to 49.2 nm; and maximum CNT yield increased from 11.37% to 14.31%. Since both the CNT diameters and the percentage yield increased following the addition of polystyrene, we conclude that carbon from PS contributes to the carbon within the MWCNTs. The electrical contact resistance of acid-washed Bucky papers produced from each loading ranged from 2.2 to 4.4 Ohm, with no direct correlation to PS loading. Due to this narrow range, materials with different loadings were mixed to create the six wires of an Ethernet cable and tested using iPerf3; the cable achieved up- and downlink speeds of ~99.5 Mbps, i.e., comparable to Cu wire with the same dimensions (~99.5 Mbps). The lifecycle assessment (LCA) of CNT wire production was compared to copper wire production for a use case in a Boeing 747-400 over the lifespan of the aircraft. Due to their lightweight nature, the CNT wires decreased the CO_2 footprint by 21 kTonnes (kTe) over the aircraft's lifespan.

Keywords: carbon nanotube; plastic; chemical recycling; life cycle assessment; ethernet; circular economy; data transmission; carbon footprint

1. Introduction

If carbon nanotubes (CNTs) are to be used for their lightweight and electrical conduction properties on a large/global scale [1], is there a scenario that justifies their use, especially given the (typically) large embodied energy requirements for their manufacture? Moreover, in the age of sudden climatic shifts linked to carbon emissions, then where would the carbon come from that would be used to make these CNTs? Additionally, can the related production method from that carbon source contribute to the goal of achieving a positive climate output? To address these questions, we suggest that both plastics and solvents be used as carbon sources for CNT manufacture, and that they can create a positive environmental impact over the lifespan of their application in the aerospace sector. For instance, one of the key attributes of CNTs is being lightweight with a density that is 1/6th that of copper. This mass decrease will result in fuel savings for the automotive and aviation sectors.

Plastic products synthesised from recycled plastics are often of inferior quality compared to freshly manufactured plastics, and are not feasible for the same applications. This results in fresh plastics having a wider range of uses, and therefore, holding greater economic value. Moreover, the recycling process is often thwarted by the inclusion of fillers, pigments, and flame retardants, for example. Additionally, mixed plastics and/or composite plastic products are also challenging or impossible to recycle unless separated. To overcome these problems, one can consider open-loop recycling by making new products other than plastic. The most prominent technique uses thermal pyrolysis to break down the long chain polymer molecules into smaller, less complex molecules [2] via the application of intense heat [3]. This is typically carried out in the absence of oxygen to avoid the formation of undesirable carbon oxides, and is done in the presence of a catalyst to increase efficiency, tailor the resulting product, and improve scalability. The product mix is typically composed of oils, chars, and gases that require subsequent refinement and separation, thereby requiring more energy. As such, the primary focus of the field has been to improve synthesis techniques with decreased energy requirements and to create more refined and homogenous products [4,5]. To that end, we have developed a novel approach towards the pyrolytic growth of carbon nanotubes by including a dissolution step prior to the high temperature cycle.

CNT growth from plastics is typically achieved using pyrolysis, whereby polymers such as polypropylene (PP) [6], polyethylene (PE) [7], polyethylene terephthalate (PET) [8] and high-density polyethylene (HDPE) [9] are heated in a solid–gas fluidized bed reactor [10]. The resultant off-gas traverses via a carrier-gas to a catalyst site to complete the conversion from vapor to the solid CNT product. The pyrolytic growth of CNTs can be improved by the addition of a thermodynamically compatible solvent [11–13] such as toluene [14,15] to dissolve plastics including polystyrene (PS). The dissolution process confers five benefits over dry plastic pyrolysis. Polymer disentanglement initiates in the solvent [16], thus increasing the reactive surface area. The polymer begins to decompose [17], and thus lowers the required energy for subsequent C–H cleavage, thereby facilitating the processing of mixed plastic products. All nonsoluble material crashes out of solution, such as flame retardants and other noncompatible additives, effectively cleaning the plastic prior to CNT growth. Also, once dissolved, mass transport at scale is readily achieved using pumps and pipes; therefore, the liquid injection method is beneficial for large-scale operations. Moreover, mixing plastics in a solvent increases the carbon density, which can lead to increased production capacity of CNTs via chemical recycling of mixed plastics and solvents.

In conducting a life-cycle analysis (LCA), one must consider the lifespan of the material from inception to application, with particular focus on its manufacturing, transformation, use, and disposal [18]. The complexity of the phenomena involved and the interactions among these steps is a source of uncertainty regarding the real value of the impacts, which is why we can only create "potential" life cycle assessments. Although considered for a long time as an experimental tool, the international standards ISO 14040 and 14044 (revised in

2006) have set the methodological and ethical bases for this type of assessment. In this age of climate uncertainty, it is imperative to adopt protocols and solutions that minimise harm compared with the problems of 'business as usual', which these protocols are intended to solve [19].

The energy cost associated with the high entropy and exergy of plastics reconstitution is often cited as the single biggest reason to avoid plastics feedstock, especially with respect to CNT growth. Moreover, this reason is used to justify the continued use of virgin hydrocarbons, despite the fact that CNT growth from these sources never accounts for the energy or material cost associate with purification and delivery of the refined feedstocks. Though this is true in terms of energy consumption, the longer-term challenge deals with the supply of virgin materials based on oil extraction before reaching a peak oil scenario. Before that point, it would be prudent to establish and develop the science and technology to use premade plastics, in all the states in which they are found [20,21], especially given that these materials will otherwise be strewn about the planet, ending up in our soil and food supply [22].

Herein, we report that carbon from plastics can act as a feedstock for carbon nanotube growth by the upcycling of plastic to high-value materials via a chemical process. This can be considered a viable alternative to landfill and incineration. Environmental challenges exist from both liquid and solid hydrocarbons, so we applied toluene and PS as model materials.

2. Materials and Methods

2.1. The Synthesis of Carbon Nanotubes

The growth of multiwalled carbon nanotubes (MWCNTs) was carried out via catalytic chemical vapour deposition (CCVD) in a two-zoned horizontal furnace (Nanotech Innovations SSP-354, Oberlin, OH, USA) liquid injection reactor (LIR), with full details described previously [23]. In summary, control CNTs were grown by injection of 1 mL (865 mg) anhydrous toluene (98% ($C_6H_5CH_3$) Sigma Aldrich (Gillingham, UK)) at 5 mL/h under a gas flow of 1 L/min using blended carrier gas having 5 vol% hydrogen in argon (BOC, Guildford, UK) into the two-zone horizontal furnace. The first zone, used for vapour formation was set to 225 °C; the second zone used for growth was set to 780 °C. MWCNTs were grown in a 100 cm long quartz tube with diameter of 38 mm (Multi-Lab, Newcastle upon Tyne, UK). All reactions were carried out using a 20 gauge needle.

Polystyrene ($(C_8H_8)_n$, with a molecular weight of 6400 (Mn 64,000, Sample#P2444-S, Polymer Source Inc., Dorval, QC, Canada), was added to toluene in concentrations of 1, 2, and 4 wt% (w/w) using PS masses of 8.75, 17.5, and 35.00 mg, respectively. All reactions were carried out with a fixed catalyst ratio of ferrocene (5 wt% w/w) (98% ($C_{10}H_{10}Fe$) Sigma Aldrich (Gillingham, UK)) with respect to the total reactant from toluene and/or toluene and polystyrene. Prior to each growth, the reactants were thoroughly mixed and degassed for 15 min using bath sonication. All materials were used as received without prior cracking or drying, and handled as described here [24]. Each concentration of PS and control was grown three times to ensure that the observed trends were valid for each series. No noticeable effect or carbothermal reduction from the aging of the quartz tube was identified [25] (Figure S1).

2.2. The Characterization and Measurement of Material Properties

High resolution transmission electron microscopy (HRTEM) was used to characterise the as-grown samples (Figure 1) using a FEI Talos 200X (FEI, Hillsboro, OR, USA) Transmission Electron Microscope (TEM) in high-resolution TEM mode, operating at 200 kV. The TEM samples were prepared by dipping holey carbon TEM grids into CNT powders. Fast Fourier transforms (FFT) of selected images were obtained to determine the materials' atomic structures.

Scanning electron microscopy (SEM) using a JEOL 7800F FEG (JEOL, Akishima, Tokyo, Japan) was used to corroborate the presence of MWCNTs (Figure 2). A small fraction of each sample was suspended in 3 mL ethanol, and 100 µL of the suspension was dried on the surface of a clean silicon wafer for imaging. The SEM was used at an operating voltage to 5 kV or below, with a working distance of ca. 10 mm. Diameters were measured using ImageJ [26].

A Renishaw inVia Raman microscope (Renishaw plc, Miskin, Pontyclun, UK) using a laser at 633 nm wavelength and 5% beam power was used for data acquisition between 100 cm^{-1} and 3200 cm^{-1} Raman shift. The laser beam was focused by maximizing the G-peak intensity to confirm best z-height alignment of the beam between sample and detector. For each CNT sample, a Raman spectrum was acquired in three separate locations. All G/D values for that series were then averaged and reported along with the maximum G/D for that series and plotted in Figure 3. The Raman spectra also show background intensity in the region from 1400 cm^{-1} to 1475 cm^{-1} that is due to fluorescence in amorphous carbon [27]. Line integration of the background intensity between that range was plotted to compare amorphous carbon content (Figure 3C). For each reaction condition, the individual CNT samples were probed in at least three separate locations in order to both overcome variance within a single sample and to create a significant quantity of data for comparative analyses between series.

Thermogravimetric analysis (TGA/DTA) (TA Instruments, Stamford Avenue, Cheshire, UK) [28] of the CNT samples was used to determine the MWCNT product yield by using ca. 10 mg of sample placed in a platinum pan and heating under active air flow up to 800 °C. The ramp rate was 5 °C/min and hold time was 30 min at 765 °C. The sampling interval was set for 3 s. MWCNT wt% is determined as the complete weight loss after full oxidation at 800 °C in air using TGA [29] (Equation (1)).

2.3. Preparation of CNTs for Testing Voltage Drop

Due to presence of residual iron catalyst witnessed in the HRTEM images, an acid wash was used to remove excess iron to establish the CNT voltage drop. The oxidising acid wash, using equimolar HNO_3 + H_2SO_4 reflux (70 °C, 24 h), stripped away the amorphous carbon in addition to helping clear out the graphitic "onion" layers from the residual catalyst materials that would otherwise hinder iron from being digested [30,31]. Note that the acid wash may have inadvertently damaged and etched some of the MWCNTs, thus increasing the measured resistance. However, it was important to prioritise iron removal, given its potential influence in electrical contact measurements.

2.4. Device Preparation and Measurement

Thin films were prepared using the "bucky-paper" [32] technique to the measure electrical conductivity of acid washed samples. After the acid wash, the CNTs were suspended in isopropanol and CNT films were made using vacuum filtration [33]. The CNT films were dried at 80 °C for 3 h prior to testing and use.

Electrical resistance values were derived using Ohm's law based on measured values of voltage drop at constant current in a range of values between 0 to 100 mA. The samples were measured along a 2 cm separation for all samples to ensure consistent path length.

The carbon nanotube cables were comprised of CNT powders firmly packed into the sheath of heat shrink tubing. Copper wire was inserted into the CNT wire ends and compressed to ensure maximum contact with the Cu lead prior to heat shrinking the outer sheath. The Cu leads were then used to crimp into pins and inserted into retail purchased RJ45 connectors (RS Components, Corby, UK) for testing as ethernet cable.

Quantification of the CNT ethernet cable was conducted using iPerf3 (iPerf3 is principally developed by ESnet/Lawrence Berkeley National Laboratory. It is released under a three-clause BSD license). The cable was directly connecting two computers, both running the same Windows 10 OS update, going from a Realtek 8125B 2.5 G LAN adapter (Realtek Semiconductor Corp., Hsinchu, Taiwan) capable for two-way traffic (server) to an Intel Killer E3100X LAN adapter (Intel, Santa Clara, CA, USA) as the client; both were capable of transfer speeds up to 2500 Mbps, ensuring that the CNT cable could perform without being bottlenecked. A standard 10 s/10 run test was performed to both authenticate the data transfer and measure the transfer speeds from the server (uplink) to the client (downlink). The test was repeated for statistical accuracy and the results recorded for further discussion.

2.5. Life Cycle Assessment(LCA) Methodolgy, Assumptions, and Boundary Conditions

An LCA was conducted using Simapro 9.1 with Ecoinvent 3.6 as database using the method "IPCC 2013 GWP 100a (incl. CO_2 uptake) V1.00". In the following LCA, the CO_2 emissions (Global Warming Potential) are expressed as CO_2 equivalent (CO_2 eq.) for the extraction, production, transport, use and end of life phases of CNT and copper wires. They were compared in the use case of electrical wiring in a standard Boeing 747-400 aircraft. This study includes analyses of both the environment impact of the manufacturing of CNT wires compared to Cu wire, and of the effect these wires have over the lifespan of a 747 in terms of CO_2 generation from fuel consumption.

Assessing the potential environmental impacts of CNT manufacturing based on life cycle assessment required the following assumptions: the mass of 141 miles [34] of copper wire is 1519.6 kg (Supplementary information); copper ore extraction was conducted in Spain; copper wire manufacturing is done in the UK within a 100 km radius of London. The mass of 141 miles of CNT wire was 356.2 kg for the same cross-section as the copper wire [35]. The raw materials for the CNTs were produced locally (within a 300 km radius of Swansea) and the CNTs were manufactured in the UK (using energy from the average UK electricity mix). The CNTs were produced on a laboratory scale as per the methods described herein; this is because no large scale or industrial process uses the LIR model for comparative purposes at this time, and a full study of a scaled reaction process is beyond the scope of this work.

To study the effective change in CO_2 emissions from an aircraft utilising CNTs wires over Cu wires during its use phase required the following assumptions: an aircraft equipped with CNT wire will be 1163.4 kg lighter. The lifespan of a Boeing 747-400 is approximately 100,000 h of flight time with an average speed of 900 km/h. Only the CO_2 emissions generated during the use phase of the aircraft were considered in this analysis; accounting for the CO_2 emissions resulting from the manufacture of the aircraft is beyond the scope of this work.

3. Results and Discussion

3.1. Material Characteristics Determind Using Microscopy

Using high resolution transmission electron microscopy, we observed the presence of both carbon nanotubes and residual metal catalyst particles (Figure 1). The number of walls ranged between 18 and 52, with CNT diameters ranging from 18 to 45 nm. There was no apparent trend associated with PS loading (within the small number of TEM images counted). Several MWCNTs displayed closed caps, and some had several catalyst particles within a single CNT (Figure 1B).

The longest single CNT observed was 13.7 μm (Figure 1E), although longer CNTs may exist. The 0.34 nm spot in the FFT is due to the d(002) lattice spacing associated with multiple graphitic walls. This is significant because it further verifies the presence of nanotubes based on their crystallography. Additionally, the 0.17 nm spot is the d(004) spacing from the second order diffraction of the d(002) lattice. However, it is more challenging to differentiate iron oxide compounds such as Fe_2O_3, Fe_3O_4 and FeOOH from FFTs, as they have very similar spots/d-spacings (Figures S2–S4).

Using scanning electron microscopy (SEM), we easily identified CNTs based on their tangled nature and long morphology (Figure 2). The catalyst content could also be identified due to higher charging density around the Fe and FeC structures. Some of the catalyst could be seen within the CNT structures, further confirming nanotube formation due to the presence of hollow cores. In some cases, the catalyst could be seen as high intensity bright spots, both along the inside walls of the CNTs and at the ends.

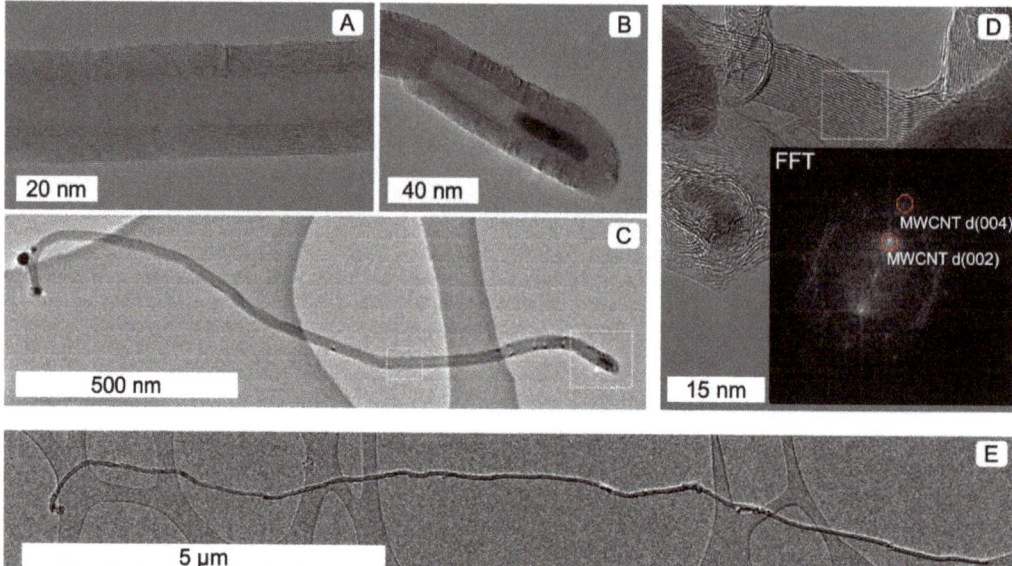

Figure 1. High resolution transmission electron microscope images showing (**A**) multiple walls and (**B**) the presence of internal catalyst particles in (**C**) long MWCNTs. (**D**) FFT analysis confirmed d(002) and d(004) line spacings. MWCNT lengths (**E**) reach over 10 μm in this case.

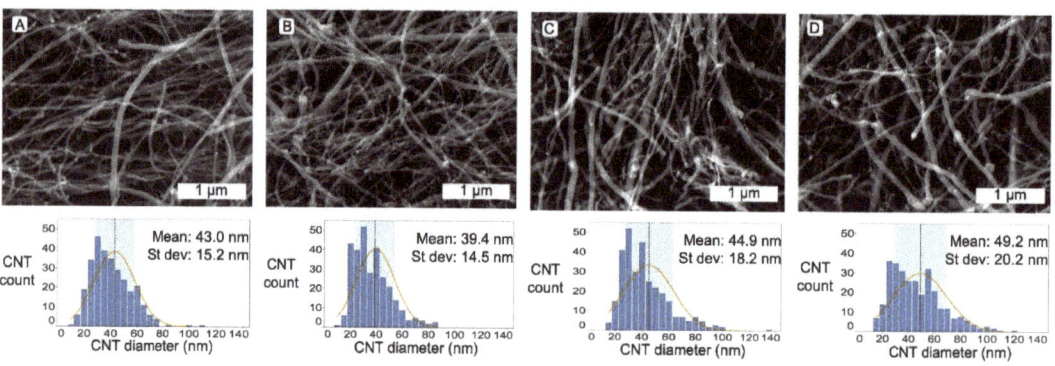

Figure 2. Representative scanning electron microscope images of carbon nanotubes made from (**A**) control with zero PS, (**B**) PS 1 wt%, (**C**) PS 2 wt% and (**D**) PS 4 wt%. Along with histogram data for each sample showing mean and standard deviation for each sample.

Most of the CNTs displayed a relatively tortuous path once deposited on the SEM stubs for imaging. In the control samples, the average diameter was found to be 43.0 nm +/−15.2 nm (Std. Dev). Interestingly, at low PS (1 wt%) concentration, the CNT diameters decreased with a concurrent smaller distribution, resulting in MWCNTs with 39.4 nm +/−14.5 nm average diameter. However, increased PS content led to increased diameters and greater standard deviations. The diameter increase can be accounted for by the increased carbon concentration in the feedstock.

3.2. Material Characteristics Determined Using Spectroscopy

Raman spectroscopy is commonly used to measure and characterise the fingerprint peaks associated with MWCNTs (Figure 3A). It can also be used to make comparative analyses of amorphous carbon contents (Figure 3B) between samples. More importantly, resonant Raman spectroscopy is used to quantify the quality of the MWCNTs by direct comparison of G- and D-peak intensities (Figure 3C).

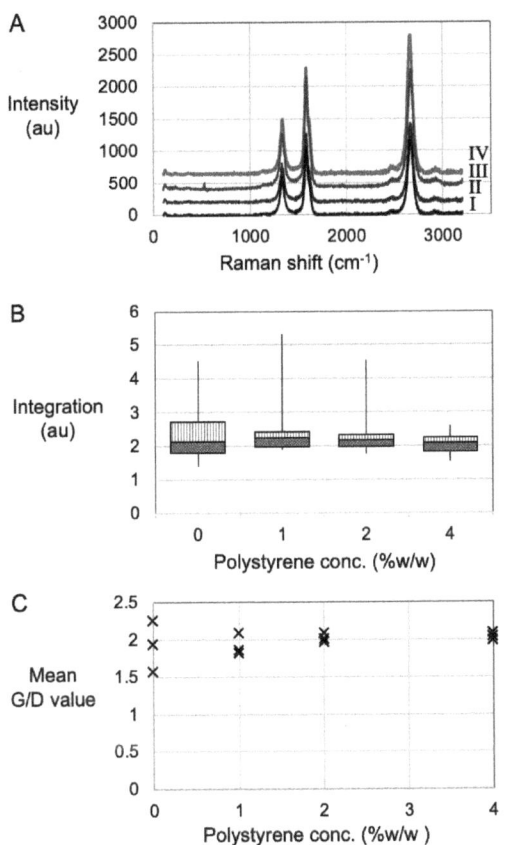

Figure 3. Raman data showing (**A**) typical spectra stacked from (I) Control, (II) PS 1 wt%, (III) PS 2 wt% and (IV) PS 4 wt%. (**B**) Box plot from line integration between 1400–1475 cm^{-1}, and (**C**) Mean G/D values. All spectra acquired using 633 nm laser.

The representative spectra for all four conditions exhibited G- and D-peaks associated with the presence of MWCNTs (Figure 3A). In the case of MWCNTs, the G-peak (also referred to as G mode, or Graphitic mode) appeared in the 1500–1600 cm^{-1} range due to the tangential displacement of C–C bond stretching, effectively indicating the density of the sp^2 hybridized carbons atoms that existed in the CNT lattice, and along the circumferential direction of the nanotube. However, the D-peak, or Disorder mode, observed at 1290–1330 cm^{-1}, represented the conversion of carbon centres from sp^2 to sp^3 hybridisation states, as can occur due to a break in the symmetry of the graphite plane [36]. Therefore, comparing the peak heights of the G- and D-peak effectively quantified the graphitic versus the defective carbon atoms within the CNT lattice. The D peaks intensities must be taken with respect to the G peak intensities. When the D peaks are high, it indicates a large portion of sp^3 carbon atoms. sp^3 carbon atoms can arise for several reasons relating to the broken symmetry of the graphitic structure. For this reason, the D-peak is used to identify defects in the CNT structure; it is not that the CNTs are themselves defective, but that they have imperfections within their body. A third fingerprint peak is also evident in Figure 3A; this G' (G prime) peak corresponds to disorder-induced carbon features arising from finite particle size distribution or lattice distortions in the CNTs. Moreover, the presence of G' peak further indicated the presence of multiple walls, which was to be expected for samples of multiwalled carbon nanotubes.

The spectral background intensity in the range 1400–1475 cm^{-1} occurred due to the fluorescence of amorphous carbon and line integration in that region was used to compare amorphous carbon concentrations between samples [27,37]. A box plot [38] (Figure 3B) of the line integration values indicates smaller and narrowing range with higher PS loading. As small integration values indicate less amorphous carbon content, we concluded that amorphous carbon decreases as a consequence of increased PS loading.

The G/D value indicated quality in bulk samples, since a large G-peak relative to the D-peak indicates a strong resonance condition of sp^2, graphitic carbon. [39] Comparisons between reaction conditions could be made in terms of range variance and minimum versus maximum values for each reaction condition. The mean G/D values for the control group suggested that, on average, this process yielded high quality materials, but they ranged from 1.5 to 2.3, suggesting they were more heterogeneous samples compared to the PS 4 wt% condition that had an average range from 1.9 to 2.1. Moreover, the PS samples generally displayed a tightening of data points which suggested greater homogeneity. The G/D values improved with a tightened range too, indicating that higher quality materials had been created by the use of PS, whereby the maximum G/D value was recorded using the PS 4 wt% (Table 1). Moreover, the increase of PS concentration may not have been detrimental, based on the steady measurement of the mean value at 2.0, even after the increase.

Table 1. Mean and maximum G/D values obtained for various carbon nanotube samples grown with incrementally higher concentrations of polystyrene feedstock.

	PS Concentration (wt%)			
	0	1	2	4
Mean G/D value	1.9	2.0	2.0	2.0
Maximum G/D value	2.3	2.3	2.3	2.5

3.3. Material Properties Determined Using Mass Balance and Thermogravimetric Analysis

The solid product was weighed using mass balance and used to determine MWCT percentage yield according to Equation (1). TGA data based on the oxidization, and subsequent mass loss at temperatures between 400–600 °C, were characteristic of MWCNTs (Figure 4). The residual mass in the TGA pan was the oxidised catalyst material, typically orange in colour and represented by a residual mass percentage which was used to calculate MWCNT percentage yield (Equation (2)). For example, using 4 wt% of PS, 130 mg was recovered from the reactor. A TGA test showed 6.53% to be residual catalyst, equating to a MWCNT mass of 121.55 mg, which, in turn, corresponded to a CNT percentage yield of 14.2% (Equation (2)). All the values were tabulated and averaged to determine the average CNT product yield (Table 2). Since we were dealing with catalytic processes where some reactions showed outstanding results compared to the norm/average, we have also reported the maximum percentage yields. Though not the norm, they represent the best-case scenario. Based on TEM observations, we assumed that negligible quantities of amorphous carbon were present; this assumption was reinforced by the line integration study of the Raman spectra.

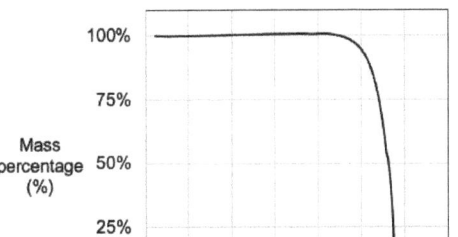

Figure 4. Thermogravimetric data of CNT growth from polystyrene 4 wt% loading, residual mass found to be 6.53%, composed of iron oxide residue from catalyst precursor.

Table 2. Mass data and percentage yield results for CNTs grown using polystyrene.

Sample	Mass of Carbon Reactant (mg)	Average CNT Product (mg)	Maximum CNT Product (mg)	Maximum CNT Yield (%)
Control	815.87	89.80	92.80	11.37%
1 wt% PS	824.20	80.47	89.36	10.84%
2 wt% PS	832.57	100.36	107.83	12.95%
4 wt% PS	849.23	106.58	121.55	14.31%

Based on the aforementioned observations that increased PS loading, and therefore higher carbon density samples, yield MWCNT with wider walls (Figure 2) with lower amorphous carbon content (Figure 3B) and higher quality due to higher maximum G/D values (Figure 3C), it was concluded that the carbon from the PS had become carbon in the MWCNTs. This correlates with the both the Baker model [40] of carbon fibre growth and the Puretzky model [41] of CNT growth, whereby as more carbon enters the reaction, more walls are created.

$$CNT\ percentage\ yield = \frac{(Mass_{Product}) - (Mass_{Fe\ residue})}{Mass_{Theoretical\ yield}} \times 100 \quad (1)$$

$$CNT\ percentage\ yield = \frac{130.00\ (mg) - 8.45\ (mg)}{849.23\ (mg)} \times 100 = 14.31\% \quad (2)$$

3.4. General Mechanism of Carbon Nanotube Growth

The CNT growth mechanism comprised a multistep process involving the formation of catalysts, the decomposition of carbon sources, and the reconstitution of carbon to a nanotube structure. This process began with the in situ decomposition of ferrocene, i.e., dissociation of the cyclopentadienyl rings sandwiched about the Fe core to render a mixture of Fe_2O_3 and Fe_3O_4 catalytic nanoparticles. The oxide form was anticipated because the reaction chamber was not first pumped down; therefore, it is likely that some amount of oxygen was resident in the chamber despite the fact the system had been flushed with concentrated argon in advance of each reaction.

The iron agglomerated and formed different sized structures ranging from nanoparticles to microparticles. These iron particles acted as catalytic surfaces known to catalytically cleave the C-H bonds in hydrocarbon reactants such as cyclopentadienyl, polystyrene plastics, and toluene. The lifetime and activity of the catalyst was maintained by using a constant flow of hydrogen in the gas stream that simultaneously reduced the metal, making it active for carbon cracking, and attracted errant carbon moieties that might otherwise have saturated the catalyst, leading to coking and eventual catalyst poisoning. Complete control over CNT products is highly challenging because of catalysts are highly structure-sensitive [42] materials, and even slight variance from the optimum condition can lead to drastic changes in the product, as determined by volcano plots. A key struggle with CNT growth is the inverse relationship between sample yield and quality—it is consistently one or the other.

A mechanistic understanding of carbon nanotube growth suggests that carbon in the form of C2 [43–45] enters the catalyst lattice and saturates the molten metal crystal. Following saturation within the catalyst, carbon precipitation occurs at a growth facet, typically a high energy facet such as <111>, exiting the catalyst in graphitic tubular form. The tubular morphology of the carbon is physically bounded by the outer edges and outer diameter of the catalyst particles and the graphitic form is adopted because of the low entropy state of graphite.

3.5. Contribution to CNT Growth from TOLUENE

Toluene decomposition compliments CNT growth from plastics in two distinct ways. Firstly, liquid phase dissolutions allows for sonication and, though resulting cavitation, assists the PS decomposition by decreasing the PS molecular weight [46]. Then, in the gas phase (at the conditions used for CNT growth), toluene readily forms C2 units, leading to CNT nucleation. CNT nucleation is notably a function of hydrocarbon thermal decomposition to suitable fragments (C2 units). C2 formation from thermal decomposition of toluene [47–49] can occur via several pathways. For example, toluene can decompose either through benzyl radical (Equation (3)) or phenyl radical (Equation (4)) formation; however, the former dominates at the temperatures applied in the present study. The benzyl radical undergoes subsequent decomposition to form C5, C3, and C4 units, that can, in turn, decompose further; see Equations (5) and (6). Toluene decomposition leads to the formation of C2, C3 and C4 units that decompose either in the gas phase due to temperature or via radical attack, or on the catalyst particle surface, to render the C2 fragments associated with nanotube growth. Moreover, the highly active C2 units or radicals mentioned herein can also initiate and advance PS decomposition.

$$C_6H_5CH_3 \rightarrow C_6H_5CH_2^* + H^* \tag{3}$$

$$C_6H_5CH_3 \rightarrow C_6H_5^* + CH_3^* \tag{4}$$

$$C_6H_5CH_2^* \rightarrow C_5H_5 + C_2H_2 \tag{5}$$

$$C_6H_5CH_2^* \rightarrow C_3H_3 + C_4H_4 \tag{6}$$

3.6. Contribution to CNT Growth from Polystyrene

PS is a conjugated aromatic polymer structure that can be decomposed by scissioning the conjugated chain into the styrene monomer units via H* attack [50]. H* originates from several sources, both in the gas and liquid phase. For example, liquid phase decomposition of toluene (Equation (3)) liberates H*, especially during the sonication used in the mixing of the liquid or at elevated temperatures. The temperature in the CVD furnace vaporizes toluene, resulting in further H* release which compliments H* originating from the thermal splitting of hydrogen in the gas flow. Further H* can be formed at elevated temperatures by catalytic cracking of hydrocarbon over an iron oxide catalyst [51]. Even without a catalyst, thermal cracking of the polymer produces an aromatic product state consisting of high concentrations of styrene (~50–79 wt%), together with the styrene dimer and trimer and other aromatic compounds including toluene, xylene, and alkylated benzenes [52]. There is a complimentary cascade of chemical reactions induced by various pathways to create H* that assist in the decomposition of polystyrene to styrene monomers. Once decomposed to styrene monomer or a similar hydrocarbon, they are further decomposed to C2 units, allowing CNT growth to occur.

3.7. Contribution to CNT Growth from Cyclopentadienyl Rings

Carbon from the cyclopentadienyl could also contribute to the carbon of the CNT [53]. Cyclopentadiene moieties, sandwiching iron in the case of ferrocene, tend to exhibit relatively high thermal stability on account of the strong binding affinity between those groups and the internal metal atoms [54]. Therefore, complete decomposition of the ferrocene complex was only expected to occur in zone two, where the temperature for growth was set at 780 °C. First, the cyclopentadienyls disassociated from the iron core and underwent further decomposition in the presence of the newly formed catalysts nanoparticles. Reaction products such as methane and ethane readily formed [55]; these compounds are widely used as carbon sources for CNT growth. Therefore, carbon from the cyclopentadiene could also potentially form part of the CNT product. In all reactions, the ferrocene concentration was fixed at 5 wt% of the mass of solvent and plastic. For example, in the control, in the absence of polystyrene plastic, 41.15 mg of ferrocene was used, thus contributing 26.34 mg of carbon. Note however that the average mass of the control reactions was 89.9 mg (Table 2, Figure S5). The carbon from the cyclopentadiene only consisted of ~30% of the average CNT mass in the control reactions, indicating the carbon from the cyclopentadiene could only form part of the CNT product; the remaining carbon must have originated from the solvent, or, when PS was added, from the PS too.

3.8. Carbon Nanotube Devices Measurement and Application

In all cases, the voltage increased linearly with current, indicating ohmic resistance of the CNT films. As a reference, the electrical resistance of a retail piece of copper tape was measured by the same procedure. The electrical resistance of the CNT samples was found to be in the range of 2.4 to 4.4 Ω. This is two orders of magnitude higher than that of the copper reference sample. On average, the use of PS increased electrical resistance (Figure 5). However, the best conductivity performance came from sample PS-2 (2 wt% w/w), i.e., 2.4 Ω compared to R = 0.6 Ω with the control. Although these values are generally higher than that of copper, the mass difference between carbon and copper make these materials an attractive alternative to copper once the I^2R losses are improved, especially since their lightweight nature is critical. Moreover, it is likely that our attempt to remove contributions from catalyst particles [56,57] (used as catalyst) by virtue of the acid wash may inadvertently have resulted in the poor electrical properties. The oxidizing acid wash may have created additional sidewall defects that ultimately hindered electron transport and this may be why no apparent trend in electrical performance was noted regarding PS loading in the feedstock.

By quantifying the uplink and downlink speed of a CNT ethernet cable (Figure 6A) using iPerf3, we concluded that the MWCNT wires were capable of data transfer rates of at least 99 Mbps. Based on our findings, the Cu wire gauge used, as well as the RJ45 connectors, were holding back the potential of the CNT wires, as determined by the fact Cu wires, once connected to identical RJ45 connectors, also maxed out at 94.7 Mbps (Figure 6B). Industrial grade RJ45 connectors certified for the CAT7 standards coupled with higher quality, thicker gauge copper wiring as well as better assembly and crimping would likely help negate the bottleneck. As it stands, the CNT ethernet cables can meet to the rated uplink and downlink speeds of the CAT5 standard, with results saturating at 100 Mbps (Video S1), as has been seen previously [58]. These speeds are adequate for UK adoption and Broadband classification, as determined by the UK Government regulator [59].

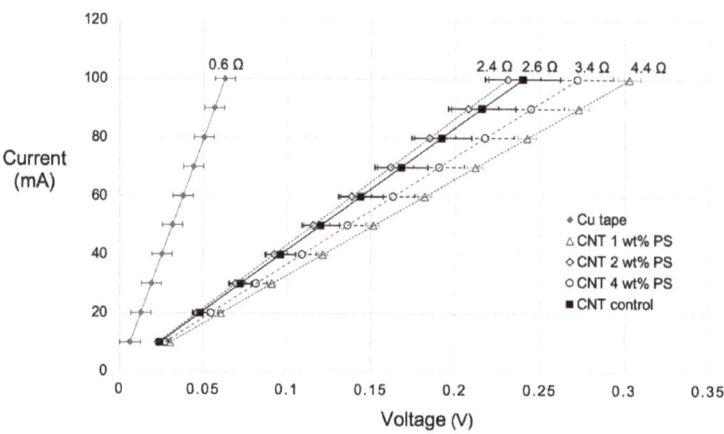

Figure 5. Voltage measurements of acid-washed carbon nanotube samples as compared with commercial copper tape.

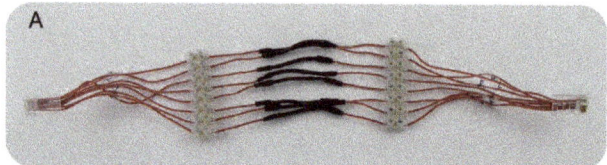

B	Maximum uplink speed (Mbps)	Maximum downlink speed (Mbps)
Cat6 cable	998	995
Copper cable	92.7	94.7
CNT cable	94.9	97.0

Figure 6. Photograph showing the CNT ethernet cable made using polystyrene-toluene feedstock (**A**), and the ethernet speed results using three devices, one (Cat6) commercial device and two lab made devices using CNTs or CU wire as active transmission component (**B**).

3.9. The Life Cycle Assessment of MWCNT Growth and Wire Production Versus Cu Wire Formation

What is the impact of making CNT wires (lab scale) compared to making an equivalent length of Cu wire from an industrial process? Moreover, what carbon saving could be achieved on account of a lighter aircraft over its typical lifespan, and where is the breakeven point for using CNT wires instead of Cu wire? For every 1 kg of CNT powder, 1.58 kg of CO_2 is created (Supplementary information), largely due to the electricity requirement for heating [60,61]. The electricity sources available in the UK did not make a significant impact on decreasing the CO_2 generation (Supplementary information). Using the CNT powder to make 141 miles of cabling, 356.2 kg of CNTs resulted in CO_2 emissions of 545 metric Tonnes (Te), compared to making 1519.6 kg of copper wire, which generated 12 Te of CO_2; this is nearly 45 times more CO_2 just from the manufacturing of CNT cables compared to Cu wires (Figure 7A). However, once economies of scale such as Wright's Law [62,63] have been considered, we anticipate CNT cable manufacturing on a larger scale to decrease CO_2 emissions [64].

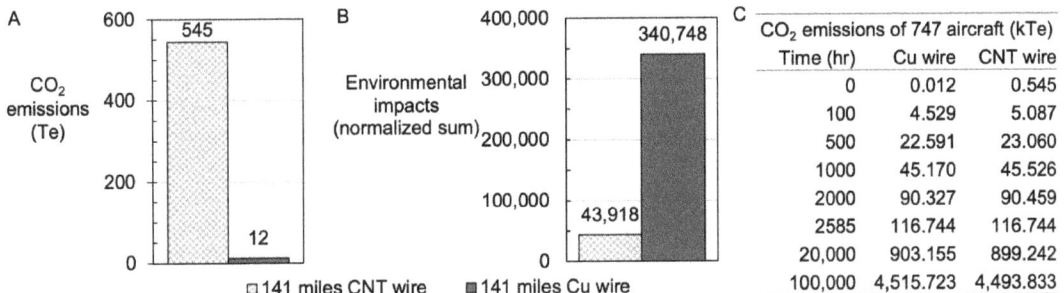

Figure 7. Life cycle analysis comparing (**A**) CO_2 emissions and (**B**) normalized sum of all environmental impacts between the production of 141 miles of CNTs wires versus Cu wire. Comparative table (**C**) showing the CO_2 emission (kTe) per copper versus CNT wires.

Other environmental factors aside from CO_2 emissions exist, such as (but not limited to) freshwater eutrophication, freshwater ecotoxicity, marine ecotoxicity and stratospheric ozone depletion. When the normalised sum [65] of these impacts is compared between CNT (at lab scale) and Cu wires, one can note an approximate eight-fold increase in the negative impact from Cu wire manufacture (Figure 7B).

3.10. The Life Cycle Assessment of Use Case of MWCNTs in a Boeing 747-400 Aircraft

Using CNT wires would make a Boeing 747-400 series 6% lighter; this saving would improve fuel efficiency, such that after just 2585 h of flight-time, overall CO_2 emissions would decrease comparatively (Figure 7C). This savings could be improved once the wire manufacturing process were scaled by virtue of scaling laws in manufacturing proficiencies. Note that almost 99% of the CO_2 eq. emissions are CO_2 emissions (of the 4520 kTe of CO_2 eq. emitted by the plane, 4450 kTe are CO_2 emissions). Due to the lightweight nature of CNT wires compared to heavier Cu wires, over the life span of the aircraft (100,000 h), the use of CNT wires decreases the carbon footprint by 21 kTe per plane. Moreover, given the fact that 694 craft were delivered between 1989 and 2009 [66] a total of 14,574 kTe CO_2 reduction could be projected for the 747-400 fleet. This projection demonstrates how the use of lightweight carbon wire technologies could create positive impact towards solving global CO_2 grand challenges.

4. Conclusions

Multiwalled carbon nanotubes (MWCNTs) were grown using liquid injection chemical vapour deposition (CVD) at 780 °C using ferrocene catalyst particles to obtain carbon from ferrocene, toluene and polystyrene (PS) at various PS concentrations, i.e., from 1 to 4 wt% (w/w). Samples were characterized using scanning electron microscopy (SEM) and transmission electron microscopy (TEM). MWCNT diameters were found to increase with greater PS concentrations due to carbon from PS incorporating into the MWCNTs. Quality was measured using Raman spectroscopy. The maximum Raman G/D values both increased and the range of mean values narrowed due to the presence of higher quality products, with greater homogeneity at greater PS concentration. This synthesis method improved MWCNT quantity without incurring a measurable loss in quality. Due to the electrical nature of MWCNTs, a MWCNT ethernet cable was produced, which found to have ~99.5 Mbps uplink and downlink speeds, i.e., comparable to those of Cu wires of similar diameter. A life cycle assessment (LCA) of the MWCNT wires made using PS suggested that the electricity powering the CVD furnace represented the largest impact. Moreover, the LCA determined that over the lifespan of a single Boeing 747-400, the use of MWCNTs wires would decrease CO_2 production by 21 kTe due to the lightweight nature of MWCNTs which is projected to reduce 14,574 kTe CO_2 footprint across the entire fleet of 747-400 aircraft. This projection demonstrates how a plastic circular economy making lightweight energy transmission carbon cables can impact global grand challenges.

5. Patents

Two patents have been filed from this work. A.H. and A.O.W. filed: PROCESS FOR REUSE OF PLASTIC THROUGH THE CONVERSION TO CARBON NANOMATERIALS United States Patent Application 20190375639; T.Y. and A.O.W. filed: CABLES AND METHODS THEREOF United States Patent Application 20210158995.

Supplementary Materials: The following are available online at https://www.mdpi.com/article/10.3390/nano12010009/s1, Figure S1: MWCNT yield versus order of the reaction, Figure S2: FFT of HRTEM images (MWCNT)., Figure S3: FFT of HRTEM images (local area), Figure S4: FFT of whole HRTEM image, Figure S5: Screenshot of mass balance calculations for reactants showing relevant carbon content, Video S1: Video showing the iPerf3 testing of ethernet speeds of CNT ethernet cable.

Author Contributions: Conceptualization, A.O.W.; methodology, A.O.W.; software, A.O.A., L.P.; validation, V.S.G., R.E.P., R.C., M.L.D.; formal analysis, A.O.W.; investigation, A.H., T.Y., Y.N., S.L., V.S.G., L.P., G.S., A.C., J.S.-C. and A.O.W.; resources, R.E.P., M.L.D. and A.O.W.; data curation, A.O.W.; writing—original draft preparation, A.O.W., A.H.; writing—review and editing, A.O.W., R.E.P., M.L.D., R.C., A.H., T.Y., Y.N., S.L., V.S.G., L.P., G.S. and I.T.; visualization, A.O.W.; supervision, A.O.W., R.E.P., M.L.D. and S.J.C.; project administration, A.O.W.; funding acquisition, A.O.W., R.E.P., M.L.D. and S.J.C. All authors have read and agreed to the published version of the manuscript.

Funding: A.O.W. is funded through Sêr Cymru II Fellowship by the Welsh Government and the European Regional Development Fund (ERDF). A.O.W. and T.Y. acknowledges funding from Welsh Government Circular Economy Capital Fund FY 2020-21. A.H. was funded by the Copper Nanotube Ultraconductive (UCC) wire project funded by Sêr Cymru National Research Network for Advanced Engineering and Materials (NRN) with contributions from E-Corp. G.S. received funding from Swansea Employability Academy (SEA) via the summer placements scheme. Thanks to funding by Welsh Government for Knowledge Economy Skills Scholarships (KESS2), part funded by the Welsh Government's European Social Fund (ESF) convergence programme for West Wales and the Valleys. L.P. and I.T. funded through KESS2 and TRIMTABS Ltd. J.S.-C. funded through KESS2 and Salts Healthcare Ltd. M.L.D and R.C. are grateful for the financial support of the EPSRC (EP/S001336/1) and for funding LCA software and database licenses. We would like to acknowledge the Life Cycle Analysis for Circular Economy (LCA4CE) Lab. J.S.-C. & V.S.G. funded thanks to Salts Healthcare Ltd. A.A. funded in part by the Swansea University Texas Strategic Partnership. The authors acknowledge access to the TEM provided by the Swansea University AIM Facility, funded in part by the EPSRC (EP/M028267/1), the European Regional Development Fund through the Welsh Government (80708) and the Welsh Government's Sêr Cymru program. The funders had no role in the design of the study; in the collection, analyses, or interpretation of data; in the writing of the manuscript, or in the decision to publish the results.

Data Availability Statement: The data presented in this study are available on request from the corresponding author.

Acknowledgments: We would like to thank Keysight Technologies for the use of a test model of the B2900A SMU. We would like to acknowledge the assistance provided by Swansea University College of Engineering AIM Facility. We would like to thank TRIMTABS Ltd. for purchasing equipment required for making ethernet cables. Thanks to Swansea Employability Academy (SEA) for the summer placements scheme. Thanks to the Swansea University Texas Strategic Partnership. R.E.P. acknowledges his work was associated with the IMPACT operation. We acknowledge pixabay for use of imagery in the graphical abstract (https://pixabay.com/vectors/airplane-boeing-747-transport-4811157/ (accessed on 1 December 2021)).

Conflicts of Interest: A.O.W. founded TRIMTABS Ltd. who co-sponsored I.T. and L.P. through the KESS2 program. All other authors declare no conflict of interest. The funders had no role in the design of the study; in the collection, analyses, or interpretation of data; in the writing of the manuscript, or in the decision to publish the results.

References

1. Smalley, R.E. Future Global Energy Prosperity: The Terawatt Challenge. *MRS Bull.* **2005**, *67*, 412–417. [CrossRef]
2. Anuar Sharuddin, S.D.; Abnisa, F.; Wan Daud, W.M.A.; Aroua, M.K. A Review on Pyrolysis of Plastic Wastes. *Energy Convers. Manag.* **2016**, *115*, 308–326. [CrossRef]
3. Khodabakhshi, S.; Kiani, S.; Niu, Y.; White, A.O.; Suwaileh, W.; Palmer, R.E.; Barron, A.R.; Andreoli, E. Facile and Environmentally Friendly Synthesis of Ultramicroporous Carbon Spheres: A Significant Improvement in CVD Method. *Carbon* **2021**, *171*, 426–436. [CrossRef]
4. Geyer, R.; Jambeck, J.R.; Law, K.L. Production, Uses, and Fate of All Plastics Ever Made. *Sci. Adv.* **2017**, *3*, 5. [CrossRef]
5. Bazargan, A.; McKay, G. A Review—Synthesis of Carbon Nanotubes from Plastic Wastes. *Chem. Eng. J.* **2012**, *195–196*, 377–391. [CrossRef]
6. Acomb, J.C.; Wu, C.; Williams, P.T. Control of Steam Input to the Pyrolysis-Gasification of Waste Plastics for Improved Production of Hydrogen or Carbon Nanotubes. *Appl. Catal. B Environ.* **2014**, *147*, 571–584. [CrossRef]
7. Aboul-Enein, A.A.; Adel-Rahman, H.; Haggar, A.M.; Awadallah, A.E. Simple Method for Synthesis of Carbon Nanotubes over Ni-Mo/Al 2 O 3 Catalyst via Pyrolysis of Polyethylene Waste Using a Two-Stage Process. *Fuller. Nanotub. Carbon Nanostruct.* **2017**, *25*, 211–222. [CrossRef]
8. El Essawy, N.A.; Konsowa, A.H.; Elnouby, M.; Farag, H.A. A Novel One-Step Synthesis for Carbon-Based Nanomaterials from Polyethylene Terephthalate (PET) Bottles Waste. *J. Air Waste Manag. Assoc.* **2017**, *67*, 358–370. [CrossRef]
9. Liu, X.; Sun, H.; Wu, C.; Patel, D.; Huang, J. Thermal Chemical Conversion of High-Density Polyethylene for the Production of Valuable Carbon Nanotubes Using Ni/AAO Membrane Catalyst. *Energy Fuels* **2018**, *32*, 4511–4520. [CrossRef]
10. Arena, U.; Mastellone, M.L.; Camino, G.; Boccaleri, E. An Innovative Process for Mass Production of Multi-Wall Carbon Nanotubes by Means of Low-Cost Pyrolysis of Polyolefins. *Polym. Degrad. Stab.* **2006**, *91*, 763–768. [CrossRef]
11. Miller-Chou, B.A.; Koenig, J.L. A Review of Polymer Dissolution. *Prog. Polym. Sci.* **2003**, *28*, 1223–1270. [CrossRef]
12. Puengjinda, P.; Sano, N.; Tanthapanichakoon, W.; Charinpanitkul, T. Selective Synthesis of Carbon Nanotubes and Nanocapsules Using Naphthalene Pyrolysis Assisted with Ferrocene. *J. Ind. Eng. Chem.* **2009**, *15*, 375–380. [CrossRef]
13. Charinpanitkul, T.; Sano, N.; Puengjinda, P.; Klanwan, J.; Akrapattangkul, N.; Tanthapanichakoon, W. Naphthalene as an Alternative Carbon Source for Pyrolytic Synthesis of Carbon Nanostructures. *J. Anal. Appl. Pyrolysis* **2009**, *86*, 386–390. [CrossRef]
14. Orbaek, A.W.; Barron, A.R. Towards a 'Catalyst Activity Map' Regarding the Nucleation and Growth of Single Walled Carbon Nanotubes. *J. Exp. Nanosci.* **2015**, *10*, 66–76. [CrossRef]
15. Hedayati, A.; Barnett, C.; Swan, G.; Orbaek White, A. Chemical Recycling of Consumer-Grade Black Plastic into Electrically Conductive Carbon Nanotubes. *C* **2019**, *5*, 32. [CrossRef]
16. Narasimhan, B.; Peppas, N.A. Disentanglement and Reptation during Dissolution of Rubbery Polymers. *J. Polym. Sci. Part B Polym. Phys.* **1996**, *34*, 947–961. [CrossRef]
17. Achilias, D.S.; Giannoulis, A.; Papageorgiou, G.Z. Recycling of Polymers from Plastic Packaging Materials Using the Dissolution–Reprecipitation Technique. *Polym. Bull.* **2009**, *63*, 449–465. [CrossRef]
18. Bjørn, A.; Owsianiak, M.; Molin, C.; Laurent, A. Main Characteristics of LCA. In *Life Cycle Assessment Theory and Practice*, 1st ed.; Hauschild, M., Rosenbaum, R., Olsen, S., Eds.; Springer: Cham, Switzerland, 2018; pp. 9–16.
19. Heidari, M.; Younesi, H. Synthesis, Characterization and Life Cycle Assessment of Carbon Nanospheres from Waste Tires Pyrolysis over Ferrocene Catalyst. *J. Environ. Chem. Eng.* **2020**, *8*, 103669. [CrossRef]
20. Rybicka, J.; Tiwari, A.; Leeke, G.A. Technology Readiness Level Assessment of Composites Recycling Technologies. *J. Clean. Prod.* **2016**, *112*, 1001–1012. [CrossRef]
21. Al-Salem, S.M.; Lettieri, P.; Baeyens, J. Recycling and Recovery Routes of Plastic Solid Waste (PSW): A Review. *Waste Manag.* **2009**, *29*, 2625–2643. [CrossRef]
22. Kwon, J.-H.; Kim, J.-W.; Pham, T.D.; Tarafdar, A.; Hong, S.; Chun, S.-H.; Lee, S.-H.; Kang, D.-Y.; Kim, J.-Y.; Kim, S.-B.; et al. Microplastics in Food: A Review on Analytical Methods and Challenges. *Int. J. Environ. Res. Public Health* **2020**, *17*, 6710. [CrossRef]
23. Orbaek, A.W.; Aggarwal, N.; Barron, A.R. The Development of a 'Process Map' for the Growth of Carbon Nanomaterials from Ferrocene by Injection CVD. *J. Mater. Chem. A* **2013**, *1*, 14122. [CrossRef]
24. Gangoli, V.S.; Raja, P.M.V.; Esquenazi, G.L.; Barron, A.R. The Safe Handling of Bulk Low-Density Nanomaterials. *SN Appl. Sci.* **2019**, *1*, 644. [CrossRef]
25. Dee, N.T.; Li, J.; White, A.O.; Jacob, C.; Shi, W.; Kidambi, P.R.; Cui, K.; Zakharov, D.N.; Janković, N.Z.; Bedewy, M.; et al. Carbon-Assisted Catalyst Pretreatment Enables Straightforward Synthesis of High-Density Carbon Nanotube Forests. *Carbon N. Y.* **2019**, *153*, 196–205. [CrossRef]
26. Rueden, C.T.; Schindelin, J.; Hiner, M.C.; DeZonia, B.E.; Walter, A.E.; Arena, E.T.; Eliceiri, K.W. ImageJ2: ImageJ for the next Generation of Scientific Image Data. *BMC Bioinform.* **2017**, *18*, 1–26. [CrossRef]
27. King, S.G.; McCafferty, L.; Stolojan, V.; Silva, S.R.P. Highly Aligned Arrays of Super Resilient Carbon Nanotubes by Steam Purification. *Carbon N. Y.* **2015**, *84*, 130–137. [CrossRef]
28. Chiang, I.W.; Brinson, B.E.; Smalley, R.E.; Margrave, J.L.; Hauge, R.H. Purification and Characterization of Single-Wall Carbon Nanotubes. *J. Phys. Chem. B* **2001**, *105*, 1157–1161. [CrossRef]

29. Mansfield, E.; Kar, A.; Hooker, S.A. Applications of TGA in Quality Control of SWCNTs. *Anal. Bioanal. Chem.* **2010**, *396*, 1071–1077. [CrossRef]
30. Khodabakhhshi, S.; Fulvio, P.F.; Sousaraei, A.; Kiani, S.; Niu, Y.; Palmer, R.E.; Kuo, W.C.H.; Rudd, J.; Barron, A.R.; Andreoli, E. Oxidative Synthesis of Yellow Photoluminescent Carbon Nanoribbons from Carbon Black. *Carbon N. Y.* **2021**, *183*, 495–503. [CrossRef]
31. Andreoli, E.; Suzuki, R.; Orbaek, A.W.; Bhutani, M.S.; Hauge, R.H.; Adams, W.; Fleming, J.B.; Barron, A.R. Preparation and Evaluation of Polyethyleneimine-Single Walled Carbon Nanotube Conjugates as Vectors for Pancreatic Cancer Treatment. *J. Mater. Chem. B* **2014**, *2*, 4740. [CrossRef]
32. Chiang, I.W.; Brinson, B.E.; Huang, A.Y.; Willis, P.A.; Bronikowski, M.J.; Margrave, J.L.; Smalley, R.E.; Hauge, R.H. Purification and Characterization of Single-Wall Carbon Nanotubes (SWNTs) Obtained from the Gas-Phase Decomposition of CO (HiPco Process). *J. Phys. Chem. B* **2001**, *105*, 8297–8301. [CrossRef]
33. López-Lorente, A.I.; Simonet, B.M.; Valcárcel, M. The Potential of Carbon Nanotube Membranes for Analytical Separations. *Anal. Chem.* **2010**, *82*, 5399–5407. [CrossRef]
34. Aircraft Electrical Wire. Available online: https://www.mitrecaasd.org/atsrac/FAA_PI-Engineer_Workshop/2001/aircraft_electrical_wire.pdf (accessed on 21 December 2021).
35. Lu, Q.; Keskar, G.; Ciocan, R.; Rao, R.; Mathur, R.B.; Rao, A.M.; Larcom, L.L. Determination of Carbon Nanotube Density by Gradient Sedimentation. *J. Phys. Chem. B* **2006**, *110*, 24371–24376. [CrossRef]
36. Dresselhaus, M.S.; Dresselhaus, G.; Jorio, A. Raman Spectroscopy of Carbon Nanotubes in 1997 and 2007. *J. Phys. Chem. C* **2007**, *111*, 17887–17893. [CrossRef]
37. Schwan, J.; Ulrich, S.; Batori, V.; Ehrhardt, H.; Silva, S.R.P. Raman Spectroscopy on Amorphous Carbon Films. *J. Appl. Phys.* **1996**, *80*, 440–447. [CrossRef]
38. Box Plot. Available online: https://www.itl.nist.gov/div898/handbook/eda/section3/boxplot.htm (accessed on 21 October 2021).
39. Meam Plot. Available online: https://www.itl.nist.gov/div898/handbook/eda/section3/meanplot.htm (accessed on 21 October 2021).
40. Baker, R.T.K. Catalytic Growth of Carbon Filaments. *Carbon N. Y.* **1989**, *27*, 315–323. [CrossRef]
41. Wood, R.F.; Pannala, S.; Wells, J.C.; Puretzky, A.A.; Geohegan, D.B. Simple Model of the Interrelation between Single- and Multiwall Carbon Nanotube Growth Rates for the CVD Process. *Phys. Rev. B* **2007**, *75*, 235446. [CrossRef]
42. Somorjai, G.A.; Carrazza, J. Structure Sensitivity of Catalytic Reactions. *Ind. Eng. Chem. Fundam.* **1986**, *25*, 63–69. [CrossRef]
43. Rao, R.; Liptak, D.; Cherukuri, T.; Yakobson, B.I.; Maruyama, B. In Situ Evidence for Chirality-Dependent Growth Rates of Individual Carbon Nanotubes. *Nat. Mater.* **2012**, *11*, 213–216. [CrossRef]
44. Marchand, M.; Journet, C.; Guillot, D.; Benoit, J.-M.; Yakobson, B.I.; Purcell, S.T. Growing a Carbon Nanotube Atom by Atom: "And Yet It Does Turn". *Nano Lett.* **2009**, *9*, 2961–2966. [CrossRef]
45. Maksimova, N.I.; Krivoruchko, O.P.; Mestl, G.; Zaikovskii, V.I.; Chuvilin, A.L.; Salanov, A.N.; Burgina, E.B. Catalytic Synthesis of Carbon Nanostructures from Polymer Precursors. *J. Mol. Catal. A Chem.* **2000**, *158*, 301–307. [CrossRef]
46. Gogate, P.R.; Prajapat, A.L. Depolymerization using sonochemical reactors: A critical review. *Ultrason. Sonochem.* **2015**, *27*, 480–494. [CrossRef]
47. Matsugi, A. Thermal Decomposition of Benzyl Radicals: Kinetics and Spectroscopy in a Shock Tube. *J. Phys. Chem. A* **2020**, *124*, 824–835. [CrossRef]
48. Brouwer, L.D.; Mueller-Markgraf, W.; Troe, J. Thermal Decomposition of Toluene: A Comparison of Thermal and Laser-Photochemical Activation Experiments. *J. Phys. Chem.* **1988**, *92*, 4905–4914. [CrossRef]
49. Oehlschlaeger, M.A.; Davidson, D.F.; Hanson, R.K. Thermal Decomposition of Toluene: Overall Rate and Branching Ratio. *Proc. Combust. Inst.* **2007**, *31*, 211–219. [CrossRef]
50. Maafa, I.M. Pyrolysisof Polystyrene Waste: A Review. *Polymers* **2021**, *13*, 225. [CrossRef]
51. Fürstner, A. Iron Catalysis in Organic Synthesis: A Critical Assessment of What It Takes to Make This Base Metal a Multitasking Champion. *ACS Cent. Sci.* **2016**, *2*, 778–789. [CrossRef] [PubMed]
52. Williams, P.T. Hydrogen and Carbon Nanotubes from Pyrolysis-Catalysis of Waste Plastics: A Review. *Waste Biomass Valorization* **2021**, *12*, 1–28. [CrossRef]
53. Rao, C.N.R.; Sen, R. Large aligned-nanotube bundles from ferrocene pyrolysis. *Chem. Commun.* **1998**, *15*, 1525–1526. [CrossRef]
54. An, J.; Choi, E.; Shim, S.; Kim, H.; Kang, G.; Yun, J. Thermal Decomposition In Situ Monitoring System of the Gas Phase Cyclopentadienyl Tris(dimethylamino) Zirconium (CpZr(NMe2)3) Based on FT-IR and QMS for Atomic Layer Deposition. *Nanoscale Res. Lett.* **2020**, *15*, 175. [CrossRef]
55. Wulan, P.P.D.K.; Rivai, G.T. Synthesis of carbon nanotube using ferrocene as carbon source and catalyst in a vertically structured catalyst reactor. *E3S Web Conf.* **2018**, *67*, 03038. [CrossRef]
56. Barnett, C.J.; McCormack, J.E.; Deemer, E.M.; Evans, C.R.; Evans, J.E.; White, A.O.; Dunstan, P.R.; Chianelli, R.R.; Cobley, R.J.; Barron, A.R. Enhancement of Multiwalled Carbon Nanotubes' Electrical Conductivity Using Metal Nanoscale Copper Contacts and Its Implications for Carbon Nanotube-Enhanced Copper Conductivity. *J. Phys. Chem. C* **2020**, *124*, 18777–18783. [CrossRef]
57. Cesano, F.; Uddin, M.J.; Lozano, K.; Zanetti, M.; Scarano, D. All-Carbon Conductors for Electronic and Electrical Wiring Applications. *Front. Mater.* **2020**, *7*, 219. [CrossRef]

58. Jarosz, P.; Schauerman, C.; Alvarenga, J.; Moses, B.; Mastrangelo, T.; Raffaelle, R.; Ridgley, R.; Landi, B. Carbon Nanotube Wires and Cables: Near-Term Applications and Future Perspectives. *Nanoscale* **2011**, *3*, 4542. [CrossRef] [PubMed]
59. House of Commons Library, Broadband. Available online: https://commonslibrary.parliament.uk/broadband-faqs/ (accessed on 27 October 2021).
60. Upadhyayula, V.K.K.; Meyer, D.E.; Curran, M.A.; Gonzalez, M.A. Life Cycle Assessment as a Tool to Enhance the Environmental Performance of Carbon Nanotube Products: A Review. *J. Clean. Prod.* **2012**, *26*, 37–47. [CrossRef]
61. Griffiths, O.G.; O'Byrne, J.P.; Torrente-Murciano, L.; Jones, M.D.; Mattia, D.; McManus, M.C. Identifying the Largest Environmental Life Cycle Impacts during Carbon Nanotube Synthesis via Chemical Vapour Deposition. *J. Clean. Prod.* **2013**, *42*, 180–189. [CrossRef]
62. Nagy, B.; Farmer, J.D.; Bui, Q.M.; Trancik, J.E. Statistical Basis for Predicting Technological Progress. *PLoS ONE* **2013**, *8*, e52669. [CrossRef]
63. Pawelke, R.H. Marrying Wright's Law to Thermodynamics for an Ideal Relative Final Cost-Predicting Model of Carbon-Fuel Substitution. *ChemRxiv* **2021**. [CrossRef]
64. Gavankar, S.; Suh, S.; Keller, A.A. The Role of Scale and Technology Maturity in Life Cycle Assessment of Emerging Technologies: A Case Study on Carbon Nanotubes. *J. Ind. Ecol.* **2015**, *19*, 51–60. [CrossRef]
65. Healy, M.L.; Dahlben, L.J.; Isaacs, J.A. Environmental Assessment of Single-Walled Carbon Nanotube Processes. *J. Ind. Ecol.* **2008**, *12*, 376–393. [CrossRef]
66. Boeing 747-400. Available online: https://en.wikipedia.org/wiki/Boeing_747-400 (accessed on 21 December 2021).

Ferric Ion Diffusion for MOF-Polymer Composite with Internal Boundary Sinks

Kirsten I. Louw, Bronwyn H. Bradshaw-Hajek * and James M. Hill

UniSA STEM, University of South Australia, Mawson Lakes, SA 5095, Australia; kirsten.louw@mymail.unisa.edu.au (K.I.L.); jim.hill@unisa.edu.au (J.M.H.)
* Correspondence: bronwyn.hajek@unisa.edu.au; Tel.: +61-8-8302-3084

Abstract: Simple and economical ferric ion detection is necessary in many industries. An europium-based metal organic framework has selective sensing properties for solutions containing ferric ions and shows promise as a key component in a new sensor. We study an idealised sensor that consists of metal organic framework (MOF) crystals placed on a polymer surface. A two-dimensional diffusion model is used to predict the movement of ferric ions through the solution and polymer, and the ferric ion association to a MOF crystal at the boundary between the different media. A simplified one-dimensional model identifies the choice of appropriate values for the dimensionless parameters required to optimise the time for a MOF crystal to reach steady state. The model predicts that a large non-dimensional diffusion coefficient and an effective association with a small effective flux will reduce the time to steady-state. The effective dissociation is the most significant parameter to aid the estimation of the ferric ion concentration. This paper provides some theoretical insight for material scientists to optimise the design of a new ferric ion sensor.

Keywords: diffusion; ferric ion sensor; MOF; finite difference; composite materials

1. Introduction

There are many situations in which it is important to monitor the concentration of ferric ions (trivalent iron cation or iron(III)) in a solution. For example, in environmental contexts, a high concentration of ferric ions promotes bacterial and algal growth, which can lead to the death of aquatic animals and plants [1,2]. In the mining industry, the concentration of ferric ions can affect copper and gold yield during mineral leaching, with both high and low concentrations producing adverse effects [3,4]. Other areas in which a knowledge of the ferric ion concentration is important include health and drinking water quality assurance [5,6].

This research is part of a larger project on lean mineral processing wherein a new sensor is being developed to detect ferric ions in a time-, cost-, and energy-efficient manner. Current ferric ion sensing methods are complex, use expensive equipment, suffer from interference from other ions, and are not adaptable to real time monitoring [7]. In addition, ferric ion concentrations are often determined indirectly [7]. As a consequence, a new ferric ion sensor that is stable, cost-effective, energy-efficient, and easy to use is desirable and would impact many sectors. In this paper, we present an idealised model of a ferric ion sensor. We propose a diffusion model of ferric ions through two different media with a sink located at the boundary separating the media.

Xu et al. [8] recently established the effectiveness of a europium-based metal organic framework (EuMOF) for ferric ion sensing. The crystalline EuMOF was suspended in an aqueous solution and tested against various concentrations of metal ions for changes in luminescence. Most MOF ferric ion sensors are intensity-based "turn-off" sensors where the intensity of light emitted by the sensor diminishes in the presence of ferric ions. However, the EuMOF [8] is bimodal, so changes in the emission ratio of two frequency peaks are linearly proportional to the ferric ion concentration, while the intensity of the emission is not as important. Metal ions such as Fe^{2+}, Ag^{2+}, Ca^{2+}, and Zn^{2+} have little effect on the emission ratio, and consequently, the ferric ion selectivity of the EuMOF has promising sensing capabilities.

In [9] we investigated the van der Waals interaction energies and forces between a hydrated ferric ion and an EuMOF crystal pore. The Coulombic forces were not considered since ferric ions exist in highly acidic solutions, and the electrostatic forces are negligible [9]. The findings suggest that the hydrated ferric ion is attracted to the pore but does not enter due to steric interactions. This is advantageous from a practical point of view because the hydrated ferric ion can be "washed" away and the sensor can be reused.

In this paper, we investigate how ferric ions diffuse through a solution and a polymer layer and interact with a MOF crystal located on the boundary between the solution and the polymer. One-dimensional diffusion through multiple layers has been studied before. Hickson et al. used a numerical approach to solving diffusion through multiple layers while considering various matching conditions at the boundary [10]. Carr and Turner [11] developed a semi-analytical method to address the complexities that arise for a large number of layers. An analytical solution exists for the one-dimensional problem with two layers (or m layers) for slabs, cylinders, and spheres [12].

A proposed new ferric ion sensor consists of a thin film of polymer coating a cut-away section of optic fibre and embedded with EuMOF crystals. In this scenario, the sensor will have MOF crystals distributed throughout the polymer composite and over its surface. In our model, we place an EuMOF crystal at the boundary between a solution and a polymer, and we investigate the changing concentration of ferric ions bound to the crystal. The paper provides important insight into ferric ion behaviour for a proposed sensor that is still under development and for which there is no readily available experimental data. Squires et al. [13] explore how the physical attributes of a sensor affect the flux onto the sensor, the rate at which substance of interest binds to the sensor, and the approximate times scales for the system to reach equilibrium. They present simple rules to equip the reader with a basic understanding of the environment and enable the design of a sensor better suited to the environment. Yariv [14] developed an advection–diffusion–reaction model of analytes binding to a sensor located on a solid surface adjacent to a shear flow of solution, using an equation first proposed in [13] to describe the concentration on the sensor. Here, we use a similar model to represent the association and disassociation of ferric ions to the EuMOF crystals.

In Section 2, we describe the geometry and the physical environment of the ferric ion sensor. We present a mathematical model that describes both the diffusion of ferric ions through an analyte solution and a polymer region, as well as the association of ferric ions to a MOF crystal located at the boundary between the two regions. Section 3 discusses outcomes of the model and the role and importance of each of the parameters.

2. Mathematical Modelling and Assumptions

One possible design for a ferric ion sensor is to suspend EuMOF crystals in a polymer composite to create a thin film on a cut away section of an optic fibre, as described by [15]. The ferric ions bind to the MOF pores on the crystal surface. A light pulse is sent through the MOF–polymer-coated optic fibre and a detector measures changes in luminosity. Changes in luminosity are indicative of the concentration of ferric ions in the solution. Figure 1 depicts a cross section of the optic fibre and MOF–polymer coating.

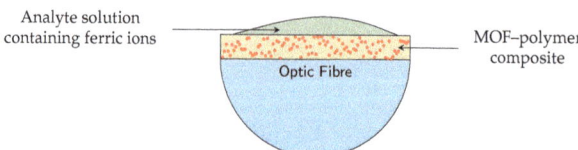

Figure 1. Two-dimensional cross-section of optic fibre with MOF–polymer thin film. Red dots represent MOF crystals in yellow polymer composite. Green represents solution being analysed.

Here, we are particularly interested in the diffusion of the ferric ions through the mixed media and their association with the MOF crystals in order to gain some insight into which physical parameters might be most important. As such, we study an idealised version of the device depicted in Figure 2. We model only the process of diffusion in the analyte solution, the association of ferric ions to a MOF crystal located at the interface between the polymer composite and the solution, and diffusion in the polymer composite. We position one MOF crystal at the interface between the solution and the polymer matrix. We do not consider the association of ferric ions with MOF crystals inside the polymer composite.

The MOF–polymer composite is exposed to a solution containing ferric ions, where the movement of ferric ions is governed by Brownian motion. We assume that the diffusivity of the ferric ions is constant through the solution, denoted by D_1. Since, in practice, the polymer is kept hydrated and the ferric ions do not enter the MOF pores, we assume that the diffusivity in the polymer is also constant, $D_2 < D_1$, and the Vrentas–Duda theory is not applicable. As a consequence, the movement of ferric ions in both the solution and the polymer is assumed to be governed by the conventional diffusion equation with distinct diffusivities,

$$\frac{\partial}{\partial T}C(T,\mathbf{X}) = D_1 \nabla^2 C(T,\mathbf{X}), \qquad X_2 \geq 0,$$
$$\frac{\partial}{\partial T}C(T,\mathbf{X}) = D_2 \nabla^2 C(T,\mathbf{X}), \qquad X_2 < 0,$$

where $C(T,\mathbf{X})$ is the concentration of ferric ions, and the origin is located at the centre of the MOF strip so that the X_1-axis coincides with the solution–polymer boundary. Here, ∇^2 is the usual Laplacian in two dimensions. The ferric ion concentration profile in the analyte solution is given by $C(T,\mathbf{X})$ for $X_2 \geq 0$, and the concentration in the polymer is given by $C(T,\mathbf{X})$ for $X_2 < 0$. The time since the sensor's exposure to the solution is given by T.

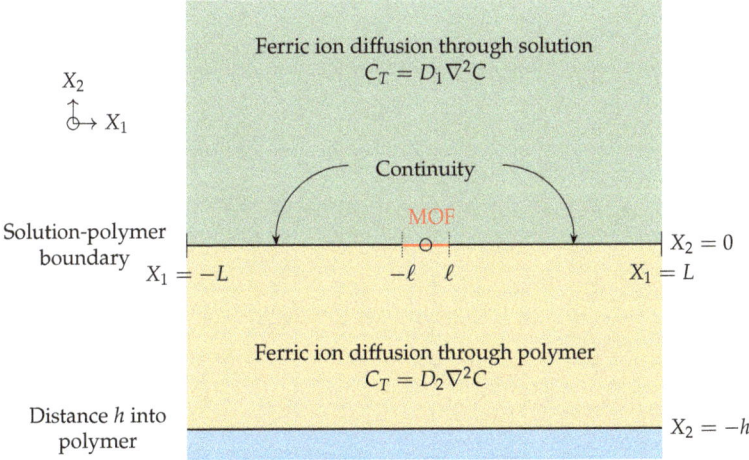

Figure 2. Two-dimensional system: X_1-axis is solution–polymer boundary, $X_2 > 0$ is distance into solution, and $X_2 < 0$ is into polymer composite.

To ensure that continuity is maintained for boundaries shared by two media, we include two additional conditions. The ferric ion concentration at the boundary is assumed to be the same when approaching from above or below the boundary, and the flux out of the solution is equivalent to the flux into the polymer [10],

$$C_+ = C_-, \qquad D_1 \frac{\partial C_+}{\partial X_2} = D_2 \frac{\partial C_-}{\partial X_2}, \qquad X_2 = 0.$$

Far from the sensor in the direction of the analyte solution, we assume a far field boundary condition, where the ferric ion concentration remains constant and equal to that of the bulk solution C_0. For numerical purposes, we set the location of this condition to be at $X_2 = L$, so that

$$C(T, X_1, L) = C_0, \qquad X_2 = L,$$

where L is assumed to be a large distance away from the solution–polymer boundary. We also assume no flux at a horizontal distance L away from the MOF centre (symmetry) and at the bottom boundary of the polymer composite,

$$\frac{\partial C}{\partial X_1} = 0, \qquad X_1 = \pm L, \qquad \frac{\partial C}{\partial X_2} = 0, \qquad X_2 = -h.$$

The flux condition at the solution–MOF interface is given by

$$D_1 \frac{\partial C_+}{\partial X_2} = a[k_{on}C(B_0 - B) - k_{off}B], \qquad |X_1| \leq \ell, \qquad X_2 = 0,$$

where $B(T, X_1)$ is the occupancy of the MOF pores by the ferric ions, k_{on} is the association rate compared to the ferric ion concentration, k_{off} is the dissociation rate, B_0 is the number of MOF pores available for ferric ion occupation, and a is a constant. Due to steric effects, the ferric ion only associates at the MOF pore's entrance and does not diffuse into the MOF pore. The first term captures the association of ferric ions to the MOF crystal, and this term vanishes when the MOF crystal is completely occupied. The second term captures the dissociation of ferric ions from the MOF pore.

We assume that ferric ions only associate and dissociate from the MOF pores in the X_2-direction and impose a no-flux boundary condition to the sides of the MOF crystal,

$$\frac{\partial C}{\partial X_1} = 0, \qquad |X_1| = \ell, \qquad X_2 = 0.$$

The occupancy of the MOF pores is governed by the pseudo-ordinary differential equation (ODE),

$$\frac{\partial B}{\partial T} = k_{on}C(B_0 - B) - k_{off}B, \qquad |X_1| \leq \ell, \qquad X_2 = 0.$$

In addition, since B_0 is the number of MOF pores available, $B(T, X_1) \in [0, B_0]$ and the total ferric ion occupancy in the MOF crystal is,

$$B_{tot}(T) = \int_{-\ell}^{\ell} B(T, X_1) \, dX_1 \leq 2\ell B_0.$$

The assumed initial conditions are

$$C(0, \mathbf{X}) = C_0, \qquad X_2 \geq 0, \qquad C(0, \mathbf{X}) = 0, \qquad X_2 < 0,$$
$$B(0, X_1) = 0, \qquad |X_1| \leq \ell, \qquad X_2 = 0.$$

2.1. Dimensionless Two-Dimensional Model

We non-dimensionalise distance with the half length of the MOF strip, ℓ, the ferric ion concentration with the bulk concentration, C_0, and the MOF binding site occupancy with the maximum occupancy, B_0. There are two options for the time scale: a diffusive time-scale, where $\tau = \ell^2/D_1$, or an association time-scale, where $\tau = 1/C_0 k_{on}$. At this stage, we non-dimensionalise time with τ without specifying which time-scale so that the system of equations can be written

$$\frac{\partial c}{\partial t} = s_1 \nabla^2 c, \quad x_2 \geq 0, \qquad \frac{\partial c}{\partial t} = s_1 D \nabla^2 c, \quad x_2 < 0, \tag{1}$$

$$\frac{\partial b}{\partial t} = s_2 [c(1-b) - s_3 b], \quad |x_1| \leq 1, \quad x_2 = 0, \tag{2}$$

where $D = D_2/D_1$, $s_1 = D_1 \tau / \ell^2$, $s_2 = C_0 k_{on} \tau$ and $s_3 = k_{off}/C_0 k_{on}$ are non-dimensional parameters. If the problem is scaled with the diffusive time-scale, then $s_1 = 1$. If the problem is scaled with the association time-scale, then $s_2 = 1$.

The continuity conditions at the solution–polymer boundary become

$$c_+ = c_-, \qquad \frac{\partial c_+}{\partial x_2} = D \frac{\partial c_-}{\partial x_2}, \qquad x_2 = 0. \tag{3}$$

Far away from the solution–polymer boundary, the boundary condition becomes

$$c(t, x_1, L/\ell) = 1, \qquad x_2 = L/\ell, \tag{4}$$

while the no-flux boundary conditions at the sides and at the polymer–optic fibre boundary become

$$\frac{\partial c}{\partial x_1} = 0, \quad x_1 = \pm L/\ell, \qquad \frac{\partial c}{\partial x_2} = 0, \quad x_2 = -h/\ell. \tag{5}$$

At the solution–MOF boundary,

$$s_1 \frac{\partial c_+}{\partial x_2} = s_4 s_2 [c(1-b) - s_3 b], \quad |x_1| \leq 1, \; x_2 = 0, \tag{6}$$

where $s_4 = a B_0 / \ell C_0$ is a non-dimensional constant, and the no-flux boundary conditions at the sides of the MOF crystal become

$$\frac{\partial c}{\partial x_1} = 0, \quad |x_1| = 1. \tag{7}$$

The initial conditions and the total MOF pore occupancy are

$$c(0, \mathbf{x}) = 1, \quad x_2 \geq 0, \qquad c(0, \mathbf{x}) = 0, \quad x_2 < 0, \tag{8}$$

$$b(0, x_1) = 0, \quad |x_1| \leq 1, \qquad x_2 = 0 \tag{9}$$

$$b_{tot}(t) = \int_{-1}^{1} b(t, x_1) \, dx_1 \leq 2. \tag{10}$$

2.2. One-Dimensional Model

To obtain a preliminary indicator of the behaviour of the system and the importance of the various parameters, we further simplify this model. Figure 3 shows a simpler one-dimensional system, where the MOF crystal is represented by a red square, and diffusion in the polymer composite region is completely ignored.

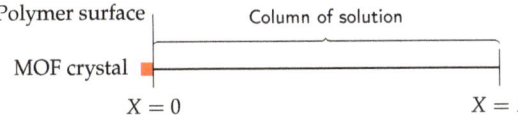

Figure 3. One-dimensional representation of ferric ions diffusing through a column of solution with MOF crystal at the left boundary.

The concentration of ferric ions is assumed to be only a function of two variables, $C(T, X)$, and the MOF pore occupancy is assumed to depend only on time, $B(T)$,

$$\frac{\partial C}{\partial T} = D_1 \frac{\partial^2 C}{\partial X^2}, \qquad X \geq 0,$$

$$\frac{\partial B}{\partial T} = k_{on} C (B_0 - B) - k_{off} B, \qquad X = 0.$$

The corresponding flux condition at the boundary $X = 0$ is given by

$$D_1 \frac{\partial C}{\partial X} = a [k_{on} C (B_0 - B) - k_{off} B],$$

and the far field condition is $C(T, L) = C_0$, with initial conditions $C(0, X) = C_0$ and $B(0) = 0$.

2.3. Dimensionless One-Dimensional Model

We non-dimensionalise concentration, occupancy, length, and time in the same way as before, where ℓ is a characteristic length representative of the MOF crystal size so that the system of equations becomes

$$\frac{\partial c}{\partial t} = s_1 \frac{\partial^2 c}{\partial x^2}, \quad x \geq 0, \qquad \frac{\partial b}{\partial t} = s_2 [c(1-b) - s_3 b], \quad x = 0, \tag{11}$$

where the s_i are as previously defined. The non-dimensional flux condition at the boundary $x = 0$ is

$$s_1 \frac{\partial c}{\partial x} = s_4 s_2 [c(1-b) - s_3 b], \tag{12}$$

while the non-dimensional far field condition at $x = L/\ell$ becomes

$$c(t, L/\ell) = 1, \tag{13}$$

and the initial conditions are given by

$$c(0, x) = 1, \quad b(0) = 0. \tag{14}$$

3. Results and Discussions

In this section, we discuss the results for the diffusion of ferric ions through the two regions and the ferric ion occupancy in the MOF pores. We vary the dimensionless constants, s_i and D, to analyse how the parameters affect ferric ion diffusion and MOF occupancy.

The models are solved numerically, as no analytical solution exists for this system, using purpose written code in MATLAB. First, we examine the dimensionless one-dimensional model where the equation for $b(t)$ is solved using a Runga–Kutta method, and the diffusion equation for $c(t, x)$ is solved with an implicit Crank–Nicolson scheme. Second, we examine the dimensionless two-dimensional model where the equation for $b(t, x_1)$ is solved using a Runga–Kutta method, and the diffusion equation for $c(t, \mathbf{x})$ is solved with a forward-time-centred-space finite difference scheme.

Further details about the numerical schemes for the dimensionless one-dimensional and two-dimensional models are presented in Appendices A and B. In addition, the authors have tested some extremely simple cases; for example, if the MOF crystal is not present. However, these results are trivial and are not included here.

3.1. Dimensionless One-Dimensional Model

Figure 4a depicts the time evolution of the ferric ion concentration in solution, and Figure 4b shows the MOF occupancy. Ferric ions in the solution initially associate rapidly to the MOF crystal, reducing the concentration in the solution near the MOF crystal. Over time, the ferric ion concentration returns to the steady-state far-field condition, as shown in

Figure 4a. We note that the MOF crystal's occupancy initially increases rapidly and then slowly increases until steady state is reached, as shown in Figure 4b.

In the following subsections, we describe the impact of varying parameters. In all cases for the one-dimensional scheme, $\Delta x = 0.1$ and $\Delta t = 0.005$, and we set $L/\ell = 10$.

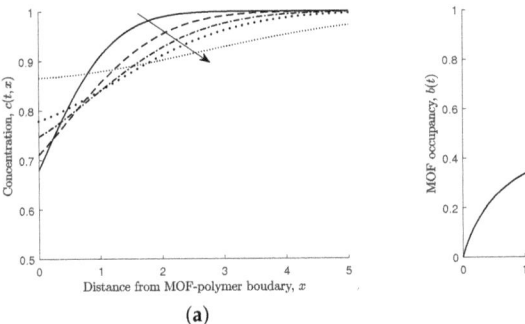
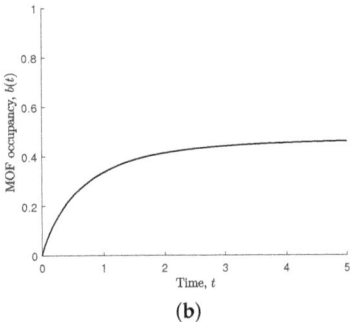

Figure 4. (a) Non-dimensional concentration of ferric ions in solution, $c(t, 0)$. Arrow indicates increasing time, $t = 0.5, 1, 1.5, 2, 5$. (b) Non-dimensional MOF occupancy, $b(t)$. Here, $s_1 = s_2 = s_3 = s_4 = 1$.

3.1.1. Varying the Non-Dimensional Diffusion Coefficient, s_1

Figure 5 shows the effect of varying the non-dimensional diffusion coefficient ($s_1 = D_1 \tau / \ell^2$) on the ferric ion concentration in the solution at the MOF crystal and the MOF occupancy. A smaller non-dimensional diffusion coefficient causes the concentration of ferric ions in solution to be more significantly reduced near the MOF boundary, and, following this initial reduction, it takes longer for the concentration in the solution to recover towards the steady-state concentration. A larger non-dimensional diffusion coefficient results in less reduction near the boundary because ferric ions can more readily diffuse from the column of solution. MOF occupancy reaches close to steady-state sooner with a high non-dimensional diffusion coefficient.

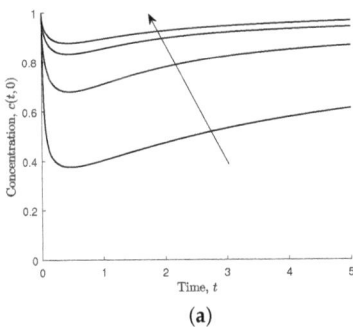
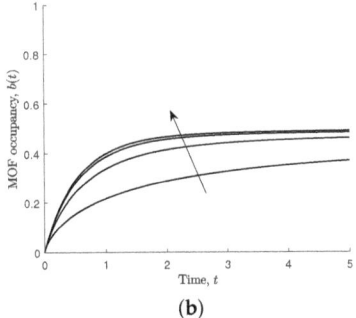

Figure 5. Results for different values of non-dimensional diffusion, $s_1 = 0.1, 1, 5, 10$, and $s_2 = s_3 = s_4 = 1$. Arrows indicate the direction of increasing s_1. (a) Non-dimensional concentration of ferric ions at MOF–polymer boundary, $c(t, 0)$. (b) Non-dimensional MOF occupancy, $b(t)$.

3.1.2. Varying the Effective Association Parameter, s_2

Increasing the effective association parameter ($s_2 = C_0 k_{on} \tau$) means that ferric ions associate to the MOF crystal faster than they diffuse through the solution. This results in the MOF occupancy reaching the steady state sooner, as shown in Figure 6b. When s_2 is large, the concentration near the solution–polymer boundary is reduced very quickly, followed by a slow increase as ferric ions diffuse from the rest of the region, as shown in Figure 6a.

Upon exposure of the MOF to the solution, the concentration at the boundary varies significantly for different values of s_2. However, at large times, the ferric ion concentration and the MOF occupancy is largely independent of s_2.

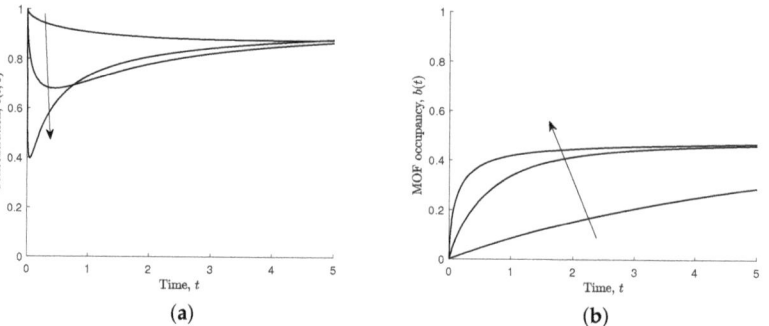

Figure 6. Results for different values of effective association, $s_2 = 0.1, 1, 10$, and $s_1 = s_3 = s_4 = 1$. Arrows indicate the direction of increasing s_2. (**a**) Non-dimensional concentration of ferric ions at MOF–polymer boundary, $c(t,0)$. (**b**) Non-dimensional MOF occupancy, $b(t)$.

3.1.3. Varying the Effective Dissociation Parameter, s_3

The effective dissociation parameter ($s_3 = k_{\text{off}}/C_0 k_{\text{on}}$) captures the ferric ions that dissociate from the MOF crystal. We have investigated various values of s_3, including $s_3 = 0$, which represents the situation when ferric ions can only associate—see Figure 7.

Increasing the effective dissociation parameter means that the ferric ions dissociate more readily and the non-dimensional steady-state MOF pore occupancy can be calculated from Equation (11),

$$b_{\text{eq}} = b(t \to \infty) = \frac{1}{1+s_3}. \tag{15}$$

A similar expression for the number of effective bound receptors at equilibrium is given in Squires et al. [13], with a different combination of parameters. (The present authors believe that the equilibrium constant K_D in Squires et al. [13] should be defined as $K_D = k_{\text{off}}/k_{\text{on}}$ rather than $K_D = k_{\text{on}}/k_{\text{off}}$ as stated. This also aligns with working in their Box 1).

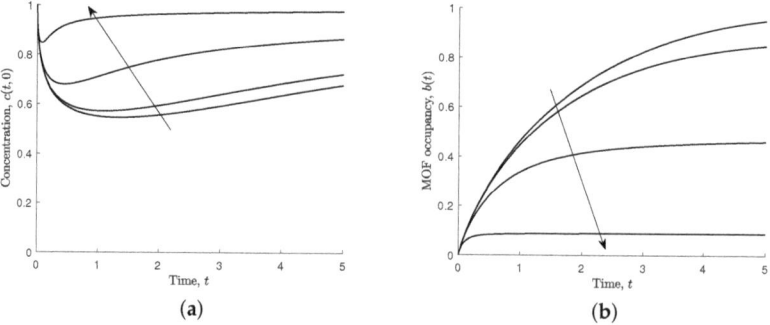

Figure 7. Results for different values of effective dissociation, $s_3 = 0, 0.1, 1, 10$, and $s_1 = s_2 = s_4 = 1$. Arrows indicate the direction of increasing s_3. (**a**) Non-dimensional concentration of ferric ions at MOF–polymer boundary, $c(t,0)$. (**b**) Non-dimensional MOF occupancy, $b(t)$.

In the steady state, the occupancy of the pores is lower for larger values of the effective dissociation parameter, and the steady state is reached more quickly. In the case when the ferric ions cannot dissociate, that is $s_3 = 0$, the steady state MOF occupancy is equal to its maximum value ($b(t \to \infty) = 1$), and the concentration of ferric ions at the MOF–solution boundary approaches the bulk concentration at large times.

3.1.4. Varying the Effective Flux Parameter, s_4

An important consequence of the flux condition is the time taken for the MOF crystal to approach steady state. Increasing the effective flux parameter ($s_4 = aB_0/\ell C_0$) increases the time taken to reach the steady state (see Figure 8b). The increase affects ferric ion concentration at the MOF crystal boundary, as ferric ions quickly associate to the MOF crystal and the ferric ions do not diffuse fast enough for the MOF crystal to reach steady state in a timely manner, as shown in Figure 8a.

The delay to reaching steady state is attributed to ferric ions associating to the MOF crystal faster and quickly depleting the ferric ions in the solution. A comparable mechanism is observed when increasing the effective association parameter, s_2. The effective association parameter is a MOF characteristic, and an increase results in a shorter time for the MOF crystal to reach steady state (see Figure 6b). The ferric ion concentration at the MOF boundary is largely unchanged after $t = 4$, as shown in Figure 6a. However, the effective flux parameter is a ferric ion characteristic, and an increase in this parameter results in a longer time for the MOF crystal to reach steady state, as shown in Figure 8b. The ferric ion concentration at the MOF boundary is very sensitive to the increase in the effective flux parameter, as seen by the long depletion time for a larger parameter value in Figure 8a.

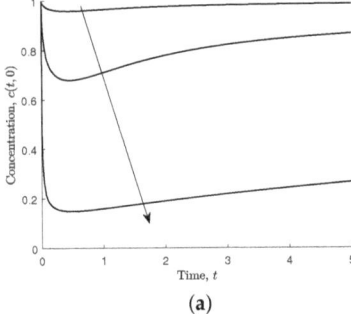
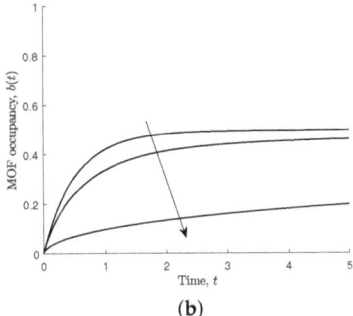

(a) (b)

Figure 8. Results for different values of effective flux, $s_4 = 0.1, 1, 10$, and $s_1 = s_2 = s_3 = 1$. Arrows indicate the direction of increasing s_4. (**a**) Non-dimensional concentration of ferric ions at MOF–polymer boundary, $c(t,0)$. (**b**) Non-dimensional MOF occupancy, $b(t)$.

3.2. Dimensionless Two-Dimensional Model

In this section, we return to the dimensionless two-dimensional model and to the primary aim of the paper, which is to investigate ferric ion diffusion through a solution into a polymer composite with a MOF sink located at the boundary between the two media.

Figure 9 shows the ferric ion concentration for the domain $-5 \leq x_1 \leq 5$, $-5 \leq x_2 \leq 5$, where $x_2 < 0$ is the concentration in the polymer and $x_2 \geq 0$ is the concentration in the solution. The MOF crystal is located at $|x_1| \leq 1$ and $x_2 = 0$. Figure 9a,b show the ferric ion profiles at time $t = 0.5$ and 5, respectively. In all cases for the numerical two-dimensional scheme, $\Delta x_1 = \Delta x_2 = 0.1$ and $\Delta t = 0.0005$, and $D = 0.5$ (unless otherwise stated) and $L/\ell = h/\ell = 10$. Figure 10a shows the ferric ion concentration along the line $x_1 = 0$, which shows the changes through the mixed media over time. In comparison to the one-dimensional model (Figure 4a), the two-dimensional model requires additional time to reach steady state. This is due to the size of the MOF crystal and the time taken to reach steady state in the two-media system. The MOF occupancy at the centre of the

MOF ($b(t,0)$) and the total MOF occupancy over the whole MOF crystal (Equation (10)) is shown in Figure 10b. By comparing Figure 10b with Figure 4b, we see that the MOF occupancy as predicted by the two-dimensional model is very similar to that predicted by the one-dimensional model. However, the two-dimensional model includes the effect of the polymer on ferric ion diffusion and greater MOF occupancy potential.

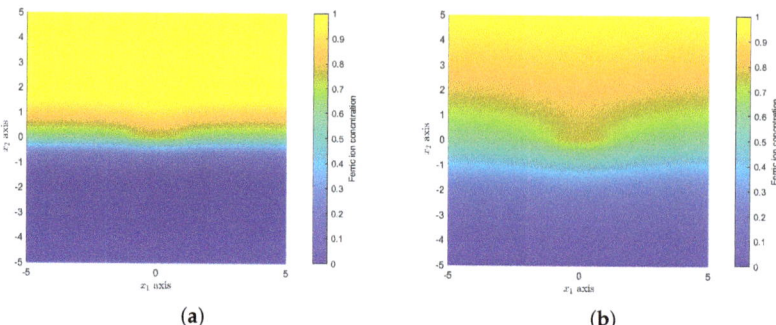

Figure 9. Non-dimensional concentration of ferric ions across two media $c(t, x_1, x_2)$: (**a**) at time $t = 0.5$; (**b**) at time $t = 5$.

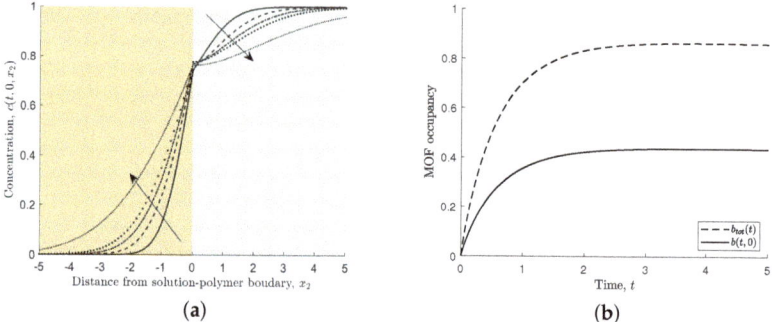

Figure 10. (**a**) Non-dimensional concentration of ferric ions through MOF-mid point across two media, $c(t, 0, x_2)$. Arrow indicates the direction of increasing time, $t = 0.5, 1, 1.5, 2, 5$. (**b**) Non-dimensional MOF occupancy, $b(t, 0)$ and $b_{tot}(t)$. Here, $s_1 = s_2 = s_3 = s_4 = 1$, and $D = 0.5$.

3.2.1. Varying the Non-Dimensional Parameters

Variation of the dimensionless constants has the same effect on the two-dimensional MOF occupancy $b(t, 0)$ profile when compared to the one-dimensional MOF occupancy $b(t)$. The two-dimensional ferric ion concentration at the centre of the MOF crystal behaves in a similar way to that of the one-dimensional ferric ion profile shown in Figure 4a. However, more time is needed for the ferric ion concentration to reach steady state due to a larger MOF crystal surface area.

Figure 11 shows the effect of varying the non-dimensional diffusion coefficient ($s_1 = 0.1, 1, 5$). For all values of s_1, the concentration in the solution near the MOF surface initially decreases before increasing towards the steady state value. For larger values of s_1, the decrease is more significant, which is in contrast to the one-dimensional results, as shown in Figure 5.

The changes in the other non-dimensional parameters (s_2, s_3, s_4) produce similar results to those discussed for the one-dimensional model. Figures showing the details can be found in Appendix C.

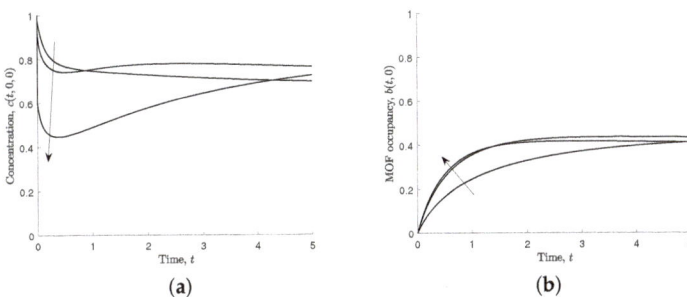

Figure 11. Results for different values of non-dimensional diffusion, $s_1 = 0.1, 1, 5$, $s_2 = s_3 = s_4 = 1$, and $D = 0.5$. Arrows indicate the direction of increasing s_1. (**a**) Non-dimensional concentration of ferric ions through MOF mid-point across two media, $c(t, 0, 0)$. (**b**) Non-dimensional MOF occupancy, $b(t, 0)$.

3.2.2. Varying the Relative Diffusion Coefficient, D

If ferric ion diffusivity in the solution and the polymer is the same, then the relative diffusion coefficient is unity, $D = D_2/D_1 = 1$. In practice, the diffusion coefficient of ferric ions in the solution will exceed that in the polymer and the larger the difference the smaller the relative diffusion coefficient. Figure 12 shows the effects of varying the relative diffusion coefficient.

Decreasing the relative diffusion coefficient means that ferric ions take longer to diffuse in the polymer and away from the solution–polymer boundary. This affords the ferric ions more opportunity to associate to the MOF crystal. Figure 12c shows that the occupancy in the crystal is highest in the case of the smallest relative diffusion coefficient.

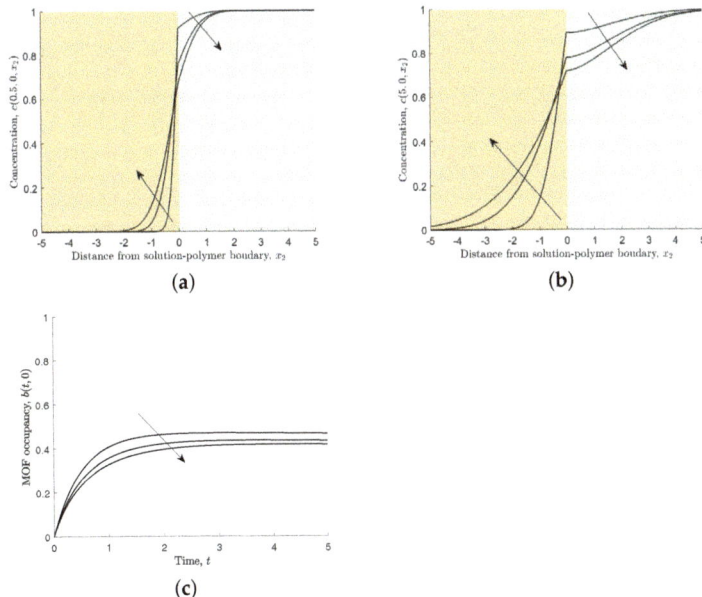

Figure 12. Results for different values of relative diffusion coefficient, $D = 0.1, 0.5, 1$, and $s_1 = s_2 = s_3 = s_4 = 1$. Arrows indicate the direction of increasing D. (**a**,**b**) Non-dimensional concentration of ferric ions through MOF mid-point across two media, $c(t, 0, x_2)$ at time $t = 0.5$ and 5, respectively. (**c**) Non-dimensional MOF occupancy, $b(t, 0)$.

4. Conclusions

In this paper, we have modelled a proposed new ferric ion sensor. We first examined a two-dimensional model to describe the diffusion of the ions through a solution and into a polymer layer, as well as their attachment to an EuMOF crystal located at the boundary of the two media. We then examined a simplified model that includes diffusion through the solution only, together with attachment at the MOF crystal. The solutions to the corresponding non-dimensional models are then analysed to investigate the importance of the various non-dimensional parameters.

The one-dimensional model indicates that at steady state, the occupancy of the MOF crystal is given by Equation (15), and this occupancy depends only on the initial concentration of ferric ions in the solution and the effective dissociation constant, $s_3 = k_{\text{off}}/C_0 k_{\text{on}}$. As a consequence, if the steady-state occupancy, $B_{\text{eq}} = B(t \to \infty)$, can be measured with the new sensor, and the association rate k_{on}, dissociation rate k_{off}, and maximum occupancy B_0 are known, then the concentration of ferric ions in solution can be calculated (in dimensional terms) using

$$C_0 = \frac{k_{\text{off}} B_{\text{eq}}}{k_{\text{on}}(B_0 - B_{\text{eq}})}. \tag{16}$$

In practice, it is most likely to be useful to minimise the time taken to reach steady state so that sensing of multiple samples can be carried out quickly and efficiently. To achieve this, it would be helpful to have a large non-dimensional diffusion coefficient s_1 and effective association constant s_2 with a small effective flux constant s_4.

While the two-dimensional model is more representative of the proposed sensor and provides more information about the behaviour of the ferric ions in the solution and the polymer, many of the broad behaviours and impacts of changing the system parameters are captured by the one-dimensional model. The relative diffusion coefficient D suggests that the polymer composite should hinder ferric ion diffusion in that region. A final sensor design is likely to include MOF crystals embedded within the polymer matrix rather than on the surface only, so the polymer composite should be designed so that the relative diffusion coefficient is close to unity in order for ferric ions to associate to MOF crystals inside the polymer composite. This situation is not examined here but will be the subject of future work. In this case, a two-dimensional model is likely to provide more insight.

Author Contributions: All authors have contributed equally to the development and execution of all aspects of this work. All authors have read and agreed to the published version of the manuscript.

Funding: This research is supported by the SA Government through the PRIF RCP Mining Consortium "Unlocking Complex Resources through Lean Processing" and the University of South Australia through the International Research Tuition Scholarship.

Data Availability Statement: Numerical details for this study are presented in the Appendices. Data sharing is not applicable for this article.

Acknowledgments: The authors wish to gratefully acknowledge Linda Rozenberga for much chemical knowledge and her considerable insight.

Conflicts of Interest: The authors declare that they do not have any conflict of interest.

Abbreviations

The following abbreviations are used in this manuscript:

MOF   Metal organic framework
EuMOF Europium based metal organic framework
ODE   Ordinary differential equation

Appendix A. Algorithm for Dimensionless One-Dimensional Model

The functions describing the ferric ion concentration $c(t,x)$ and MOF occupancy $b(t)$ are discretised, and the following definitions are used,

$$c(t_n, x_i) = c_i^n, \quad c(t,0) = c_0^n, \quad c(t, L/\ell) = c_l^n, \quad b(t_n) = b^n.$$

We first update the value of the MOF occupancy b^{n+1} so that we can use the updated value to calculate the ferric ion concentration c_0^{n+1} at $x = 0$. A Runge–Kutta 4 scheme is used

$$f(y) = s_2 c_0^n (1-y) - s_2 s_3 y,$$

$$k_1 = \Delta t f(b^n), \quad k_2 = \Delta t f\left(b^n + \frac{k_1}{2}\right),$$

$$k_3 = \Delta t f\left(b^n + \frac{k_2}{2}\right), \quad k_4 = \Delta t f(b^n + k_3).$$

At time step $n+1$, the MOF occupancy is

$$b^{n+1} = b^n + \frac{1}{6}(k_1 + 2k_2 + 2k_3 + k_4),$$

where $b(0) = b^0 = 0$.

Next, we update the ferric ion concentration throughout the solution using the Crank–Nicolson central forwards time, centred space. The following difference formulas for the first and second derivatives were used

$$c(t,x) = \frac{1}{2}(c_i^n + c_i^{n+1}),$$

$$\frac{\partial}{\partial x} c(t,x) = \frac{1}{2}\left[\frac{c_{i+1}^n - c_{i-1}^n}{2\Delta x} + \frac{c_{i+1}^{n+1} - c_{i-1}^{n+1}}{2\Delta x}\right], \quad \frac{\partial}{\partial t} c(t,x) = \frac{c_i^{n+1} - c_i^n}{\Delta t}$$

$$\frac{\partial^2}{\partial x^2} c(t,x) = \frac{1}{2}\left[\frac{c_{i+1}^n - 2c_i^n + c_{i-1}^n}{\Delta x^2} + \frac{c_{i+1}^{n+1} - 2c_i^{n+1} + c_{i-1}^{n+1}}{\Delta x^2}\right].$$

Substituting these formulas into the ferric ion diffusion equation gives

$$\frac{c_i^{n+1} - c_i^n}{\Delta t} = \frac{s_1}{2}\left[\frac{c_{i+1}^n - 2c_i^n + c_{i-1}^n}{\Delta x^2} + \frac{c_{i+1}^{n+1} - 2c_i^{n+1} + c_{i-1}^{n+1}}{\Delta x^2}\right].$$

We collect terms with time step $n+1$ on the left and n on the right hand side of the equation so that the general scheme is

$$-\alpha c_{i+1}^{n+1} + (2+2\alpha)c_i^{n+1} - \alpha c_{i-1}^{n+1} = \alpha c_{i+1}^n + (2-2\alpha)c_i^n + \alpha c_{i-1}^n,$$

where $\alpha = s_1 \Delta t / \Delta x^2$. The boundary conditions concerning the ferric ion concentration need to be incorporated, and so, the flux at the MOF–polymer boundary (Equation (12)) becomes

$$s_1\left[c_1^n - c_{-1}^n + c_1^{n+1} - c_{-1}^{n+1}\right] = 2s_4 s_2 \Delta x (c_0^n + c_0^{n+1})(1 - b^{n+1}) - 4\Delta x s_4 s_2 s_3 \, b^{n+1}.$$

Now, adjust the general scheme at and near the boundaries:
for $i = 0$, the MOF–polymer boundary,

$$-2\alpha c_1^{n+1} + [2 + 2\alpha + 2\alpha s_4 s_2 \Delta x (1 - b^{n+1})/s_1] c_0^{n+1}$$
$$= 2\alpha c_1^n + [2 - 2\alpha - 2\alpha s_4 s_2 \Delta x (1 - b^{n+1})/s_1] c_0^n + 4\Delta x s_2 s_3 s_4 \, b^{n+1}/s_1,$$

for $i = I$, far-field boundary (Equation (13)),
$$c_I^n = 1, \quad \forall n,$$
for $i = I - 1$,
$$(2 + 2\alpha)c_{I-1}^{n+1} - \alpha\, c_{I-2}^{n+1} = (2 - 2\alpha)c_{I-1}^n + \alpha\, c_{I-2}^n + 2\alpha.$$

The finite difference scheme can be written as a linear system,
$$L\underline{c}^{n+1} = R\underline{c}^n + A,$$
where
$$L = \begin{bmatrix} 2 + 2\alpha + 2\alpha s_4 s_2 \Delta x(1 - b^{n+1})/s_1 & -\alpha & 0 & 0 & \cdots & 0 & 0 & 0 & 0 \\ -\alpha & 2 + 2\alpha & -\alpha & 0 & \cdots & 0 & 0 & 0 & 0 \\ \ddots & & \ddots & \ddots & & 0 & 0 & 0 & 0 \\ 0 & & 0 & 0 & 0 & \cdots & -\alpha & 2 + 2\alpha & -\alpha & 0 \\ 0 & & 0 & 0 & 0 & \cdots & 0 & -\alpha & 2 + 2\alpha & 0 \\ 0 & & 0 & 0 & 0 & \cdots & 0 & 0 & 0 & 1 \end{bmatrix},$$

$$R = \begin{bmatrix} 2 - 2\alpha - 2\alpha s_4 s_2 \Delta x(1 - b^{n+1})/s_1 & \alpha & 0 & 0 & \cdots & 0 & 0 & 0 & 0 \\ \alpha & 2 - 2\alpha & \alpha & 0 & \cdots & 0 & 0 & 0 & 0 \\ \ddots & & \ddots & \ddots & & 0 & 0 & 0 & 0 \\ 0 & & 0 & 0 & 0 & \cdots & \alpha & 2 - 2\alpha & \alpha & 0 \\ 0 & & 0 & 0 & 0 & \cdots & 0 & \alpha & 2 - 2\alpha & 0 \\ 0 & & 0 & 0 & 0 & \cdots & 0 & 0 & 0 & 0 \end{bmatrix},$$

$$A = [4\Delta x s_2 s_3 s_4\, b^{n+1}/s_1, 0, \cdots, 0, 2\alpha, 1]^{\mathrm{T}},$$
and
$$\underline{c}^n = [c_0^n, c_1^n, c_2^n, \cdots, c_{I-2}^n, c_{I-1}^n, c_I^n]^{\mathrm{T}},$$

where L and R are real $(I+1) \times (I+1)$ matrices, and A and $\underline{c}^n$ are $(I+1)$-vectors.

The system describing the concentration of ferric ions through the solution can be evaluated as
$$\underline{c}^{n+1} = L^{-1}(R\underline{c}^n + A),$$
where $c(0, x) = c_i^0 = 1$. Table A1 provides the numerical values for the graphs shown in the main text. Tests were performed for various timesteps and mesh sizes to ensure numerical accuracy.

Table A1. Numerical values for one-dimensional simulation.

Δt	Δx	L/ℓ
0.005	0.1	10

Appendix B. Algorithm for Dimensionless Two-Dimensional Model

The functions describing the ferric ion concentration $c(t, \mathbf{x})$ and MOF occupancy $b(t, x_1)$ are discritised, and the following definitions are used:

$$c(t, x_1, x_2) = c(t_n, x_i, x_j) = c_{ij}^n, \quad c(t, -L/\ell, x_2) = c_{0j}^n, \quad c(t, L/\ell, x_2) = c_{Ij}^n,$$
$$c(t, -1, x_2) = c_{il\,j}^n, \quad c(t, 1, x_2) = c_{ir\,j}^n, \quad c(t, x_1, 0) = c_{i\,jm}^n,$$
$$c(t, x_1, -h/\ell) = c_{i0}^n, \quad c(t, x_1, L/\ell) = c_{iJ}^n, \quad b(t, x_1) = b(t_n, x_i) = b_i^n.$$

In the same way as the one-dimensional model was solved, the MOF occupancy is first updated for the next time step b_i^{n+1} and later used to update the ferric ion concentration c_{ij}^{n+1} at $|x_1| \leq 1$ and $x_2 = 0$. A Runge–Kutta 4 scheme is used to discretise $b(t, x_1)$ at $|x_1| \leq 1$,

$$f(y) = s_2 \, c_{ijm}^n (1-y) - s_2 s_3 y, \quad \text{for } i = il, \ldots, rl,$$
$$k_1 = \Delta t \, f(b_i^n), \quad k_2 = \Delta t \, f\left(b_i^n + \frac{k_1}{2}\right),$$
$$k_3 = \Delta t \, f\left(b_i^n + \frac{k_2}{2}\right), \quad k_4 = \Delta t \, f(b_i^n + k_3).$$

At the next time step $n+1$ for $i = il, \ldots, ir$, the MOF occupancy is

$$b_i^{n+1} = b_i^n + \frac{1}{6}(k_1 + 2k_2 + 2k_3 + k_4),$$

where $b_i^0 = 0$.

We update the ferric ion concentration and use a forwards time, two-dimensional centred space scheme. The following difference formulas for the first and second derivatives were used:

$$\frac{\partial}{\partial t} c(t, \mathbf{x}) = \frac{c_{ij}^{n+1} - c_{ij}^n}{\Delta t},$$
$$\frac{\partial}{\partial x_1} c(t, \mathbf{x}) = \frac{c_{i+1j}^n - c_{i-1j}^n}{2\Delta x_1}, \quad \frac{\partial}{\partial x_2} c(t, \mathbf{x}) = \frac{c_{ij+1}^n - c_{ij-1}^n}{2\Delta x_2},$$
$$\nabla^2 c(t, \mathbf{x}) = \frac{c_{i+1j}^n - 2c_{ij}^n + c_{i-1j}^n}{\Delta x_1^2} + \frac{c_{ij+1}^n - 2c_{ij}^n + c_{ij-1}^n}{\Delta x_2^2}.$$

Substituting the formulas into the ferric ion diffusion equation in the solution for $x_2 \geq 0$ gives

$$\frac{c_{ij}^{n+1} - c_{ij}^n}{\Delta t} = s_1 \left[\frac{c_{i+1j}^n - 2c_{ij}^n + c_{i-1j}^n}{\Delta x_1^2} + \frac{c_{ij+1}^n - 2c_{ij}^n + c_{ij-1}^n}{\Delta x_2^2} \right].$$

We collect terms with time step $n+1$ on the left and n on the right hand side of the equation so that the general scheme $\forall i$ and $j = jm + 1 : J - 1$ is

$$c_{ij}^{n+1} = (1 - 2s_x - 2s_y)c_{ij}^n + s_x(c_{i+1j}^n + c_{i-1j}^n) + s_y(c_{ij+1}^n + c_{ij-1}^n).$$

Similarly, diffusion in the polymer $\forall i$ and $j = 1 : jm - 1$ is

$$c_{ij}^{n+1} = (1 - 2s_x D - 2s_y D)c_{ij}^n + s_x D(c_{i+1j}^n + c_{i-1j}^n) + s_y D(c_{ij+1}^n + c_{ij-1}^n),$$

where $s_x = s_1 \Delta t / \Delta x_1^2$ and $s_y = s_1 \Delta t / \Delta x_2^2$. We need to account for the boundary conditions of the system and update the numerical scheme accordingly:
for $\forall i, n$, and $j = J$, far-field boundary condition (Equation (4)) at $x_2 = L/\ell$,

$$c_{iJ}^n = 1,$$

for $i = 0$ and $j = jm + 1 : J - 1$, no-flux boundary condition (Equation (5)) for solution medium at $x_1 = -L/\ell$,

$$c_{0j}^{n+1} = (1 - 2s_x - 2s_y)c_{0j}^n + s_x(c_{1j}^n + c_{1j}^n) + s_y(c_{0j+1}^n + c_{0j-1}^n),$$

for $i = 0$ and $j = 1 : jm - 1$, no-flux boundary condition (Equation (5)) for polymer medium at $x_1 = -L/\ell$,

$$c_{0j}^{n+1} = (1 - 2s_xD - 2s_yD)c_{0j}^n + s_xD(c_{1j}^n + c_{1j}^n) + s_yD(c_{0j+1}^n + c_{0j-1}^n),$$

for $i = I$ and $j = jm + 1 : J - 1$, no-flux boundary condition (Equation (5)) for solution medium at $x_1 = L/\ell$,

$$c_{Ij}^{n+1} = (1 - 2s_x - 2s_y)c_{Ij}^n + s_x(c_{I-1j}^n + c_{I-1j}^n) + s_y(c_{Ij+1}^n + c_{Ij-1}^n),$$

for $i = I$ and $j = 1 : jm - 1$, no-flux boundary condition (Equation (5)) for polymer medium at $x_1 = L/\ell$,

$$c_{Ij}^{n+1} = (1 - 2s_xD - 2s_yD)c_{Ij}^n + s_xD(c_{I-1j}^n + c_{I-1j}^n) + s_yD(c_{Ij+1}^n + c_{Ij-1}^n),$$

for $i = 1, ..., I - 1$ and $j = 0$, no-flux boundary condition (Equation (5)) for polymer–optic fibre at $x_2 = -h/\ell$,

$$c_{i0}^{n+1} = (1 - 2s_xD - 2s_yD)c_{i0}^n + s_xD(c_{i+10}^n + c_{i-10}^n) + s_yD(c_{i1}^n + c_{i1}^n),$$

for $i = 0$ and $j = 0$, no-flux boundary condition (Equation (5)) at $x_1 = -L/\ell$ and $x_2 = -h/\ell$,

$$c_{00}^{n+1} = (1 - 2s_xD - 2s_yD)c_{00}^n + s_xD(c_{10}^n + c_{10}^n) + s_yD(c_{01}^n + c_{01}^n),$$

for $i = I$ and $j = 0$, no-flux boundary condition (Equation (5)) at $x_1 = L/\ell$ and $x_2 = -h/\ell$,

$$c_{I0}^{n+1} = (1 - 2s_xD - 2s_yD)c_{I0}^n + s_xD(c_{I-10}^n + c_{I-10}^n) + s_yD(c_{I1}^n + c_{I1}^n),$$

for $i = 1, ..., I - 1$ and $j = jm$, solution–polymer boundary continuity condition (Equation (3)) at $x_2 = 0$,

$$\frac{\partial c}{\partial t} = s_1 \frac{\partial^2 c}{\partial x_1^2} + s_1 \frac{\partial}{\partial x_2}\left[D_i \frac{\partial c}{\partial x_2}\right] = s_1 \frac{\partial^2 c}{\partial x_1^2} + \frac{s_1}{\Delta x_2}\left[\frac{\partial c_+}{\partial x_2} - D\frac{\partial c_-}{\partial x_2}\right],$$

$$\frac{c_{ijm}^{n+1} - c_{ijm}^n}{\Delta t} = s_1 \left[\frac{c_{i+1jm}^n - 2c_{ijm}^n + c_{i-1jm}^n}{\Delta x_1^2}\right] + \frac{s_1}{\Delta x_2}\left[\frac{c_{ijm+1}^n - c_{ijm}^n}{\Delta x_2}\right] - \frac{s_1 D}{\Delta x_2}\left[\frac{c_{ijm}^n - c_{ijm-1}^n}{\Delta x_2}\right],$$

$$c_{ijm}^{n+1} = (1 - 2s_x - s_y(1+D))c_{ijm}^n + s_x(c_{i+1jm}^n + c_{i-1jm}^n) + s_y c_{ijm+1}^n + s_y D c_{ijm-1}^n,$$

for $i = 0$ and $j = jm$, solution–polymer boundary continuity condition (Equation (3)) and no-flux boundary condition (Equation (5)) at $x_1 = -L/\ell$ and $x_2 = 0$,

$$c_{0jm}^{n+1} = (1 - 2s_x - s_y(1+D))c_{0jm}^n + s_x(c_{1jm}^n + c_{1jm}^n) + s_y c_{0jm+1}^n + s_y D c_{0jm-1}^n,$$

for $i = I$ and $j = jm$, solution–polymer continuity condition (Equation (3)) and no-flux boundary condition (Equation (5)) at $x_1 = L/\ell$ and $x_2 = 0$,

$$c_{Ijm}^{n+1} = (1 - 2s_x - s_y(1+D))c_{Ijm}^n + s_x(c_{I-1jm}^n + c_{I-1jm}^n) + s_y c_{Ijm+1}^n + s_y D c_{Ijm-1}^n,$$

for $i = il - 1$ and $j = jm$, solution–polymer continuity condition (Equation (3)) and no-flux boundary condition (Equation (7)) at $x_1 = -1_-$ and $x_2 = 0$,

$$c_{il-1jm}^{n+1} = (1 - 2s_x - s_y(1+D))c_{il-1jm}^n + s_x(c_{il-2jm}^n + c_{il-2jm}^n) + s_y c_{il-1jm+1}^n + s_y D c_{il-1jm-1}^n,$$

for $i = ir + 1$ and $j = jm$, solution–polymer continuity condition solution–polymer continuity condition (Equation (3)) and no-flux boundary condition (Equation (7)) at $x_1 = 1_+$ and $x_2 = 0$,

$$c_{ir+1\,jm}^{n+1} = (1 - 2s_x - s_y(1+D))c_{ir+1\,jm}^n + s_x(c_{ir+2\,jm}^n + c_{ir+2\,jm}^n) + s_y c_{ir+1\,jm+1}^n + s_y D c_{ir+1\,jm-1}^n,$$

At the solution–MOF–polymer boundary, we use the continuity condition (Equation (3)), to find

$$c_{i\,jm-1}^n = \frac{-c_{i\,jm+1}^n + (1+D)c_{i\,jm}^n}{D}.$$

For $i = il, ..., ir$ and $j = jm$, MOF–solution flux condition (Equation (6)) at $|x_1| \leq 1$ and $x_2 = 0$ gives

$$s_1 \left[\frac{c_{i\,jm+1}^{n+1} - c_{i\,jm-1}^{n+1}}{2\Delta x_2} \right] = s_4 s_2 (c_{i\,jm}^{n+1}(1 - b_i^{n+1}) - s_3 b_i^{n+1}),$$

$$c_{i\,jm}^{n+1} = \frac{s_1(1+D)c_{i\,jm+1}^{n+1} + 2\Delta x_2 D s_4 s_2 s_3 b_i^{n+1}}{s_1(1+D) + 2\Delta x_2 D s_4 s_2 (1 - b_i^{n+1})}.$$

Table A2 provides the numerical values for the graphs shown in the main text. Tests were performed for various timesteps and mesh sizes to ensure numerical accuracy.

Table A2. Numerical values for two-dimensional simulation.

Δt	Δx_1	Δx_2	L/ℓ	h/ℓ	D
0.0005	0.1	0.1	10	10	0.5

Unless otherwise stated, D takes the value found in Table A2.

Appendix C. Results of Varying Parameters in the Two-Dimensional Model

This section contains the figures that show the effects of variation of the remaining non-dimensional parameters for the two-dimensional model.

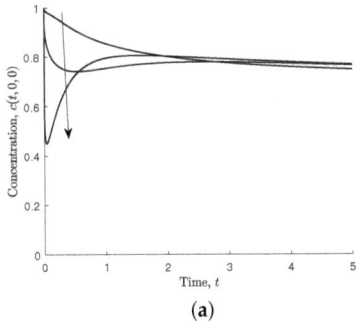
(a)

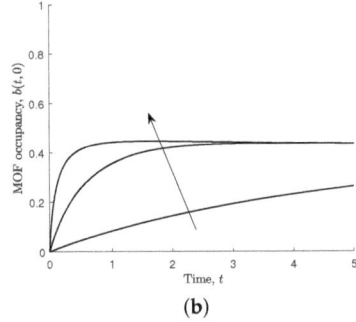
(b)

Figure A1. Results for different values of effective association, $s_2 = 0.1, 1, 10$, $s_1 = s_3 = s_4 = 1$, and $D = 0.5$. Arrows indicate the direction of increasing s_2. (a) Non-dimensional concentration of ferric ions through MOF-mid point across two media, $c(t, 0, x_2)$. (b) Non-dimensional MOF occupancy, $b(t, 0)$.

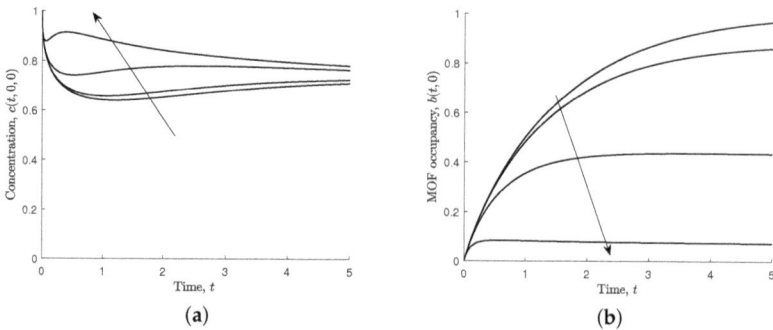

Figure A2. Results for different values of effective dissociation, $s_3 = 0, 0.1, 1, 10$, $s_1 = s_2 = s_4 = 1$, and $D = 0.5$. Arrows indicate the direction of increasing s_3. (**a**) Non-dimensional concentration of ferric ions through MOF-mid point across two media, $c(t, 0, x_2)$. (**b**) Non-dimensional MOF occupancy, $b(t, 0)$.

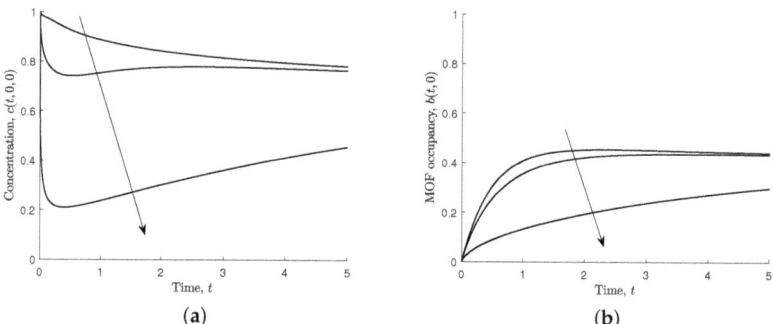

Figure A3. Results for different values of effective capacity, $s_4 = 0.1, 1, 10$, $s_1 = s_2 = s_3 = 1$, and $D = 0.5$. Arrows indicate the direction of increasing s_4. (**a**) Non-dimensional concentration of ferric ions through MOF-mid point across two media, $c(t, 0, x_2)$. (**b**) Non-dimensional MOF occupancy, $b(t, 0)$.

References

1. Hogsden, K.L.; Harding, J.S. Consequences of acid mine drainage for the structure and function of benthic stream communities: A review. *Freshw. Sci.* **2012**, *31*, 108–120. [CrossRef]
2. Vajargah, M.F. A review on the effects of heavy metals on aquatic animals. *J. Biomed. Res. Environ. Sci.* **2021**, *2*, 865–869. [CrossRef]
3. Vuori, K. Direct and indirect effects of iron on river ecosystems. *Ann. Zool. Fenn.* **1995**, *32*, 317–329.
4. Thomas, J.E.; Smart, S.S.C.; Skinner, W.M. Kinetic factors for oxidative and non-oxidative dissolution of iron sulfides. *Miner. Eng.* **2000**, *13*, 1149–1159. [CrossRef]
5. Wang, C.; Babitt, J.L. Liver iron sensing and body iron homeostasis. *Blood, J. Am. Soc. Hematol.* **2019**, *133*, 18–29. [CrossRef]
6. Chaturvedi, S.; Dave, P.N. Removal of iron for safe drinking water. *Desalination* **2012**, *303*, 1–11. [CrossRef]
7. Sahoo, S.K.; Sharma, D.; Bera, R.K.; Crisponi, G.; Callan, J.F. Iron (III) selective molecular and supramolecular fluorescent probes. *Chem. Soc. Rev.* **2012**, *41*, 7195–7227. [CrossRef] [PubMed]
8. Xu, H.; Dong, Y.; Wu, Y.; Ren, W.; Zhao, T.; Wang, S.; Gao, J. An–OH group functionalized MOF for ratiometric Fe^{3+} sensing. *J. Solid State Chem.* **2018**, *258*, 441–446. [CrossRef]
9. Louw, K.I.; Bradshaw-Hajek, B.H.; Hill, J.M. Interaction of Ferric Ions with Europium Metal Organic Framework and Application to Mineral Processing Sensing. *Philos. Mag.* **2022**, submitted.
10. Hickson, R.I.; Barry, S.I.; Mercer, G.N.; Sidhu, H.S. Finite difference schemes for multilayer diffusion. *Math. Comput. Model.* **2011**, *54*, 210–220. [CrossRef]
11. Carr, E.J.; Turner, T.W. A semi-analytical solution for multilayer diffusion in a composite medium consisting of a large number of layers. *Appl. Math. Model.* **2016**, *40*, 7034–7050. [CrossRef]
12. Hahn, D.W.; Özisik, M.N. *Heat Conduction*; John Wiley & Sons: Hoboken, NJ, USA, 2012.

13. Squires, T.M.; Messinger, R.J.; Manalis, S.R. Making it stick: Convection, reaction and diffusion in surface-based biosensors. *Nat. Biotechnol.* **2008**, *26*, 417–426. [CrossRef] [PubMed]
14. Yariv, E. Small Péclet-number mass transport to a finite strip: An advection–diffusion–reaction model of surface-based biosensors. *J. Appl. Math.* **2020**, *31*, 763–781. [CrossRef]
15. Rivero, P.J.; Goicoechea, J.; Arregui, F.J. Optical fiber sensors based on polymeric sensitive coatings. *Polymers* **2018**, *10*, 280. [CrossRef] [PubMed]

Communication

Luminescence of SiO_2-BaF_2:Tb^{3+}, Eu^{3+} Nano-Glass-Ceramics Made from Sol–Gel Method at Low Temperature

Natalia Pawlik [1,*], Barbara Szpikowska-Sroka [1], Tomasz Goryczka [2], Ewa Pietrasik [1] and Wojciech A. Pisarski [1,*]

[1] Institute of Chemistry, University of Silesia, 40-007 Katowice, Poland; barbara.szpikowska-sroka@us.edu.pl (B.S.-S.); ewa.pietrasik@us.edu.pl (E.P.)
[2] Institute of Materials Engineering, University of Silesia, 41-500 Chorzow, Poland; tomasz.goryczka@us.edu.pl
* Correspondence: natalia.pawlik@us.edu.pl (N.P.); wojciech.pisarski@us.edu.pl (W.A.P.); Tel.: +48-32-349-7658 (N.P.); +48-32-349-7678 (W.A.P.)

Abstract: The synthesis and characterization of multicolor light-emitting nanomaterials based on rare earths (RE^{3+}) are of great importance due to their possible use in optoelectronic devices, such as LEDs or displays. In the present work, oxyfluoride glass-ceramics containing BaF_2 nanocrystals co-doped with Tb^{3+}, Eu^{3+} ions were fabricated from amorphous xerogels at 350 °C. The analysis of the thermal behavior of fabricated xerogels was performed using TG/DSC measurements (thermogravimetry (TG), differential scanning calorimetry (DSC)). The crystallization of BaF_2 phase at the nanoscale was confirmed by X-ray diffraction (XRD) measurements and transmission electron microscopy (TEM), and the changes in silicate sol–gel host were determined by attenuated total reflectance infrared (ATR-IR) spectroscopy. The luminescent characterization of prepared sol–gel materials was carried out by excitation and emission spectra along with decay analysis from the $^5D_4$ level of Tb^{3+}. As a result, the visible light according to the electronic transitions of Tb^{3+} ($^5D_4 \rightarrow {}^7F_J$ (J = 6–3)) and Eu^{3+} ($^5D_0 \rightarrow {}^7F_J$ (J = 0–4)) was recorded. It was also observed that co-doping with Eu^{3+} caused the shortening in decay times of the $^5D_4$ state from 1.11 ms to 0.88 ms (for xerogels) and from 6.56 ms to 4.06 ms (for glass-ceramics). Thus, based on lifetime values, the Tb^{3+}/Eu^{3+} energy transfer (ET) efficiencies were estimated to be almost 21% for xerogels and 38% for nano-glass-ceramics. Therefore, such materials could be successfully predisposed for laser technologies, spectral converters, and three-dimensional displays.

Keywords: BaF_2 nanophase; oxyfluoride nano-glass-ceramics; Tb^{3+}/Eu^{3+} energy transfer; sol–gel chemistry

Citation: Pawlik, N.; Szpikowska-Sroka, B.; Goryczka, T.; Pietrasik, E.; Pisarski, W.A. Luminescence of SiO_2-BaF_2:Tb^{3+}, Eu^{3+} Nano-Glass-Ceramics Made from Sol–Gel Method at Low Temperature. *Nanomaterials* 2022, 12, 259. https://doi.org/10.3390/nano12020259

Academic Editors: Federico Cesano, Simas Rackauskas and Mohammed Jasim Uddin

Received: 17 December 2021
Accepted: 12 January 2022
Published: 14 January 2022

Publisher's Note: MDPI stays neutral with regard to jurisdictional claims in published maps and institutional affiliations.

Copyright: © 2022 by the authors. Licensee MDPI, Basel, Switzerland. This article is an open access article distributed under the terms and conditions of the Creative Commons Attribution (CC BY) license (https://creativecommons.org/licenses/by/4.0/).

1. Introduction

Barium fluoride, BaF_2, belongs to the group of attractive nanoparticles, produced using different preparation methods and applied in numerous multifunctional applications. Nd^{3+}:BaF_2 nanocrystals synthesized by the reverse microemulsion technique present interesting luminescence properties [1]. Indeed, the quenching of fluorescence intensity (λ_{em} = 1052 nm) in nanosized Nd^{3+}:BaF_2 domains was not observed even under very high dopant levels (~45 mol.% of Nd^{3+}). Further experiments revealed the crystallization of cubic and orthorhombic BaF_2 nanoparticles, and it was proven that such fluoride crystals could be quite easily transformed from the orthorhombic phase to the more thermodynamically stable cubic phase under certain preparation conditions. This effect was confirmed by X-ray diffraction (XRD) and high-resolution transmission electron microscopy (HR-TEM) used for self-assembled monodisperse BaF_2 nanocrystals accomplished by the liquid–solid-solution (LSS) approach [2]. BaF_2 nanocrystals were also fabricated from precursor Na_2O-K_2O-BaF_2-Al_2O_3-SiO_2 glasses via their controlled heat treatment. Their self-organized nanocrystallization processes [3] and size distribution [4] have been presented and discussed in detail. Luminescence properties of nanosized Eu^{3+}-doped

BaF$_2$ synthesized via an ionic liquid-assisted solvothermal method in different solvents (e.g., DMSO, water, or water with PVP solution) confirmed that these fluoride nanoparticles can be effectively used for bioimaging applications [5].

From the accumulated experience and literature data, it is known that RE^{3+} ions can be introduced into the fluoride nanocrystals dispersed within the transparent glassy host. Indeed, several precursor glasses doped with RE^{3+} ions were heat treated to fabricate RE^{3+}:BaF$_2$ nanocrystals and obtain transparent glass-ceramics with enhanced luminescence properties. Nano-glass-ceramics with RE^{3+}:BaF$_2$ have been examined for visible [6] and near-infrared [7,8] luminescence as well as white up-conversion applications [9]. Special attention has been devoted to the structure and luminescent properties of BaF$_2$ nanocrystals in glass-ceramics singly doped with Er^{3+} [10,11] and co-doped with Er^{3+}/Yb^{3+} [12,13]. Among RE^{3+}, trivalent europium ions are commonly used as a spectroscopic probe, indicating structural changes around the optically active ions and their surrounding environment [14]. Additionally, europium ions in a divalent oxidation state can also exist, thus, the silicate glasses containing EuF$_3$ synthesized by melt-quenching in reducing atmosphere tend to form Eu^{2+}-doped glass-ceramics after the heat-treatment process. The prepared glass-ceramic system with Eu^{2+}:BaF$_2$ nanocrystals could be potentially utilized as a blue phosphor for UV–LED applications [15]. Divalent europium ions in fluorosilicate glass-ceramics can be well stabilized via lattice site substitution [16]. In the field of preparation of the RE^{3+}-doped glass-ceramics containing BaF$_2$ nanocrystals, particular attention should also be focused on the sol–gel method. The first synthesis of 95SiO$_2$–5BaF$_2$ (mol.%) nano-glass-ceramics via the sol–gel technique was reported and described in work by D. Chen et al. [17]. The authors proved that the size of precipitated BaF$_2$ nanocrystals (2–15 nm) and the luminescence of Er^{3+} ions are strictly dependent on heat-treatment conditions of initially obtained xerogels. C.E. Secu et al. [18,19] presented the fabrication, structure, and luminescence of 95SiO$_2$–5BaF$_2$ (mol.%) nano-glass-ceramics singly doped with Pr^{3+}, Ho^{3+}, Dy^{3+}, Sm^{3+}, and Eu^{3+} ions. Except for Eu^{3+}-doped samples, the emission bands of other active dopants were revealed after controlled heat treatment of precursor xerogels, which was explained by incorporating RE^{3+} into BaF$_2$ crystals (3–7 nm) and removing residual OH groups from the silicate sol–gel host. The recently published work by M. Hu et al. [20] was concentrated on properties of 95SiO$_2$–5BaF$_2$ (mol.%) glass-ceramics singly and doubly doped by Tb^{3+}, Eu^{3+}, and Dy^{3+} ions, containing fluoride nanocrystals with an average size of ~5 nm. The authors verified the thermal stability of generated luminescence in a range from 30 °C to 290 °C, proving that synthesized sol–gel nano-glass-ceramics could be utilized as color and white light emitters. The properties of RE^{3+}-doped sol–gel glass-ceramics containing BaF$_2$ nanocrystals were compared with other oxyfluoride systems in an extensive review published recently by Secu et al. [21]. This class of RE^{3+}-doped materials is widely considered as a promising candidate for selected applications, e.g., three-dimensional displays, flat color screens, spectral converters, light-emitting diodes (LEDs), etc. [21].

Our previously published work [22] was concerned with sol–gel SiO$_2$-BaF$_2$ nano-glass-ceramic systems doped with europium ions in a trivalent oxidation state. Their structural and optical properties have been studied using various experimental techniques, such as differential scanning calorimetry (DSC), X-ray diffraction (XRD), transmission electron microscopy (TEM) coupled with the energy-dispersive X-ray spectroscopy (EDS), infrared (ATR-IR), and luminescence spectroscopy. The properties of Tb^{3+}, Eu^{3+} co-doped glass-ceramic systems containing BaF$_2$ nanocrystals made from the sol–gel method at a low temperature are communicated here. To the best of our knowledge, these aspects for SiO$_2$-BaF$_2$:Tb^{3+}, Eu^{3+} nano-glass-ceramics have not yet been examined.

2. Materials and Methods

The xerogels singly doped with Tb^{3+} and co-activated with Tb^{3+}, Eu^{3+} ions were prepared via the previously described sol–gel synthesis [22]. The reagents from Sigma-Aldrich (St. Louis, MO, USA) were applied for the fabrication the samples. In the first step of preparation, precursor (TEOS), ethanol, deionized water, and acetic acid (AcOH) were mixed (in a molar ratio 1:4:10:0.5) in round-bottom flasks for 30 min. TEOS, $Si(OC_2H_5)_4$, was used as a precursor for creating the SiO_2 silicate host, water was necessary to perform the hydrolysis reaction of TEOS, and AcOH played a role as a catalyst. Due to the significantly limited solubility of TEOS in water, ethyl alcohol was introduced into the reaction systems, enabling the hydrolysis reaction by increasing the TEOS–water contact surface. The hydrolysis could be expressed by the following reaction:

$$Si(OC_2H_5)_4 + nH_2O \rightleftarrows Si(OH)_n(OC_2H_5)_{(4-n)} + nC_2H_5OH,$$

in which $n \leq 4$. Simultaneously with the hydrolysis reaction, the condensation begins, which allows for the creation of a silicate network through the formation of siloxane bridges, Si–O–Si. The homocondensation could be given by the following chemical reaction:

$$(OC_2H_5)_n(OH)_{(3-n)}Si-OH + HO-Si(OC_2H_5)_n(OH)_{(3-n)} \rightleftarrows$$
$$(OC_2H_5)_n(OH)_{(3-n)}Si-O-Si(OC_2H_5)_n(OH)_{(3-n)} + H_2O,$$

and the heterocondensation could be expressed by:

$$(OC_2H_5)_n(OH)_{(3-n)}Si-OH + C_2H_5O-Si(OC_2H_5)_n(OH)_{(3-n)} \rightleftarrows$$
$$(OC_2H_5)_n(OH)_{(3-n)}Si-O-Si(OC_2H_5)_n(OH)_{(3-n)} + C_2H_5OH,$$

in which $n \leq 3$. The mechanisms of hydrolysis and condensation reactions of alkoxides were discussed in detail in the paper [23].

After pre-hydrolysis and pre-condensation, the solutions of $Ba(AcO)_2$ and $RE(AcO)_3$ (RE = Tb or Tb/Eu) in trifluoroacetic acid (CF_3COOH, TFA) and deionized water were added dropwise, and the obtained mixtures were stirred for the next 60 min. Since the electrolytic dissociation of TFA acid ($K_a = 5.9 \times 10^{-1}$) is greater than for AcOH ($K_a = 1.8 \times 10^{-5}$), TFA is a much stronger acid than AcOH, and the following reaction occurs:

$$Ba(AcO)_2 + 2TFA \rightarrow Ba(TFA)_2 + 2AcOH.$$

For Tb^{3+}-doped samples, the molar ratio of $TFA:Ba(AcO)_2:Tb(AcO)_3$ was equal to 5:0.95:0.05, and for Tb^{3+}, Eu^{3+} co-doped materials, the molar ratio of $TFA:Ba(AcO)_2:Tb(AcO)_3:Eu(AcO)_3$ was equal to 5:0.9:0.05:0.05. The mass of TEOS, ethanol, deionized water, and acetic acid reached 90 wt.% of each sample, and the mass of the remaining part containing TFA, $Ba(AcO)_2$, and $RE(AcO)_3$ (RE = Tb or Tb/Eu) equaled 10 wt.%. The liquid sols were dried at 35 °C for several weeks and then heat treated at 350 °C per 10 h in a muffle furnace (Czylok, Jastrzębie-Zdrój, Poland). The thermal treatment of xerogels at 350 °C aims to transform them into SiO_2-BaF_2 nano-glass-ceramics. Indeed, TFA was introduced as a fluorination reagent, allowing for successful crystallization of BaF_2 fraction inside the silicate sol–gel host. The fabricated xerogels were denoted as XG_{Tb} and $XG_{Tb/Eu}$ (for singly and doubly doped xerogels), as well as nGC_{Tb} and $nGC_{Tb/Eu}$ (for singly and co-doped nano-glass-ceramics).

The thermogravimetry and differential scanning calorimetry (TG/DSC) were carried out using a Labsys Evo system with a heating rate of 10 °C/min in argon atmosphere (SETARAM Instrumentation, Caluire, France). To verify the formation of fluoride nanocrystals within the silicate sol–gel host at 350 °C, the X-ray diffraction was performed using an X'Pert Pro diffractometer equipped by PANalytical with CuK_α radiation (Almelo, the Netherlands). Additionally, the fluoride nanocrystals were observed by a JEOL JEM 3010 transmission electron microscope operated at 300 kV (JEOL, Tokyo, Japan). The structural characterization was supplemented by infrared spectroscopy (IR). The experiment was performed with the use of the Nicolet iS50 ATR spectrometer (Thermo Fisher Scientific Instruments, Waltham, MA, USA), and the spectra were collected in attenuated total reflectance (ATR) configuration within the 4000–400 cm^{-1} as well as 500–200 cm^{-1} ranges (64 scans, 4 cm^{-1} resolution).

The luminescence measurements were performed on a Photon Technology International (PTI) Quanta-Master 40 (QM40) UV/VIS Steady State Spectrofluorometer (Photon Technology International, Birmingham, NJ, USA) supplied with a tunable pulsed optical parametric oscillator (OPO) pumped by the third harmonic of a Nd:YAG laser (Opotek Opolette 355 LD, OPOTEK, Carlsbad, CA, USA). The laser system was coupled with a 75 W xenon lamp, a double 200 mm monochromator, and a multimode UV/VIS PMT (R928) (PTI Model 914) detector. The excitation and emission spectra were recorded with a resolution of 0.5 nm. The luminescence decay curves were recorded by a PTI ASOC-10 (USB-2500) oscilloscope. All structural and optical measurements were carried out at room temperature.

3. Results and Discussion

3.1. Thermal Behavior of Synthesized Xerogels

Figure 1 presents the TG/DSC curves recorded for fabricated xerogels in an inert gas atmosphere in a temperature range from 45 °C to 475 °C (the heating rate during measurement was 10 °C/min). According to TG curves, there are two distinguishable degradation steps for both fabricated samples: first, identified at 45–(~205) °C, and second, observed in the temperature range (~205) °C–(~320) °C. A slight weight loss, about 2.75% (XG_{Tb}) and 3.56% ($XG_{Tb/Eu}$), is associated with the elimination of residual solvents (ethyl alcohol, acetic acid) and water from the porous sol–gel host. At higher temperatures, strong exothermic peaks with maxima at 305 °C (XG_{Tb}) and 306 °C ($XG_{Tb/Eu}$) were identified, which appear along with ~17.55% weight loss. Generally, trifluoroacetates tend to decompose at temperatures near ~300 °C, which is well documented and described in the current literature [24–26]. Thus, recorded exothermic DSC peaks are clearly correlated with thermal decomposition of $Ba(TFA)_2$ and crystallization of BaF_2, which could be given by the chemical reaction:

$$Ba(TFA)_2 \xrightarrow{T} BaF_2 + CF_3CFO + CO_2 + CO.$$

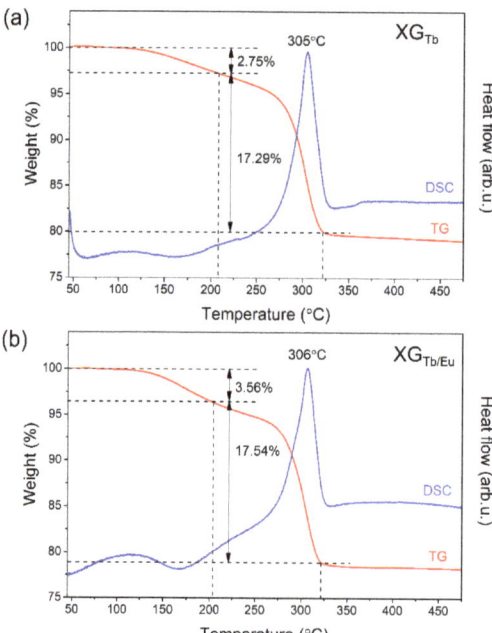

Figure 1. DSC and TG curves of prepared xerogels singly doped with Tb^{3+} (**a**), and co-doped with Tb^{3+}/Eu^{3+} ions (**b**).

The thermolysis led to cleavage of C–F bonds from –CF_3 groups, and the resultant fluorine ions (F^-) tend to react with Ba–O bonds, forming BaF_2 phase [27]. The heat exchanged during degradation of $Ba(TFA)_2$ in studied sol–gel materials is close to −118 J/g. J. Farjas et al. [28] pointed out that the denoted heat exchange during the degradation of trifluoroacetates depends on atmosphere (air or ambient gas) and the presence of vapored water. Our obtained value is comparable with DSC results obtained for pure $Ba(TFA)_2$ salt in argon atmosphere [28]. The data obtained from TG/DSC analysis for the studied sol–gel samples are shown in Tables 1 and 2.

Table 1. The parameters from TG analysis for studied sol–gel materials.

Sample	Number of Degradation Steps	Temperature Range (°C)	Weight Loss (%)
XG_{Tb}	1st	45–208	2.75
	2nd	208–322	17.56
$XG_{Tb/Eu}$	1st	45–204	3.56
	2nd	204–321	17.54

Table 2. The parameters from DCS curves recorded for fabricated silicate sol–gel samples.

Sample	Peak Maximum (°C)	Exchanged Heat (J/g)
XG_{Tb}	305	−118.3
$XG_{Tb/Eu}$	306	−117.9

3.2. Structural Characterization by XRD, TEM, and ATR-IR

Figure 2 presents the X-ray diffraction (XRD) patterns of the xerogels and nano-glass-ceramics fabricated at 350 °C. The diffractograms collected for Tb^{3+} singly doped samples are depicted in Figure 2a, meanwhile, the data for Tb^{3+}, Eu^{3+} co-doped materials are shown in Figure 2b. The XRD patterns of the precursor xerogels revealed any sharp diffraction lines, but only a broad hump with a maximum at ~25°, indicating their amorphous nature without long-range order [29]. Conversely, the intense diffraction lines were observed only after thermal treatment of xerogels at 350 °C for 10 h. The XRD patterns of prepared glass-ceramics are in accordance with the standard diffraction lines of cubic BaF_2 crystallized in the Fm3m space group (ICDD card no. 00-004-0452), confirming the precipitation of fluoride crystals inside the silicate sol–gel matrix. The crystalline size of BaF_2 in fabricated glass-ceramics was evaluated by calculations with the Scherrer formula given below [30]:

$$D = \frac{K\lambda}{B \cos \theta} \quad (1)$$

in which K is a shape factor (in our calculations, K = 1 was taken), λ is a wavelength of X-rays (0.154056 nm, Kα line of Cu), B is a broadening of the diffraction peak at half the maximum intensity, and θ is Bragg's angle. The average crystal sizes of BaF_2 were calculated to be 5 nm ± 0.1 nm for both nGC_{Tb} and $nGC_{Tb/Eu}$ samples. The average size of BaF_2 nanocrystallites was also calculated from a Williamson–Hall plot as follows [31]:

$$D = \frac{K\lambda}{\beta \cos \theta} \quad (2)$$

where β is half of the width of the diffraction line, whereas (Δa/a) refers to the lattice deformation.

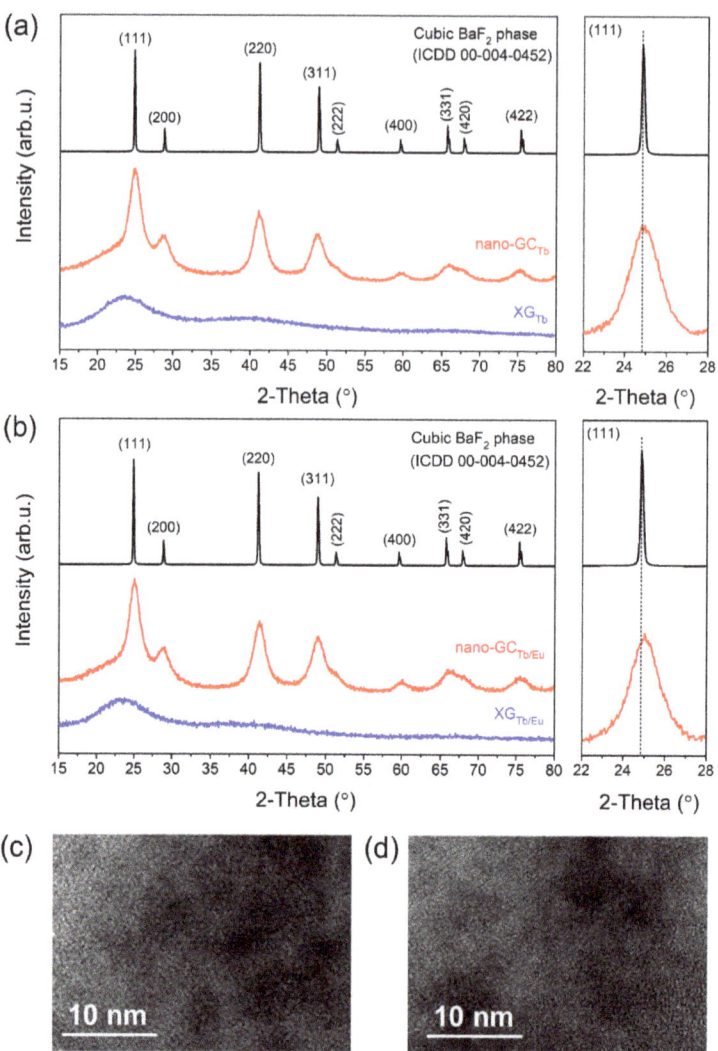

Figure 2. XRD patterns of prepared sol–gel samples: Tb^{3+} singly doped materials (**a**) and Tb^{3+}, Eu^{3+} co-doped specimens (**b**). The standard data for the BaF_2 cubic phase (ICDD card no. 00-004-0452) are also shown for comparison. TEM images revealed the presence of fluoride crystals in glass-ceramics singly doped with Tb^{3+} (**c**) and co-doped with Tb^{3+}, Eu^{3+} (**d**).

The Scherrer method makes the half-width of the diffraction line dependent only on the size of the crystallites. On the other hand, in the Williamson–Hall analysis, the internal stresses reflected by the lattice deformation are additionally taken into account in the broadening of the diffraction line. The mean crystal sizes, calculated with the Williamson–Hall method, were estimated to be 4.3 nm ± 0.1 nm for nGC_{Tb}, and 4.8 nm ± 0.1 nm for $nGC_{Tb/Eu}$. Moreover, the lattice deformation was negligible and less than 0.1%. The obtained results of the average crystallite size, from both methods, reveal good agreement. They proves the lack of internal stresses in the formed BaF_2 particles. The crystal lattice parameters for BaF_2 phase were determined to be 6.188 (8) Å (Tb^{3+}-doped sample) and 6.169 (7) Å (Tb^{3+}, Eu^{3+} co-doped sample), which are slightly smaller than the lattice parameter for undoped barium fluoride (a_0 = 6.2001 Å). Indeed, both Tb^{3+} (1.04 Å) and

Eu^{3+} (1.07 Å) ions [32] with smaller ionic radii could substitute Ba^{2+} (1.35 Å) [33] cations in BaF_2 crystal lattice, resulting in a decrease in the unit cell volume. The indicated changes in the lattice parameter are also noticeable as a slight shift of the recorded diffraction lines towards higher values of the 2θ angle (an enlargement within a 22–28° angle, in which the (111) diffraction line was detected; shown in Figure 1). Additionally, it was observed that the shift of diffraction lines is more clearly visible for the $nGC_{Tb/Eu}$ sample, which evidences that the total incorporation of RE^{3+} ions inside BaF_2 nanocrystals is higher than for nGC_{Tb}. Similar results from XRD measurements were described in the literature for other oxyfluoride optical systems, e.g., SiO_2-LaF_3:Er^{3+} sol–gel nano-glass-ceramics [34] and germano-gallate glass-ceramics containing BaF_2:Er^{3+} nanocrystals [11]. Figure 2c,d display the TEM images of prepared nano-glass-ceramic samples singly doped with Tb^{3+} and co-doped with Tb^{3+}, Eu^{3+} ions, respectively. The size of BaF_2 nanocrystals was average, estimated from the Scherrer equation and the Williamson–Hall method.

The ATR-IR spectrum within the 4000–400 cm^{-1} range for a representative $XG_{Tb/Eu}$ sample is shown in Figure 3a, and the assignment of individual IR peaks was carried out based on the literature [35,36]. The recorded infrared signals confirmed the formation of a polycondensed silicate network created by Q^2 (949 cm^{-1}), Q^3 (1045 cm^{-1}), and Q^4 (1134 cm^{-1}) units of SiO_2 tetrahedrons as well as Si–O–Si siloxane bridges (1192 cm^{-1}, 801 cm^{-1}). On the other hand, the signals recorded at ~3665 cm^{-1} and ~3400 cm^{-1}, according to vicinal/geminal and hydrogen-bonded Si–OH groups, respectively, clearly pointed to the presence of unreacted silanol groups. Indeed, xerogels are highly porous materials [37], hence, the IR bands originating from Si–OH groups are expected. Indeed, the next recorded infrared band, 1659 cm^{-1}, revealed the vibrations of Si–OH groups, but also oscillations within C=O carbonyl groups (from residual AcOH and unreacted TFA), as well as molecular water. A signal at ~3200 cm^{-1} was interpreted as vibrations from hydrogen-bonded OH groups in organic compounds and water, and may confirm that pores inside the silicate network are filled with liquids. It should be noted that peaks located near ~1134 cm^{-1} and ~1192 cm^{-1} could be assigned, despite Q^4 units and Si–O–Si bridges, to oscillations of C–F bonds in $Ba(TFA)_2$ and unreacted TFA. Indeed, a comparison of ATR-IR spectra in this region for $XG_{Tb/Eu}$ xerogel and an analogous sample prepared without the addition of $Ba(AcO)_2$ and TFA revealed that the signals are more intense for $XG_{Tb/Eu}$ (inset of Figure 3a). This could point to the presence of additional oscillators that contribute to overall signals recorded at ~1134 cm^{-1} and ~1192 cm^{-1}. The signal recorded at ~420 cm^{-1} was assigned to the O–Si–O bending vibration.

The ATR-IR spectrum within the 4000–400 cm^{-1} region registered for a representative $nGC_{Tb/Eu}$ sample obtained at 350 °C is shown in Figure 3b. Compared with the ATR-IR spectrum for $XG_{Tb/Eu}$, the intensities of signals at >3000 cm^{-1} and 1659 cm^{-1} weakened significantly, which allows us to make conclusions about evaporation of volatile chemical components from the sol–gel host and progressive reactions between unreacted Si–OH groups. Additionally, it was also observed that for the $nGC_{Tb/Eu}$ nano-glass-ceramic sample, the intensity of the IR signal near ~949 cm^{-1} is weaker than for the $XG_{Tb/Eu}$ xerogel sample. An indicated effect may also suggest a continuation of polycondensation because Q^2 units probably transformed into Q^3 and Q^4 ones, which favor the creation of a more cross-linked sol–gel host. It was also observed that intensities of IR signals near ~1134 cm^{-1} and ~1192 cm^{-1} weakened compared with those recorded for the xerogel. This effect could be explained by thermal decomposition of $Ba(TFA)_2$ compound into BaF_2 crystals within the prepared silicate sol–gel host in the proposed heat-treatment conditions. Indeed, the Ba–F vibrations might be observed at lower frequencies (inset of Figure 3b), which agrees with the IR spectrum recorded for pure BaF_2 [38]. The peak with a maximum at ~440 cm^{-1} was recorded for both $nGC_{Tb/Eu}$ nano-glass-ceramics and an analogous sample prepared without the addition of $Ba(AcO)_2$ and TFA, confirming that such a band is not related to the fluoride fraction but to the oscillations within the silicate host (O–Si–O vibration). It should be noticed that this band shifts toward a higher frequency for nano-glass-ceramics in comparison with the xerogel. The reason for such spectral behavior could be explained by differences in the inter-tetrahedra angle of SiO_4 units in xerogels and glass-ceramics, as was stated in the literature [39].

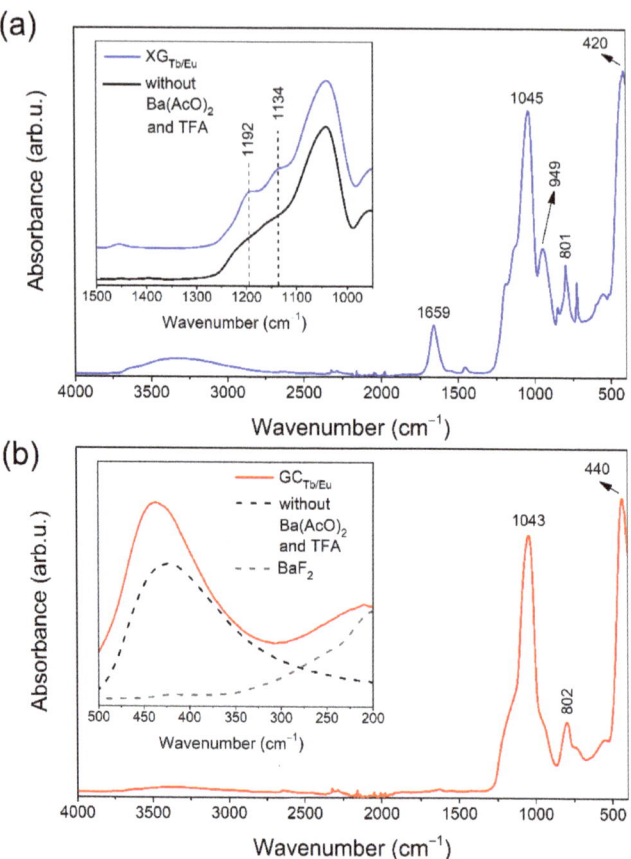

Figure 3. ATR-IR spectra recorded for xerogel $XG_{Tb/Eu}$ (**a**) and nano-glass-ceramic $nGC_{Tb/Eu}$ (**b**) co-doped with Tb^{3+}, Eu^{3+} ions.

3.3. Luminescence of Amorphous Silicate Xerogels

Figure 4a shows the excitation spectra of the prepared $XG_{Tb/Eu}$ samples. The spectra were registered within the 340–520 nm spectral range on collecting the luminescence at 541 nm and 612 nm wavelengths. The excitation spectrum, while monitoring the green emission line at 541 nm, revealed the characteristic bands for Tb^{3+} ions according to the following transitions within the near-UV and VIS scope: $^7F_6 \rightarrow {}^5L_9$ (352 nm), $^7F_6 \rightarrow {}^5L_{10}$ (370 nm), $^7F_6 \rightarrow {}^5D_3$ (379 nm), and $^7F_6 \rightarrow {}^5D_4$ (488 nm). Meanwhile, the spectrum recorded by collecting the red luminescence at 612 nm showed the excitation lines of Eu^{3+} related to the electronic transitions from the $^7F_0$ ground level into the following excited states: ^5G_J (376 nm), $^5L_7$ (384 nm), $^5L_6$ (394 nm), $^5D_3$ (418 nm), and $^5D_2$ (464 nm). However, it was observed that the spectrum recorded for $XG_{Tb/Eu}$ contains some additional weak bands, which did not appear for the sample singly doped with Eu^{3+} (for better visibility, an enlargement of the 340–390 nm scope is presented in the inset of Figure 2a, and the bands are marked by asterisks). It should be noted that the recorded additional bands correspond to the contribution of excitation lines originating from Tb^{3+} ions ($^7F_6 \rightarrow {}^5L_9$ (352 nm), $^7F_6 \rightarrow {}^5L_{10}$ (379 nm), and $^7F_6 \rightarrow {}^5D_4$ (488 nm)). Moreover, a slight shift of the $^7F_0 \rightarrow {}^5L_7$ band (from 384 nm to 382 nm) was also denoted, which could be related to its overlapping with the $^7F_6 \to {}^5D_3$ excitation line originating from Tb^{3+} co-dopant. Hence, the obtained results could suggest the occurrence of $Tb^{3+} \to Eu^{3+}$ ET. A similar interpretation of excitation spectra was described for lead borate glasses co-doped with Tb^{3+} and Eu^{3+} ions [40].

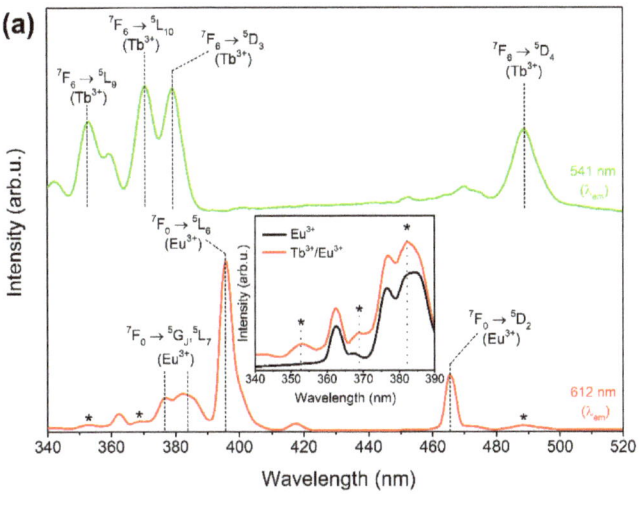

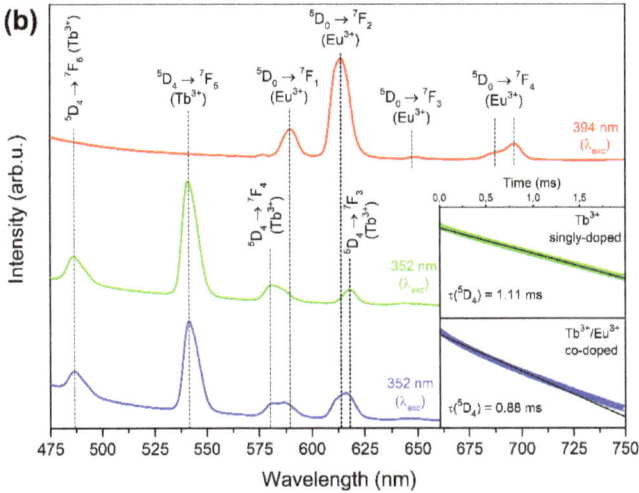

Figure 4. The excitation spectra recorded for Tb^{3+} (λ_{em} = 541 nm) and Eu^{3+} (λ_{em} = 612 nm) ions in fabricated amorphous xerogels. For the latter, the additional lines originated from Tb^{3+} ions were marked by asterisks (**a**). The registered luminescence spectra collected for XG_{Tb} (green line, λ_{exc} = 352 nm) and $XG_{Tb/Eu}$ samples (blue line, λ_{exc} = 352 nm; red line, λ_{exc} = 394 nm). Inset shows the decay curves recorded for the $^5D_4$ state of Tb^{3+} ions (**b**).

The fluorescence spectra of prepared sol–gel specimens are displayed in Figure 4b. The emission spectrum recorded for the $XG_{Tb/Eu}$ sample under excitation at 394 nm (presented as a red line) consisted of several emission lines at 574 nm ($^5D_0 \to {}^7F_0$), 590 nm ($^5D_0 \to {}^7F_1$), 612 nm ($^5D_0 \to {}^7F_2$), 648 nm ($^5D_0 \to {}^7F_3$), and 696 nm ($^5D_0 \to {}^7F_4$) within the reddish-orange light area. It was observed that the $^5D_0 \to {}^7F_2$ red emission band is the most prominent luminescence line, and the spectrum is similar to other Eu^{3+}-doped typical glassy-like optical materials described in the literature [41,42]. Based on the collected spectrum, the R/O ratio (red-to-orange) was calculated using the areas of the $^5D_0 \rightarrow {}^7F_2$ (R) and the $^5D_0 \rightarrow {}^7F_1$ (O) bands. The R/O ratio value estimated for precursor silicate xerogel is relatively high and equals 3.92. It indicates that Eu^{3+} ions are far from an inversion center, which is characteristic for amorphous materials. In the luminescence spectrum of the XG_{Tb} sample (marked as a green line), the bands centered at 486 nm, 541 nm, 580 nm, and 618 nm were attributed to the $^5D_4 \rightarrow {}^7F_J$ (J = 6–3) electronic transitions, respectively.

To verify the occurrence of ET between Tb^{3+} and Eu^{3+} ions in the studied silicate xerogels, the emission spectrum for the $XG_{Tb/Eu}$ sample was recorded upon excitation at a 352 nm wavelength (shown as a blue line). The spectrum consisted of the following emission bands in the VIS spectral range: blue (486 nm), an intense green (541 nm), yellowish-orange (584 nm), and red (616 nm). The same bands within the blue–green light area were detected for the XG_{Tb} xerogel, and the mentioned emission lines were ascribed to the $^5D_4 \rightarrow {}^7F_6$ and the $^5D_4 \rightarrow {}^7F_5$ electronic transitions, respectively. Although the positions of these emission bands are the same, their intensity is slightly lower for the co-doped $XG_{Tb/Eu}$ sample than for the singly doped XG_{Tb} one. Simultaneously, an increase in luminescence intensity within the yellowish-orange as well as red ranges was observed, and—compared with emissions recorded for the XG_{Tb} xerogel—the maxima of these bands were slightly shifted (from 580 nm to 584 nm, and from 618 nm to 616 nm). Thus, based on this observation, we could conclude that the indicated shift is a result of the superimposition of the yellow ($^5D_4 \rightarrow {}^7F_4$, 580 nm) and red band ($^5D_4 \rightarrow {}^7F_3$, 618 nm) of Tb^{3+} ions with orange ($^5D_0 \rightarrow {}^7F_1$, 590 nm) and red ($^5D_0 \rightarrow {}^7F_2$, 612 nm) luminescence originating from Eu^{3+}.

Hence, our experimental results indicate the occurrence of Tb^{3+}/Eu^{3+} ET upon excitation at a 352 nm wavelength when Tb^{3+} ions are excited from the $^7F_6$ ground state. Then, the electrons at the $^5L_9$ level decay rapidly through the $^5G_5$, $^5L_{10}$, and $^5D_3$ states by the multiphonon relaxation process until the $^5D_4$ level is populated. Since there is the energetical resemblance of the $^5D_4$ (Tb^{3+}) and the $^5D_1/^5D_0$ (Eu^{3+}) levels, the Tb^{3+}/Eu^{3+} energy migration is feasible, and the excitation energy is transferred from Tb^{3+} to the adjacent Eu^{3+} ion. The acceptor ions relax from the $^5D_0$ state to the ^7F_J levels, promoting the light emission within the reddish-orange spectral region [43]. The ET is schematized in the level diagram presented in Figure 5.

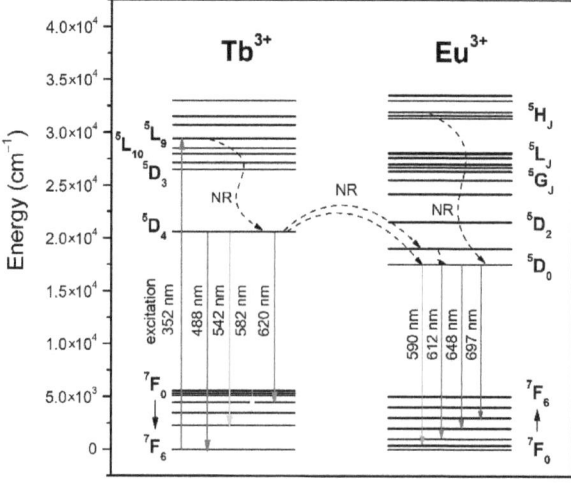

Figure 5. Energy level scheme of Tb^{3+} and Eu^{3+} ions.

The decay curves were registered for the green light at 541 nm, upon excitation at 352 nm from the near-UV range (inset in Figure 4b). For xerogels, a mono-exponential fit was used to evaluate the lifetimes of Tb^{3+}, and the fitted curves are marked with a black line, while the collected experimental data are shown as green and blue lines for XG_{Tb} and $XG_{Tb/Eu}$, respectively. A slight shortening in the decay time of the $^5D_4$ (Tb^{3+}) state from 1.11 ms (XG_{Tb}) to 0.88 ms ($XG_{Tb/Eu}$) was identified. An indicated decline in a lifetime for co-doped xerogel could be explained by introducing an additional decay pathway via Eu^{3+} ions. Indeed, the ET from Tb^{3+} to Eu^{3+} enhances the decay rate of the excited Tb^{3+} ions, resulting in the shortening of the $^5D_4$ (Tb^{3+}) lifetime. Hence, the analysis of luminescence decay curves also enables calculation of the efficiency of Tb^{3+}/Eu^{3+} ET, based on the following equation [44]:

$$\eta_{ET} = \left(1 - \frac{\tau}{\tau_0}\right) \cdot 100\%. \quad (3)$$

where τ_0 and τ are the lifetimes of the $^5D_4$ (Tb^{3+}) state for sample singly doped with Tb^{3+}, and the sample co-doped with Tb^{3+}, Eu^{3+} ions, respectively. In the case of the studied xerogels, the efficiency of Tb^{3+}/Eu^{3+} ET was estimated to be about 21%, and the comparable values were denoted for, e.g., fluoroborate glass (η_{ET} = 20%) [45].

3.4. Luminescence of SiO_2-BaF_2 Nano-Glass-Ceramics

The excitation spectra recorded for the $nGC_{Tb/Eu}$ sample are shown in Figure 6a. The spectra emerged by monitoring the green luminescence characteristic for Tb^{3+} (541 nm), and the red emission originating from Eu^{3+} ions (612 nm). The luminescence of Tb^{3+} ions (541 nm) could be efficiently excited by the following wavelengths from the near-UV scope: 352 nm ($^7F_6 \rightarrow {}^5L_9$), 369 nm ($^7F_6 \rightarrow {}^5L_{10}$), and 377 nm ($^7F_6 \rightarrow {}^5D_3$), as well as from the VIS range: 485 nm ($^7F_6 \rightarrow {}^5D_4$). In the case of the excitation spectrum recorded at a 612 nm emission wavelength, an intense line appeared at 394 nm ($^7F_0 \rightarrow {}^5L_6$, Eu^{3+}), but a few weaker bands at 376 nm ($^7F_0 \rightarrow {}^5G_J$), 384 nm ($^7F_0 \rightarrow {}^5L_7$), 418 nm ($^7F_0 \rightarrow {}^5D_3$), and 465 nm ($^7F_0 \rightarrow {}^5D_2$) were also detected. Similarly, as for xerogels, the recorded additional excitation lines at 352 nm, 369 nm, and 485 nm—marked in Figure 6a by asterisks—are typical for the $^7F_6 \rightarrow {}^5L_{9,10}, {}^5D_4$ transitions of Tb^{3+} ions, which could suggest the occurrence of Tb^{3+}/Eu^{3+} ET in the studied nano-glass-ceramic samples. Similar results were found for other Tb^{3+}, Eu^{3+} co-doped fluoride-based optical systems, e.g., pure CaF_2 nanocrystals [46], and glass-ceramics containing SrF_2 [47], as well as $NaYF_4$ nanocrystals [48].

Figure 6b depicts the emission spectra collected for $nGC_{Tb/Eu}$ and nGC_{Tb} samples, recorded upon excitation at 352 nm (a blue line for $nGC_{Tb/Eu}$, and a green line for nGC_{Tb}) and 394 nm (a red line) wavelengths. An excitation of the nGC_{Tb} sample using 352 nm results in registration of the visible emissions ascribed to the $^5D_4 \rightarrow {}^7F_6$ (487 nm), $^5D_4 \rightarrow {}^7F_5$ (541 nm), $^5D_4 \rightarrow {}^7F_4$ (580 nm, 587 nm), and $^5D_4 \rightarrow {}^7F_3$ (619 nm) transitions characteristic for Tb^{3+} ions. Subsequently, when the $nGC_{Tb/Eu}$ co-doped sample was excited by a 394 nm wavelength, the luminescence bands originating from Eu^{3+} ions centered at 589 nm ($^5D_0 \rightarrow {}^7F_1$), 611 nm/614 nm ($^5D_0 \rightarrow {}^7F_2$), 648 nm ($^5D_0 \rightarrow {}^7F_3$), and 688 nm/696 nm ($^5D_0 \rightarrow {}^7F_4$) were observed. One can see that, in contrast to xerogel, the $^5D_0 \rightarrow {}^7F_1$ magnetic dipole transition dominates the spectrum, which indicates that Eu^{3+} ions are placed at sites close to an inversion symmetry [49]. According to the calculated R/O ratio value (3.92 for $XG_{Tb/Eu}$ and 0.51 for $nGC_{Tb/Eu}$) and the literature [18], the observed change in emission profile clearly suggests that Eu^{3+} ions tend to embed into the BaF_2 fluoride nanocrystal lattice by substituting Ba^{2+} cations. The decrease in the R/O ratio value was denoted for other Eu^{3+}-doped oxyfluoride glass-ceramic systems described in the literature [50–52].

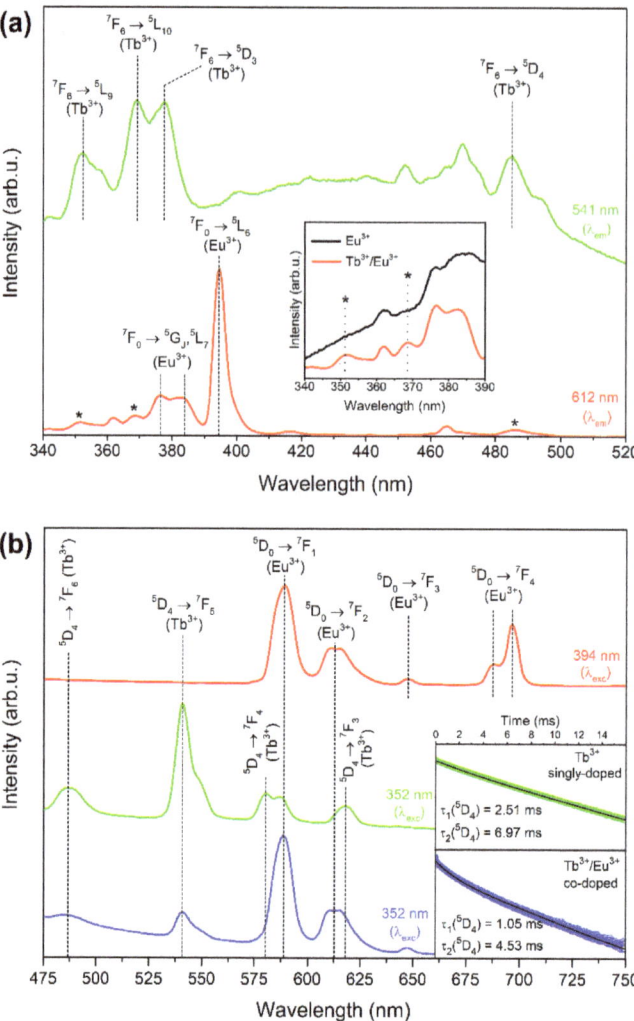

Figure 6. The excitation spectra recorded for Tb^{3+} (λ_{em} = 541 nm) and Eu^{3+} (λ_{em} = 612 nm) ions in prepared SiO_2-BaF_2 nano-glass-ceramics. For the latter, the additional lines originated from Tb^{3+} ions were marked by asterisks (**a**). The registered emission spectra for nGC_{Tb} (green line, λ_{exc} = 352 nm) as well as $nGC_{Tb/Eu}$ glass-ceramics (blue line, λ_{exc} = 352 nm; red line, λ_{exc} = 394 nm). Inset shows the decay curves recorded for the $^5D_4$ (Tb^{3+}) state in nano-glass-ceramics (**b**).

The emission spectrum of the $nGC_{Tb/Eu}$ sample, collected upon 352 nm excitation, revealed an intense orange (589 nm) and red (611 nm/615 nm, and 647 nm) luminescence corresponding to the transitions of Eu^{3+} from the $^5D_0$ level. Along with those bands, two emission lines with relatively low intensity were found within the blue–green scope and were assigned to the emissions originating from the $^5D_4$ state of Tb^{3+} ions. Therefore, compared with $XG_{Tb/Eu}$, the luminescence in the reddish-orange spectral range is particularly enhanced for $nGC_{Tb/Eu}$. Based on this observation, we could conclude that the distance between interacting Tb^{3+} and Eu^{3+} ions in the prepared nano-glass-ceramics might be significantly shorter than in xerogels. Such shortening in the inter-ionic distance, strictly related to the segregation of rare earths inside BaF$_2$ nanocrystals precipitated at 350 °C, could be responsible for a more efficient transfer of excitation energy from Tb^{3+} to Eu^{3+} ions.

For the SiO$_2$-BaF$_2$ nano-glass-ceramics, the luminescence decay from the ^{5}D$_4$ level follows a double-exponential function with two different decay lifetimes. It results from the distribution of RE^{3+} ions between either the sol–gel host (described by faster τ_1 component) and BaF$_2$ nanocrystals (described by longer τ_2 lifetime). The results are presented in the inset of Figure 6b, and the fitted decay curves are labeled with a black line, whereas the experimental data are tagged as green and blue lines for nGC$_{Tb}$ and nGC$_{Tb/Eu}$, respectively. For the sample singly doped with Tb^{3+} ions, the lifetime components are equal to τ_1 = 2.51 ms and τ_2 = 6.97 ms, while for the sample co-doped with Tb^{3+}, Eu^{3+} the decay times are equal to τ_1 = 1.05 ms and τ_2 = 4.53 ms. Based on lifetime components, the average decay times, τ_{avg}, were calculated from the following formula [53]:

$$\tau_{avg} = \frac{A_1\tau_1^2 + A_2\tau_2^2}{A_1\tau_1 + A_2\tau_2}. \tag{4}$$

Thus, the average luminescence lifetime of the ^{5}D$_4$ (Tb^{3+}) state for nGC$_{Tb/Eu}$ was determined to be τ_{avg} = 4.06 ms, and for nGC$_{Tb}$ it equaled τ_{avg} = 6.56 ms. The analysis of luminescence decay curves showed a noticeable prolongation in lifetimes for SiO$_2$-BaF$_2$ nano-glass-ceramics compared with xerogels. It suggests that the amount of OH groups characterized by high vibrational energy (>3000 cm^{-1}) should be significantly reduced in glass-ceramics. Moreover, the dopant ions tend to enter into the BaF$_2$ nanocrystals characterized by low phonon energy (~319 cm^{-1} [54]), making the radiative relaxation from the ^{5}D$_4$ level more prominent compared with xerogels.

Additionally, based on luminescence lifetimes, the calculated ET efficiency for prepared SiO$_2$-BaF$_2$ nano-glass-ceramic exceeds 38%. In such a case, the distance between interacting RE^{3+} ions entering into BaF$_2$ nanocrystals decreased, resulting in a reinforced transfer of energy absorbed by Tb^{3+} to Eu^{3+}. Indeed, it is related to creating an energy transfer net among the donor and acceptor ions, causing the ET to become more frequent. Comparable values of ET efficiency were described for glass-ceramics containing YF$_3$:1Tb^{3+}, 0.5Eu^{3+} (mol.%) nanophase ($\eta_{ET} \approx$ 39%) [55].

Summarizing, due to the unique properties of BaF$_2$, e.g., a broad region of transparency from 0.14 µm up to 14 µm, wide bandgap (11 eV), and low maximum phonon energy (~319 cm^{-1}), the oxyfluoride glass-ceramics containing BaF$_2$ nanophase are extensively applied to generate an efficient up- [11] and down-conversion luminescence [9], or white light emission [20]. Therefore, such materials could be successfully used for laser technologies, spectral converters, and three-dimensional displays [10]. Since Eu^{3+} ions emit within the red or reddish-orange light area, and Tb^{3+} ions are well known as green emitters, the fabricated SiO$_2$-BaF$_2$:Tb^{3+}, Eu^{3+} nano-glass-ceramics are able to generate multicolor luminescence. Thus, sol–gel materials might be considered for use as optical elements in RGB lighting optoelectronic devices operating upon near-UV excitation.

4. Conclusions

This work presented the fabrication of Tb^{3+}, Eu^{3+} co-doped oxyfluoride glass-ceramics at 350 °C from xerogels prepared via the sol–gel technique. The analysis of the thermal behavior of xerogels was performed using TG/DSC measurements, and the structural properties were determined based on ATR-IR spectroscopy. The crystallization of BaF$_2$ at the nanoscale was confirmed by XRD and TEM measurements. The characterization of sol–gel samples involved an excitation of the prepared sol–gel materials upon near-UV irradiation at 352 nm which showed the Tb^{3+}/Eu^{3+} energy transfer, resulting in strengthening the luminescence within the reddish-orange light scope due to additional emission from Eu^{3+} ions. Nevertheless, for xerogels, the blue–green luminescence (^{5}D$_4 \to {}^7$F$_{5,6}$ of Tb^{3+}) dominated, meanwhile, the reddish-orange emission (^{5}D$_0 \to {}^7$F$_{0-4}$ of Eu^{3+} overlapped with ^{5}D$_4 \to {}^7$F$_{4,3}$ bands of Tb^{3+}) was particularly enhanced for SiO$_2$-BaF$_2$ nano-glass-ceramics. The luminescence decay kinetics showed that in the co-doped sol–gel materials, the energy transfer from Tb^{3+} to Eu^{3+} ions occurred with an efficiency that varied from 21% for xerogels to 38% for nano-glass-ceramics. An indicated increase in energy transfer efficiency for prepared nano-glass-ceramics could be explained by shortening the distance between interacting Tb^{3+} and Eu^{3+} ions embedded into the BaF_2 nanocrystal lattice. The obtained results suggest that the fabricated SiO_2-BaF_2:Tb^{3+}, Eu^{3+} nano-glass-ceramics could be predisposed to application in selected technologies, e.g., three-dimensional displays and color screens.

Author Contributions: Conceptualization, N.P.; methodology, N.P. and B.S.-S.; software, N.P.; validation, N.P.; formal analysis, N.P.; investigation, N.P., T.G. and E.P.; resources, W.A.P.; data curation, N.P.; writing—original draft preparation, N.P. and W.A.P.; writing—review and editing, N.P. and W.A.P.; visualization, N.P.; supervision, N.P.; project administration, N.P.; funding acquisition, W.A.P. All authors have read and agreed to the published version of the manuscript.

Funding: The research activities are co-financed by the funds granted under the Research Excellence Initiative of the University of Silesia in Katowice.

Institutional Review Board Statement: Not applicable.

Informed Consent Statement: Not applicable.

Data Availability Statement: The data presented in this study are available on request from the corresponding authors.

Conflicts of Interest: The authors declare no conflict of interest.

References

1. Bender, C.M.; Burlitch, J.M.; Barber, D.; Pollock, C. Synthesis and fluorescence of neodymium-doped barium fluoride nanoparticles. *Chem. Mater.* **2000**, *12*, 1969–1976. [CrossRef]
2. Xie, T.; Li, S.; Peng, Q.; Li, Y. Monodisperse BaF_2 Nanocrystals: Phases, Size Transitions, and Self-Assembly. *Angew. Chem. Int. Ed.* **2009**, *48*, 196–200. [CrossRef]
3. Bocker, C.; Rüssel, C. Self-organized nano-crystallisation of BaF_2 from $Na_2O/K_2O/BaF_2/Al_2O_3/SiO_2$ glasses. *J. Eur. Ceram. Soc.* **2009**, *29*, 1221–1225. [CrossRef]
4. Bocker, C.; Bhattacharyya, S.; Höche, T.; Rüssel, C. Size distribution of BaF_2 nanocrystallites in transparent glass ceramics. *Acta Mater.* **2009**, *57*, 5956–5963. [CrossRef]
5. Sharma, R.K.; Nigam, S.; Chouryal, Y.N.; Nema, S.; Bera, S.P.; Bhargava, Y.; Ghosh, P. Eu-Doped BaF_2 Nanoparticles for Bioimaging Applications. *ACS Appl. Nano Mater.* **2019**, *2*, 927–936. [CrossRef]
6. Huang, L.; Jia, S.; Li, Y.; Zhao, S.; Deng, D.; Wanh, H.; Jia, G.; Hua, W.; Xu, S. Enhanced emissions in Tb^{3+}-doped oxyfluoride scintillating glass ceramics containing BaF_2 nanocrystals. *Nucl. Instrum. Methods Phys. Res. Sect. A* **2015**, *788*, 111–115. [CrossRef]
7. Zhang, W.-J.; Chen, Q.-J.; Qian, Q.; Zhang, Q.-Y. The 1.2 and 2.0 μm emission from Ho^{3+} in glass ceramics containing BaF_2 nanocrystals. *J. Am. Ceram. Soc.* **2012**, *95*, 663–669. [CrossRef]
8. Zhao, Z.; Liu, C.; Xia, M.; Yin, Q.; Zhao, X.; Han, J. Intense ~1.2 μm emission from Ho^{3+}/Y^{3+} ions co-doped oxyfluoride glass-ceramics containing BaF_2 nanocrystals. *J. Alloys Compd.* **2017**, *701*, 392–398. [CrossRef]
9. Li, C.; Xu, S.; Ye, R.; Deng, D.; Hua, Y.; Zhao, S.; Zhuang, S. White up-conversion emission in $Ho^{3+}/Tm^{3+}/Yb^{3+}$ tri-doped glass ceramics embedding BaF_2 nanocrystals. *Phys. B* **2011**, *406*, 1698–1701. [CrossRef]
10. Zhao, Z.; Ai, B.; Liu, C.; Yin, Q.; Xia, M.; Zhao, X.; Jiang, Y. Er^{3+} Ions-Doped Germano-Gallate Oxyfluoride Glass-Ceramics Containing BaF_2 Nanocrystals. *J. Am. Ceram. Soc.* **2015**, *98*, 2117–2121. [CrossRef]
11. Lesniak, M.; Zmojda, J.; Kochanowicz, M.; Miluski, P.; Baranowska, A.; Mach, G.; Kuwik, M.; Pisarska, J.; Pisarski, W.A.; Dorosz, D. Spectroscopic Properties of Erbium-Doped Oxyfluoride Phospho-Tellurite Glass and Transparent Glass-Ceramic Containing BaF_2 Nanoparticles. *Materials* **2019**, *12*, 3429. [CrossRef]
12. Qiao, X.; Fan, X.; Wang, M.; Zhang, X. Spectroscopic properties of Er^{3+}–Yb^{3+} co-doped glass ceramics containing BaF_2 nanocrystals. *J. Non-Cryst. Solids* **2008**, *354*, 3273–3277. [CrossRef]
13. Dan, H.K.; Zhou, D.; Wang, R.; Jiao, Q.; Yang, Z.; Song, Z.; Yu, X.; Qiu, J. Effects of gold nanoparticles on the enhancement of upconversion and near-infrared emission in Er^{3+}/Yb^{3+} co-doped transparent glass–ceramics containing BaF_2 nanocrystals. *Ceram. Int.* **2015**, *41*, 2648–2653. [CrossRef]
14. Qiao, X.; Luo, Q.; Fan, X.; Wang, M. Local vibration around rare earth ions in alkaline earth fluorosilicate transparent glass and glass ceramics using Eu^{3+} probe. *J. Rare Earths* **2008**, *26*, 883–888. [CrossRef]
15. Luo, Q.; Fan, X.; Qiao, X.; Yang, H.; Wang, M.; Zhang, X. Eu^{2+}-doped glass ceramics containing BaF_2 nanocrystals as a potential blue phosphor for UV-LED. *J. Am. Ceram. Soc.* **2009**, *92*, 942–944. [CrossRef]
16. Wang, C.; Chen, X.; Luo, X.; Zhao, J.; Qiao, X.; Liu, Y.; Fan, X.; Qian, G.; Zhang, X.; Han, G. Stabilization of divalent Eu^{2+} in fluorosilicate glass-ceramics via lattice site substitution. *RSC Adv.* **2018**, *8*, 34536–34542. [CrossRef]

17. Chen, D.; Wang, Y.; Yu, Y.; Ma, E.; Zhou, L. Microstructure and luminescence of transparent glass ceramics containing Er^{3+}:BaF_2 nano-crystals. *J. Solid State Chem.* **2006**, *179*, 532–537. [CrossRef]
18. Secu, C.E.; Secu, M.; Ghica, C.; Mihut, L. Rare-earth doped sol-gel derived oxyfluoride glass-ceramics: Structural and optical characterization. *Opt. Mater.* **2011**, *33*, 1770–1774. [CrossRef]
19. Secu, C.E.; Bartha, C.; Polosan, S.; Secu, M. Thermally activated conversion of a silicate gel to an oxyfluoride glass ceramics: Optical study using Eu^{3+} probe ion. *J. Lumin.* **2014**, *146*, 539–543. [CrossRef]
20. Hu, M.; Yang, Y.; Min, X.; Liu, B.; Wu, Y.; Wu, Y.; Yu, L. Rare earth ion (RE = Tb/Eu/Dy) doped nanocrystalline oxyfluoride glass-ceramic $5BaF_2-95SiO_2$. *J. Am. Ceram. Soc.* **2021**, *104*, 5317–5327. [CrossRef]
21. Secu, M.; Secu, C.; Bartha, C. Optical Properties of Transparent Rare-Earth Doped Sol-Gel Derived Nano-Glass Ceramics. *Materials* **2021**, *14*, 6871. [CrossRef]
22. Pawlik, N.; Szpikowska-Sroka, B.; Pisarska, J.; Goryczka, T.; Pisarski, W.A. Reddish-Orange Luminescence from BaF_2:Eu^{3+} Fluoride Nanocrystals Dispersed in Sol-Gel Materials. *Materials* **2019**, *12*, 3735. [CrossRef]
23. Danks, A.E.; Hall, S.R.; Schnepp, Z. The evolution of 'sol–gel' chemistry as a technique for materials synthesis. *Mater. Horiz.* **2016**, *3*, 91–112. [CrossRef]
24. Mosiadz, M.; Juda, K.L.; Hopkins, S.C.; Soloducho, J.; Glowacki, B.A. An in-depth in situ IR study of the thermal decomposition of yttrium trifluoroacetate hydrate. *J. Therm. Anal. Calorim.* **2012**, *107*, 681–691. [CrossRef]
25. Yoshimura, Y.; Ohara, K. Thermochemical studies on the lanthanoid complexes of trifluoroacetic acid. *J. Alloys Compd.* **2006**, *408–412*, 573–576. [CrossRef]
26. Kemnitz, E.; Noack, J. The non-aqueous fluorolytic sol–gel synthesis of nanoscaled metal fluorides. *Dalton Trans.* **2015**, *44*, 19411–19431. [CrossRef]
27. Sun, X.; Zhang, Y.W.; Du, Y.P.; Yan, Z.G.; Si, R.; You, L.P.; Yan, C.H. From Trifluoroacetate Complex precursors to Monodisperse Rare-Earth Fluoride and Oxyfluoride Nanocrystals with Diverse Shapes through Controlled Fluorination in Solution Phase. *Chem. Eur. J.* **2007**, *13*, 2320–2332. [CrossRef]
28. Farjas, J.; Camps, J.; Roura, P.; Ricart, S.; Puig, T.; Obradors, X. The thermal decomposition of barium trifluoroacetate. *Thermochim. Acta* **2012**, *544*, 77–83. [CrossRef]
29. Khan, A.F.; Yadav, R.; Singh, S.; Dutta, V.; Chawla, S. Eu^{3+} doped silica xerogel luminescent layer having antireflection and spectrum modifying properties suitable for solar cell applications. *Mater. Res. Bull.* **2010**, *45*, 1562–1566. [CrossRef]
30. De Pablos-Martín, A.; Mather, G.C.; Muñoz, F.; Bhattacharyya, S.; Höche, T.; Jinschek, J.R.; Heil, T.; Durán, A.; Pascual, M.J. Design of oxy-fluoride glass-ceramics containing $NaLaF_4$ nano-crystals. *J. Non-Cryst. Solids* **2010**, *356*, 3071–3079. [CrossRef]
31. Holder, C.F.; Schaak, R.E. Tutorial on Powder X-ray Diffraction for Characterizing Nanoscale Materials. *ACS Nano* **2019**, *13*, 7359–7365. [CrossRef]
32. Qin, D.; Tang, W. Energy transfer and multicolor emission in single-phase $Na_5Ln(WO_4)_{4-z}(MoO_4)_z$:$Tb^{3+}$,$Eu^{3+}$ (Ln = La, Y, Gd) phosphors. *RSC Adv.* **2016**, *6*, 45376–45385. [CrossRef]
33. Hameed, A.S.H.; Karthikeyan, C.; Sasikumar, S.; Kumar, V.S.; Kumaresan, S.; Ravi, G. Impact of alkaline metal ions Mg^{2+}, Ca^{2+}, Sr^{2+} and Ba^{2+} on the structural, optical, thermal and antibacterial properties of ZnO nanoparticles prepared by the co-precipitation method. *J. Mater. Chem. B* **2013**, *1*, 5950–5962. [CrossRef]
34. Gorni, G.; Velázquez, J.J.; Mosa, J.; Mather, G.C.; Serrano, A.; Vila, M.; Castro, G.R.; Bravo, D.; Balda, R.; Fernández, J.; et al. Transparent Sol-Gel Oxyfluoride Glass-Ceramics with High Crystalline Fraction and Study of RE Incorporation. *Nanomaterials* **2019**, *9*, 530. [CrossRef]
35. Innocenzi, P. Infrared spectroscopy of sol-gel derived silica-based films: A spectra-microstructure overview. *J. Non-Cryst. Solid.* **2003**, *316*, 309–319. [CrossRef]
36. Aguiar, H.; Serra, J.; González, P.; León, B. Structural study of sol-gel silicate glasses by IR and Raman spectroscopies. *J. Non-Cryst. Solid.* **2009**, *355*, 475–480. [CrossRef]
37. Yamasaki, S.; Sakuma, W.; Yasui, H.; Daicho, K.; Saito, T.; Fujisawa, S.; Isogai, A.; Kanamori, K. Nanocellulose Xerogels With High Porosities and Large Specific Surface Areas. *Front. Chem.* **2019**, *7*, 316. [CrossRef]
38. Richman, I. Longitudinal Optical Phonons in CaF_2, SrF_2, and BaF_2. *J. Chem. Phys.* **1964**, *41*, 2836–2837. [CrossRef]
39. Gorni, G.; Pascual, J.M.; Caballero, A.; Velázquez, J.J.; Mosa, J.; Castro, Y.; Durán, A. Crystallization mechanism in sol-gel oxyfluoride glass-ceramics. *J. Non-Cryst. Solids* **2018**, *501*, 145–152. [CrossRef]
40. Pisarska, J.; Kos, A.; Sołtys, M.; Żur, L.; Pisarski, W.A. Energy transfer from Tb^{3+} to Eu^{3+} in lead borate glass. *J. Non-Cryst. Solids* **2014**, *388*, 1–5. [CrossRef]
41. Shinozaki, K.; Honma, T.; Komatsu, T. High quantum yield and low concentration quenching of Eu^{3+} emission in oxyfluoride glass with high BaF_2 and Al_2O_3 contents. *Opt. Mater.* **2014**, *36*, 1384–1389. [CrossRef]
42. Deng, C.-B.; Zhang, M.; Lan, T.; Zhou, M.-J.; Wen, Y.; Zhong, J.; Sun, X.-Y. Spectroscopic investigation on Eu^{3+}-doped $TeO_2-Lu_2O_3-WO_3$ optical glasses. *J. Non-Cryst. Solids* **2021**, *554*, 120565. [CrossRef]
43. Xie, F.; Li, J.; Dong, Z.; Wen, D.; Shi, J.; Yan, J.; Wu, M. Energy transfer and luminescent properties of $Ca_8MgLu(PO_4)_7$:Tb^{3+}/Eu^{3+} as a green-to-red color tunable phosphor under NUV excitation. *RSC Adv.* **2015**, *5*, 59830–59836. [CrossRef]
44. Wang, J.; Peng, X.; Cheng, D.; Zheng, Z.; Guo, H. Tunable luminescence and energy transfer in $Y_2BaAl_4SiO_{12}$:Tb^{3+},Eu^{3+} phosphors for solid-state lighting. *J. Rare Earths* **2021**, *39*, 284–290. [CrossRef]

45. Gopi, S.; Jose, S.K.; Sreeja, E.; Manasa, P.; Unnikrishnan, N.V.; Joseph, C.; Biju, P.R. Tunable green to red emission via Tb sensitized energy transfer in Tb/Eu co-doped alkali fluoroborate glass. *J. Lumin.* **2017**, *192*, 1288–1294. [CrossRef]
46. Back, M.; Marin, R.; Franceschin, M.; Hancha, N.S.; Enrichi, F.; Trave, E.; Polizzi, S. Energy transfer in color-tunable water-dispersible Tb-Eu codoped CaF_2 nanocrystals. *J. Mater. Chem. C* **2016**, *4*, 1906–1913. [CrossRef]
47. Kłonkowski, A.M.; Wiczk, W.; Ryl, J.; Szczodrowski, K.; Wileńska, D. A white phosphor based on oxyfluoride nano-glass-ceramics co-doped with Eu^{3+} and Tb^{3+}: Energy transfer study. *J. Alloys Compd.* **2017**, *724*, 649–658. [CrossRef]
48. Li, X.; Peng, Y.; Wei, X.; Yuan, S.; Zhu, Y.; Chen, D. Energy transfer behaviors and tunable luminescence in Tb^{3+}/Eu^{3+} codoped oxyfluoride glass ceramics containing cubic/hexagonal $NaYF_4$ nanocrystals. *J. Lumin.* **2019**, *210*, 182–188. [CrossRef]
49. Binnemans, K. Interpretation of europium(III) spectra. *Coord. Chem. Rev.* **2015**, *295*, 1–45. [CrossRef]
50. Wang, X.; Chen, J.; Li, J.; Guo, H. Preparation and luminescent properties of Eu-doped transparent glass-ceramics containing SrF_2 nanocrystals. *J. Non-Cryst. Solids* **2011**, *357*, 2290–2293. [CrossRef]
51. Yanes, A.C.; Santana-Alonso, A.; Méndez-Ramos, J.; del-Castillo, J.; Rodríguez, V.D. Novel Sol-Gel Nano-Glass-Ceramics Comprising Ln^{3+}-Doped YF_3 Nanocrystals: Structure and High Efficient UV Up-Conversion. *Adv. Funct. Mater.* **2011**, *21*, 3136–3142. [CrossRef]
52. Velázquez, J.J.; Mosa, J.; Gorni, G.; Balda, R.; Fernández, J.; Pascual, L.; Durán, A.; Castro, Y. Transparent SiO_2-GdF_3 sol-gel nano-glass ceramics for optical applications. *J. Sol-Gel Sci. Technol.* **2019**, *89*, 322–332. [CrossRef]
53. Bao, W.; Yu, X.; Wang, T.; Zhang, H.; Su, C. Tb^{3+}/Eu^{3+} co-doped Al_2O_3-B_2O_3-SrO glass ceramics: Preparation, structure and luminescence properties. *Opt. Mater.* **2021**, *122*, 111772. [CrossRef]
54. Ritter, B.; Haida, P.; Fink, F.; Krahl, T.; Gawlitza, K.; Rurack, K.; Scholz, G.; Kemnitz, E. Novel and easy access to highly luminescent Eu and Tb doped ultra-small CaF_2, SrF_2 and BaF_2 nanoparticles—structure and luminescence. *Dalton Trans.* **2017**, *46*, 2925–2936. [CrossRef]
55. Chen, D.; Wang, Z.; Zhou, Y.; Huang, P.; Ji, Z. Tb^{3+}/Eu^{3+}:YF_3 nanophase embedded glass ceramics: Structural characterization, tunable luminescence and temperature sensing behavior. *J. Alloys Compd.* **2015**, *646*, 339–344. [CrossRef]

Article

Influence of Nanoparticles and Metal Vapors on the Color of Laboratory and Atmospheric Discharges

Victor Tarasenko [1,2,*], Nikita Vinogradov [1], Dmitry Beloplotov [1], Alexander Burachenko [1], Mikhail Lomaev [1] and Dmitry Sorokin [1]

[1] Institute of High-Current Electronics SB RAS, 634055 Tomsk, Russia; vinikitavin@mail.ru (N.V.); rff.qep.bdim@gmail.com (D.B.); bag@loi.hcei.tsc.ru (A.B.); lomaev@loi.hcei.tsc.ru (M.L.); sdma-70@loi.hcei.tsc.ru (D.S.)

[2] Department of Quantum Electronics and Photonics, National Research Tomsk State University, 634050 Tomsk, Russia

* Correspondence: vft@loi.hcei.tsc.ru; Tel.: +7-903-9539631

Abstract: Currently, electrical discharges occurring at altitudes of tens to hundreds of kilometers from the Earth's surface attract considerable attention from researchers from all over the world. A significant number of (nano)particles coming from outer space burn up at these altitudes. As a result, vapors of various substances, including metals, are formed at different altitudes. This paper deals with the influence of vapors and particles released from metal electrodes on the color and shape of pulse-periodic discharge in air, nitrogen, argon, and hydrogen. It presents the results of experimental studies. The discharge was implemented under an inhomogeneous electric field and was accompanied by the generation of runaway electrons and the formation of mini-jets. It was established that regardless of the voltage pulse polarity, the electrode material significantly affects the color of spherical- and cylindrical-shaped mini jets formed when bright spots appear on electrodes. Similar jets are observed when the discharge is transformed into a spark. It was shown that the color of the plasma of mini-jets is similar to that of atmospheric discharges (red sprites, blue jets, and ghosts) at altitudes of dozens of kilometers and differs from the color of plasma of pulsed diffuse discharges in air and nitrogen at the same pressure. It was revealed that to observe the red, blue and green mini-jets, it is necessary to use aluminum, iron, and copper electrodes, respectively.

Keywords: mini-jets; diffuse discharge; spark discharge; red sprites; blue jets; ghosts

Citation: Tarasenko, V.; Vinogradov, N.; Beloplotov, D.; Burachenko, A.; Lomaev, M.; Sorokin, D. Influence of Nanoparticles and Metal Vapors on the Color of Laboratory and Atmospheric Discharges. *Nanomaterials* **2022**, *12*, 652. https://doi.org/10.3390/nano12040652

Academic Editors: Federico Cesano, Simas Rackauskas and Mohammed Jasim Uddin

Received: 14 December 2021
Accepted: 11 February 2022
Published: 15 February 2022

Publisher's Note: MDPI stays neutral with regard to jurisdictional claims in published maps and institutional affiliations.

Copyright: © 2022 by the authors. Licensee MDPI, Basel, Switzerland. This article is an open access article distributed under the terms and conditions of the Creative Commons Attribution (CC BY) license (https://creativecommons.org/licenses/by/4.0/).

1. Introduction

At present, nanopowders are widely used in various fields. Micro- and nanoparticles are used in cases of exposure to solids and liquids [1–3], as well as in biology [4], agriculture [5], and other fields [6]. These highly demanded areas have already become traditional; therefore, scientific teams from different countries continue to conduct intensive research aimed at studying the properties of such powders (see, for example, [7–10]). It is known that metal vapors and various particles affect the optical properties of gas-discharge plasma. In [11,12], it was demonstrated that the plasma of a repetitively pulsed spark discharge containing metal vapors can be used as a source of spontaneous radiation in the UV region of the spectrum. On the other hand, it is of interest to establish the degree of influence of micro- and nanoparticles, as well as metal vapors, on the properties of discharges in the upper layers of the atmosphere of our planet. It is known that in the upper layers of the atmosphere, at an altitude of tens of kilometers, a large number of micrometeorites burn down and evaporate [13]. As a result, vapors of different materials, including metals, appear at different altitudes. It is of interest to determine whether vapors of meteorites burning in the Earth's atmosphere affect the properties of high-altitude discharges.

High-altitude discharges in the Earth's atmosphere are studied by many scientific groups (see [14–19]). In recent years, the improvement of instruments for detecting various types of radiation, as well as means for capturing images from the International Space Station (ISS), have contributed to obtaining new results [19,20]. Recently, a large number of color photographs of high-altitude transient luminous events (TLEs; red sprites, elves, ghosts, blue jets, and their analogues (starters, giant jets) are referred to as TLEs) have appeared [14–21]. Figure 1 demonstrates a collage of images of TLEs, composed from photographs obtained using various sources [21].

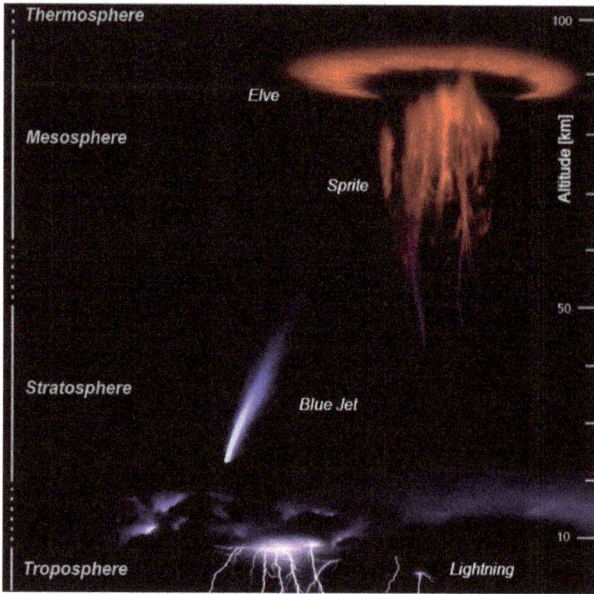

Figure 1. Different types of electrical phenomena in the atmosphere (TLE) [21]. Reproduced from https://en.wikipedia.org; Author: Abestrobi; accessed on 14 February 2022. The image is licensed under the Creative Commons Attribution-Share Alike 3.0 (CC BY-SA 3.0) Unported license).

One issue requiring more research is the determination of the color source for each of these phenomena. It is believed that the color of red sprites that initiate at altitudes of about 70 km and are visible at altitudes of 40 to 90 km is determined by the radiation of the first positive 1P system of a nitrogen N_2 molecule [15,17,18]. This fact is also confirmed by spectral studies of sprite emission [22]. The color of blue jets, which originate from the tops of clouds at about 18 km and reach an altitude of 50 km or more, according to [15,17,19,23], is associated with the radiation of the second positive 2P system of a nitrogen molecule N_2 and the first negative 1N system of a nitrogen N_2^+ ion. There is no information in the available sources about the nature of the green color of phenomena such as ghosts. However, the color of TLEs of the same type, described in different sources, may differ significantly. Thus, the elve in Figure 1 has a red color [21], while on the video taken from the ISS, it is blue-pink [20].

In laboratory experiments, with decreasing air pressure, the color of the plasma of pulsed discharges usually changes from blue to pink, while the color of the sprites observed in natural conditions is bright red, and that of jets is blue [14–21,23]. However, both sprites and jets (or their parts) can be observed at the same altitude (see [16]). Consequently, lines and bands of radiation of a different nature can determine the color of observed TLEs. On the other hand, it was shown in [24,25] that in diffuse and spark discharges in air, mini-jets are observed near electrodes made of aluminum, stainless steel, and copper, the colors of which are red, blue, and green, respectively.

The aim of this work is to study the optical properties of plasma of diffuse and spark discharges in an inhomogeneous electric field at different pressures of air, argon, nitrogen, and hydrogen with the injection of electrode material into the discharge region due to the explosion of microprotrusions on the electrode surface and/or transition to a spark. The color of mini-jets observed in such discharges is compared with that of high-altitude atmospheric discharges (blue jets, red sprites, and ghosts).

2. Experimental Setup and Methods

The study of the discharge modes and optical characteristics of the formed plasma was carried out using electrodes of various shapes, made of different materials. A cone-shaped cathode and a plane anode ("point-plane" gap geometry) were used in most experiments. This gap geometry provides the formation of diffuse discharges in various gases at a relatively low amplitude of voltage pulse due to the strongly non-uniform electric field strength distributions and the generation of runaway electrons [26,27]. It is known that high-altitude discharges are also accompanied by the generation of runaway electrons and other high-energy particles [23].

A sketch of an experimental setup, consisting of a voltage-pulsed generator (NPG-15/2000N or NPG-18/3500N, generators differ in the ranges of adjustment of the amplitude and repetition rate of voltage pulses), a 3 m long coaxial cable with a wave impedance of 75 Ω, and a discharge chamber, is presented in Figure 2.

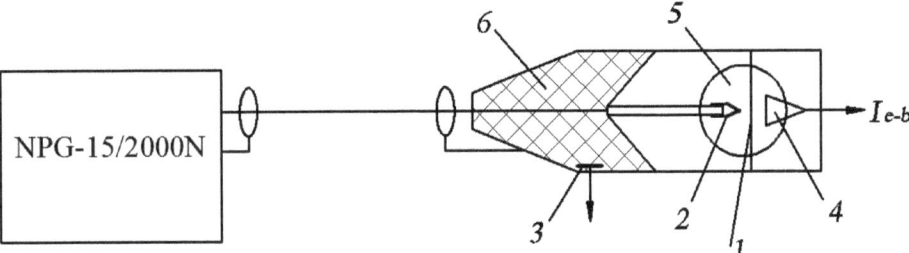

Figure 2. Sketch of the experimental setup. 1—plane anode, 2—cone-shaped cathode, 3—capacitive voltage divider, 4—collector, which was used to measure the runaway electron beam current, 5—discharge chamber with quartz side windows for capturing the discharge plasma glow, 6—insulator.

The NPG generators form voltage pulses U_g with amplitudes ranging from 12 to 18 kV (the voltage across the discharge gap doubled due to its reflection from a "cold" gap) with a full width at half maximum (FWHM) and a rise time of ≈6 and ≈3 ns, respectively. The studies were carried out at a pulse repetition rate f from 60 to 1000 Hz. Some experiments were performed in single-pulse mode. In addition, a RADAN-220 generator with a voltage pulse amplitude in an incident wave U_g of 120 kV and a duration at a matched load of 2 ns was used.

Aluminum, copper, stainless steel, and tungsten were used as electrode materials. The high-voltage, cone-shaped electrode (2 in Figure 2) was made of D14 aluminum, copper, or stainless steel. It had an apex angle of 40 degrees. However, the radius of its tip rounding varied from 70 to 380 μm in the experiments. In several experiments, the cone-shaped electrode was replaced by a bundle of 0.2-mm-diameter tungsten wires with sharp ends. The plane electrode (1 in Figure 2) was made of aluminum, stainless steel, or copper. A gas (air, nitrogen, argon, and hydrogen) pressure varied from 0.1 to 760 Torr. The gap width varied from 1 to 12 mm.

To increase the concentration of the sputtered and evaporated electrode material, as well as the intensity of the discharge plasma radiation, the experiments were carried out under the ignition of the discharge in a repetitively pulsed mode at a voltage of tens of kilovolts. In some experiments, the gas pressure corresponded to that of air at altitudes where blue jets and red sprites are observed.

The use of an electrode with a small radius of curvature as a high-voltage cathode made it possible to determine the conditions under which the generation of runaway electrons takes place. When registering the runaway electron beam current, a plane anode made of a thin aluminum foil or a grid with a light transmission of 67% was used. The results of studies of the properties of a beam of runaway electrons generated in a repetitively pulsed discharge ignition mode on a similar experimental setup are described in detail in [26].

The discharge glow images were photographed with a Sony A100 digital camera (Kuala Lumpur, Malaysia). The colors of red sprites, blue jets, and other TLEs in the images given in [17–19,23] and [20,21] were compared with the color of the repetitively pulsed discharge plasma.

Emission spectra from the different discharge zones were recorded with an EPP-2000C spectrometer (StellarNet Inc. (Tampa, Fl, USA), λ = 192–854.5 nm). Waveforms of the voltage from a capacitive voltage divider and electron beam current from a collector were recorded with a Tektronix MDO 3104 oscilloscope (1 GHz, sampling rate 5 GS/s).

3. Results

3.1. Pulse Breakdown Conditions for Laboratory and Atmospheric Discharges

It is well known that the gap width, the shape and material of the electrodes, the amplitude and rise time of the voltage pulse, the gas pressure and gas type, as well as the method of the gas pre-ionization, determine the breakdown characteristics. When studying laboratory discharges, experimental conditions are relatively easy to control, including measuring the current of runaway electrons [25,26]. In addition, the amplitude and duration of a voltage pulse, the current density, and the energy stored in a high-voltage generator can be varied over a wide range.

The situation with discharges in the upper layers of the Earth's atmosphere is much more complicated. In the case of TLEs, it is necessary to determine not only the altitude of their formation and air composition in this region, but also the electrical properties of the atmosphere (electric field strength, discharge current density, and pulse duration). However, high-altitude atmospheric discharges are transient and electrical parameters are very difficult to measure. The initiation of TLEs depends on the air temperature, as well as the height of the clouds, their density, and the distribution of charges in them. In addition, the formation of these discharges is associated with the appearance of lightning in the lower layers of the Earth's atmosphere. High-energy particles from space, from the Sun, and generated in strong electric fields, are important for charge accumulation and discharge initiation. VUV/UV radiation from the Sun plays an important role in the ionization of the Earth's atmosphere. All these factors can affect the type of discharge and its color. In this paper, the main focus is on the study of the influence of metal vapors and particles appearing in the gap on the coloration of various discharge regions. Data on the appearance of micro- and nanoparticles in the gap, as well as metal vapors, that affect the color of the discharge plasma, are given below.

3.2. Formation of Micro- and Nanoparticles during Spark and Diffuse Discharges

It should be noted that the formation of particles and electrode material evaporation have previously been studied in detail for vacuum discharges (see, for example, [28–30]). The presence of particles in those experiments was observed under various discharge modes. Double diffuse jets (streamers) from the ends of particle tracks were detected for the first time. Below are photographs showing the appearance of particles in the discharge. The cone-shaped, high-voltage electrode used in most of the experiments is placed at the bottom of the images. In the case of electrodes that are poorly distinguishable against the background of the glow of the discharge plasma, their outlines are shown by white lines.

First, the appearance of particles in a spark discharge is demonstrated. The photographs in Figure 3 show the spark discharge glow, the cone-shaped electrode with a small radius of the tip rounding, luminous tracks of particles, and diffuse jets at the ends of the tracks.

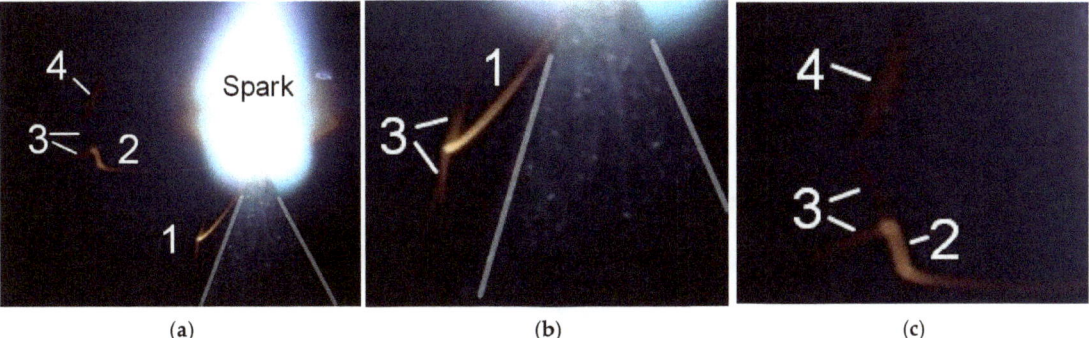

Figure 3. Integral images of the discharge in air at a pressure of 760 Torr with stainless steel electrodes. Single-pulse mode. NPG-18/3500N voltage pulse generator. The voltage pulse amplitude is 18 kV. The height of photograph (**a**) is 2 mm. The plane electrode is at the top and the cone-shaped electrode is at the bottom (outlines of the cone-shaped electrode in (**a**,**b**) are marked with white lines). The gap length is $d = 1$ mm. Zoomed images of the bottom 1 (**b**) and left 2 (**c**) tracks, which end with diffuse jets 3 and 4, are also shown.

The photograph was captured in the single-pulse mode of the spark discharge in the 1 mm long gap filled with atmospheric-pressure air. A short gap was used to increase the current density during the spark stage of the discharge. The spark phase of discharge combustion provides the greatest contribution to the intensity of radiation from the discharge gap. Figure 3a demonstrates a bright spot (the glow from the spark channel), which touches the tip of the cone-shaped electrode with its bottom part and adjoins the flat electrode with its top part. However, even with intense radiation from the spark channel, the tracks of two microparticles are visible. They were emitted from the point of contact between the spark and the cone-shaped electrode. Under these conditions, the brightness of the tracks increases with distance from the electrode. The particle trajectories are different: one of them (2) abruptly changes the direction of its motion.

Figure 3b,c shows zoomed images of the particles' tracks, which, as already noted, can change direction and end in jets. The smooth change in the direction of movement of the microparticles can be explained by the influence of an electric field in this direction. The nature of the glow of the particles in Figure 3 corresponds to the glow of a micrometeorite that burns down in the Earth's atmosphere [13]. The brightness of the micrometeorite (the particle) glow increases towards the track's end. This cannot be explained by an increase in the particle velocity, since the particle stops. This occurs after the voltage pulse action. We believe that the increase in the radiation intensity of the particle is due to its heating during deceleration on gas particles.

Let us note an important feature in the formation of tracks, shown in Figure 3. At the ends of the tracks, jets (streamers) are visible. Usually, they are oriented in opposite directions. The appearance of these diffuse jets at the ends of the tracks is due to the formation of streamers from the plasma surrounding the particle surface. As is known, streamers are formed when the threshold plasma concentration for the electric field at a given point is reached. The formation of streamers is confirmed by the formation of two rectilinearly propagating jets from the same region (3), as well as an abrupt change in the direction of propagation of the upper jet (4) (see Figure 3c). Streamers appear to carry away a significant part of the charge, which leads to a deceleration of the microparticle. In this case, the discharge was formed at the relatively high air pressure (760 Torr) and the small interelectrode distance, which made it possible to observe tracks of individual particles. The observation of microparticle tracks is facilitated in the case of the ignition of a spark discharge with the high current density on the electrodes.

With decreasing gas pressure and increasing interelectrode distance, the current density decreased; sparks do not have time to form at short voltage pulse durations. Accordingly, particles forming tracks (as in Figure 3) were practically not formed, and, therefore, were not registered. The investigations showed that at a pressure of 30 Torr or less, it is easier to register particle tracks if an electrode with a small radius of curvature made of a heavy metal, such as tungsten, is used. It was found that each of the gases has its optimal pressure for observing the tracks. An image of several tracks from the high-voltage anode made of pointed tungsten wires 0.2 mm in diameter, tied into a bundle, is shown in Figure 4.

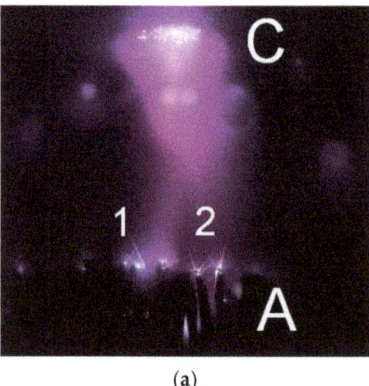
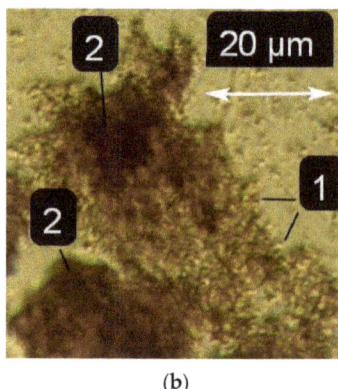

(a) (b)

Figure 4. (a) Image of a discharge in hydrogen at ta pressure of 30 Torr with a tungsten wire anode (bottom) and a plane cathode made of aluminum (top), obtained during one voltage pulse (U_g~120 kV, FWHM at matched load is 2 ns, rise time is 0.5 ns) from the RADAN-220 generator). C—cathode, A—anode. d = 5 mm. (b) Photograph of a part of the surface of a caprolon plate located at the sidewall of the discharge chamber opposite the discharge gap. Air pressure is 100 Torr, d = 2 mm. 1—nanoparticles with a transverse size of ~500 nm or less. 2—clusters of nanoparticles. NPG-15/2000N voltage pulse generator. U_g = 13 kV, f = 60 Hz. The cone-shaped electrode is made of copper, and the plane electrode is made of stainless steel.

The image shows several tracks (see 1 and 2 in Figure 4a) formed by moving microparticles, starting from bright spots on the tungsten anode. Bright spots are also visible on the plane aluminum cathode, but there are no tracks under these conditions. This experiment shows that particles can appear even in the absence of a spark channel. The formation of bright spots on the electrode is sufficient for their initiation. Moreover, the electrode made of heavy metal turned out to be the most suitable for obtaining tracks with a large gap length. At pressures of different gases from ones to tens of Torr and with the tungsten electrode with a small radius of curvature, the intensity of the track glow decreased with distance from the electrode. In addition, in contrast to the conditions in Figure 3, they had a direct trajectory. The length of the tracks depended on the pressure and decreased with increasing pressure. At low pressures (less than 1 Torr), bright spots on were not formed the electrodes and no tracks were observed. The appearance of the tracks was affected by both the pressure and the type of gas. During a discharge in argon with the RADAN-220 generator, bright spots were formed on the tungsten electrode in a narrow pressure range compared to hydrogen; accordingly, this narrowed the range of conditions under which particle tracks were observed. The results shown in Figures 3 and 4 show that pulsed discharges produce particles that, under certain conditions, can be observed from the tracks they leave behind. The particles appear due to the rapid heating of the electrodes at local points by the flowing discharge current.

Figure 4b shows a photograph in which the particles formed in the discharge can be seen. They were scraped off the side surface of the discharge chamber and placed on a microscope slide. Metal particles and their compounds with oxygen and nitrogen with a size of ~500 nm and less are non-uniformly distributed over the surface of the slide. In some regions, clusters of particles are visible (see region 2 in Figure 4b). The size of the particles depended on the discharge mode, the type of gas and its pressure, and the electrode material.

In addition to the formation of particles, spark discharges, and the appearance of bright spots on the electrodes, the evaporation and sputtering of the electrode material occurs. Moreover, when operating in a repetitively pulsed mode, the concentration of metal atoms increases [25]. Below are the results showing the effect of vapors of the electrode material on the discharge color, including at low pressures.

3.3. Effect of Electrode Material on the Color of Pulsed Diffuse Discharges

As is known, the color of a discharge plasma depends on the type of discharge and its operation mode, as well as on the composition of a gas mixture and pressure. In turn, the discharge formation process is determined by the amplitude and duration of a voltage pulse, the current density, the shape of the electrodes, and the interelectrode distance. In the pulse-periodic mode, the intensity of the discharge plasma glow increased. Under these conditions, this made it easier to photograph the discharges at low power consumption. Figure 5 shows photographs of air plasma glow during the discharge between two aluminum electrodes at different air pressures.

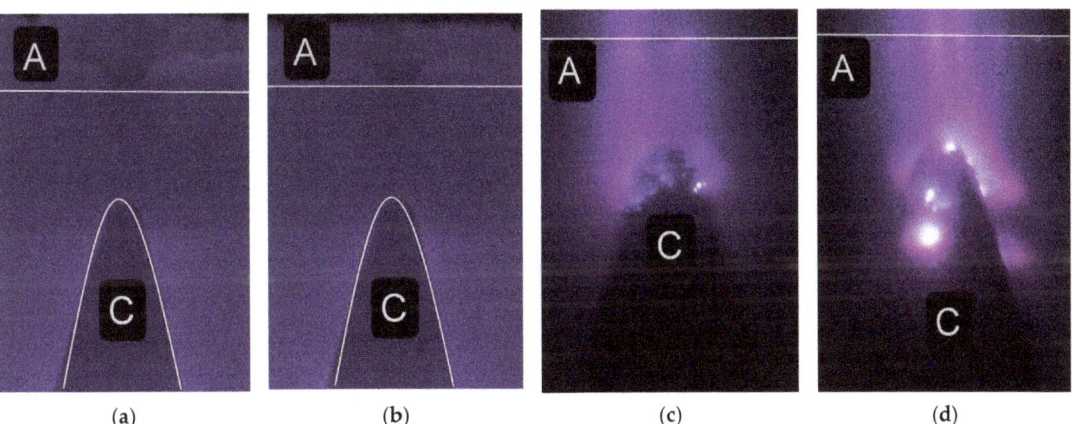

Figure 5. Photographs of the discharge in air at pressures of 1.5 (**a**), 3 (**b**), 10 (**c**), and 30 Torr (**d**) between two aluminum electrodes, obtained at U_g = 12 kV and f = 77 Hz. The plane electrode is located at the top and the cone-shaped electrode is located at the bottom. C—cathode, A—anode. d = 2 mm. Cathode (**a**,**b**) and anode (**a**–**d**) surfaces are marked with white lines. NPG-15/2000N voltage pulse generator.

The discharge has a diffuse form at pressures of fractions of Torr and covers the entire volume of the discharge chamber, and its plasma glow has low-intensity radiation (Figure 5a,b). Its color is far from the bright red of the sprites and elves shown in Figure 1. However, it is closer to the color of the elves on the video from the ISS [15]. Due to the low density of the discharge current under these conditions, bright spots were not observed on the electrodes. Accordingly, the density of the metal vapors from the electrodes at air pressures of 1.5 and 3 Torr was low.

The color of the discharge in the gap became reddish when the gas pressure increased to 10 and 30 Torr, but it also did not correspond to the typical color of sprites (see Figure 1). So, as shown in Figure 5b,c, red spherical jets appeared near the bright spots on the electrodes. The color of these jets matches the color of the sprites. The differences in the color of the discharge plasma in the gap filled with air and near the bright spot on the cone-shaped aluminum electrode is also clearly seen in Figure 6a.

(a)

(b)

(c)
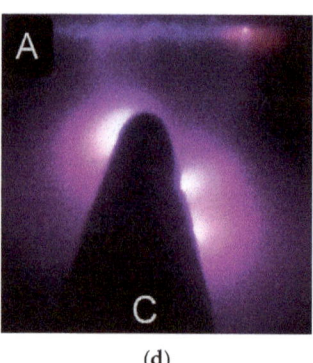
(d)

Figure 6. Images of the discharge in air at a pressure of 30 Torr (**a**) and in argon at pressures of 30 (**b**), 60 (**c**) and 10 Torr (**d**), obtained at $U_g = 12$ kV. $f = 620$ (**a**–**c**), and 77 Hz (**d**). In all photographs, the plane electrode made of aluminum (**a**,**d**) or stainless steel (**b**,**c**) is at the top, and the cone-shaped electrode made of aluminum is at the bottom. C—cathode, A—anode. $d = 3$ (**a**–**c**) and 2 mm (**d**). NPG-18/3500N voltage pulse generator.

For the discharge in air with the cone-shaped aluminum electrode (Figure 6a), as in Figure 5c,d, the discharge color at the bright spot on the electrode changed and a red spherical jet appeared. When the gap width was decreased and the chamber was filled with argon, a bright spot also appeared on the plane electrode (Figure 6d). The color of the spherical jet near it was red, as was that near the cone-shaped electrode. Filling the discharge chamber with argon eliminates the influence of N_2 (1P) radiation on the discharge color.

Experiments were also carried out with electrodes made of different metals. Photographs of the discharge plasma glow in the gap formed by a plane stainless steel anode and a cone-shaped aluminum anode are shown in Figure 6b,c. The use of aluminum electrodes changed the color of the discharge in argon, air, and nitrogen near the electrodes to red. On the surface of the cone-shaped electrode, bright spots surrounded by red halo can be seen. The number of bright spots depends on the discharge current density and gas pressure, as well as on the shape and polarity of the electrodes. As the interelectrode distance increases, the number of bright spots on the plane electrode usually decreases. The voltage pulse repetition rate also affects the number of bright spots. However, the color of the discharge plasma inside them mainly depends on the electrode material. Figure 6b,c demonstrates photographs of the discharge in the gap with a plane stainless steel electrode. The glow color of the plasma cloud (spherical jets) on the plane electrode in these figures is blue.

The use of copper electrodes resulted in an intense green color around the bright spots on the high-voltage electrode (Figure 7).

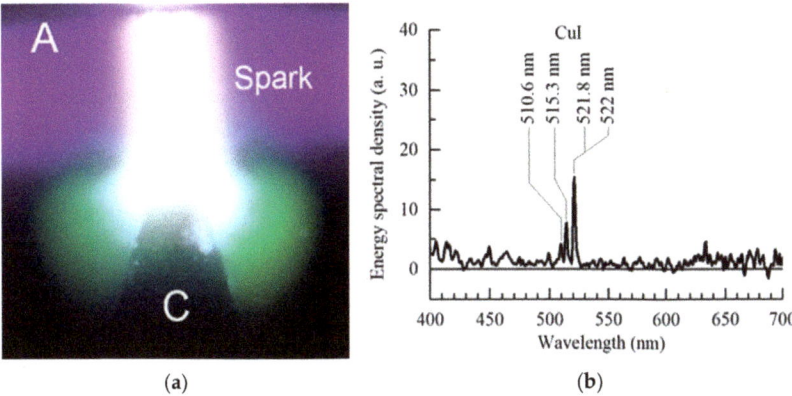

Figure 7. Photograph of the discharge in air at the pressure of 100 Torr (**a**) and the spectrum of the green discharge region near the cone-shaped copper cathode (**b**). The plane stainless steel electrode is located at the top. f = 60 Hz. C—cathode, A—anode. d = 2 mm. NPG-18/3500N voltage pulse generator. U_g = 13 kV.

The lines of the copper atom dominate in the visible region of the emission spectrum (Figure 7b). An important role in the excitation of the above spectral transitions in copper atoms is played by the transfer of energy from the metastable level $A^3\Sigma_u^+$ of molecular nitrogen [25]. A green glow was also observed during high-altitude discharges in the Earth's atmosphere (ghosts in [16]), but the probability of their observation is negligible. In [16], these green-colored areas are located above the red sprites. They have a spherical shape and occur at an altitude of about 80 km.

It should also be noted that for the conditions in Figures 5–7, as in [26], runaway electrons were registered behind a foil or mesh anode at pressures from ones to tens of Torr. These electrons pre-ionize air and other gases and contribute to the formation of diffuse discharges [21,27]. When TLEs appear in the Earth's atmosphere, high-energy particles, including electrons, are also detected [23].

4. Discussion

Usually, the region emitting a certain color near bright spots on a copper, aluminum, or steel electrode (see Figures 5–7) has a spherical shape. However, sometimes, it can also have the shape of a cylindrical jet. Figure 8 shows blue jets obtained in laboratory discharges. These jets correspond to ones in the video from the ISS [20].

Although the sizes of blue jets in the Earth's atmosphere and in laboratory experiments differ by several orders of magnitude, their color and shape are similar. Atmospheric blue jets are usually in the form of a cylinder or a cone, from which thinner jets can propagate. The blue jets near the stainless electrode obtained at the discharge in the air (Figure 8b–c) had a similar shape.

The shape of red sprites is more complex. They are formed in two directions. Initially, one or several jets start from a plasma formation at an altitude of about 75 km [18,31] from the Earth's surface to the ground. Next, the jet(s) or diffuse cloud propagate in the opposite direction. A similar picture was observed in our experiments (Figure 3). Double diffuse jets, starting from one region, propagated in opposite directions. To initiate a streamer (ionization wave) in an electric field, it is necessary to create plasma with a sufficiently high concentration of charged particles. This can be facilitated by, for example, the development of an electron avalanche or a preliminary discharge. An experiment on the registration of particle tracks showed that jets (streamers) could also be initiated at the ends of tracks. We assume that micrometeorites, entering the dense layers of the atmosphere, also create plasma trails with high concentrations of charged particles and initiate some TLEs. Primarily, they are ghosts, and, under certain conditions, they are red sprites.

A micrometeorite containing copper, sputtering and burning, can create a vapor cloud, as well as initiating an ionization wave (streamer), which, at high altitudes has a spherical shape. Discharges in copper vapor are green, while in nitrogen and oxygen there are no intense lines (bands) in this region of the spectrum. The experiments in [32] showed that the presence of copper vapor during an apokampic discharge increases the length of the plasma jet (streamer) and decreases the voltage at which it appears.

Figure 8. Photos of jets near the cone-shaped stainless steel cathode (**a**–**c**) at an air pressure of 760 Torr captured in one pulse. d = 2 (**a**) and 3 (**b**,**c**) mm. NPG-18/3500N voltage pulse generator.

The red color of the positive column, similar to that of a sprite, was recorded in [33] during a continuous-glow discharge in air at a pressure of 1.2 Torr. However, at the cathode, the discharge plasma color had a blue tint. The sprites spread in both directions from the place of their initiation at an altitude of about 75 km; most of them were red along their entire length. Therefore, micrometeorites, and not only ionospheric/mesospheric inhomogeneities, can also contribute to the initiation of red sprites and influence their color [34].

5. Conclusions

This study showed that the vapors of the electrode material can significantly affect the color of the plasma glow of pulsed and pulse-periodic discharges. The largest change in color is observed near bright spots on the electrodes, primarily near electrodes with a small radius of curvature, as well as in the area where the spark channels are adjacent to the electrodes. Based on these results, it is possible to put forward a hypothesis about the influence of cosmic dust [13] on the color of parts of transient luminous events and their initiation.

Author Contributions: Conceptualization, writing—original draft, V.T.; methodology, N.V., D.B. and A.B.; supervision, M.L.; review & editing, D.S. All authors have read and agreed to the published version of the manuscript.

Funding: This research was performed within the framework of the State assignment of the IHCE SB RAS, project No. FWRM-2021-0014.

Data Availability Statement: Data are contained within the paper.

Acknowledgments: The authors are thankful to Evgenii Baksht for help with the experiments.

Conflicts of Interest: The authors declare no conflict of interest. The funders had no role in the design of the study; in the collection, analyses, or interpretation of data; in the writing of the manuscript, or in the decision to publish the results.

References

1. Yu, G.; Wang, X.; Liu, J.; Jiang, P.; You, S.; Ding, N.; Guo, Q.; Lin, F. Applications of Nanomaterials for Heavy Metal Removal from Water and Soil: A Review. *Sustainability* **2021**, *13*, 713. [CrossRef]
2. Pang, W.; Li, Y.; DeLuca, L.T.; Liang, D.; Qin, Z.; Liu, X.; Xu, H.; Fan, X. Effect of Metal Nanopowders on the Performance of Solid Rocket Propellants: A Review. *Nanomaterials* **2021**, *11*, 2749. [CrossRef]
3. Santhosh, C.; Velmurugan, V.; Jacob, G.; Jeong, S.K.; Grace, A.N.; Bhatnagar, A. Role of nanomaterials in water treatment applications: A review. *Chem. Eng. J.* **2016**, *306*, 1116–1137. [CrossRef]
4. Yaqoob, A.A.; Ahmad, H.; Parveen, T.; Ahmad, A.; Oves, M.; Ismail, I.M.I.; Qari, H.A.; Umar, K.; Ibrahim, M.N.M. Recent Advances in Metal Decorated Nanomaterials and Their Various Biological Applications: A Review. *Front. Chem.* **2020**, *8*, 341. [CrossRef] [PubMed]
5. Khot, L.R.; Sankaran, S.; Maja, J.M.; Ehsani, R.; Schuster, E.W. Applications of nanomaterials in agricultural production and crop protection: A review. *Crop. Prot.* **2012**, *35*, 64–70. [CrossRef]
6. Sun, C.; Qin, C.; Zhai, H.; Zhang, B.; Wu, X. Optical Properties of Plasma Dimer Nanoparticles for Solar Energy Absorption. *Nanomaterials* **2021**, *11*, 2722. [CrossRef]
7. Bréchignac, C.; Houdy, P.; Lahmani, M. (Eds.) *Nanomaterials and Nanochemistry*; Springer: Berlin/Heidelberg, Germany, 2008; p. 747.
8. *Metal Nanoparticles in Microbiology*; Rai, M.; Duran, N., Eds.; Springer: Berlin/Heidelberg, Germany, 2008; p. 303.
9. Carpenter, M.A.; Mathur, S.; Kolmakov, A. (Eds.) *Metal Oxide Nanomaterials for Chemical Sensors*; Springer: New York, NY, USA, 2012; p. 548.
10. *Nanomaterials Handbook*; Gogotsi, Y., Ed.; CRC Press: Boca Raton, FL, USA, 2017; p. 712.
11. Baksht, E.K.; Tarasenko, V.F.; Shut'Ko, Y.V.; Erofeev, M.V. Point-like pulse-periodic UV radiation source with a short pulse duration. *Quantum Electron.* **2012**, *42*, 153–156. [CrossRef]
12. Shuaibov, A.K.; Minya, A.Y.; Chuchman, M.P.; Malinina, A.A.; Malinin, A.N.; Gomoki, Z.T.; Kolozhvari, Y.C. Optical characteristics of overstressed nanosecond discharge in atmospheric pressure air between chalcopyrite electrodes. *Plasma Res. Express* **2018**, *1*, 015003. [CrossRef]
13. Plane, J.M.; Flynn, G.J.; Määttänen, A.; Moores, J.E.; Poppe, A.R.; Carrillo-Sanchez, J.D.; Listowski, C. Impacts of cosmic dust on planetary atmospheres and surfaces. *Space Sci. Rev.* **2018**, *21*, 23. [CrossRef]
14. Sentman, D.D.; Wescott, E.M. Red sprites and blue jets: Thunderstorm-excited optical emissions in the stratosphere, mesosphere, and ionosphere. *Phys. Plasmas* **1995**, *2*, 2514–2522. [CrossRef]
15. Pasko, V.P.; Inan, U.; Bell, T.F.; Taranenko, Y.N. Sprites produced by quasi-electrostatic heating and ionization in the lower ionosphere. *J. Geophys. Res. Earth Surf.* **1997**, *102*, 4529–4561. [CrossRef]
16. Flickr.com. Available online: https://www.flickr.com/photos/frankie57pr/49610428072/ (accessed on 10 February 2022).
17. Gordillo-Vázquez, F.J.; Luque, A.; Simek, M. Spectrum of sprite halos. *J. Geophys. Res. Earth Surf.* **2011**, *116*, 093919. [CrossRef]
18. Huang, A.; Lu, G.; Yue, J.; Lyons, W.; Lucena, F.; Lyu, F.; Cummer, S.A.; Zhang, W.; Xu, L.; Xue, X.; et al. Observations of Red Sprites Above Hurricane Matthew. *Geophys. Res. Lett.* **2018**, *45*, 13–158. [CrossRef]
19. Chanrion, O.; Neubert, T.; Mogensen, A.; Yair, Y.; Stendel, M.; Singh, R.; Siingh, D. Profuse activity of blue electrical discharges at the tops of thunderstorms. *Geophys. Res. Lett.* **2017**, *44*, 496–503. [CrossRef]
20. YouTube. Available online: https://youtu.be/4VR3yBlKsFM (accessed on 1 November 2021).
21. Wikipedia.org. Available online: https://en.wikipedia.org/wiki/Sprite_(lightning)#/media/File:Upperatmoslight1.jpg (accessed on 14 February 2022).
22. Hampton, D.L.; Heavner, M.J.; Wescott, E.M.; Sentman, D.D. Optical spectral characteristics of sprites. *Geophys. Res. Lett.* **1996**, *23*, 89–92. [CrossRef]
23. Heumesser, M.; Chanrion, O.; Neubert, T.; Christian, H.J.; Dimitriadou, K.; Gordillo-Vazquez, F.J.; Luque, A.; Pérez-Invernón, F.J.; Blakeslee, R.J.; Østgaard, N.; et al. Spectral Observations of Optical Emissions Associated with Terrestrial Gamma-Ray Flashes. *Geophys. Res. Lett.* **2021**, *48*, 2020GL090700. [CrossRef]
24. Tarasenko, V.F.; Beloplotov, D.V.; Lomaev, M.I. Colored Diffuse Mini Jets in Runaway Electrons Preionized Diffuse Discharges. *IEEE Trans. Plasma Sci.* **2016**, *44*, 386–392. [CrossRef]
25. Beloplotov, D.V.; Lomaev, M.I.; Sorokin, D.A.; Tarasenko, V.F. Blue and green jets in laboratory discharges initiated by runaway electrons. *J. Phys. Conf. Ser.* **2015**, *652*, 012012. [CrossRef]
26. Baksht, E.K.; Burachenko, A.G.; Erofeev, M.V.; Tarasenko, V.F. Pulse-periodic generation of supershort avalanche electron beams and X-ray emission. *Plasma Phys. Rep.* **2014**, *40*, 404–411. [CrossRef]
27. Tarasenko, V.F. Runaway electrons in diffuse gas discharges. *Plasma Sources Sci. Technol.* **2019**, *29*, 034001. [CrossRef]
28. Mesyats, G.A. Ecton mechanism of the vacuum arc cathode spot. *IEEE Trans. Plasma Sci.* **1995**, *23*, 879–883. [CrossRef]
29. Proskurovsky, D.I.; Popov, S.A.; Kozyrev, A.V.; Pryadko, E.L.; Batrakov, A.V.; Shishkov, A.N. Droplets Evaporation in Vacuum Arc Plasma. *IEEE Trans. Plasma Sci.* **2007**, *35*, 980–985. [CrossRef]

30. Anders, A. A review comparing cathodic arcs and high power impulse magnetron sputtering (HiPIMS). *Surf. Coatings Technol.* **2014**, *257*, 308–325. [CrossRef]
31. Williams, E.R. Sprites, elves, and glow discharge tubes. *Phys. Today* **2001**, *54*, 41–47. [CrossRef]
32. Tarasenko, V.F.; Kuznetsov, V.S.; Panarin, V.A.; Skakun, V.S.; Sosnin, E.A. Whether and how the vapors of Al, Cu, Fe, and W influence the dynamics of apokamps. *J. Phys. Conf. Ser.* **2020**, *1499*, 012051. [CrossRef]
33. Williams, E.; Valente, M.; Gerken, E.; Golka, R. *Sprites, Elves and Intense Lightning Discharges*; Springer: Dordrecht, The Netherlands, 2006; pp. 237–251.
34. Liu, N.; Dwyer, J.R.; Stenbaek-Nielsen, H.C.; McHarg, M.G. Sprite streamer initiation from natural mesospheric structures. *Nat. Commun.* **2015**, *6*, 7540. [CrossRef] [PubMed]

MDPI
St. Alban-Anlage 66
4052 Basel
Switzerland
Tel. +41 61 683 77 34
Fax +41 61 302 89 18
www.mdpi.com

Nanomaterials Editorial Office
E-mail: nanomaterials@mdpi.com
www.mdpi.com/journal/nanomaterials